Amorphous and Polycrystalline Thin-Film Silicon Science and Technology—2010

MATERIALS RESEARCH SOCIETY
SYMPOSIUM PROCEEDINGS VOLUME 1245

Amorphous and Polycrystalline Thin-Film Silicon Science and Technology—2010

Symposium held April 5–9, 2010, San Francisco, California

EDITORS:

Qi Wang
The National Renewable Energy Laboratory
Golden, Colorado, U.S.A.

Baojie Yan
United Solar Ovonic LLC
Troy, Michigan, U.S.A.

Andrew Flewitt
University of Cambridge
Cambridge, United Kingdom

Chuang-Chuang Tsai
National Chiao Tung University
Hsinchu, Taiwan

Seiichiro Higashi
Hiroshima University
Hiroshima, Japan

Materials Research Society
Warrendale, Pennsylvania

CAMBRIDGE
UNIVERSITY PRESS

University Printing House, Cambridge CB2 8BS, United Kingdom

One Liberty Plaza, 20th Floor, New York, NY 10006, USA

477 Williamstown Road, Port Melbourne, VIC 3207, Australia

314-321, 3rd Floor, Plot 3, Splendor Forum, Jasola District Centre, New Delhi - 110025, India

103 Penang Road, #05-06/07, Visioncrest Commercial, Singapore 238467

Cambridge University Press is part of the University of Cambridge.

It furthers the University's mission by disseminating knowledge in the pursuit of education, learning and research at the highest international levels of excellence.

www.cambridge.org
Information on this title: www.cambridge.org/9781605112220

Materials Research Society
506 Keystone Drive, Warrendale, PA 15086
http://www.mrs.org

First published 2010
First paperback edition 2012

Single article reprints from this publication are available through University Microfilms Inc., 300 North Zeeb Road, Ann Arbor, MI 48106

CODEN: MRSPDH

A catalogue record for this publication is available from the British Library

ISBN 978-1-605-11222-0 Hardback
ISBN 978-1-107-40800-5 Paperback

CONTENTS

*Invited Paper

SOLAR CELL: LIGHT TRAPPING

LOW GAP MATERIALS

*Invited Paper

POSTER SESSION: CRYSTALLIZATION

POSTER SESSION: SOLAR CELLS

NOVEL DEVICES

NANOSTRUCTURED SILICON

*Invited Paper

FILMS AND GROWTH

FLEXIBLE ELECTRONICS

THIN FILM TRANSISTORS AND MATERIALS

CHARACTERIZATION

*Invited Paper

DEFECTS AND METASTABILITY

CARRIER TRANSPORT

POSTER SESSION: FILMS AND GROWTH

*Invited Paper

POSTER SESSION: CHARACTERIZATION

POSTER SESSION: NOVEL DEVICES

POSTER SESSION: THIN FILM TRANSISTORS

CRYSTALLINE SI FILM

SOLAR CELL: FUNDAMENTAL

*Invited Paper

PREFACE

Symposium A, "Amorphous and Polycrystalline Thin-Film Silicon Science and Technology," held April 5–9 at the 2010 MRS Spring Meeting in San Francisco, California, is the latest in an annual series. Thin-film silicon, including hydrogenated amorphous silicon, nano- and microcrystalline silicon, and poly-crystalline silicon materials, is one of the long-standing attractive fields of science and technology. Along the way, it has experienced the initial invention of amorphous silicon with effective doping, invention of amorphous silicon solar cells and thin film transistors (TFT), improvements in material quality and device performance, and scale-up to massive production of hundreds of megawatts of thin film silicon solar panels and millions of a-Si:H TFT active-matrix liquid crystal displays for TVs and computer monitors. For nearly 30 years, this series has played an important role in the scientific research and technology development of thin-film silicon materials and devices. It has provided a platform for scientists and engineers around the world to share progress, exchange ideas, and collaborate to resolve various challenges. The spectrum of the symposium has been changed from the original a-Si:H based materials to include structured materials such as nanocrystalline, microcrystalline and epitaxial growth of crystal silicon thin films. The scope of applications has also expanded to a wide range of electronics and optoelectronics, with many kinds of novel devices.

Along with the progress in the science and technology of thin-film silicon, new challenges evolved with the expansion of the thin-film silicon family from amorphous to nano-structured and macro-structured materials. From a scientific point of view, the understanding of thin-film silicon still needs to be improved. Because of the complexity in material structures, such as disordered chemical bonds, grain boundaries, and alloys, many fundamental properties have not been fully characterized, such as defect distribution, carrier transport, and metastability. The insufficient knowledge in basic material properties limits the device modeling, design and optimization. Additionally, many engineering challenges remain related to making the products more functional, reliable and cost effective. Currently, large-area, high-rate, and uniform depositions of thin-film silicon materials and devices are the major areas that need to be improved for thin-film silicon solar panels to compete with other solar technologies. Also, research on high-mobility TFTs is desirable for high-speed, high-resolution displays. To address these issues, various new materials and fabrication methods have been developed. It is clear that thin-film silicon has been, and will continue to be, an important field of scientific research and technology development.

This symposium started off on April 5 with an extremely well-attended full-day tutorial lectured by Profs. Andrew Flewitt and Arokia Nathan, aimed at young researchers and people new to the field. During the four days of 14 oral sessions and two evenings of poster sessions, 15 invited talks reviewed the recent progress; and 56 oral presentations and 58 poster presentations reported new results in various areas, covering fundamental studies and technology advances. 60 percent of the papers were from 15 countries outside of the United States. The focus sessions on thin-film silicon solar cells were well attended. Remarkable milestones of 10% and 12.5% stable total-area efficiencies of a-Si:H single-junction and a-Si:H/nc-Si:H/nc-Si:H triple-junction solar cells were presented by Oerlikon Solar-Lab and by United Solar Ovonic LLC, respectively. Light trapping with advanced nano-technologies was an emerging topic with a great expectation to enhance solar cell efficiency. Very-high-rate depositions of microcrystalline silicon were demonstrated by AIST and Osaka University. High-performance thin-film silicon TFTs were made by several institutions, such as Hitachi Ltd. Studies on scaling up to large-area substrates and microscopic modifications of materials for TFTs also drew special attention. Fundamental studies on long outstanding issues in amorphous silicon and alloys were reported. Characterizations of new materials, such as nano-, micro- and polycrystalline silicon thin films and alloys were hot topics in the symposium. Studies on carrier transport and defect structures in nanocrystalline silicon provided important information on material optimization, device simulation, and device design. One highlight of the symposium was the invited talk given by Prof. Hellmut Fritzsche, a pioneer of the field. He reviewed the progress and remaining issues in the metastability of amorphous silicon alloys.

The success of Symposium A reflects the great needs in the research and development of thin-film silicon materials and devices. The high-quality presentations ensured the high quality of the symposium. As the organizers of Symposium A, we greatly acknowledge the invaluable contributions of the authors of oral and poster presentations, especially those who made written contributions to this volume.

Many people were involved in the symposium before, during and after the conference. Special appreciation goes to all of the referees for their careful review of papers in the proceedings and valuable feedback given to the authors. We sincerely thank Mary Ann Woolf, who supervised and managed the abstract and manuscript reviewing process. Her experience and hard work allowed for smooth and timely production of this volume. The MRS staff provided friendly and professional support throughout the organization of the Symposium and Proceedings.

On behalf of all the participants, we thank the generous financial support of our corporate sponsors:

- Photovoltaics Technology Center
- ITRI
- NREL
- ULVAC Inc.
- United Solar Ovonic LLC

Qi Wang
Baojie Yan
Andrew Flewitt
Chuang-Chuang Tsai
Seiichiro Higashi

July 2010

MATERIALS RESEARCH SOCIETY SYMPOSIUM PROCEEDINGS

Volume 1245 — Amorphous and Polycrystalline Thin-Film Silicon Science and Technology—2010, Q. Wang, B. Yan, C.C. Tsai, S. Higashi, A. Flewitt, 2010, ISBN 978-1-60511-222-0
Volume 1246 — Silicon Carbide 2010—Materials, Processing and Devices, S.E. Saddow, E.K. Sanchez, F. Zhao, M. Dudley, 2010, ISBN 978-1-60511-223-7
Volume 1247E —Solution Processing of Inorganic and Hybrid Materials for Electronics and Photonics, 2010, ISBN 978-1-60511-224-4
Volume 1248E —Plasmonic Materials and Metamaterials, J.A. Dionne, L.A. Sweatlock, G. Shvets, L.P. Lee, 2010, ISBN 978-1-60511-225-1
Volume 1249 — Advanced Interconnects and Chemical Mechanical Planarization for Micro- and Nanoelectronics, J.W. Bartha, C.L. Borst, D. DeNardis, H. Kim, A. Naeemi, A. Nelson, S.S. Papa Rao, H.W. Ro, D. Toma, 2010, ISBN 978-1-60511-226-8
Volume 1250 — Materials and Physics for Nonvolatile Memories II, C. Bonafos, Y. Fujisaki, P. Dimitrakis, E. Tokumitsu, 2010, ISBN 978-1-60511-227-5
Volume 1251E —Phase-Change Materials for Memory and Reconfigurable Electronics Applications, P. Fons, K. Campbell, B. Cheong, S. Raoux, M. Wuttig, 2010, ISBN 978-1-60511-228-2
Volume 1252— Materials and Devices for End-of-Roadmap and Beyond CMOS Scaling, A.C. Kummel, P. Majhi, I. Thayne, H. Watanabe, S. Ramanathan, S. Guha, J. Mannhart, 2010, ISBN 978-1-60511-229-9
Volume 1253 — Functional Materials and Nanostructures for Chemical and Biochemical Sensing, E. Comini, P. Gouma, G. Malliaras, L. Torsi, 2010, ISBN 978-1-60511-230-5
Volume 1254E —Recent Advances and New Discoveries in High-Temperature Superconductivity, S.H. Wee, V. Selvamanickam, Q. Jia, H. Hosono, H-H. Wen, 2010, ISBN 978-1-60511-231-2
Volume 1255E —Structure-Function Relations at Perovskite Surfaces and Interfaces, A.P. Baddorf, U. Diebold, D. Hesse, A. Rappe, N. Shibata, 2010, ISBN 978-1-60511-232-9
Volume 1256E —Functional Oxide Nanostructures and Heterostructures, 2010, ISBN 978-1-60511-233-6
Volume 1257 — Multifunctional Nanoparticle Systems—Coupled Behavior and Applications, Y. Bao, A.M. Dattelbaum, J.B. Tracy, Y. Yin, 2010, ISBN 978-1-60511-234-3
Volume 1258 — Low-Dimensional Functional Nanostructures—Fabrication, Characterization and Applications, H. Riel, W. Lee, M. Zacharias, M. McAlpine, T. Mayer , H. Fan, M. Knez, S. Wong, 2010, ISBN 978-1-60511-235-0
Volume 1259E —Graphene Materials and Devices, M. Chhowalla, 2010, ISBN 978-1-60511-236-7
Volume 1260 — Photovoltaics and Optoelectronics from Nanoparticles, M. Winterer, W.L. Gladfelter, D.R. Gamelin, S. Oda, 2010, ISBN 978-1-60511-237-4
Volume 1261E —Scanning Probe Microscopy—Frontiers in NanoBio Science, C. Durkan, 2010, ISBN 978-1-60511-238-1
Volume 1262 — *In-Situ* and Operando Probing of Energy Materials at Multiscale Down to Single Atomic Column—The Power of X-Rays, Neutrons and Electron Microscopy, C.M. Wang, N. de Jonge, R.E. Dunin-Borkowski, A. Braun, J-H. Guo, H. Schober, R.E. Winans, 2010, ISBN 978-1-60511-239-8
Volume 1263E —Computational Approaches to Materials for Energy, K. Kim, M. van Shilfgaarde, V. Ozolins, G. Ceder, V. Tomar, 2010, ISBN 978-1-60511-240-4
Volume 1264 — Basic Actinide Science and Materials for Nuclear Applications, J.K. Gibson, S.K. McCall, E.D. Bauer, L. Soderholm, T. Fanghaenel, R. Devanathan, A. Misra, C. Trautmann, B.D. Wirth, 2010, ISBN 978-1-60511-241-1
Volume 1265 — Scientific Basis for Nuclear Waste Management XXXIV, K.L. Smith, S. Kroeker, B. Uberuaga, K.R. Whittle, 2010, ISBN 978-1-60511-242-8
Volume 1266E —Solid-State Batteries, S-H. Lee, A. Hayashi, N. Dudney, K. Takada, 2010, ISBN 978-1-60511-243-5
Volume 1267 — Thermoelectric Materials 2010—Growth, Properties, Novel Characterization Methods and Applications, H.L. Tuller, J.D. Baniecki, G.J. Snyder, J.A. Malen, 2010, ISBN 978-1-60511-244-2
Volume 1268 — Defects in Inorganic Photovoltaic Materials, D. Friedman, M. Stavola, W. Walukiewicz, S. Zhang, 2010, ISBN 978-1-60511-245-9
Volume 1269E —Polymer Materials and Membranes for Energy Devices, A.M. Herring, J.B. Kerr, S.J. Hamrock, T.A. Zawodzinski, 2010, ISBN 978-1-60511-246-6

MATERIALS RESEARCH SOCIETY SYMPOSIUM PROCEEDINGS

Volume 1270 — Organic Photovoltaics and Related Electronics—From Excitons to Devices, V.R. Bommisetty, N.S. Sariciftci, K. Narayan, G. Rumbles, P. Peumans, J. van de Lagemaat, G. Dennler, S.E. Shaheen, 2010, ISBN 978-1-60511-247-3

Volume 1271E —Stretchable Electronics and Conformal Biointerfaces, S.P. Lacour, S. Bauer, J. Rogers, B. Morrison, 2010, ISBN 978-1-60511-248-0

Volume 1272 — Integrated Miniaturized Materials—From Self-Assembly to Device Integration, C.J. Martinez, J. Cabral, A. Fernandez-Nieves, S. Grego, A. Goyal, Q. Lin, J.J. Urban, J.J. Watkins, A. Saiani, R. Callens, J.H. Collier, A. Donald, W. Murphy, D.H. Gracias, B.A. Grzybowski, P.W.K. Rothemund, O.G. Schmidt, R.R. Naik, P.B. Messersmith, M.M. Stevens, R.V. Ulijn, 2010, ISBN 978-1-60511-249-7

Volume 1273E —Evaporative Self Assembly of Polymers, Nanoparticles and DNA , B.A. Korgel, 2010, ISBN 978-1-60511-250-3

Volume 1274 — Biological Materials and Structures in Physiologically Extreme Conditions and Disease, M.J. Buehler, D. Kaplan, C.T. Lim, J. Spatz, 2010, ISBN 978-1-60511-251-0

Prior Materials Research Society Symposium Proceedings available by contacting Materials Research Society

Solar Cell: From Research to Manufacture

Mater. Res. Soc. Symp. Proc. Vol. 1245 © 2010 Materials Research Society 1245-A01-01

Thin Film Silicon Photovoltaic Technology – From Innovation to Commercialization

Subhendu Guha and Jeffrey Yang
United Solar Ovonic LLC, 1100 W. Maple Road, Troy, MI 48084

ABSTRACT

The last decade has witnessed tremendous progress in the science and technology of thin film silicon (amorphous and nanocrystalline) photovoltaic. The shipment of solar panels using this technology was about 200 MW in 2009; based on announcement of new or expanded production capacity, the shipment is projected to grow ten-times in the next 3-5 years. The key factor that will determine the wide-scale acceptance of the products will be the cost of solar electricity achieved using this technology. Efficiency of solar modules and throughput of production equipment will play a key role. In this paper, we discuss our roadmap to improve the product efficiency and machine throughput.

INTRODUCTION

The world market for photovoltaic (PV) has been growing at an annual rate of about 40% over the last five years. In spite of the economic downturn in the global economy, the shipment for 2009 exceeded 6,000 MW, a gain of 10% over 2008. The technology mix was dominated by single crystal and polycrystalline silicon with thin film silicon (both amorphous and nanocrystalline) accounting for about 3% of the total market [1]. The low material cost and ease of large-scale manufacturing have attracted many companies to thin film silicon, and over the last two years several companies have announced new or expanded production capacity. It has been projected that thin film silicon technology may capture 30% of the global PV market within the next five years. This will of course be determined by how competitive the thin film silicon products will be in comparison to other contenders. The key driver for large-scale deployment of PV is the levelized cost of electricity that depends on the energy yield in terms of kWh produced per kW installed and the installed system cost. Thin film silicon alloy solar cells are less sensitive to temperature than the crystalline counterpart, and it has been demonstrated that they produce more kWh/kW under real world conditions [2]. Flexible solar laminates have lower installation cost for rooftop application that reduces the overall system cost. In order to reduce module cost further, one must increase the efficiency and also improve production throughput so that the capital cost can be lowered. The goal is to offer solar electricity at grid parity. Innovation will play an important role in achieving this target. In this paper, we shall review the technology that we have been developing to address efficiency and throughput, and will discuss our roadmap to reduce cost further to reach grid parity.

UNITED SOLAR OVONIC (USO) TECHNOLOGY

The three key components of USO technology are: 1. Roll-to-roll production, 2. Multi-junction thin film silicon solar cell structure, and 3. Flexible solar laminates. As shown in Fig. 1, innovation has played a key role in taking the laboratory results to production. In 1981, we built our first roll-to-roll machine with the web transported through just a single chamber. A pilot plant producing same-gap amorphous silicon (a-Si:H/a-SiH) double-junction cells was installed in 1991. With advances in the laboratory demonstrating the advantages of the triple-junction cell technology, we built our first triple-junction processor with an annual capacity of 5 MW in 1996. It was at that time that we recognized the advantage of flexible products for the rooftop market and we introduced our first building-integrated photovoltaic (BIPV) product in 1997. With increasing acceptance of our products, we embarked on an aggressive expansion plan and today we have about 150 MW of annual capacity.

The commercial laminate fabrication process consists of three basic steps [3, 4]. In the first, a proprietary roll-to-roll deposition technology is used to deposit an amorphous silicon/amorphous silicon-germanium/amorphous silicon-germanium (a-Si:H/a-SiGe:H/a-SiGe:H) triple-junction solar cell on a flexible and lightweight stainless steel substrate using a radio frequency *(rf)* glow-discharge system. Six rolls of stainless steel, each 1.5 mile long, are loaded into the triple-junction processor and nine miles of solar cells are produced in 62 hours. The bottom sub-cell absorbs the red light, the middle cell the yellow/green light, and the top cell

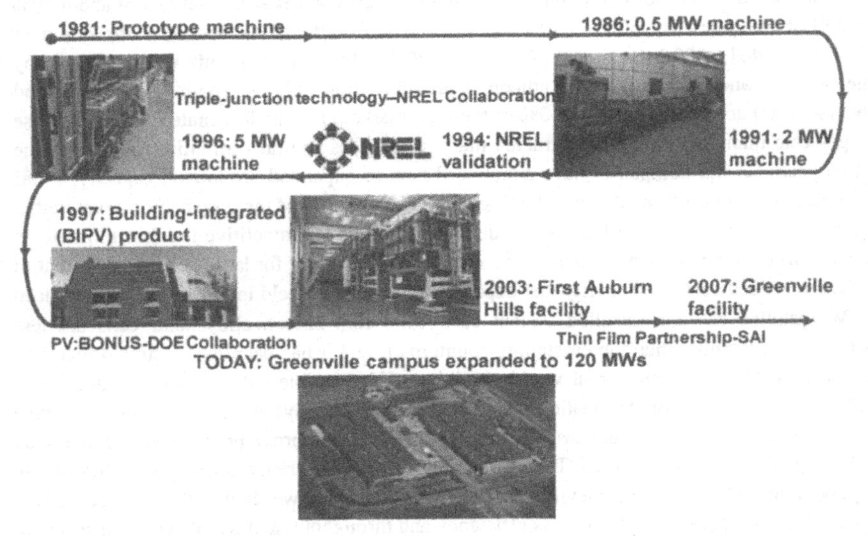

Figure 1. From innovation to commercialization. The history of ECD-USO flexible product development.

the blue light. An Al/ZnO back reflector at the bottom of the solar cell improves the reflectivity and texture of the substrate, which results in improved light trapping and enhancement in conversion efficiency. The second fabrication step consists of cutting the roll of solar cell into smaller pieces, and processing them for cell delineation, short passivation, and top and bottom current collection bus bar application. The third and final step consists of interconnecting the individual solar cells into a series string and encapsulating the string in UV-stabilized and weather-resistant polymers to form the final product. Figure 2 shows schematics of: (a) spectrum-splitting triple-junction solar cell structure, (b) cross section of the module, and (c) roll-to-roll a-Si:H alloy processor and triple-junction structure formation.

USO offers a unique and differentiated product (Fig. 3) compared to any of its competitors. The lightweight and flexible products come with an adhesive and a release paper at the back and can be easily integrated with the roof [4]. This reduces the installation cost significantly.

The current laminate has a total-area efficiency of 6.7% and an aperture-area efficiency of 8.2%. USO has an aggressive plan to increase the aperture-area efficiency first to 10% and later to 12% within the next several years. In the next section, we shall discuss our approach to meet these goals.

EFFICIENCY AND THROUGHPUT IMPROVEMENT

We are using two parallel approaches to improve efficiency. One is replacing the Al/ZnO back reflector (BR) with a superior Ag/ZnO BR; the other is the use of an improved deposition process to develop high quality a-Si:H, a-SiGe:H and nanocrystalline silicon (nc-Si:H) alloys. The improved deposition process results in higher throughput from the deposition machine as well.

Back reflector

In the simplest case, the cell can be deposited on a specular surface resulting in just two optical passes in the cell. Multiple passes (Fig. 4) within the cell by the use of a textured BR increase the optical path and improve photon absorption, especially for the long-wavelength photons [5, 6]. Yablonovitch and Cody have shown that for random scattering, the path length can be increased by a factor of $4n^2$, where n is the refractive index of the material [7, 8]. For a-Si:H alloy, this amounts to about 50 passes or 25 reflections. Figure 5 shows the calculated value of the fraction of absorbed light as a function of wavelength for different number of internal reflections [5, 6]. The calculation assumes an a-Si:H solar cell with an optical gap of 1.68 eV and *i*-layer thickness of 500 nm. The series of curves show the significant enhancement in photon absorption that can be obtained in the red/infrared part of the spectrum with an increasing number of reflections within the cell. This of course will lead to higher short-circuit current density. The enhancement in short-circuit current density is larger when one uses a lower bandgap material.

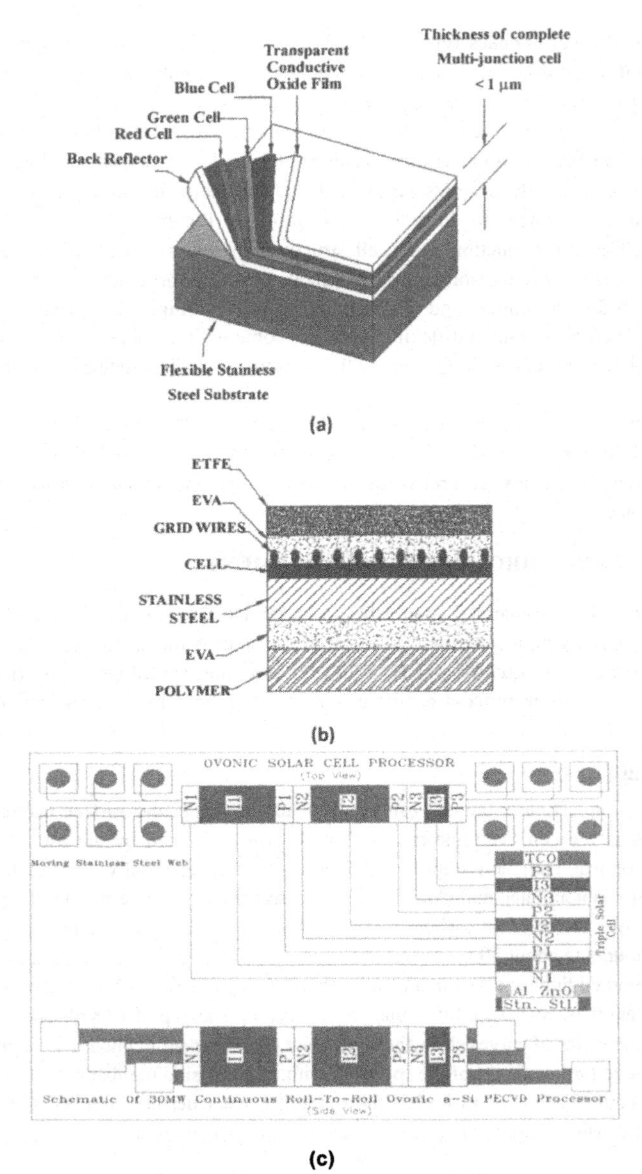

Figure 2. Schematics of (a) triple-junction device structure, (b) cross section of module, and (c) roll-to-roll thin film silicon processor and triple-junction structure formation.

Conventional Solar Panels ***UNI-SOLAR® Laminates***

Figure 3. Comparison of (left) conventional rigid solar panels with (right) United Solar's flexible solar laminate.

We have been using a bi-layer of Ag and ZnO in the laboratory to obtain efficient light trapping. The silver layer can be textured by sputtering at a high temperature. In some cases, ZnO is also textured either by depositing at a high temperature or by subsequent etching. Several tools are used to analyze the quality of the BR. Atomic Force Microscopy (AFM) and Scanning Electron Microscopy (SEM) are used routinely to evaluate the surface roughness. Increasing surface roughness leads to larger scattering, and this is analyzed by measuring the angular dependence of the reflected light when the light is incident normally. Losses at the reflecting surface are determined by measuring total and diffused reflection from the surface. Finally, the quantum efficiency of the finished cell determines the short-circuit current density.

Figure 6 shows AFM and SEM photographs of transparent conductive oxides (TCO) and back reflectors made under various conditions and chemical treatments. It is clear that a wide

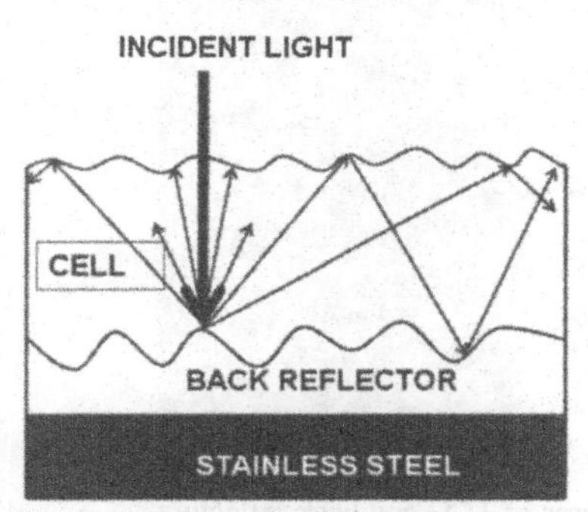

Figure 4. Schematic diagram of a textured back reflector.

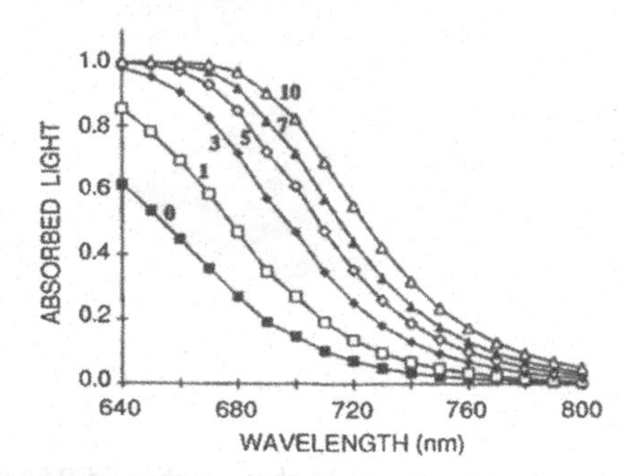

Figure 5. Improvement in absorption with multiple trapping.

variety of textures can be obtained by changing deposition parameters or subsequent processing. Surface roughness ranging from 30 nm to 80 nm can easily be obtained.

Typical angular scattering from various textured substrates is shown in Fig. 7 [9]. It is clear that scattering at an angle to the normal can be increased by increasing the texture of the metal or the ZnO layer. The largest scattering is seen when both the Ag and ZnO is textured. Theoretically, the perfect random surface for light scattering should produce a scattering light intensity proportional to $\cos(\alpha)$, where α is the angle between the scattered light and the normal line to the surface (in Fig. 7 (b), $\theta = 90°-\alpha$). The factor of cosine takes into account the effective area of the light spot on the surface to the viewer at a given angle. Figure 7 (b) plots the measured scattering light intensity divided by $\cos(\alpha)$. It shows that the sample with textured Ag and thick ZnO layer has quite a flat response as a function of the viewing angle. Even though the scattering is adequate, one must make sure that both diffused and total reflections from the BR surface are adequate. Figure 8 (a) shows typical reflection plots for Al/ZnO and Ag/ZnO BR. It is clear from Fig. 8 (b) that higher reflection results in larger red response giving rise to higher short-circuit current density. Theoretically, the best back reflector is with flat Ag and textured ZnO, because the flat Ag reduces the plasmon loss at the Ag/ZnO interface and the textured ZnO provides the required light scattering [10]. Yan *et al.* have shown that it is true for ZnO thicker than 2 μm for a-SiGe:H solar cells [11]. With a thinner ZnO layer, a textured Ag is still necessary because the light scattering is not sufficient at the ZnO and a-Si:H interface. We should mention that recent experiments on nc-Si:H solar cells have shown that thinner ZnO gives rise to better cell performance [12]. This is attributed to poorer quality of nc-Si:H material when grown on a highly textured substrate.

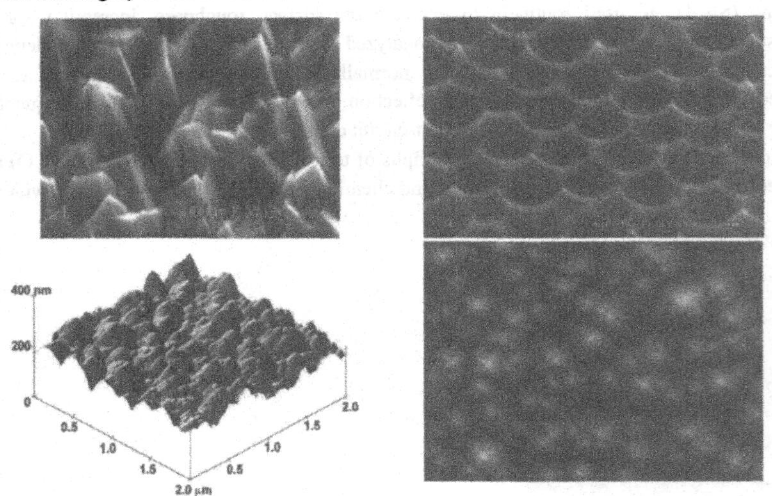

Figure 6. AFM and SEM photographs of different textures of TCO and back reflectors.

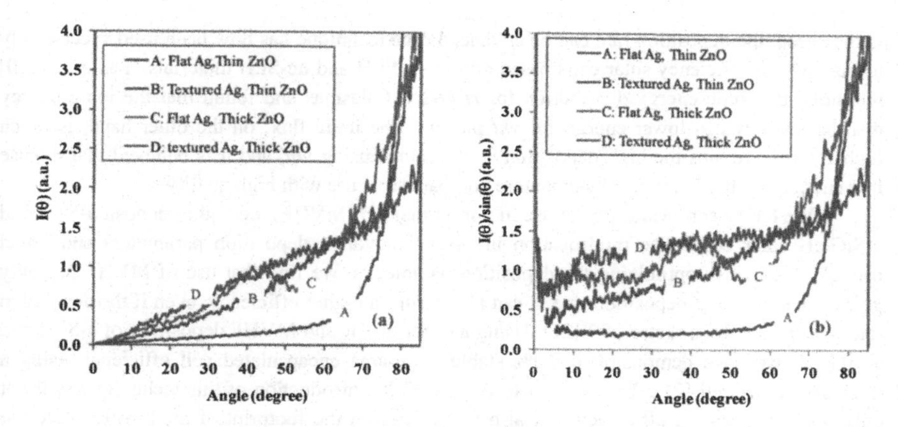

Figure 7. (a) Angular distribution of scattering light intensity I(θ) and (b) the scattering light intensity with correction of viewing area.

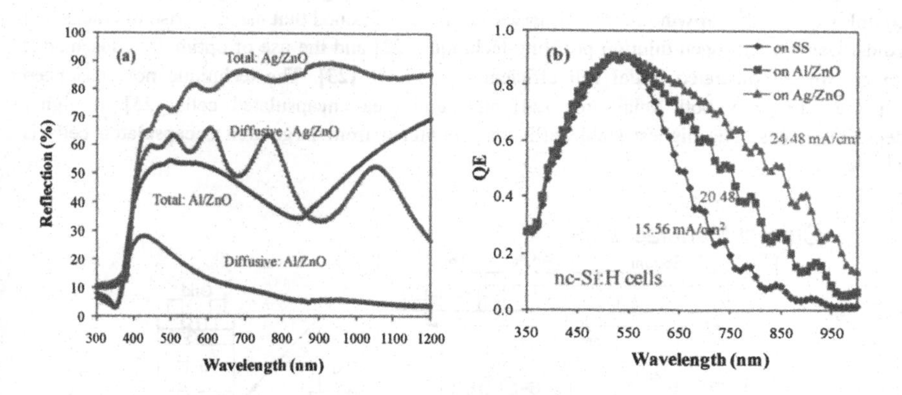

Figure 8. (a) Wavelength dependence of reflectance of Al/ZnO and Ag/ZnO BRs and (b) corresponding quantum efficiency plots of nc-Si:H solar cells.

Many new methods of efficient light trapping are now being explored. They include the use of optical confinement by a grating structure [13, 14] nano-particles [15] and photonic structures [16]. Comprehensive reviews can be found in the literature [17, 18].

New Deposition methods

Most of the reported glow-discharge systems use an *rf* frequency of 13.56 MHz. Very high frequency (*vhf*) has recently been used successfully to grow both a-Si:H and nc-Si:H materials. It was first shown by the Neuchatel group [19] that as the plasma excitation frequency is increased, the deposition rate can be enhanced. The technique has now been used successfully to obtain high efficiency solar cells based on both a-Si:H and nc-Si:H materials. Yan *et al.* [20] measured the ionic energy distribution for *rf* and *vhf* plasmas and found that the ionic energy distribution shifts to lower energy for *vhf* plasma. The ionic flux, on the other hand, is much higher. This explains the high deposition rate obtained using *vhf*, and it is believed that intense low energy bombardment results in a more compact structure with high quality.

We have been exploring the use of a modified *vhf* (MVHF) method to deposit a-Si:H and a-SiGe:H solar cells. The modification involves innovative deposition parameters and novel cathode design. By optimizing the deposition parameters, we find that use of MVHF not only gives rise to a higher deposition rate, it can also result in higher efficiency, even if the deposition rate is increased by a factor of 2 [21]. Using a superior BR and MVHF deposition of a-Si:H and a-SiGe:H, we have demonstrated 9.5% stable large-area encapsulated cell efficiency using a double-junction cell [21]. The immediate impact of the introduction of this technology will not only be an increase in efficiency, but also a reduction in the footprint of the a-Si:H processor (Fig. 9) leading to lower capital cost.

The MVHF technique has also been applied successfully to make nc-Si:H single- and multi-junction cells. Several barriers need to be crossed. The growth morphology of nc-Si:H is complex and as the growth progresses, larger grains are obtained that can give rise to cracks and voids. Using a hydrogen dilution profiling technique [22] and the use of optimally chosen seed layers, we demonstrated initial cell efficiency of 15.4% [23]. The technique now has been expanded to grow both small-area [24] and large-area encapsulated cells [25] at higher deposition rates. The highest stable efficiency obtained from large-area encapsulated cells is 11.2% [25].

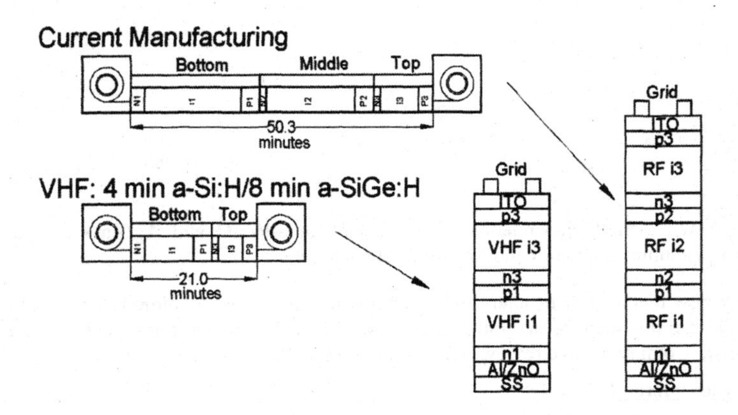

Figure 9. Reduction in footprint of deposition system with MVHF as compared to current *rf* manufacturing.

MANUFACTURING ROADMAP

As mentioned earlier, the current product has an aperture-area efficiency of 8.2% and a total-area efficiency of 6.7%. We shall reduce the inactive area in our laminate to obtain a higher total-area efficiency for the same aperture-area efficiency. The introduction of MVHF and the superior BR in the production line will also increase the efficiency. We follow a stage gate approach to introduce a process in production. The stage gates involve demonstration of efficiency in small- (0.25 cm^2) and large-area (> 400 cm^2), and demonstration of film thickness uniformity over even larger area (> 3500 cm^2). We have crossed these stage gates now and are ready to introduce the two approaches involving MVHF a-SiGe:H and nc-Si:H in a staggered manner in production. The effect on product efficiency is shown in Fig. 10. From the current 8.2% efficiency, introduction of MVHF process will result in an aperture-area efficiency of 10% for a-SiGe:H and 12% for nc-Si:H. This will result in a near-term improvement of total-area efficiency by 70%!

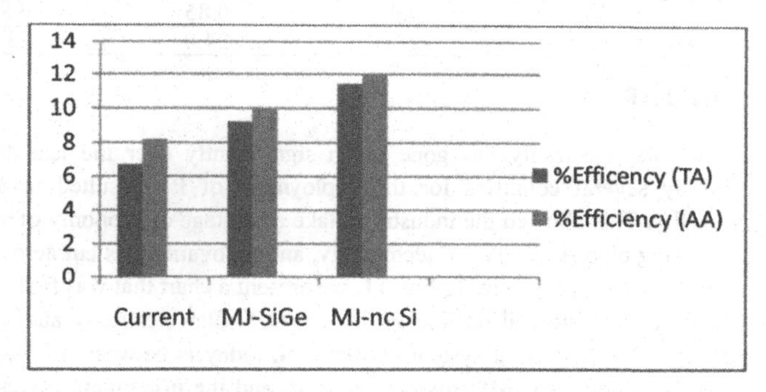

Figure 10. Near-term improvement in total-area (TA) and aperture-area (AA) efficiency.

While we have been discussing the near-term roadmap, it is important to keep some aspirational but achievable goal in front of us. We have already demonstrated a small-area initial cell efficiency of 15.4% using a triple-junction structure incorporating nc-Si:H. In Table I, we show the cell parameters that need to be achieved to obtain 25% cell efficiency for different cell structures. The major improvements needed are in fill factor and short-circuit current density. We should mention that our understanding of an inhomogeneous material like nc-Si:H is still in its infancy. New deposition methods such as rapid thermal annealing, epitaxy using hot-wire CVD, hollow cathode discharge, etc. [26] are being explored. Using a multi-junction approach, an increase in fill factor from 0.75 to 0.85 is not a daunting challenge. We have already shown a fill factor of 0.75 in a triple-junction structure with the component nc-Si:H cell having a fill factor around 0.65. Crystalline silicon solar cells with 200 μm thickness show fill factors exceeding 0.75. We need to improve the nc-Si:H cell of about 2 μm thickness to have a fill factor of 0.75. To obtain short-circuit current density exceeding 36 mA/cm^2 of course will need much superior light trapping. We have not even reached the maximum current density available from a randomized BR, and there is a potential to obtain further gain using more innovative BR's. These include periodic, plasmonic, and photonic BR's [14, 15]. Early results on the use of periodic grating to diffract light at the BR surface did not give higher current density than the random BR [27]. Our understanding was not as advanced at that time, and with the renewed attention that light trapping is now getting, there is a good chance to break the random scattering barrier [28]. The use of up-conversion to convert unabsorbed infra-red light to visible light is also being pursued [29].

Table I. Cell parameters needed to obtain initial 25% cell efficiency for different structures.

	Current	Option 1	Option 2	Option 3
Cell structure	a-Si/a-SiGe/nc-Si	a-Si/a-Si/nc-Si	a-Si/a-SiGe/nc-Si	a-Si/nc-Si/nc-Si
V_{oc} (V)	2.24	2.50	2.30	2.15
J_{sc} (mA/cm^2)	9.13	11.8	12.3	12.6
FF	0.75	0.85	0.85	0.85
Eff (%)	15.4	25.1	24.0	23.0

REACHING GRID PARITY

The cost of solar electricity has gone down significantly over the last decade. The incentives offered by several countries for the deployment of PV resulted in addition of manufacturing capacity. This allowed the industry to take advantage of economy of scale. There is also innovation taking place in every PV technology, and innovation has cut across the entire value chain from module to deployment. In Fig. 11, we present a chart that was first provided by DOE in 2007 showing the historical decline in the price of solar electricity and the goals to achieve. The installed cost of rooftop systems in the U.S. today is between $3.5-$4/watt, and using the Solar Advisor Model (SAM) provided by DOE and the investment tax credit that is available today, solar electricity is already lower than 10 c/kWh in a sunny place like Phoenix, Arizona. This is below the projection made by DOE just two years ago. There has been a notion that we need a revolution to reach grid parity. Actually, the revolution has already taken place in many different technologies, and we need to keep the momentum going in R&D to reduce the cost further.

CONCLUSION

Significant advances have been made in our understanding of a-Si:H, a-SiGe:H and nc-Si:H solar cells. Many manufacturers are establishing new facilities or expanding their manufacturing capacity. In order for this technology to remain competitive in the market place, increase in product efficiency and improvement in machine throughput are needed. Innovation will play a key role in addressing these goals. In this paper, we have outlined our technology roadmap using improved photon harvesting and new deposition methods to increase efficiency and improve throughput of our production machine.

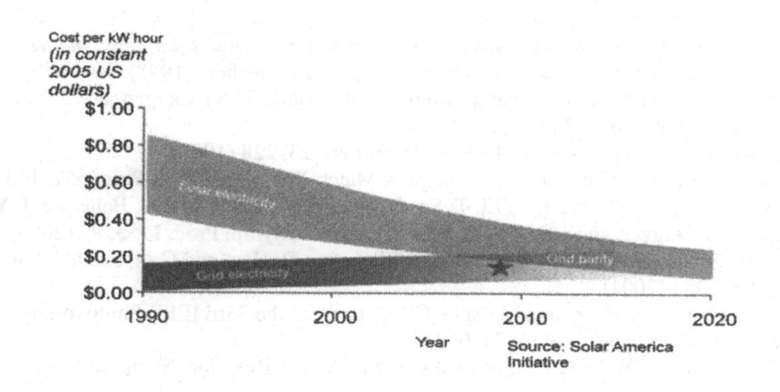

Figure 11. Path toward grid parity.

ACKNOWLEDGEMENT

This work was partially supported by DOE under the Solar America Initiative Program Contract No. DE-FC36-07 GO 17053. The authors thank the entire R&D group at USO for the great team work in the technology development.

REFERENCES

1. Paula Mints, Navigant Consulting (Private Communication).
2. A. Gregg, T. Parker, and R. Swenson, Proc. of the 31st IEEE Photovoltaic Specialists Conf. 1587 (2005).
3. J. Yang, A. Banerjee, and S. Guha, Sol. Energy Mater. Sol. Cells **78**, 597 (2003).
4. S. Guha, Proc. of the 31st IEEE Photovoltaic Specialists Conf. 12 (2005).
5. A. Banerjee and S. Guha, J. Appl. Phys. **69**, 1030 (1991).
6. A. Banerjee, J. Yang, K. Hoffman, and S. Guha, Appl. Phys. Lett. **65**, 472 (1994).
7. E. Yablonovitch and G. D. Cody, IEEE Trans. Electron Devices ED-**29**, 300 (1982).
8. E. Yablonovitch, J. Opt. Soc. Am. **72**, 899 (1982).
9. J. Yang, B. Yan, G. Yue, and S. Guha, Mater. Res. Soc. Symp. Proc. **1153**, 247 (2009).
10. B. Sopori, J. Madjdpour, Y. Zhang, W. Chen, S. Guha, J. Yang, A. Banerjee, and S. Hegedus, Mater. Res. Soc. Symp. Proc. **557**, 755 (1999).
11. B. Yan, G. Yue, C.-S. Jiang, Y. Yan, J. M. Owens, J. Yang, and S. Guha, Mater. Res. Soc. Symp. Proc. **E1101**, on line at www.mrs.org.
12. G. Yue, L. Sivec, J. M. Owens, B. Yan, J. Yang, and S. Guha, Appl. Phys. Lett. **95**, 263501 (2009).
13. C. Haase and H. Stiebig, Prog. Photovolt: Res. Appl. **14**, 629 (2006).
14. H. Sai, Y. Kanamori, and M. Kondo, Mater. Res. Soc. Symp. Proc. **1153**, 29 (2009).
15. S. Pillai, K. R. Catchpole, T. Trupke, and M. A. Green, J. Appl. Phys. **101**, 093105 (2007).
16. D. Zhou and R. Biswas, J. Appl. Phys. **103**, 093102 (2008).

17. R. I. Schropp and M. Zeman, *Amorphous and microcrystalline solar cells: modeling, materials, and device technology*, Kluwer Academic Publishers (1998).
18. M. Zeman, O. Isabella, K. Jaeger, S. Solntsev, R. Lyang, R. Santbergen, and J. Krc, Mater. Res. Soc. Symp. Proc. **1245**, A3.3 (2010).
19. H. Curtins, N. Wyrsch, and A. Shah, Electron Lett. **23**, 228 (1987).
20. B. Yan, J. Yang, S. Guha, and A. Gallagher, Mater. Res. Soc. Symp. Proc. **557**, 115 (1999).
21. X. Xu, D. Beglau, S. Ehlert, Y. Li, T. Su, G. Yue, B. Yan, K. Lord, A. Banerjee, J. Yang, S. Guha, P.G. Hugger, and J.D. Cohen, Mater. Res. Soc. Symp. Proc. **1153**, 99 (2009).
22. B. Yan, G. Yue, J. Yang, S. Guha, D.L. Williamson, D. Han, and C.-S. Jiang, Appl. Phys. Lett. **85**, 1955 (2004).
23. B. Yan, G. Yue, J. Yang, and S. Guha, CD of Proc. of the 33rd IEEE Photovoltaic Specialists Conf., paper No. 257 (2008).
24. G. Yue, L. Sivec, B. Yan, J. Yang, and S. Guha, Mater. Res. Soc. Symp. Proc. **1245**, A21.1 (2010).
25. X. Xu, T. Su, S. Ehlert, D. Bobela, D. Beglau, G. Pietka, Y. Li, J. Zhang, G. Yue, B. Yan, G. DeMaggio, C. Worrel, K. Lord, A. Banerjee, J. Yang, and S. Guha, Mater. Res. Soc. Symp. Proc. **1245**, A2.1 (2010).
26. C. W. Teplin, K. Alberi, M. J. Romero, R. C.Reedy, D. L. Young, I. T. Martin, M. Shub, E. Iwaniczko, C. L. Beall, P. Stradins, and H. M. Branz, Mater. Res. Soc. Symp. Proc. **1245**, A5.10 (2010).
27. S. Guha, Optoelectronics **5**, 201 (1990).
28. H. Zhao, B. Ozturk, E. Schiff, B. Yan, J. Yang, and S. Guha, Mater. Res. Soc. Symp. Proc. **1245**, A3.2 (2010).
29. D. Hreniak, P. Gluchowski, W. Strek, M. Bettinelli, A. Kozlowska, and M. Kozlowski, Mater. Science-Poland **24**, 405 (2006).

Mater. Res. Soc. Symp. Proc. Vol. 1245 © 2010 Materials Research Society 1245-A01-02

From R&D to Large-Area Modules at Oerlikon Solar

J. Meier, S. Benagli, J. Bailat, D. Borello, J. Steinhauser, J. Hötzel, L. Castens, J-B. Orhan, Y. Djeridane, E. Vallat-Sauvain, U. Kroll,
Oerlikon Solar-Lab SA, Puits-Godet 12a, CH-2000 Neuchâtel, Switzerland

ABSTRACT

Amorphous silicon single-junction p-i-n and Micromorph tandem solar cells are deposited in KAI-M reactors on in-house developed LPCVD ZnO front TCO's. An a-Si:H p-i-n cell with a stabilized efficiency of 10.09 % on 1 cm^2 has been independently confirmed by NREL. An alternative ZnO/a-Si:H cell process with an intrinsic absorber of only 180 nm has reached 10.06 % NREL confirmed stabilized efficiencies as well. Up-scaling of such thin cells to 10x10 cm^2 mini-modules has led to an aperture module efficiency of stabilized 9.20 ± 0.19 % as well independently confirmed by ESTI of JRC Ispra.

Micromorph tandem cells with stabilized efficiencies of 11.0 % have been achieved on as-grown LPCVD ZnO front TCO at bottom cell thickness of just 1.3 μm in combination with the in-house developed AR concept. Applying an advanced LPCVD ZnO front TCO stabilized tandem cells of 10.6 % have been realized at a bottom cell thickness of only 0.8 μm. Implementing in-situ intermediate reflectors in Micromorph tandems on LPCVD ZnO reached in a stabilized cell efficiency of 11.3 % with a bottom cell thickness of 1.6 μm.

INTRODUCTION

On the path towards achieving grid parity, thin film silicon solar modules offer a significant potential for manufacturing cost reduction. The challenge of amorphous and microcrystalline silicon based technology is the improvement of module performance towards crystalline silicon technology. While at Oerlikon Solar's customers sites several manufacturing lines based on amorphous and microcrystalline silicon have now been installed, the need for higher efficiencies is the major interest beside cost decrease. Therefore, Oerlikon Solar is concentrating with its R&D group to challenge improved device efficiencies. In this paper we report on the status of amorphous p-i-n single-junction and Micromorph tandems cells using industrial PECVD KAI equipment and in-house developed LPCVD (Low Pressure Chemical Vapor Deposition) ZnO as TCO technology. As light-trapping is one of the most important keys to improve performance, special care on the development of LPCVD ZnO tailored to amorphous or Micromorph tandems has been taken. In addition Oerlikon has developed an in-house AR concept that allows further reducing losses of light coupling into the absorber.

EXPERIMENT

The heart of Oerlikon Solar's thin film PV technology is the KAI PECVD reactor (Plasma Enhanced Chemical Vapor Deposition). To improve deposition rates for solar device-quality amorphous, and especially microcrystalline silicon [1-3], the flat panel display-type reactors were adapted to run at a higher excitation frequency of 40.68 MHz. In this study results were obtained in KAI-M (520x410 mm^2) reactors. More details regarding PECVD processes are given in [4-7].

In order to improve light-trapping the tuning of the LPCVD front ZnO contact layer for optimized a-Si:H single-junction, respectively Micromorph tandem solar cells, was in the focus. Therefore, different types of front TCO's (as-grown type-A, and type-B with haze over 40 % at 600 nm) have been developed and adjusted for very efficient light-scattering. In addition an in-house AR (Anti-Reflecting) concept has been found that allows for further enhanced light coupling into the device.

Recently [8] we have developed the intermediate reflector concept based on PECVD processes in combination with commercial SnO_2 as front TCO. This, however, leads to enhanced optical losses in the microcrystalline silicon bottom cell compared to LPCVD ZnO as front TCO [9]. Consequently intermediate reflectors have been implemented in Micromorph tandems on LPCVD ZnO improving every interface and taking into account the advantage of the enhanced optical light-management of this type of front TCO.

ZnO back contacts in combination with a white reflector reveal excellent light-trapping properties and have been systematically applied in all cells presented here. The test cells were laser scribed to areas of well-defined 1 cm^2. Mini-modules were patterned by laser-scribing to monolithic series connection.

In order to evaluate the stabilized performance the tandem cells were light-soaked at 50°C under 1 sun illumination for 1000 hours. The devices were characterized under AM 1.5 illumination delivered from double-source sun simulators.

Spectral data of transmission were analyzed by a Perkin-Elmer lambda 950 spectrometer.

RESULTS AND DISCUSSION

LPCVD ZnO front contacts

The ZnO front contact layers have been developed in our R&D LPCVD reactor system resulting in improved optical transmission characteristics as given in Fig. 1. This ZnO film represents type-A material, whereas a type-B ZnO is differently processed to achieve very high haze of ~ 40%.

Amorphous single-junction p-i-n cells on ZnO substrates

In previous studies the influence of thickness of the intrinsic a-Si:H absorber layer on the initial and stabilized efficiencies has carefully been investigated especially for SnO_2 and on in-house developed LPCVD ZnO [6, 7]. Whereas for commercially available SnO_2 the properties of the TCO are fixed and determined by the supplier, the in-house LPCVD process allows a further

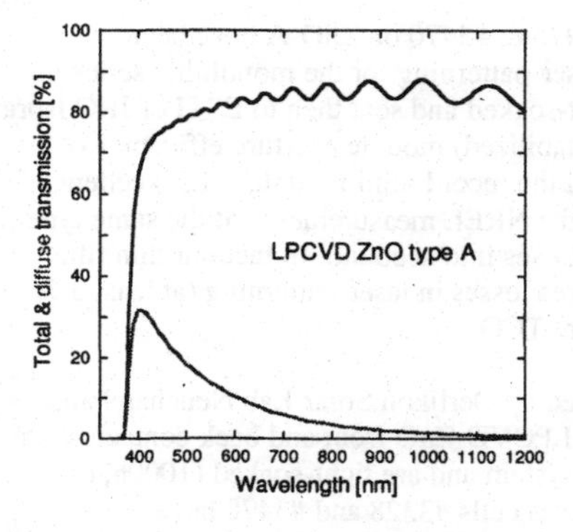

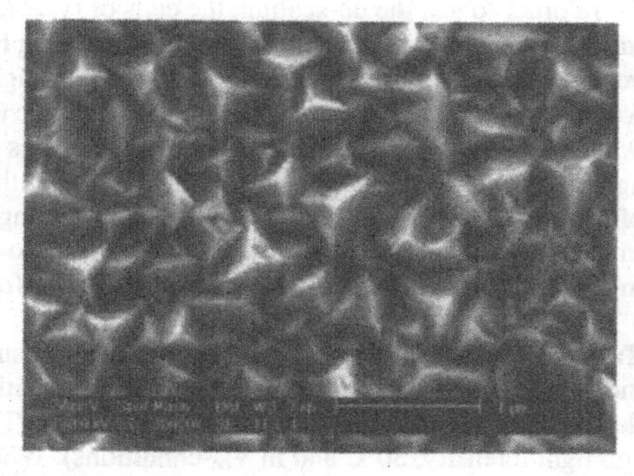

Figure 1. Left: Total and diffuse transmission of LPCVD ZnO of type-A on a Schott Borofloat 33 glass substrate. Right: The ZnO is as-grown and applied in our Micromorph devices. Note the haze of type-A ZnO is about 12 %.

improvement with respect to the optical and structural features of the front TCO. Therefore, tuning both the PECVD cell deposition and the front TCO process opens a new window for high efficient amorphous cells at rather thin intrinsic absorbers. Thus, stabilized cell efficiencies in several runs and on two different types of TCO have achieved over the 10 % barrier. In order to verify our own measurements cells were sent immediately after light-soaking to NREL. The comparison between our cells characterization and those of NREL are given in Tab. 1.

The record cell #3497 measured by NREL is as well shown in Fig. 2. The remarkably high stabilized efficiency of 10.09 ± 0.3 % has been confirmed. Furthermore, Tab. 1 confirms that the Oerlikon characterization is very close to the one of NREL. To note is that our I-V characterization is based on reference cells calibrated by ESTI of JRC Ispra. Compared to the previous record (η = 9.47 ± 0.3 % obtained by IMT Neuchâtel [10, 11]) a significant improvement of 0.6 % absolute could be attained. The initial efficiency based on our AM1.5 I-V characterization and area measurement has reached for cell #3497 11.42 % with a V_{oc}-value of 885 mV, a short-circuit current density (J_{sc}) of 17.94 mA/cm^2 and a FF of 71.92%.

The absolute external QE characteristics of these cells are remarkable high throughout the range of absorption of amorphous silicon. Fig. 3 represents the data deduced from the NREL measurements. Even in the light-soaked state the cells reach 90 % QE and 80 % at short wavelength of 400 nm. This result could finally be attained thanks to an optimization of all involved layers and interfaces forming the cell. In particular, apart of the high quality and standard band gap i-layer (deposited in the single-chamber KAI reactor) the excellent optical and light-scattering properties of our LPCVD-ZnO are one of the main key elements for the enhanced performance.

The 10.09 % stabilized cell is a remarkable new result for amorphous silicon technology, however, the cells achieving 10.06 % at an i-layer thickness of 180 nm only is even more striking. This rather thin cell is based on a front ZnO process that is already in mass production at Oerlikon's customer lines. Thus, the 10.06 % cell on type-A ZnO is very close to present industrialized mass processes, however, at remarkable reduce cell device thickness which allows for further reduction of fabrication cost.

In order to test the up-scaling, the cells of type #3473 & #3470 on ZnO-A have been implemented in 10x10 cm^2 mini-modules applying laser-patterning for the monolithic series connection. As well the mini-modules were fully light-soaked and sent then to ESTI of JRC Ispra for independent characterization. Fig. 4 reflects the (stabilized) module aperture efficiency of 9.20 ± 0.19 % as certified. The ESTI characteristics of the record mini-module is in excellent agreement within the given measurement errors with the NREL measurements of the same type of device (#3473 & #3470) taking typical up-scaling losses into account. In fact our thin film mini-module efficiencies are mainly reduced due to area losses in laser-patterning (at least 3 % in our case) and series resistance losses due to the front TCO.

Table 1: Review of record cells prepared and measured by Oerlikon Solar-Lab Neuchâtel and independently characterized by NREL. All cells with LPCVD-ZnO front and back contacts were deposited in a R&D single-chamber KAI-M PECVD system and are light-soaked (1000h, one sun light intensity, 50°C and in V_{oc}-conditions). Whereas cells #3328 and #3470 have commercial AR coating, on cells #3497 and #3473 our in-house AR was applied. Between both measurements is time gap of about 9 days due to transport.

#Sample / Measurement	AR	I_{sc} [mA]	V_{oc} [mV]	FF [%]	Eff. [%]	Aera [cm^2]
ZnO type-B, p-i(250nm)-n cell						
#3328 / Oerlikon	com.	18.070	878	65.68	9.93	1.05
#3328 / NREL	**com.**	**18.110**	**875.6**	**65.91**	**9.94**	**1.051**
#3497 / Oerlikon	Oerl.	18.040	879	66.39	10.03	1.05
#3497 / NREL	**Oerl.**	**18.098**	**876.7**	**66.58**	**10.09**	**1.047**
ZnO type-A, p-i(180nm)-n cell						
#3473 / Oerlikon	Oerl.	17.310	885	68.61	10.01	1.05
#3473 / NREL	**Oerl.**	**17.480**	**883.8**	**68.33**	**10.06**	**1.049**
#3297 / Oerlikon	No	16.680	881	67.65	9.47	1.05
#3297 / NREL	**No**	**16.708**	**878.2**	**67.57**	**9.55**	**1.038**
#3470 / Oerlikon	com.	17.290	882	68.15	9.90	1.05
#3470 / NREL	**com.**	**17.360**	**885.6**	**67.85**	**10.06**	**1.036**

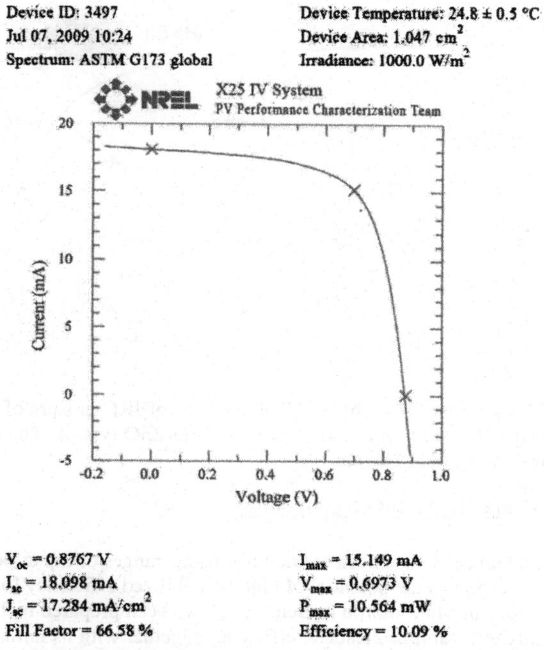

Figure 2: NREL I(V) plot of the stabilized record efficiency of 10.09 ± 0.3 % for an a-Si:H single-junction solar cell. The i-layer thickness of this cell is 250 nm. The used substrate is a 1 mm Schott Borofloat 33 glass on which LPCVD-ZnO with high haze factor (ZnO type-B) was deposited. On this cell the in-house AR was applied as well.

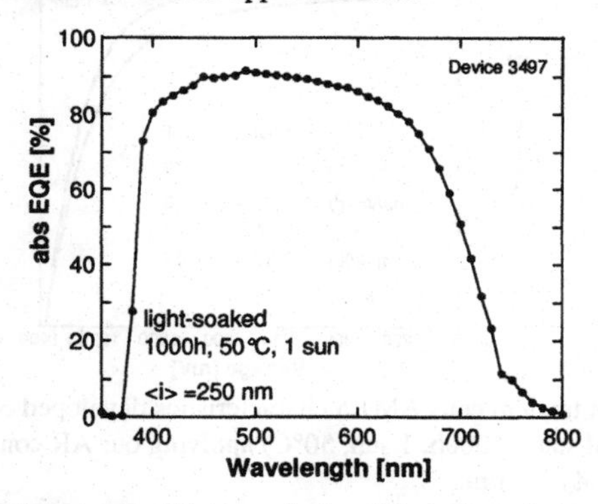

Figure 3: Absolute external quantum efficiency deduced from the relative QE of NREL and the short-circuit current density under AM1.5 measured at NREL for the record cell #3497 (Fig. 2).

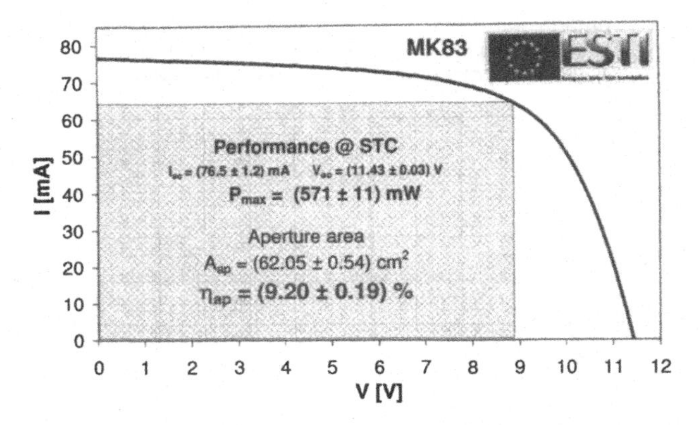

Figure 4: AM1.5 I-V characteristics by ESTI laboratories of JRC in Ispra of the best p-i-n a-Si:H (light-soaked) 10x10 cm^2 mini-module on LPCVD-ZnO type-A. The intrinsic a-Si:H absorber has a thickness of only 180 nm.

Micromorph tandem cells on ZnO type-A substrates

Micromorph tandem cells have been studied in various ranges of top & bottom cell thickness configurations with respect to the potential of highest stabilized efficiency [6, 12]. In addition a range of configurations of Micromorph tandem cells have been prepared including our AR. In Fig. 5 the present highest stabilized test cell efficiency together with its initial characteristics are given. The cell reaches an initial efficiency well above 12 % with rather high short-circuit current densities of 12.6 mA/cm^2 thanks to a very efficient light-trapping as the bottom cell is only 1.3 μm thick.

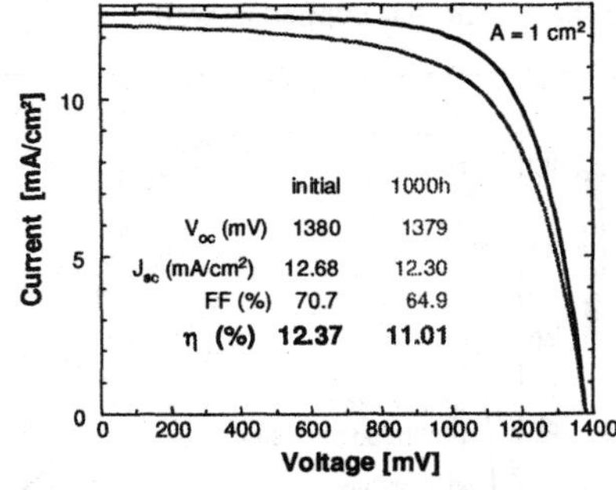

Figure 5. Micromorph tandem cells AM1.5 characteristics developed on type-A ZnO in the initial and light-soaked state (1000h, 1 sun, 50°C) applying our AR concept. The μc-Si:H bottom cell has thickness of only 1.3 μm.

Micromorph tandem cells on ZnO type-B substrates

The effect of the enhanced haze of ZnO type-B is compared with ZnO type-A in Fig. 6 by the QE of Micromorph tandem cells with similar top and identical bottom cell thicknesses. The enhanced light-trapping capability of type-B ZnO leads to a remarkable improvement in the bottom cell current.

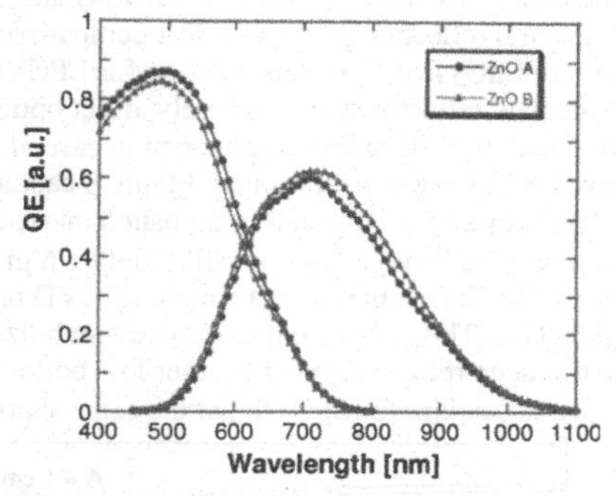

Figure 6. QE of Micromorph tandem cells on type-A & -B front ZnOs. The bottom cells have a thickness of 1.2 μm, top cells have comparable thicknesses.

Due to the very efficient light-trapping of the μc-Si:H bottom cell on type-B ZnO, the microcrystalline silicon intrinsic absorber layer thickness could remarkably be reduced. In Fig. 7 the AM1.5 I-V characteristics of a tandem cell with a microcrystalline bottom cell of only 0.8 μm is shown in the initial and light-soaked state. The 10.6 % stabilized efficiency is a remarkable result as the total silicon absorber layer (top & bottom) is only about 1 μm thick. Regarding manufacturing cost this very thin but efficient device represents a very interesting option.

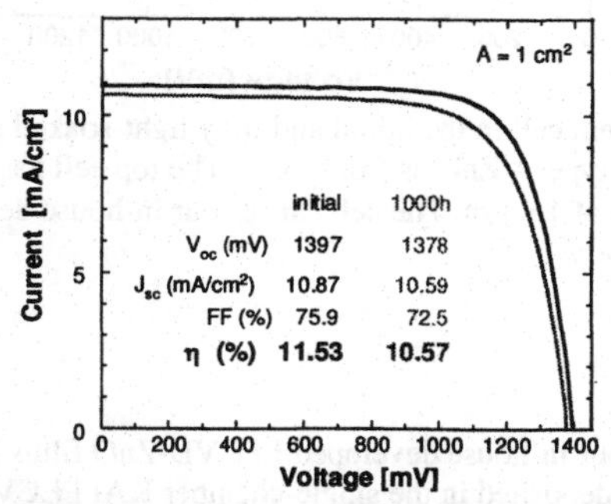

Figure 7. Micromorph tandem cell in the initial and light-soaked state on type-B front ZnO with μc-Si:H layer thickness of only 0.8 μm. The relative degradation achieved in 8.3 %.

Development of intermediate reflectors in Micromorph tandems

Intermediate reflectors based on silicon have been developed in the KAI-M reactors to enhance the light-trapping in the amorphous silicon top cell. Refractive indexes of down to 1.68 could be prepared for these layers so far [8, 9]. Such intermediate reflectors have been implemented in Micromorph tandems and studied for LPCVD ZnO and SnO_2 as front TCO windows with respect to its spectral reflection properties. The comparison indicates directly a more pronounced loss in case of SnO_2 front contacts whereas for LPCVD ZnO the implementation of the intermediate reflector seems to barely affect optical losses.

The high current potential and the reduce loss mechanism in case of ZnO motivated to further improve the device with intermediate layer incorporated. Figure 8 captures our highest stabilized test cell device of 11.3 % efficiency so far. This cell is deposited on type-A front ZnO and has a top cell thickness of 160 nm combined with a bottom cell of only 1.6 μm.

To mention is that type-A front ZnO is based on a simple LPCVD process as it is industrially already applied in mass production. Thus, at present our highest stabilized Micromorph tandem cell is achieved with an intermediate reflector and at a rather low bottom cell thickness of 1.6 μm, much thinner than one would require for SnO_2 to get the same short-circuit current level.

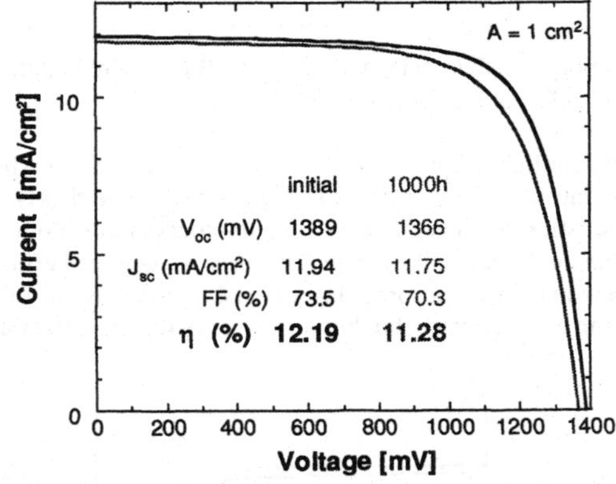

Figure 8. Micromorph tandem cell in the initial and fully light-soaked state with incorporated intermediate reflector using type-A ZnO as front TCO. The top cell has a thickness of 160 nm whereas the bottom cell one of 1.6 μm. The cell carries our in-house developed AR. Note the relative degradation is only 8 %.

CONCLUSIONS

Excellent properties of our in-house developed LPCVD-ZnO films in combination with high quality of the silicon layers deposited in the single-chamber KAI PECVD reactor have demonstrated to be very important in achieving high efficiency levels. Our ZnO layers present high transmission, high conductivity, excellent light-scattering capabilities and a surface morphology suited for the growth of high quality a-Si:H solar cell devices. A record stabilized a-Si:H cell efficiency of 10.09 ± 0.3 % on 1 cm^2 is independently confirmed by NREL. This result presents to our knowledge the highest confirmed stabilized cell efficiency for the a-Si:H single-junction solar cell technology. To note is that NREL confirmed for two other test cells with intrinsic absorber layer thickness of only 180 nm stabilized efficiencies above 10 % too.

The 180 nm a-Si:H p-i-n cell process has been transferred to mini-modules of 10 x 10 cm^2 using the monolithic series connection by laser patterning. Measurements at ESTI laboratories of JRC in Ispra on light-soaked mini-modules confirmed a module aperture area efficiency of 9.20 ± 0.19 %. This high stabilized module efficiency is coherent with the NREL cell efficiency measurements, as these mini-modules efficiencies are mainly reduced due to scribe and series resistance losses.

Micromorph tandem cells have been successfully optimized on in-house ZnO at rather thin μc-Si:H bottom cell thickness. On standard as-grown type-A ZnO stabilized efficiencies of 11.0% have been obtained with a microcrystalline bottom cell of only 1.3 μm thickness. On advanced front ZnO substrates stabilized efficiencies of 10.6 % have been reached using a bottom cell of just 0.8 μm thickness.

Applying the intermediate reflector in Micromorph tandems reveal more favorable light-trapping characteristics for LPCVD ZnO as front contact compared to commercial SnO_2 shown by a reduced spectral reflection loss. Based on this advantage Micromorph tandem cells of 11.3 % stabilized efficiencies with incorporated intermediate reflector have been yet attainted on LPCVD ZnO. Hereby the bottom cell has a thickness of only 1.6 μm.

At present up-scaling of Micromorph tandems to industrial size substrate areas of 1.4 m^2 has led in Oerlikon Solar's pilot line to 151 W initial module power [13, 14]. This corresponds to a module aperture efficiency of 11 % (initial). On several sites of Oerlikon customers Micromorph tandem module ramp-up and production phases have started [15].

ACKNOWLEDGMENTS

The authors would like to thank T. Moriarty and Dr. K. Emery of the NREL laboratories for helpful characterization of our cells. Furthermore, we wish to thank Dr. H. Müllejans and his team from the ESTI Laboratories of JRC (Ispra) for the device characterization and helpful discussions. As well we thank the PV research group of Prof. C. Ballif of the IMT Neuchâtel for scientific and technical support.

REFERENCES

1. F. Finger et al., Appl. Phys. Lett. 65(20), (1994) 2588.
2. H. Takatsuka et al., Solar Energy 77, (2004) 951.
3. T. Matsui, A. Matsuda, M. Kondo, Mat. Res. Symp. Proc. 88, (2004) A8.1.1
4. U. Kroll et al., Thin Solid Films 451-452 (2004), p. 525-530.
5. U. Kroll et al., Proc 19th EU PVSEC (Paris 2004), paper 3AO.8.1, p. 1374-1377.
6. S. Benagli et al., J. Meier et al., in Proc. of 21st & 22nd & 23rd E-PVSEC.
7. S. Benagli et al., Proc. of 24th E-PVSEC (Hamburg), p. 2293.
8. J. Bailat et al., Techn. Digest of PVSEC-18 (Kolkata, India, 2009).
9. L. Castens et al., to be publ. Physics Procedia (E-MRS 2009 spring meeting).

10. J. Meier et al., Proc. 3rd WCPEC (Osaka 2003) session S2.
11. M. A. Green, Keith Emery, Yoshihiro Hishikawa and Wilhelm Warta, Progress in Photovoltaics: Research and Applications 2009; 17:320-326.
12. J. Meier et al., Proc. of 24th E-PVSEC (Hamburg), p. 2398.
13. see press release of Oerlikon Solar of May 27, 2009. www. Oerlikon/Solar.com
14. O. Kluth et al., Proc. of 24th E-PVSEC (Hamburg), p. 2715.
15. N. Papathanasiou et al., presented 24th E-PVSEC (Hamburg), paper 3CO.12.2, or N. Papathanasiou et al., Proc. OTTI (Feb. 2010), 6th User Forum, Thin-Film Photovoltaics.

Mater. Res. Soc. Symp. Proc. Vol. 1245 © 2010 Materials Research Society 1245-A01-03

Protocrystalline Silicon for Micromorph Tandem Cells on Gen. 5 Size

Gijs van Elzakker[1], Daniel Sixtensson[1], Niklas Papathanasiou[1], Klaus Neubeck[1] and Roland Sillmann[1]
[1]Inventux Technologies AG, Wolfener Str. 23, 12681 Berlin, Germany

ABSTRACT

Inventux Technologies AG is a high volume producer of Micromorph (a-Si:H/μc-Si:H) tandem modules. The light-induced degradation of hydrogenated amorphous silicon (a-Si:H), called Staebler-Wronski effect (SWE), limits the stabilized efficiency of a-Si:H-based solar cells. Several laboratories have reported on the development of a-Si:H with increased resistance against light-soaking. This so-called 'protocrystalline' silicon can be grown with plasma-enhanced chemical vapor deposition (PECVD) by diluting the silane source gas with hydrogen.

The aim of the work presented in this paper was to scale-up the laboratory results on protocrystalline silicon to a size of 1.43 m^2 (Gen. 5) using a process that is suitable for high volume production. We demonstrate that the strict boundary conditions regarding uniformity and growth rate, which are necessary for a production process, can be met. The reduced light-induced degradation of protocrystalline solar cells fabricated with the newly developed process is confirmed by a light-soaking experiment. As an outlook towards future work, we discuss issues related to the implementation of a protocrystalline top cell in the Micromorph tandem configuration. The challenge of choosing the right top-cell thickness is illustrated by experimental results on two tandem cells. The top cells of these tandems contain protocrystalline i-layers of different thicknesses.

INTRODUCTION

To date, the most promising silicon-based thin film solar-cell configuration is the so-called Micromorph tandem cell. A Micromorph tandem cell consists of a microcrystalline silicon (μc-Si:H) bottom cell and an amorphous silicon (a-Si:H) top cell. Since the a-Si:H top cell of the a-Si:H/μc-Si:H tandem can be made thinner, and because μc-Si:H is hardly affected by light-soaking, the light-induced degradation in such a cell structure is significantly reduced in comparison to a single-junction a-Si:H cell. This is one of the reasons why Inventux focuses exclusively on the production of Micromorph modules. Despite the success of the Micromorph concept, the influence of the light-induced degradation in the top cell, which generates approximately 2/3 of the power, on the overall performance is still significant. Therefore, stability against light-soaking of the top cell material itself should be addressed.

Several research laboratories have shown that hydrogen dilution of silane in the PECVD deposition of a-Si:H can lead to an enhanced stability against SWE [1]. The a-Si:H deposited under these conditions is called protocrystalline silicon. It has been shown that by using protcrystalline silicon the light-induced degradation can be reduced to 10% for a single-junction cell with a 300-nm thick absorber [2]. The enhanced stability is due to an improved structural order, which is deduced from the narrowing of the XRD scattering peak as shown in Figure 1. We have scaled-up the deposition process for protocrystalline silicon for high volume production on a size of 1.43 m^2. The aim was to achieve a growth rate that is at least as high as

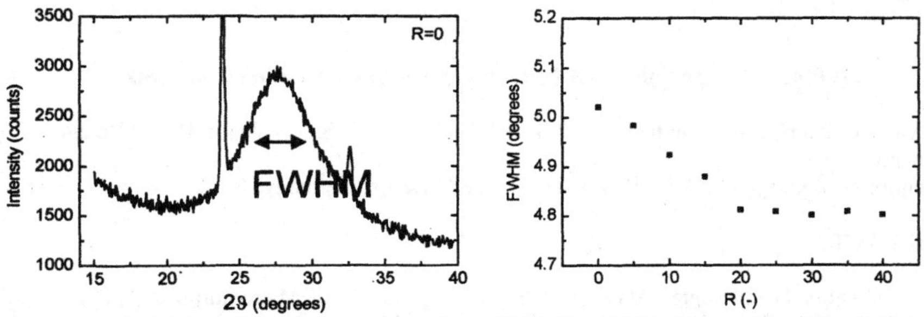

Figure 1. Left: X-Ray diffraction pattern of an a-Si:H film. Right: The width of the XRD peak decreases when the hydrogen dilution R (=$[H_2]/[SiH_4]$) is increased. Data is from reference [2]

that for standard a-Si:H (3.6 Å/s) while maintaining a good thickness uniformity over the entire area. The obtained process is evaluated using optical measurements on the deposited films and the light-soaking of single-junction solar cells.

Since the optical bandgap of protocrystalline silicon is larger than that of standard a-Si:H, the implementation of protocrystaline silicon in the top cell of a Micromorph tandem is not straightforward. We will outline the parameters influencing the design of a Micromorph tandem structure for high stabilized efficiency. We illustrate the challenge of choosing the right top-cell thickness by fabricating two tandem cells for which the top cells contain protocrystalline i-layers of different thicknesses. The stability of these tandem cells is compared to a reference Micromorph cell with a standard a-Si:H i-layer in the top cell.

EXPERIMENT

Thin-film silicon layers are deposited with PECVD in a KAI Plasmabox reactor. The plasma is maintained at a VHF frequency of 40 MHz. Individual films are deposited directly on a 1.43 m^2 glass substrate for characterization with optical measurements. Optical characterization of the films on glass is performed using a Perkin-Elmer UV-VIS spectrometer and a Woollam spectroscopic ellipsometer (SE). For solar cells (single-junction a-Si:H or a-Si:H/μc-Si-H tandems) the p-i-n structures are deposited on glass (1.43 m^2) that is coated with boron-doped LPCVD-grown ZnO. The back contact also consists of ZnO. Finally, white paint is applied as back reflector for the a-Si:H/μc-Si:H tandem structure. Single junction a-Si:H cells with an area of 1 cm^2 are defined by laser scribing. For these test cells no back reflector is applied. Tandem cells (a-Si:H /μc-Si:H) are evaluated on a "mini-module" layout with an area of 6 x 6 cm^2, that contains 10 individual cells connected in series. White paint is used as back reflector for the mini-modules.

Illuminated current-voltage characteristics of solar cells and mini modules are acquired using a Wacom dual-source sun simulator. The measurements are carried out under Standard Test Conditions (STC): a cell temperature of 25°C, irradiance of 1000 W/m^2 and AM1.5 spectral distribution. Light soaking of cells and modules is done under open-circuit conditions in a Solaronix light soaking bench, at a temperature of 50°C and an irradiance of 1000 W/m^2.

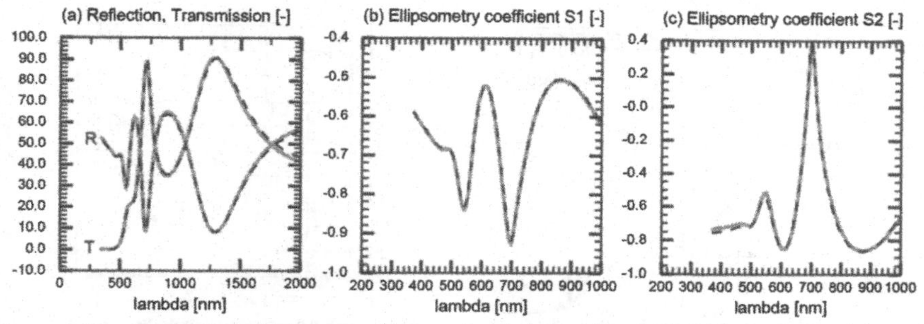

Figure 2. Fit of optical measurements with Tauc-Lorentz model in RIG-VM. (a) Reflection and Transmission data (grey) and fit (dashed). (b), (c): Spectroscopic Ellipsometry data and fit.

DISCUSSION

Protocrystalline silicon on Gen. 5 size in a KAI Plamabox deposition system

After optimizing the deposition of protocrystalline silicon for the VHF plasma process in the KAI Plasmabox reactor, the resulting film properties were characterized using reflection/transmission measurements as well as SE. The optical data is fit using the RIG-VM optical modeling program using a Tauc-Lorentz description for the band gap of a-Si:H [3]. Figure 2 shows the results of the fitting procedure (protocrystalline film). Parameters that are obtained from the fitting include the film thickness and the optical band gap. These parameters are listed in table 1 and compared to those of a standard a-Si:H layer. The comparison demonstrates that our process for the deposition of protocrystalline silicon results in a higher growth rate than achieved thus far for standard a-Si:H. A characteristic feature of protocrystalline silicon is the larger band gap than that of standard a-Si:H, which was also observed for our material. This larger band gap has consequences for the eventual application of protocrystalline silicon in Micromorph tandem cells, as we will discuss later.

Single junction cells with protocrystalline silicon i-layers

We applied protocrystalline i-layers in solar cells to confirm the improved resistance to light-induced degradation. Single-junction a-Si:H solar cells were fabricated with the newly developed protocrystalline silicon as i-layer material. For comparison, reference cells were fabricated with a standard a-Si:H absorber layer with the same thickness as that of the protocrystalline absorber layers. Figure 3(a) shows a comparison of the quantum efficiency (QE) measured for both cell types. The increased band gap of the protocrystalline silicon results in a

Table 1. Properties of a-Si:H films obtained from the fitting of optical measurement data

	Deposition time (s)	Thickness (nm)	Tauc-lorentz gap (eV)	Growth rate (Å/s)
Standard	500	180	1.68	3.6
Protocrystalline	400	180	1.71	4.5

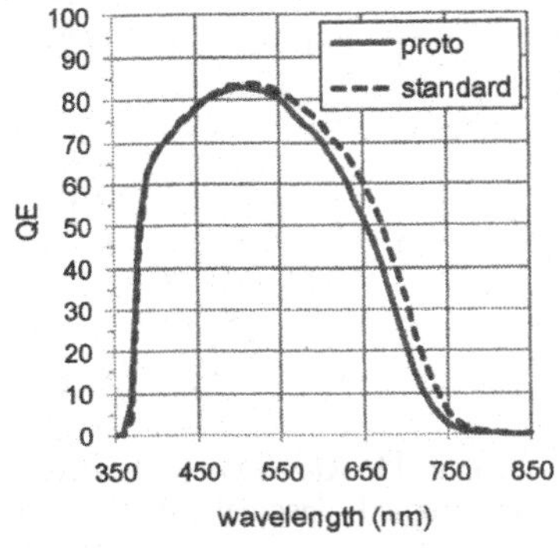

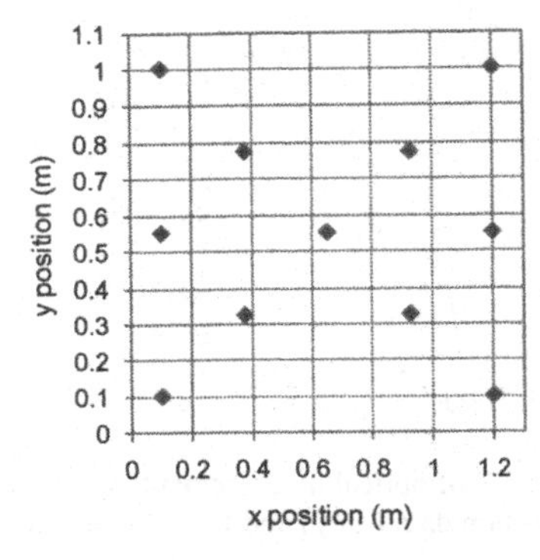

Figure 3(a) The larger band gap of the top cell results in a lower QE for longer wavelengths. (b) Distribution of 1 cm^2 test cells over a 1.43 m^2 glass (used for both cell types).

reduction of the QE in the longer wavelength region. As a result, the short-circuit current of the solar cells is lower than for standard a-Si:H cells with the same thickness. To obtain good experimental statistics and at the same time evaluate the uniformity of the deposition process, a large number of test cells were taken from different positions on the 1.43 m^2 glass, as shown in figure 3 (b). The standard deviation of the efficiency measured on these cells is less than 0.2 in absolute percent, confirming the good uniformity over the entire glass area. All test cells underwent a light-soaking test. The results (average over all cells) are shown in figure 4(a). The light-soaking test clearly confirms the improved stability of the protocrystalline a-Si:H. Figure 4(b) shows the J-V curves after ~400 hours of light-soaking. The protocrystalline cells have a higher efficiency despite the lower current due to the improved Voc and FF. The external parameters after light soaking are listed in table 2. We conclude that the up-scaling of the deposition process for protocrystalline silicon was successful.

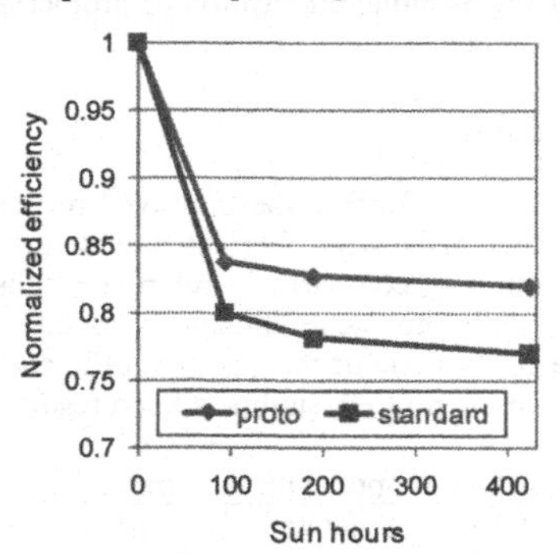

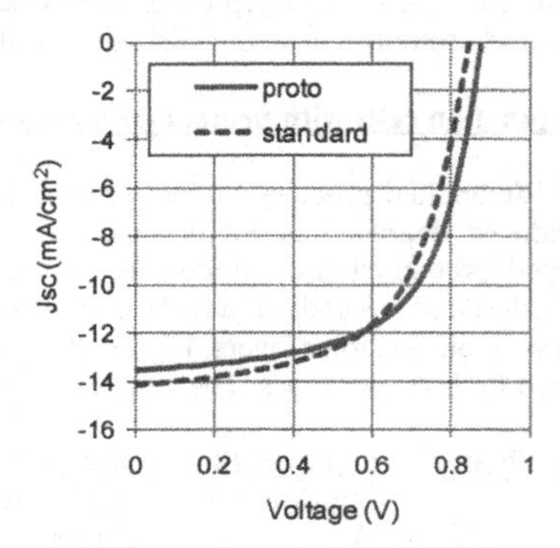

Figure 4. Stability of single junction cells with protocrystalline and standard a-Si:H i-layer. (a) Efficiency normalized to the initial value as a function of light-soaking time. (Each data point represents the average of 20 cells). (b) Typical J-V curves after ~400 hours of light soaking.

Table 2. Properties of a-Si:H solar cells after 400 hours of light soaking. The values represent the average of 20 cells. Note that these cells have no back reflector.

	Jsc (mA/cm^2)	Voc (V)	FF	Efficiency (%)
Standard	14.2 ± 0.10	0.85 ± 0.005	0.59 ± 0.012	7.0 ± 0.19
Protocrystalline	13.5 ± 0.13	0.88 ± 0.007	0.61 ± 0.007	7.3 ± 0.15

Outlook: application in Micromorph tandem cells

Because of the larger band gap, a protocrystalline silicon top cell absorber of the same thickness as a standard a-Si:H absorber would generate less current and therefore shift the current balance in a tandem towards top cell limitation. This could be overcome by increasing the thickness of the top cell, but this has a negative influence on the stability of the top cell. The stability of a single-junction a-Si:H cell becomes worse when the thickness is increased, due to the reduced strength of the electric field in the i-layer which makes carrier transport more difficult [4]. Another aspect to the optimization of the stability of the tandem is the current matching. Bottom cell limitation (i.e. having a bottom cell that produces less current than the top cell) is favorable for high stability. This is due to the fact that the initial degradation of the top cell current is not seen in the output of the tandem which is determined by the bottom cell. In summary, the use of protocrystalline silicon in micromorph tandems requires tuning of many parameters. It is a complex optimization problem, that we will address in future work using numerical optoelectrical solar-cell simulations.

To illustrate the trade-offs, we show results on two tandem cells with a protocrystalline absorber layer in the top cell using two different thicknesses. The results were obtained on mini modules with a white paint back reflector. As a reference, Micromorph tandem cells with a standard a-Si:H absorber layer for the top cell were fabricated. The thickness of the top cell is chosen such that the tandem is bottom limited (we will refer to this reference cell as 'standard'). In the first tandem with protocrystalline silicon, the top cell thickness is kept the same (called sample_A), and in the second sample the absorber was made 20 % thicker (called sample B).

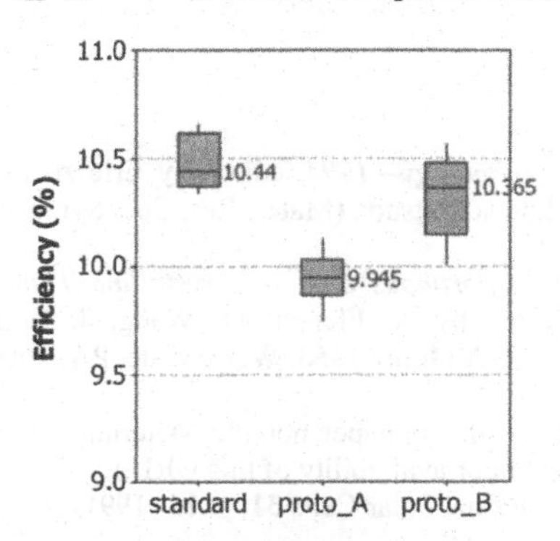

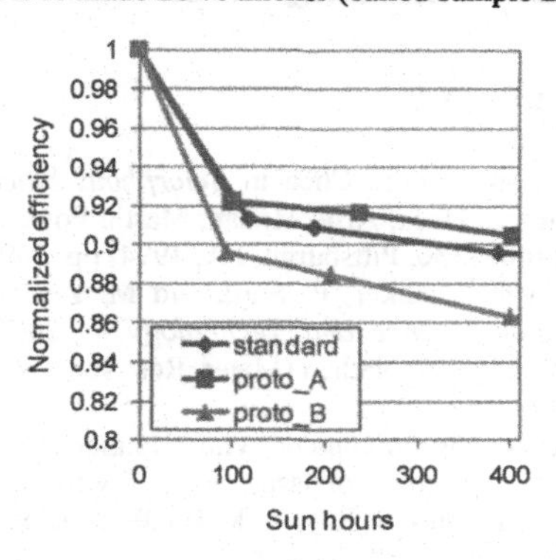

Figure 5. (a) Initial efficiency of mini modules with protocrystalline top cells (b) Light-induced degradation of the mini modules (both graphs show average data from 10 mini-modules).

As shown in figure 5(a) the proto_A tandem has a lower efficiency which is due to the lower current as a consequence of strong top cell limitation. Proto_B has nearly the same initial efficiency as the 'standard' cell. In figure 5(b) the light-soaking of the mini modules is shown. With respect to the stability against light soaking, proto_A shows a much better result and proto_B has worse properties than the 'standard' cell. The opposite results with respect to initial efficiency and stability demonstrate the trade-offs encountered in implementing protocrystalline silicon in Micromorph tandem cells.

CONCLUSIONS

We have successfully scaled the deposition process for protocrystalline silicon to Gen. 5 dimensions (1.43m^2). The strict requirements for uniformity and deposition rate are met by the newly developed process. The increased stability of the protocrystalline silicon was confirmed in a light-soaking test of single-junction a-Si:H solar cells.

We have outlined the complexity of the trade-offs that are relevant to the implementation of a protocrystalline silicon i-layer in the top cell of Micromorph tandem cells. These are illustrated by the results of two fabricated tandem cells. In one case, the thickness of the i-layer is kept the same as the reference cell, which reduces the initial efficiency due to current limitation by the top cell. In the second case, the i-layer thickness is increased, which reduces the resistance to light-soaking. Our future efforts towards a proof of concept for the use of protocrystalline silicon as the i-layer of the top cell in Micromorph tandem cells will be carried out in conjunction with numerical simulations.

ACKNOWLEDGMENTS

The authors would like to acknowledge the support from the Competence Centre Thin-Film- and Nanotechnology for Photovoltaics Berlin (PVcomB).

REFERENCES

1. L. Yang and L. Chen in *Amorphous Silicon Technology—1994*, edited by Eric A. Schiff, Michael Hack, Arun Madan, Martin Powell, Akihisa Matsuda (Mater. Res. Soc. Symp. Proc. **Volume 336**, Pittsburgh, PA, 1994) pp. 669-674.
2. G. van Elzakker, P. Šutta and M. Zeman, in *Amorphous and Polycrystalline Thin-Film Silicon Science and Technology — 2009*, edited by A. Flewitt, Q. Wang, J. Hou, S. Uchikoga, A. Nathan (Mater. Res. Soc. Symp. Proc. **Volume 1153**, Warrendale, PA, 2009) p. A18-02.
3. http://www.simkopp.de/rvm/ (Neither the authors of this paper nor the Materials Research Society warrants or assumes liability for the content or availability of this URL.)
4. P. Chaudhuri, S. Ray, A. K. Batabyal, and A. K. Barua, Solar Cells **31**, p. 13 (1991).

Mater. Res. Soc. Symp. Proc. Vol. 1245 © 2010 Materials Research Society 1245-A01-04

Uniformity and quality of monocrystalline silicon passivation by thin intrinsic amorphous silicon in a new generation plasma-enhanced chemical vapor deposition reactor

B. Strahm[1], Y. Andrault[1], D. Bätzner[1], D. Lachenal[1], C. Guérin[1], M. Kobas[1], J. Mai[2], B. Mendes[1], T. Schulze[2], G. Wahli[1], A. Buechel[1]
[1]Roth & Rau Switzerland SA, Rue de la Maladière 23, CH-2000 Neuchâtel, Switzerland.
[2]Roth & Rau AG, An der Baumschule 6-8, D-09337 Hohenstein-Ernstthal, Germany.

ABSTRACT

This work reports the first results of a new generation plasma-enhanced chemical vapor deposition (PECVD) reactor manufactured by Roth and Rau. This large area parallel plate reactor has been especially designed for the manufacturing of silicon heterojunction solar cells which are made of very thin amorphous silicon films over monocrystalline silicon substrates. Layer thickness uniformity below ± 3 % is reported for both intrinsic and doped layer over a 400 x 400 mm^2 area. Moreover, it is shown that the passivation quality is excellent with life-times up to 4.15 ms on *n*-type FZ silicon substrates. A ± 0.6 % uniformity in open circuit voltage (mean value of 701.4 mV) is achieved over 32 devices having a 4 cm^2 area and an average conversion efficiency of 19.5 %.

INTRODUCTION

Silicon heterojunction technology (Si-HJT) made of thin silicon layers on monocrystalline silicon substrate is an excellent candidate to reach photovoltaic solar cells with light conversion efficiencies above 20 %, even at the industrial production level[1]. High efficiency, low temperature processing, simple device structure, compatibility with thin wafers (< 100 µm) or small temperature coefficient are some of the advantages of Si-HJT that enable a considerable reduction of the manufacturing costs per Watt peak.

As shown in Figure 1, Si-HJT, cells built on *n*-type substrates, are constituted on the font (illumination) side successively by an intrinsic amorphous silicon passivation layer and a *p*-doped amorphous silicon emitter both deposited by plasma enhanced chemical vapor deposition (PECVD). On top of the silicon layers, an antireflective transparent conductive oxide (TCO) is deposited by physical vapor deposition (PVD) and the charge collection is made by a screen printed metallic contacting grid. On the back side, the stack is realized of an intrinsic amorphous silicon passivation layer, a back-surface field (BSF) layer made of *n*-type amorphous silicon both deposited by PECVD, a TCO layer and a metallic contacting layer.

For a successful mass production of Si-HJT cells, PECVD and PVD large area deposition reactors have to fulfill several criteria. The two most important are:

- A very high uniformity to reduce the intra-run spread in passivation quality in order to achieve narrow distribution in final cell performances.
- A very well controlled discharge ignition to guarantee inter-run reproducibility and a thickness control at the sub-nanometer level, since the layers are extremely thin (10 nm range).

In contrast to reactors used for the production of thin film solar cells, their primary objectives are to reach high deposition rates with acceptable uniformity and a thickness control in the range of 10-50 nm. This is due to the much thicker layers deposited in the case of thin film solar cells (≈ 250 nm for a-Si and ≈ 2500 nm for microcrystalline silicon cells[2]) compared to the ones needed for Si-HJT (≈ 10 nm) cells.

The present work reports about the uniformity of layers deposited in a new generation large area PECVD reactor especially designed by Roth & Rau for silicon heterojunction solar cell manufacturing. It is shown that thickness uniformity as good as ± 3 % can be achieved, whereas the uniformity in open circuit voltage is less than ± 1 % around 701.4 mV.

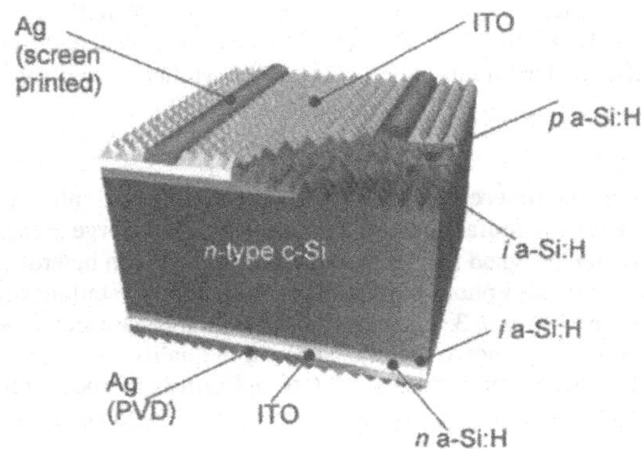

Figure 1 : Structure of a silicon heterojunction (Si-HJT) solar cells made from *n*-type monocrystalline silicon substrate.

EXPERIMENT

Solar cells have been produced following a standard amorphous/crystalline silicon heterojunction (Si-HJT) structure as described in Fig. 1. Textured *n*-type monocrystalline float zone silicon wafers (2-3 Ωcm) have been used as substrates. Prior to processing, the native silicon oxide has been removed from the wafer' surface by immersion in hydrofluoric acid before loading in the first vacuum deposition reactor. Hydrogenated amorphous silicon (a-Si:H) layers have been deposited by plasma enhanced chemical vapor deposition (PECVD). The used reactor is a large area capacitively-coupled parallel plate reactor manufactured by Roth & Rau and especially designed for Si-HJT manufacturing. This reactor is able to process 9 substrates per batch allowing considerable throughput for R&D purpose. Silicon layers have been manufactured from silane (SiH_4) and hydrogen (H_2) source gases. *p*- and *n*-doped layers have been obtained by adding trimethyl-boron ($B(CH_3)_3$) and phosphine (PH_3), respectively, to SiH_4 and H_2. Indium tin oxide (ITO) has been used as TCO both at the front and back sides. It has been deposited in a Roth & Rau manufactured magnetron sputtering system using a planar ITO ceramic target by pulsed direct current (DC) discharge. Sputtered silver has been deposited using the same tool with a DC source. Finally, the front contacting grid has been deposited using an Essemtec screen printer while using a silver containing paste and cured at low temperature (< 200 °C) in a multi-temperature belt furnace.

Devices (4 cm^2 area) have been characterized by current-voltage (IV) measurement under AM 1.5 illumination using a Vacom class A sun simulator by averaging 5 measurements.

Surface passivation and implied open circuit voltage have been determined by Sinton measurement in transient mode[3]. The layer uniformity over the reactor deposition area (500 x 500 mm^2) has been measured on a large (470 x 470 mm^2) glass substrate using white light interferometry technique. The overall uniformity is defined by :

$$U(\%) = \frac{d_{max} - d_{min}}{\bar{d}/2} \cdot 100, \quad (1)$$

where d_{max} and d_{min} are respectively the maximum and minimum measured thicknesses, and $\bar{d}$ is the mean value of the thickness over the whole deposition area.

RESULTS

Figure 2 presents mapping of the relative thickness (d/d_{max}, where d is the measured thickness) for intrinsic and *p*-doped a-Si:H deposited in the *S-cube* reactor. It is shown that the uniformity over the whole glass substrate is of ± 5.8 % and ± 9.0 % for intrinsic and doped a-Si:H, respectively, which is in the range of uniformities achieved in industrial state-of-the-art reactors, i.e. < 10 %. However, the uniformity measurements presented in Fig. 2 have been performed on a 470 x 470 mm^2 area, therefore only with a 15 mm wide frame exclusion from the reactor side-walls, which is much smaller than the usual 50 mm edge exclusion generally taken into account. If a 50 mm wide frame exclusion is considered in the *S-cube* reactor, the uniformity over the remaining 400 x 400 mm^2 is less than ± 3 % for both intrinsic (± 2.9 %) and *p*-doped (± 2.6 %) layers, which is in the range of the thickness measurement error.

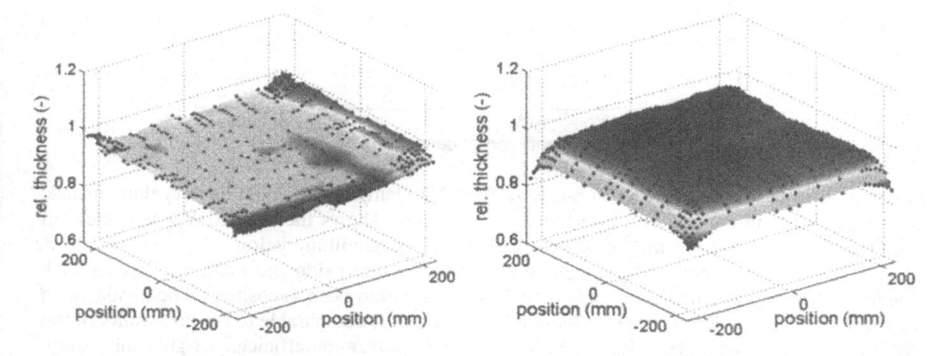

Figure 2 : Uniformity of intrinsic (left) and *p*-doped (right) a-Si:H layer deposited in the *S-cube* PECVD reactor over a 470 x 470 m^2 area. Uniformity is less than ± 3 % over 400 x 400 mm^2 for both type of layers.

Surface passivation is tested on polished *n*-type <100> FZ substrate (2-3 Ωcm) free of native oxide. Very thin (< 10 nm) amorphous silicon layers are deposited on both sides consecutively. As shown in Fig. 3, life-time up to 4.15 ms (at a minority carrier density of 10^{15} cm^{-3}) is reached after annealing on a hot plate, whereas it was already at 3.2 ms for as-deposited layers. These results are comparable with the highest reported life-times, even using other type of passivation material such as aluminum oxide [4]. Note that, such high passivations were already reported for a-Si:H on FZ c-Si, but using thick (40 nm) layers[5]. For the used substrates, a life-time of 4.15 ms corresponds to an implied V_{oc} of 730 mV at one sun illumination as shown in the inset of Fig. 3. These excellent results can be attributed to a very sharp interface between crystalline and amorphous silicon[6] even if the layers are very thin as determined by ellipsometric spectroscopy and transmission electron microscopy (not shown here).

To summarize, these results show that the *S-cube* reactor is able to deposit very uniform layers both for doped and undoped silicon with thickness uniformity better than ± 3 % over a 400 x 400 mm^2 area. Moreover, the quality of the passivation layer made of very thin amorphous silicon layers is of very high quality with life-times up to 4.15 ms, corresponding to implied V_{oc} of 730 mV.

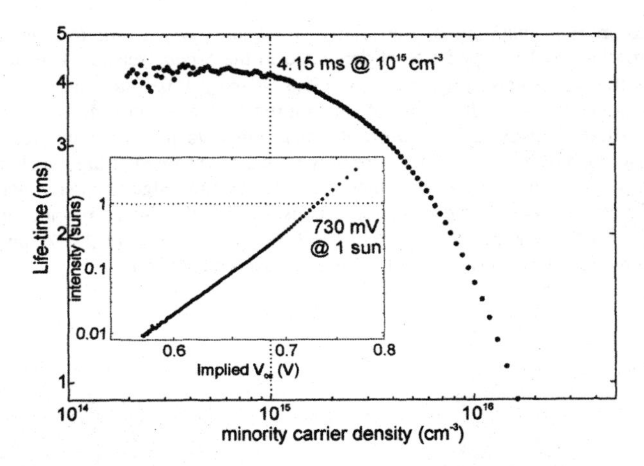

Figure 3 : Life-time measurement on *n*-type <100> (2-3 Ωcm) silicon wafer by very thin intrinsic a-Si:H deposited in the *S-cube* PECVD reactor. The inset shows the implied V_{oc} as calculated from the life-time curve reaching a value of 730 mV at one sun illumination.

Passivations using *i* and *p* layers on front (illumination) side and *i* and *n* layers on back side have been performed simultaneously on 32 cells. ITO has been deposited on both sides with a uniformity of ± 4.6 % over a 400 x 400 mm^2 area, which is comparable to the uniformity of the PECVD layers. Figure 4 presents the cell V_{oc} (left) and conversion efficiency (right) uniformity. The mean values are as high as 701.4 mV and 19.5 % for the V_{oc} and the efficiency, respectively. The use of textured substrate is at the origin of the reduction in V_{oc} as compared to the 730 mV presented in Fig. 3 which has been measured on polished substrate. The relative uniformities as determined by using Equation 1 are of ± 0.6 % and ± 2.6 %, respectively for V_{oc} and the efficiency. Even with these excellent uniformities, a larger dispersion is observed for efficiency than for V_{oc}. The almost perfect uniformity in V_{oc} confirms the excellent uniformity of the deposited layers, both silicon and ITO. Visual inspection of the cells presenting the lowest efficiencies shows some imperfections of the silver screen printed grid (finger interruption) impacting the final device performance because of losses in fill factor. Out of these results, the best 4 cm^2 cell has an efficiency of 19.7 %, with a V_{oc} of 703.6 mV, a I_{sc} of 36.2 mA/cm^2 and a fill factor of 77.3 % as summarized in Fig. 5. Note that these results include a 6.3 % loss due to illumination shadowing because of the front contact screen printed grid.

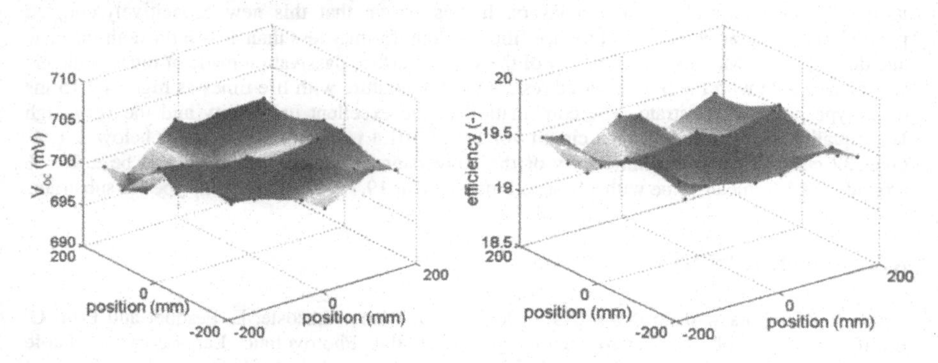

Figure 4 : Open circuit voltage (left) and light conversion efficiency (right) uniformity. Mean V_{oc} is of 701.4 mV and mean efficiency of 19.5 %, whereas the uniformity are respectively of 0.6 % and 2.6 % based on Equation 1.

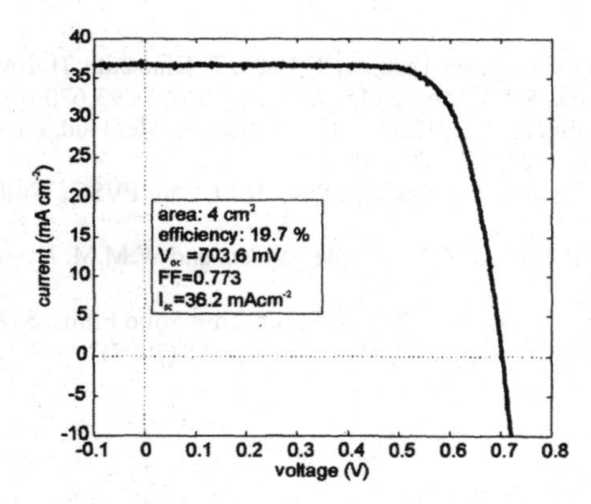

Figure 5 : Light IV measurement of a 2 x 2 cm^2 cell showing a conversion efficiency of 19.7 %.

CONCLUSIONS

A new generation large area plasma-enhanced chemical vapor deposition reactor was used to deposit amorphous silicon layers. It was shown that this new capacitively-coupled parallel plate reactor was able to produce films with uniformity less than ± 3 % for both intrinsic and doped layers. Moreover, the quality of the silicon surface passivation made of intrinsic layers was comparable with the best reported results in the literature with life-times as high as 4.15 ms on *n*–type polished substrate. The combination of the excellent uniformity and the very high layer quality resulted in cell open circuit voltage of 701.4 mV with a uniformity below ± 1 %. Over 32 cells (4 cm^2), the uniformity of the conversion efficiency was shown to be ± 2.6 % around a 19.5 % mean value with a best cell measured at 19.7 % on textured n-type FZ substrate.

ACKNOWLEDGMENTS

The authors wish to thank Drs S. de Wolf, J. Dâmon-Lacoste, L. Fesquet and Prof. C. Ballif, members of the heterojunction group of the Photovoltaic Laboratory at Ecole Polytechnique Fédérale de Lausanne (EPFL), as well as the Sächsische Aufbaubank for financial support (project No. 12898).

REFERENCES

1. Y. Tsunomura, Y. Yoshimine, M. Taguchi, T. Baba, T. Kinoshita, H. Kanno, H. Sakata, E. Maruyama and M. Tanaka, Solar Energy Mater. and Solar Cells **93**,670-673 (2009).
2. A. Shah, J. Meier, A. Buechel, U. Kroll, J. Steinhauser, F. Meillaud, H. Schade and D. Dominé, Thin Solid Films **502**, 292-299 (2006).
3. R. A. Sinton, A. Cuevas and M. Stuckings, Proc. IEEE 25th PVSC, Washington (USA), 457-460 (1996).
4. B. Hoex, J. Schmidt, P. Pohl, M.C.M. van de Sanden, and W.M.M. Kessels, J. Appl. Phys. **104**, 044903 (2008).
5. J. Damon-Lacoste, L. Fesquet, S. Olibet, C. Ballif, Thin Solid Films, **517**, 6401-6404 (2009).
6. S. De Wolf and M. Kondo, Appl. Phys. Lett. **90**, 042111 (2007)

Solar Cell: Nanocrystalline Si

Mater. Res. Soc. Symp. Proc. Vol. 1245 © 2010 Materials Research Society 1245-A02-01

High-efficiency Large-area Nanocrystalline Silicon Solar Cells Using MVHF Technology

Xixiang Xu[1], Tining Su[1], Scott Ehlert[1], David Bobela[2], Dave Beglau[1], Ginger Pietka[1], Yang Li[1], Jinyan Zhang[1], Guozhen Yue[1], Baojie Yan[1], Greg DeMaggio[1], Chris Worrel[1], Ken Lord[1], Arindam Banerjee[1], Jeff Yang[1], and Subhendu Guha[1]
[1]United Solar Ovonic LLC, 1100 West Maple Road, Troy, MI, 48084, U.S.A.
[2]National Renewable Energy Laboratory, 1617 Cole Blvd, Golden, CO, 80401, U.S.A.

ABSTRACT

We present the progress made in attaining high-efficiency large-area nc-Si:H based multi-junction solar cells using Modified Very High Frequency technology. We focused our effort on improving the spatial uniformity and homogeneity of nc-Si:H film growth and cell performance. We also conducted both indoor and outdoor light soaking studies and achieved 11.2% stabilized efficiency on large-area (≥400 cm^2) encapsulated a-Si:H/nc-Si:H/nc-Si:H triple-junction cells.

INTRODUCTION

Hydrogenated nanocrystalline silicon (nc-Si:H), with its superior long-wavelength response and light-soaking stability, has become a promising candidate to replace hydrogenated amorphous silicon-germanium alloy (a-SiGe:H) in multi-junction thin-film silicon solar cells [1,2]. However, because of its indirect bandgap, the thickness of nc-Si:H material must be larger than that of a-Si:H material in order to achieve sufficient light absorption at longer wavelengths. Typical thicknesses of the nc-Si:H based multi-junction cells are 2 to 5 μm, while the a-Si:H based cells are ≤0.5 μm thick. Therefore, the main challenges for using nc-Si:H in thin-film PV manufacturing are to (i) attain high deposition rates and (ii) achieve homogenous properties over a large deposition area [3]. In this work, we present the progress made in attaining over 11% stabilized large-area (≥400 cm^2) encapsulated nc-Si:H based triple-junction cells at a high deposition rate.

EXPERIMENTAL

a-Si:H and nc-Si:H multi-junction solar cells were deposited by PECVD in batch deposition systems on Ag/ZnO coated stainless steel substrates. The substrate dimension was 15" x 14". Intrinsic a-Si:H and nc-Si:H layers were deposited using Modified Very High Frequency (MVHF) excited plasma, while *n* and *p* doped layers were deposited with a radio frequency (RF) plasma. Typical deposition rates were 5-15 Å/s for the intrinsic a-Si:H and nc-Si:H layers.

Most studies were conducted on fully encapsulated large-area (≥ 400 cm^2) solar cells. However, small-area cells with an active area of 0.25 cm^2 were first selected from various locations of the large area to study uniformity across the large area. Current-density versus voltage (J-V) and quantum efficiency (QE) measurements were performed for solar cell characterization.

We studied both a-Si:H/nc-Si:H/nc-Si:H triple-junction and a-Si:H/nc-Si:H double-junction structures deposited by MVHF technology. Additionally, a-Si:H/a-SiGe:H/a-SiGe:H triple-junction cells deposited by RF were used as reference samples in light soaking tests.

RESULTS AND DISCUSSION

At the MRS2009 Spring Meeting, we reported initial aperture-area (400 cm^2) cell efficiency of 10 - 10.5% for encapsulated a-Si:H/nc-Si:H double-junction and a-Si:H/nc-Si:H/nc-Si:H triple-junction cells [3]. We subsequently conducted light soaking tests on these cells. As illustrated in Table I, the triple-junction cell degraded by less than 5%, while the double-junction cell degraded by ~15%. We should point out that there was no Intermediate Reflector (IR) inserted between the a-Si:H top cell and nc-Si:H bottom cell [4]. In this work, we report the progress made in developing and optimizing the a-Si:H/nc-Si:H/nc-Si:H triple-junction solar cells.

Table I. Initial and stabilized J-V characteristics of encapsulated a-Si:H/nc-Si:H double-junction and a-Si:H/nc-Si:H/nc-Si:H triple-junction cells. The aperture area is 400 cm^2.

Cell#	Cell Structure	Light Soak Time [hrs]	J_{sc} [mA/cm^2]	V_{oc} [V]	FF	Efficiency [%]
14337	a-Si/nc-Si double-junction	0	11.6	1.39	0.63	10.2
		1000	11.5	1.37	0.54	8.6
		Degradation	1.9%	0.9%	14.0%	**15.6%**
14707	a-Si/nc-Si/nc-Si triple-junction	0	7.5	1.86	0.73	10.2
		1000	7.5	1.86	0.69	9.7
		Degradation	0.9%	-0.4%	4.4%	**4.5%**

nc-Si:H material optimization

We investigated the correlation between nc-Si:H solar cell performance and nc-Si:H film properties using hydrogen effusion (H-effusion), Raman scattering, and XRD. The H-effusion spectra of good device quality nc-Si:H films have a distinct peak between 350 and 400 °C, which is significantly different from that of a-Si:H or mixed-phase films. We also observed the interrelation between features of H-effusion spectra, crystal-volume fraction from Raman spectra, and nc-Si:H cell performance. Details of the results and their potential correlation with the hydrogenation of the grain boundaries are presented elsewhere [5].

nc-Si:H component cell development

Increasing the quantum efficiency, particularly in the long wavelength range (>600 nm), of nc-Si:H component cells is a key factor in improving the a-Si:H/nc-Si:H/nc-Si:H triple-junction cell performance. We have made significant improvements in the nc-Si:H cell's long wave length response by optimizing the initial nc-Si:H layer (seed layer) deposition and effectively controlling the vertical nc-Si:H growth with a process parameter profile. As shown in Figure 1, the total current density obtained by integrating the QE (300-1200 nm) has increased from 25.7 mA/cm^2 to 27.3 mA/cm^2, with most of the gain in the bottom nc-Si:H cell.

We also modified the cathode configuration and power coupling based on extensive modeling and simulation to improve the homogeneity of the deposited nc-Si:H film. Prior to the modification, the central region of the deposited nc-Si:H showed a higher crystal volume fraction than at the edges, resulting in lower open-circuit voltage for the nc-Si:H cells, even though the films showed good spatial thickness uniformity. The modified cathode not only expands the area of homogeneous nc-Si:H, but also creates optimum plasma conditions to deposit nc-Si:H with improved properties.

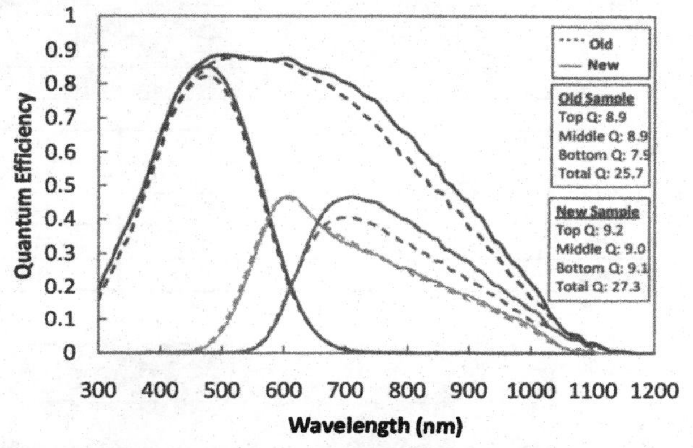

Figure 1. Quantum efficiency of an improved a-Si:H/nc-Si:H/nc-Si:H triple-junction cell in comparison with a previous cell.

The characteristics of the triple-junction solar cells made with optimized hardware and process parameters are listed in Table II and compared with a set of previous samples. These are fully encapsulated large-area cells, having an aperture area of ≥400 cm^2. The new samples show about 10% higher efficiency than the previous set, 11.5% vs. 10.4%, with almost all the gain obtained from the short-circuit current density.

Light induced degradation

We investigated light-induced degradation of the large-area (≥400 cm^2) encapsulated solar cells by conducting both indoor and outdoor tests. For the indoor light soaking test, the solar cells were exposed to white light (100 mW/cm^2) for 1100 hours under open circuit condition at a temperature of 50 °C. As shown in Fig. 2, the a-Si:H/nc-Si:H/nc-Si:H triple-junction cells exhibit less degradation and reach a saturated state much sooner than the a-Si:H based cells. We also conducted light soaking on small-area cells (1 cm^2) encapsulated in the same way as the large- area cells to monitor the change in QE of each component cell in a triple-junction cell structure. As expected, most degradation arose from the a-Si:H top cells (1-4%), while the nc-Si:H middle and bottom cells show almost no degradation (0-2%). We also fabricated sub-modules, each consisting of five interconnected cells having an aperture area of ~400 cm^2, and subjected them to outdoor testing. The degradation of these interconnected modules was in range of 3-6% after being exposed to 150 days of sunlight.

Table II. Initial J-V characteristics of a group of new a-Si:H/nc-Si:H/nc-Si:H triple-junction cells, compared with the previous group. All samples are fully encapsulated, with an aperture area ≥400 cm^2.

Cell #	V_{oc} (V)	J_{sc} (mA/cm^2)	FF	Efficiency (%)
14710	1.88	7.84	0.71	10.5
14715	1.88	7.81	0.70	10.4
14716	1.88	7.95	0.70	10.4
Previous average	1.88	7.87	0.70	10.4
15357	1.89	8.57	0.70	11.3
15359	1.88	8.56	0.71	11.5
15387	1.90	8.68	0.70	11.6
New average	1.89	8.61	0.70	11.5
New/Previous	**1.00**	**1.09**	**1.00**	**1.10**

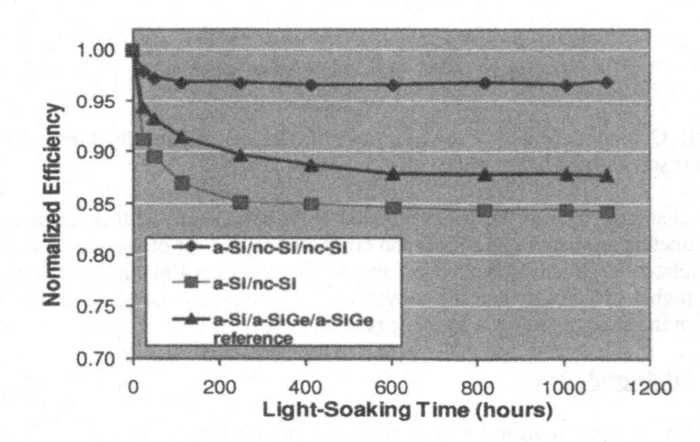

Figure 2. Normalized efficiency plotted against light soaking time, where a-Si:H/nc-Si:H/nc-Si:H triple-junction cells show a better stability than the a-Si:H/nc-Si:H double-junction cells and the a-Si:H/a-SiGe:H/a-SiGe:H reference cell.

We also compared two groups of a-Si:H/nc-Si:H/nc-Si:H triple-junction cells with different nc-Si:H component cells, with the nc-Si:H component cells characterized by different values of V_{oc} and J_{sc}, as listed in Table III. Group A (higher V_{oc} and lower J_{sc}) showed slightly better stability than Group B (lower V_{oc} and higher J_{sc}), 3% vs. 5% degradation. The best stabilized efficiency was 11.2%.

Table III. Initial and stable J-V characteristics of two groups of a-Si:H/nc-Si:H/nc-Si:H triple-junction cells. Group A, having higher V_{oc} and lower J_{sc} than Group B, yields slightly better stability. All samples are fully encapsulated with an aperture area ≥ 400 cm^2.

Sample Group	Light Soak State	V_{oc} (V)	J_{sc} (mA/cm^2)	FF	Efficiency (%)
Group A	Initial	1.89	8.60	0.70	11.4
	Stable	1.88	8.59	0.69	11.1
	Stable/Initial	1.00	1.00	0.97	0.97
Group B	Initial	1.84	8.80	0.70	11.3
	Stable	1.83	8.73	0.67	10.8
	Stable/Initial	1.00	0.99	0.96	0.95

We fabricated two sub-modules, each consisting of five interconnected cells having an aperture area of ~400 cm^2, and subjected them to outdoor testing. One of these interconnected modules was exposed to 75 days of sunlight and showed ~4% degradation, while the other showed ~6% degradation after 30 days of insolation. The modules subjected to outdoor light soaking displayed greater variability in measured degradation due to the lower winter temperatures and large variance in day-to-day sunlight intensities in Michigan.

NREL and USO measurements

Table IV lists characteristics of 4 stabilized triple-junction cells measured at both USO and NREL. All four cells are fully encapsulated, have an aperture area of ≥400 cm^2, and have been light-soaked indoors for 1000 hours. As illustrated in Table IV, USO and NREL's measurement of two previous samples (14710, 14713) agreed within 2%. However, NREL's measurement of two recent samples (15359 and 15920) was about 5% lower than USO's, mainly in the J_{sc} value. Part of difference between NREL and USO measurements on the recent samples is attributed to the updated AM1.5 spectrum adopted by NREL [6].

Table IV. Stabilized J-V characteristics of encapsulated a-Si:H/nc-Si:H/nc-Si:H triple-junction cells with aperture area ≥ 400 cm^2. Cells were measured at both USO and NREL. The agreement is within 5%.

Cell #	Measured by	V_{oc} (V)	J_{sc} (mA/cm^2)	FF	Efficiency (%)
14710	USO	1.86	7.90	0.68	10.0
	NREL	1.88	8.01	0.67	10.1
14713	USO	1.88	7.67	0.68	9.8
	NREL	1.89	7.96	0.66	9.9
15359	USO	1.88	8.54	0.69	11.1
	NREL	1.87	8.38	0.67	10.6
15920	USO	1.90	8.88	0.66	11.1
	NREL	1.89	8.50	0.66	10.5
	USO average	1.88	8.25	0.68	10.5
	NREL average	1.88	8.21	0.67	10.3
	USO/NREL	**1.00**	**1.00**	**1.02**	**1.02**

SUMMARY

We have improved the MVHF high-rate deposition process for nc-Si:H material and nc-Si:H component cells mainly by optimizing the nc-Si:H initial seed layer and controlling nc-Si:H vertical grain evolution, resulting in about a 10% improvement in long wavelength photocurrent. We also conducted both indoor and outdoor light soaking tests on large-area encapsulated a-Si:H/nc-Si:H/nc-Si:H triple-junction cells and achieved a 11.2% stabilized efficiency.

ACKNOWLEDGMENTS

The authors thank H. Fritzsche for valuable discussions, and S. Almutawalli, K. Younan, D. Wolf, T. Palmer, L. Sivec, N. Jackett, B. Sivec, J. Noch, J. Pfeiffer, G. St. John, B. Hang, B. Hartman, J. Piner, R. Capangpangan, R. Caraway, D. Tran, C. Steffes, S. Liu, J. Owens, as well as E. Chen for sample preparation and measurements. The work was supported by US DOE under the Solar America Initiative Program Contract No. DE-FC36-07 GO 17053.

REFERENCES

1. J. Meier, R. Flückiger, H. Keppner, and A. Shah, Appl. Phys. Lett. **65**, 860 (1994).
2. B. Yan, G. Yue, and S. Guha, Mat. Res. Soc. Symp. Proc. 989, 335 (2007).
3. X. Xu, Y. Li, S. Ehlert, T. Su, D. Beglau, D. Bobela, G. Yue, B. Yan, J. Zhang, A. Banerjee, J. Yang, and S. Guha, Mat. Res. Soc. Symp. Proc. Vol. **1153** (2009).
4. P. Buehlmann, J. Bailat, D. Domine, A. Billet, F. Meillaud, A. Feltrin, and C. Ballif, Appl. Phys. Lett. **91**, 143505 (2007).
5. T. Su, D. Bobela, X. Xu, S. Ehlert, D. Beglau, G. Yue, B. Yan, A. Banerjee, J. Yang, and S. Guha, to be presented at 2010 MRS Spring Meeting, Symp. A.
6. Keith Emery, private communication.

Mater. Res. Soc. Symp. Proc. Vol. 1245 © 2010 Materials Research Society 1245-A02-03

OXYGENATED PROTOCRYSTALLINE SILICON THIN FILMS FOR WIDE BANDGAP SOLAR CELLS

R.E.I. Schropp, J.A. Schüttauf, C.H.M. van der Werf
Utrecht University, Faculty of Science, Debye Institute for Nanomaterials Science, Department of Physics and Astronomy, Section Nanophotonics – Physics of Devices, P.O. Box 80.000, 3508 TA Utrecht, The Netherlands
[1]Present address: Eindhoven University of Technology, Eindhoven, The Netherlands

ABSTRACT

Protocrystalline silicon, which is a material that has enhanced medium range order (MRO), can be prepared by using high hydrogen dilution in PECVD, or, alternatively, using high atomic H production from pure silane in HWCVD. We show that this material can accommodate percentage-level concentrations of oxygen without deleterious effects. The advantage of protocrystalline SiO:H for application in multijunction solar cells is not only that it has an increased band gap, providing a better match with the solar spectrum, but also that the solar cells incorporating this material have a reduced temperature coefficient. Further, protocrystalline materials have a reduced susceptibility to light-induced defect creation. We present the unique result in the PV field that these oxygenated protocrystalline silicon solar cells have an efficiency temperature coefficient (TCE) that is virtually zero (TCE is between -0.08%/°C and 0.0/°C). It is thus beneficial to make this cell the current limiting cell in multibandgap cells, which will lead to improved annual energy yield.

INTRODUCTION

Oxygen and carbon are well known as bandgap widening constituents in hydrogenated amorphous silicon (a-Si:H). Both additions have been used primarily in p-type a-Si:H layers to make them more transparent for blue light and thus achieve higher photocurrent generation in the intrinsic layer of p-i-n type solar cells. The use of oxygenated p-layers was for instance reported by Fuji Electric, Co. [1]. The addition of oxygen to a-Si:H i-layers has been studied before [2], but the use of oxygenated i-layers in practical solar cells has not been investigated much [3,4]. Oxygen is commonly considered to lead to donor centers in silicon, accompanied with mid-gap defects particularly in amorphous silicon. In this paper, we show that the situation is completely different for *protocrystalline* silicon (proto-Si:H), a material much like amorphous silicon except for an enhanced medium range order (MRO). A fingerprint for enhanced MRO is the reduced width (FWHM < 5.5° [5]) of the first scattering peak in X-ray scattering measurements. In this FWHM the partial pair distribution functions based on next-nearest neighbor separations play an important role. A higher correlation length obtained from the narrowed FWHM then means that the bonds are more ordered. Enhanced MRO is usually achieved in Plasma Enhanced Chemical Vapor Deposition (PECVD) by strong dilution of the silane (SiH_4) with H_2. A disadvantage is that the deposition rate is only 1-3 Å/s. Alternatively, it can be obtained from pure SiH_4 at 10 Å/s using Hot Wire CVD [6].

The enhanced MRO implies that the material has a higher density and has on average stronger bonds. In this paper we show that the incorporation of oxygen at small concentrations in proto-Si thin films does not necessarily lead to deterioration of the semiconductor purity but instead helps to increase the optical band gap and thus, the open circuit voltage in thin film solar cells. Under such conditions, not leading to an increase in the density of recombination centers, oxygen alloying is a good method to form high bandgap materials.

A higher band gap has many advantages: (a) the conversion efficiency of tandem and triple cells can be improved due to better spectral splitting of solar illumination; (b) the temperature coefficient of single junction and multijunction solar cells can be reduced. A lower temperature coefficient is of importance for optimizing the annual yield of solar cells because of the high operating temperatures (up to 70°C) often encountered in outdoor applications. It is of interest to note that the undesirable high operating temperatures occur exactly when the expected energy yield is highest (at high irradiation intensities); (c) a higher open circuit voltage can be obtained, which is beneficial as it reduces series resistance losses in thin film electrodes, such as in the transparent conducting oxide (TCO) electrodes.

EXPERIMENTAL RESULTS AND DISCUSSION

Preparation

All silicon layers in this study were deposited by regular rf PECVD at 13.56 MHz in the PASTA multichamber deposition system at the Utrecht Solar Energy Laboratory (USEL). The i-layers were made using SiH_4 and CO_2, using a SiH_4/CO_2 gas flow ratio in the range from 4 to 8, under strong H_2 dilution. The i-layer deposition temperature was 140 °C. Also the p-layers were oxygenated with CO_2.

We have studied oxygenated protocrystalline silicon p-i-n solar cells, either with a standard a-SiC:H p-layer or with an a-SiO:H p-layer. The oxygenated p-layers were deposited using a CO_2/SiH_4 gas flow ratios in the range from 0.4 to 0.6 (typically 30 sccm SiH_4 and 50 sccm CO_2), and $B(CH_3)_3$ (also called TMB) as the dopant gas. The cells have the structure: glass/Asahi U-type SnO_2:F/a-SiO:H p-layer/SiO:H i-layer/a-Si:H n-layer/Ag/Al (see Fig. 1). It is of interest to note that the i-layers were made in the p-chamber; therefore only two chambers of the system were needed.

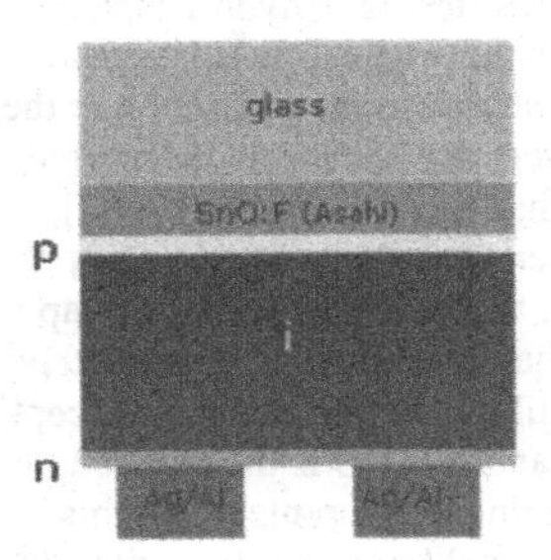

Figure 1. Schematic cross section of the experimental p-i-n solar cells in this study. The SnO_2:F is textured (not shown). The intrinsic protocrystalline silicon layer is oxygenated using CO_2 in the 13.56 MHz PECVD gas mixture. The thickness of this layer is kept at 200 nm, similar to that used in a-Si/nc-Si tandem and a-Si/nc-Si/nc-Si triple cells. No reflection enhancement layer is used together with the Ag/Al back contact.

O and C content of intrinsic silicon layers

In order to confirm the presence of O in i-layers we performed Elastic Recoil Detection (ERD) analysis on layers with the same thickness as used in the solar cells. The determination of

Si-O bonds using FTIR would have been difficult on these thin layers (~200 nm) due to the native oxide that is always immediately formed.

In ERD analysis, accelerated heavy ions are incident on the thin film, and the lighter elements in the silicon (such as H, C, N, O) are recoiled from the material under a forward angle. The ERD measurements were carried out with a 50 MeV Cu^{8+} ion beam supplied by a tandem Van de Graaff accelerator. The angle of incidence with the sample surface was 27°. Particles recoiled at an angle of 30° with the incoming beam were collected by an ionization chamber. For films thinner than a few 100 nm, the depth profiles of Si, O, N and C can be measured in this way. The hydrogen recoils were detected separately by a Si-detector at the rear end of the ionization chamber.

As can be seen in Fig. 2, we find (from right to left) an unavoidably high surface concentration of oxygen and carbon (native oxide and surface contamination), and a finite and almost constant concentration of both elements in the bulk of the layer. From the spectra in Fig. 2 we deduce that a typical i-layer made with 0.8 sccm CO_2 and 5 sccm H_2 under strong H_2 dilution (200 sccm) contains ~0.5 at.-% O and about ~0.06 at.-% C. This shows that the O from the CO_2 gas is preferentially built in during deposition. This sample had an E_{04} band gap of 2.05 eV and an optical absorption coefficient at 600 nm of 1.1×10^4 cm^{-1}.

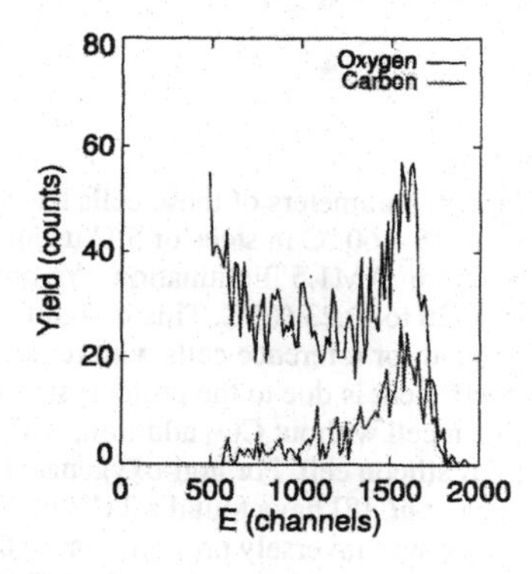

Figure 2. ERD spectra of oxygen and carbon in a 200 nm thick intrinsic protocrystalline silicon layer.

Thin film and cell properties

The a-SiO:H p-layers have an E_{04} of 2.14 eV and are thus very suitable as window layer. The conductivity activation energy of this material is 0.501 eV, which is in the right range as well [7]. The i-layers made without CO_2 addition have an E_{04} band gap of 2.00 eV and an absorption coefficient at 600 nm of 1.7×10^4 cm^{-1}. Equivalent i-layers deposited with CO_2 addition have an E_{04} band gap of 2.05 eV and an absorption coefficient at 600 nm of 1.1×10^4 cm^{-1}, consistent with the higher band gap.

As the back contact for the solar cells we used a Ag/Al double layer; there is no reflection enhancement layer at the back contact such as ZnO. Our best results so far show a V_{oc} of 0.962 V, while the fill factor is as high as 0.721 and J_{sc} is 9.00 mA/cm^2 for cells that are only 200 nm thick. The efficiency thus is 6.24%. The solar cell parameters averaged over 10 cells are 6.2% efficiency, FF = 0.714, J_{sc} = 9.04 mA/cm^2, and V_{oc} = 0.961 V, showing that the properties do not diverge much. The FF does not deteriorate due to adding CO_2. The high FF shows that the

CO_2 addition to the gas flow during deposition of the i-layer does not lead to increased recombination. In fact, the reference cell without CO_2 addition to the i-layer has a *lower* FF of 0.669. As can be seen in Fig. 3, the effect of adding CO_2 during deposition of the i-layer increases V_{oc} by ~10 mV. This is not a large increase because the i-layer of the reference cells already have a high V_{oc} of 0.953 V as they are protocrystalline. Due to the low FF of the reference cell, the efficiency of the cell with CO_2 addition is higher than that made without CO_2.

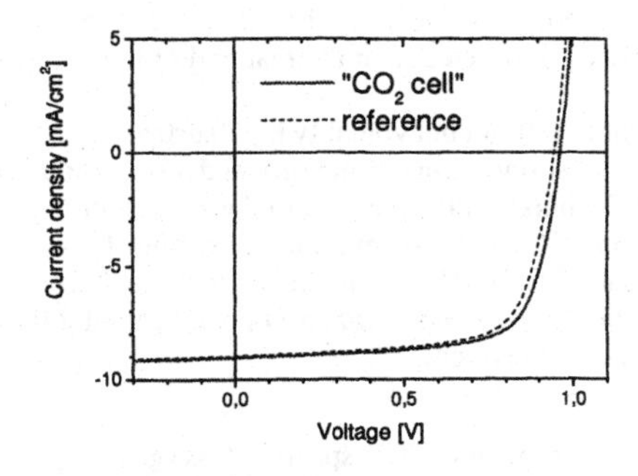

Figure 3. The J-V curves under AM1.5 100 mW/cm^2 illumination, at 25 °C, for an oxygenated protocrystalline cell compared with a reference cell made under the same conditions without the addition of CO_2 to the gas mixture during the i-layer.

Temperature coefficients

The temperature coefficient of the typical solar cell parameters of these cells has been recorded indoors in the laboratory for temperatures from 25 – 60 °C in steps of 5°C using a controllable temperature stage under calibrated 100 mW/cm^2 AM1.5 illumination. We observed a temperature coefficient for V_{oc} (TCV) ranging from -0.20 to -0.23 %/°C. This is significantly better than the coefficient of -0.26%/°C that we have found for reference cells with regular a-Si:H i-layers. A part of the improved temperature coefficient is due to the protocrystalline nature of the material: a reference protocrystalline silicon cell without CO_2 addition, with a TCV of -0.23%/°C, already behaves better than an amorphous silicon cell. For non-oxygenated (regular) protocrystalline silicon solar cells K. Sriprapha *et al.* [8] have found a TCV of -0.27 %/°C. In their article, it was also noted that the TCV was inversely proportional to the initial V_{oc}, as expected. Therefore, the improved values of the TCV, as small as -0.20 %/°C, observed in this work are probably due to the presently higher initial V_{oc} values.

In Fig. 4 we show the V_{oc} versus T of a typical oxygenated protocrystalline cell, with a TCV of -0.21 mV/°C. Both J_{sc} and FF increase as the T increases. This counteracts the V_{oc} decrease to a large extent. In particular, for cells with oxygenated i-layers (thus having a large band gap), the V_{oc} decrease is almost completely cancelled by the increases in J_{sc} and FF. This leads to the advantageous effect that the efficiency is almost insensitive to the ambient temperature. This is truly unique and, to our knowledge, not observed for any other cell technology, be it c-Si, CIGS, CdTe, III-V, organic or dye cells. For silicon thin films, the overall temperature coefficient of the efficiency (TCE) appears dependent on the absolute efficiency. This is shown in Fig. 5. The explanation is that for thinner cells, the V_{oc} is higher and thus the TCV is lower, while the temperature coefficient for the J_{sc} is not changing. In all, thinner cells have a lower overall temperature coefficient.

Previously, Yanagisawa *et al.* [9] have found TCEs for thin film *amorphous* silicon solar cells of -0.4 - -0.5 %/°C. They further noted that the TCE was dependent on the degradation state, with lower TCE's for degraded cells. In their publication, however, all TCE's were negative, as they studied the 'classical' amorphous silicon cells. For our *protocrystalline, oxygenated* cells, we have found TCEs ranging from -0.08%/°C for the cells with highest

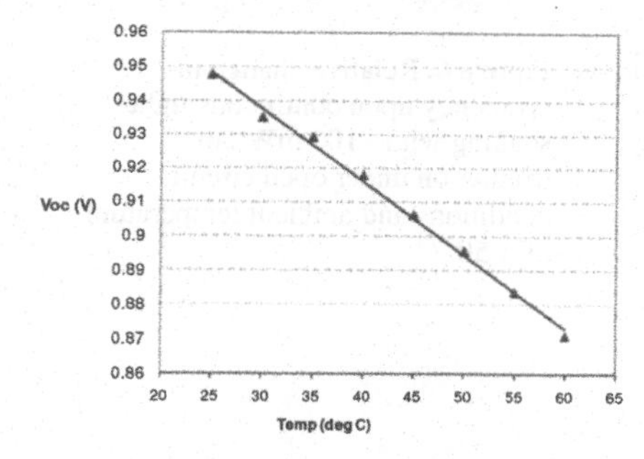

Figure 4. The V_{oc} versus T of a typical oxygenated protocrystalline silicon cell. The TCV is 0.21%/°C.

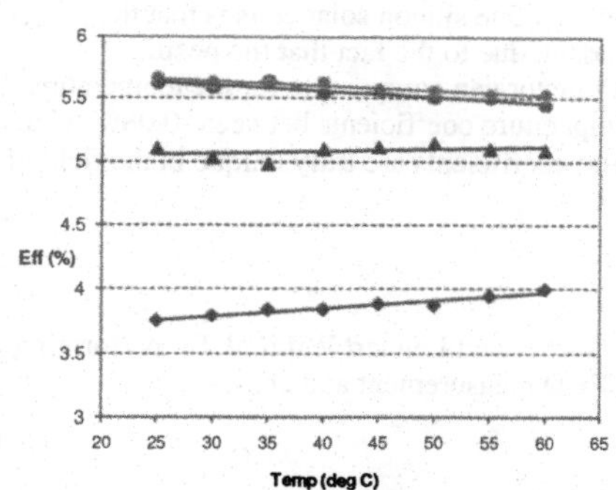

Figure 5. The dependence of the efficiency of various oxygenated protocrystalline SiO:H p-i-n solar cells. The temperature coefficient of the efficiency is very small or even positive for cells with an efficiency of ~5% and lower.

efficiencies to +0.19%/°C for cells with lower efficiencies. The cells with STC efficiency between 5.0 to 5.5% thus have a temperature coefficient of zero.

Stability

As these cells do not only have a large band gap, but also have the protocrystalline network structure, it is expected that they behave relatively stable compared to standard amorphous silicon cells. The cells with oxygenated i-layers were exposed to continuous 1-sun illumination under open circuit conditions (worst case) and at a temperature < 50°C, for a prolonged time. The decrease in efficiency is only 8 to 9% relative (Fig. 6), which is very small as compared to the 18 to 25% degradation in 'classic' amorphous silicon.

CONCLUSIONS

We have studied the temperature dependent behavior of oxygenated protocrystalline silicon solar cells. Protocrystalline silicon is of interest as the absorber layer in thin film solar cells because of its inherently improved stability. The already high open circuit voltage of such cells is advantageous to the energy yield in outdoor applications, mainly due to the reduced temperature coefficient of the open circuit voltage. The V_{oc} can be further increased by alloying with oxygen through addition of CO_2 to the feedstock gas mixture.

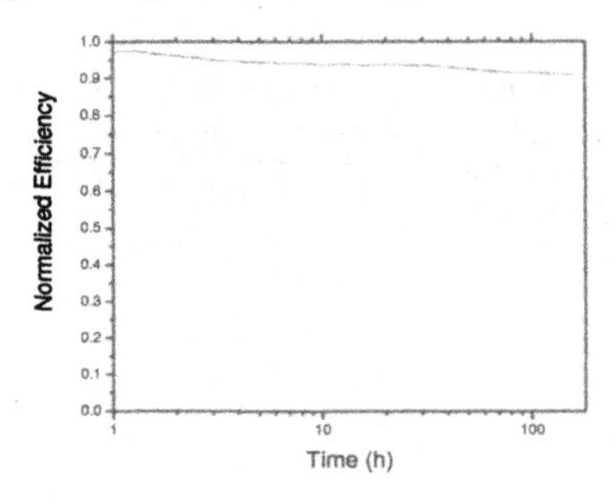

Figure 6. Relative change in efficiency upon continuous light soaking with ~100 mW/cm^2 irradiation under open circuit conditions and ambient temperature of < 50 °C.

A great advantage of the oxygenated protocrystalline silicon solar cells is that the power output (efficiency) is virtually temperature independent, due to the fact that the positive temperature coefficient for the photocurrent and fill factor can compensate for the temperature effect on the output voltage. We find efficiency temperature coefficients between -0.08%/°C to +0.19%/°C. The thus obtained near-zero temperature coefficients are truly unique in the field of solar cells.

ACKNOWLEDGMENTS

We thank Joachim Binnekade, Hans van Franeker and Lennart Wilmink for performing the characterizations and Wim Arnold Bik for the ERD measurement and analysis.

REFERENCES

1. S. Fujikake, H. Ohta, P. Sichanugrist, M. Ohsawa, Y. Ichikawa and H. Sakai: Optoelectronics—Devices and Technologies **3** (1994) 379.
2. D. Das, S.M. Iftiquar, D. Das, A.K. Barua, J. Mater. Sci. **34** (1999) 1051.
3. P. Sichanugrist, N. Pingate, and C. Piromjit, Mat. Res. Soc. Proc. **989** (2007) A18.08.
4. S. Inthisang, K. Sripraphа, A. Yamada, and M. Konagai, 33rd IEEE PV Specialists Conference, San Diego, May 11-16, 2008.
5. D.L. Williamson, Mat. Res. Soc. Proc. **557** (1999) 251.
6. R.E.I. Schropp, M.K. van Veen, C.H.M. van der Werf, D.L. Williamson, A.H. Mahan, Mat. Res. Soc. Proc. **808** (2004) A8.4.1.
7. R.E.I. Schropp and M. Zeman, *Amorphous and Microcrystalline Silicon Solar Cells: Modeling, Materials, and Device Technology*, ISBN: 978-0-7923-8317-8, Series: Electronic Materials: Science and Technology (Springer, 1998).
8. K. Sripraphа, I.A. Yunaz, S.Y. Myong, A. Yamada, and M. Konagai, Jpn. J. Appl. Phys. **46** (2007) 7212.
9. T. Yanagisawa, T. Kojima, T. Koyanagi, K. Takahisa, K. Nakamura, Solar Energy Materials & Solar Cells **69** (2001) 287.

Mater. Res. Soc. Symp. Proc. Vol. 1245 © 2010 Materials Research Society 1245-A02-04

Critical Concentrations of Atmospheric Contaminants in a-Si:H and μc-Si:H Solar Cells

T. Merdzhanova[1], J. Woerdenweber[1], T. Kilper[1], H. Stiebig[2], W. Beyer[1,2], A. Gordijn[1]
[1]IEF5-Photovoltaik, Forschungszentrum Jülich GmbH, D-52425 Jülich, Germany
[2]Malibu GmbH & Co. KG, Böttcherstr. 7, D-33609 Bielefeld, Germany

ABSTRACT

We report on a direct comparison of the effect of the atmospheric contaminants on a-Si:H and μc-Si:H p-i-n solar cells deposited by plasma-enhanced chemical vapor deposition (PECVD) at 13.56 MHz. Nitrogen and oxygen were inserted by two types of controllable contamination sources: (i) directly into the plasma through a leak at the deposition chamber wall or (ii) into the process gas supply line. Similar critical concentrations in the range of 4-6x10^{18} cm^{-3} for nitrogen and 1.2-5x10^{19} cm^{-3} for oxygen were observed for both a-Si:H and μc-Si:H cells for the chamber wall leak. Above these critical concentrations the solar cell efficiency decreases for a-Si:H solar cells due to losses in the fill factor under red light illumination (FF_{red}). For μc-Si:H cells the losses in FF_{red} and in short-circuit current density deteriorate the device performance. Only for a-Si:H the critical oxygen concentration is found to depend on the contamination source. Conductivity measurements suggest that at the critical oxygen concentration the Fermi level is located about 0.05 eV above midgap for both a-Si:H and μc-Si:H.

INTRODUCTION

An important requirement for the fabrication of thin-film a-Si:H and μc-Si:H p-i-n solar cells with high efficiency is a low contamination level of oxygen and nitrogen impurities in the intrinsic (i-) layer. However, in practice, atmospheric contaminants are always present in some extent as they are introduced by leaks in/ out-gassing of deposition chamber walls [1] on the one hand and by the impurities from the source gas system on the other hand. It has been shown that for a constant impurity flow, the incorporation of oxygen or nitrogen from a chamber leak depends on the deposition rate [2] and the deposition conditions [3]. The presence of oxygen and nitrogen influences the electrical properties (i.e. increase of dark conductivity) of single films of a-Si:H [3-6] and μc-Si:H [7-10]. Impurities are also found to influence the structural properties of μc-Si:H solar cells [10]. The μc-Si:H deposition regime is shifted towards a more amorphous growth by the incorporation of nitrogen [8] and oxygen [9]. A model describing the incorporation of impurities into the i-layer based on conductivity and electron spin resonance (ESR) measurements was proposed for a-Si:H [4,5] and μc-Si:H [8,9]. For both materials, oxygen and nitrogen impurities are assumed to lead (in part) to donor like states shifting the Fermi level towards the conduction band. By the use of ESR measurements in [11] it was found that the defect density in a-Si:H films is increased when exceeding an impurity concentration of ~ 10^{20} cm^{-3} for both, oxygen and nitrogen. Furthermore, for μc-Si:H films similar observations were measured by ESR, i.e. oxygen impurities lead to an increase of the defect density [12]. In the present work, we study the critical impurity concentrations for a-Si:H and μc-Si:H solar cells. In particular, the influence of the position of the contamination source (leak at the deposition chamber wall or at process gas supply line) on the critical concentrations is investigated.

EXPERIMENT

The thin-film silicon solar cells were fabricated by plasma-enhanced chemical vapour deposition (PECVD) at 13.56 MHz. As substrate for a-Si:H and μc-Si:H p-i-n solar cells, Asahi U-Type (SnO_2:F) and aluminium doped zinc oxide (ZnO:Al) transparent conductive oxide (TCO) coated glass were used, respectively. Thermally evaporated silver contact pads (1 cm^2) were used as back reflector. For deposition of all intrinsic Si:H films a hydrogen flow of ~ 360 sccm, a deposition pressure of ~ 10 Torr and a heater temperature of ~ 200 °C was applied. For growth of a-Si:H, a SiH_4 flow of 7.8 sccm was used and a discharge power of about 0.2 W/cm^2 giving a deposition rate of about 0.45 nm/s. For growth of μc-Si:H i-layer a discharge power of 0.53 W/cm^2 was employed. Depending on the applied silane flow the μc-Si:H i-layer deposition rate was about 0.26 nm/s (1.02 sccm) or about 0.5 nm/s (2.1 sccm). The i-layer thickness was about 350 nm for a-Si:H and about 1.2 - 1.3 μm for μc-Si:H. For intentional contamination of the i-layer with oxygen and nitrogen, these gases (purity 5.0) were inserted by a controllable leak (needle valve) at the chamber wall (i.e. the impurities were introduced directly into plasma) or into the process gases supply line. By varying the impurity flow, base pressures in the range between 10^{-8} and 10^{-4} Torr were obtained. The unintended contamination of the silane (purity 6.0) and hydrogen (purity 6.0) gasses was minimized by the use of an Entegris/Aeronex CE-35KF-SK-4R gas purifier. The impurity concentration in the i-layer was measured by secondary ion mass spectrometry (SIMS) using a quadrupole instrument (Atomika 4000). The solar cell performances (I/V curves) were measured under AM1.5 illumination at 100 mW/cm^2 and 25 °C. Dark conductivity measurements were done in a coplanar 2-point configuration.

RESULTS AND DISCUSSION

A comparison of I/V characteristics for a-Si:H and μc-Si:H p-i-n solar cells as a function of impurity (oxygen and nitrogen) concentration in the i-layer is shown in figure 1. The solar cells are deposited at optimized deposition conditions with optimized intrinsic absorber layer thicknesses (i.e. ~ 350 nm for a-Si:H and ~ 1.2 – 1.3 μm for μc-Si:H). The performance at low impurity concentrations corresponds to state-of-the-art cell performance. The impurities were inserted directly into the plasma trough a controllable leak at the deposition chamber wall. For both a-Si:H and μc-Si:H, the efficiency and fill factor under red light illumination, FF_{red}, decreases by increasing the impurity concentration when the critical impurity concentrations are exceeded. The critical concentration of contaminants (defining the minimum impurity concentration which causes a deterioration of solar cell performance) can be determined from the FF_{red}. The FF_{red} shows the quality of carrier extraction from the bulk and it is known to be very sensitive to the incorporation of defects, donors and acceptors in the i-layer [1]. We observed for a-Si:H (figure 1 (a)) and μc-Si:H (figure 1 (b)) similar critical nitrogen and oxygen concentrations in the range of ~ 4-6x10^{18} cm^{-3} and ~ 1.2-5x10^{19} cm^{-3}, respectively. However, the critical oxygen concentration for a-Si:H (2-5x10^{19} cm^{-3}) is slightly higher compared to that determined for μc-Si:H (1.2-2x10^{19} cm^{-3}). Differences between the two materials were also observed in terms of the short-circuit current density, J_{SC}, and open-circuit voltage, V_{OC}, as a function of impurities concentration. For 350 nm thick a-Si:H solar cells, nitrogen and oxygen impurities do not significantly influence the J_{SC} (~ 13 mA/cm^2) and the V_{OC} (~ 900 mV) except for the highest concentration in the examined range, as it is shown in figure 1(a).

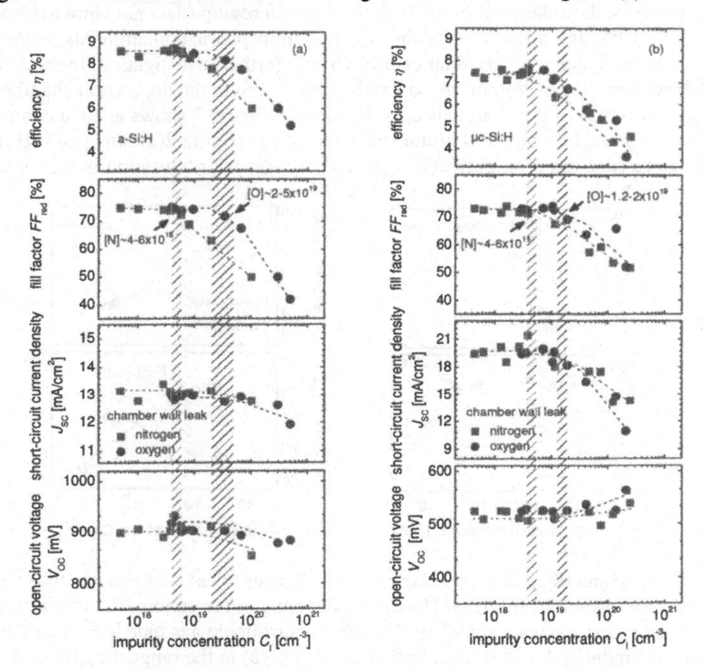

Figure 1. I/V characteristics (η, FF_{red}, J_{SC} and V_{OC}) of p-i-n solar cells with i-layer of (a) a-Si:H and (b) μc-Si:H vs. impurity concentration (oxygen and nitrogen) varied by a controllable chamber wall leak. The dotted lines are guides for the eyes, while the critical concentrations are indicated with dashed lines and arrows.

In contrast, for μc-Si:H solar cells the short-circuit current density is strongly affected by the impurities (see figure 1 (b)), which is partly attributed to a charge carrier collection problem. A previous study [10] shows that μc-Si:H solar cells thinner than 1 μm are less sensitive to oxygen and nitrogen impurities than thick cells (~ 3 μm). Furthermore, for μc-Si:H solar cells the V_{OC} increases with increasing oxygen and nitrogen concentration due to the decrease in crystallinity [8-10]. In order to prevent structural changes from a microcrystalline towards a more amorphous growth at high oxygen and nitrogen concentrations, the silane flow during deposition was slightly reduced. In conclusion, by inserting oxygen and nitrogen impurities into the plasma by a controllable chamber wall leak we observed for both a-Si:H and μc-Si:H solar cells similar critical concentrations in the range of 4-6x10^{18} cm^{-3} for nitrogen and 1.2-5x10^{19} cm^{-3} for oxygen. Above such critical oxygen and nitrogen concentrations the solar cell efficiency deteriorates due to losses in the FF_{red} for a-Si:H solar cells. For μc-Si:H cells the losses in FF_{red} and in J_{SC} decrease the device performance. In contrast to a-Si:H solar cells, structural changes related to a decrease of the crystalline volume fraction at high oxygen and nitrogen concentrations were reported for μc-Si:H solar cells [10]. For both materials the decrease of the efficiency is due to reduced sensitivity of the solar cells in the long wavelength region (data not shown) [3-5].

The similarities and differences of the effect of atmospheric contaminants on the performance of a-Si:H and μc-Si:H solar cells deserving further investigation. In the next, we study the influence of the position of the contamination source in the deposition chamber on the critical concentrations for oxygen and nitrogen impurities. Figure 2 shows again a comparison of solar cell efficiency and FF under red illumination for a-Si:H (figure 2(a)) and μc-Si:H (figure 2 (b)) solar cells as a function of impurity (oxygen and nitrogen) concentration in the i-layer.

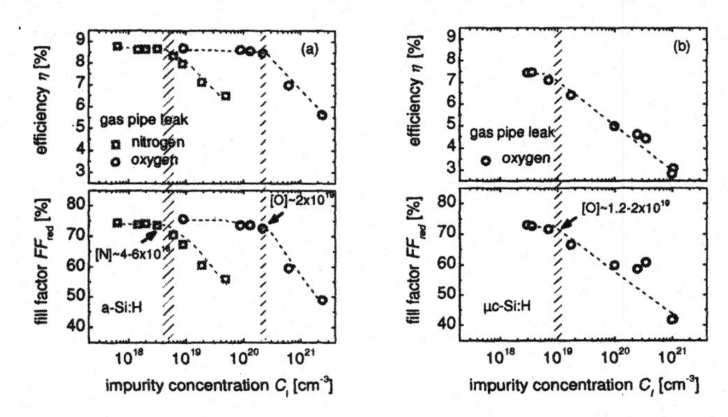

Figure 2. Efficiency and FF_{red} of p-i-n solar cells with i-layer of (a) a-Si:H solar cells vs. impurity concentration and of (b) μc-Si:H cells vs. oxygen concentration. The impurities were inserted into the process gasses supply line through a controllable gas pipe leak. The critical concentrations are indicated with dashed lines and arrows: (a) in the range $4\text{-}6\text{x}10^{18}$ cm^{-3} for nitrogen and at $2\text{x}10^{20}$ cm^{-3} for oxygen (for a-Si:H); (b) for μc-Si:H the critical concentration is in the range $1.2\text{-}2\text{x}10^{19}$ cm^{-3} for oxygen.

In this case, the impurities were inserted together with the deposition gasses silane and hydrogen through a controllable leak at the process gas supply line. Nitrogen impurities introduced in the deposition process through a gas pipe leak lead to a decrease of η and FF_{red} of a-Si:H solar cells at ~ $4\text{-}6\text{x}10^{18}$cm^{-3} (see figure 2 (a)). The critical nitrogen concentration for a-Si:H is found to be independent on the contamination source (compare figure 1 (a)). For a-Si:H solar cells contaminated with oxygen by the chamber wall leak, η and FF_{red} start to decrease at ~ 2-$5\text{x}10^{19}$ cm^{-3}. Figure 2 (a) shows a shift of the critical oxygen contamination level up to ~ $2\text{x}10^{20}$ cm^{-3} when the gas pipe leak was applied for a-Si:H solar cells (compare to figure 1 (b)). In contrast for μc-Si:H solar cells contaminated with oxygen by a gas pipe leak no difference to chamber wall leak was observed. The critical oxygen concentration for μc-Si:H is at ~ $1.2\text{-}2\text{x}10^{19}$ cm^{-3} for both type of leaks (compare figure 1 (b) and 2 (b)).

Thus, it has to be concluded that there is not one unambiguous critical oxygen concentration for a-Si:H above which the solar cell performances deteriorate. The critical oxygen concentration depends on the type of the contamination source (chamber wall and gas pipe leak). The application of the gas pipe leak leads to an increased a critical concentration of $2\text{x}10^{20}$cm^{-3}

for oxygen in a-Si:H. A high reaction probability of oxygen and SiH_4 in the gas pipe outside of the plasma is proposed to lead to a formation of electrically inactive twofold-coordinated oxygen (formation of atom groups with Si-O-Si bonding configuration) in the i-layer [13]. Due to the much lower reactivity of nitrogen with SiH_4 or H_2 [14], no difference for nitrogen incorporation from different types of leaks is expected, as observed experimentally. In case of μc-Si:H solar cells the critical oxygen concentration is found to be independent on the type of the contamination source and it is at ~ 1.2-2x10^{19} cm^{-3}. A possible explanation is that the higher discharge power (0.53 W/cm^2) for the deposition of μc-Si:H leads to the decomposition of atom groups with Si-O-Si bonding and the incorporation of oxygen in a doping-like configuration.

For a further understanding of the I/V parameter deterioration above the critical impurity concentrations for both a-Si:H and μc-Si:H, electrical properties of the corresponding intrinsic films were studied. The thicknesses were about the same as the i-layers in the solar cells. Figure 3 shows a comparison of the dark conductivity of a-Si:H (figure 3 (a)) and μc-Si:H (figure 3 (b)) intrinsic films as a function of the impurity concentration. At the critical nitrogen and oxygen concentrations, a similar dark conductivity of ~ 3-5x10^{-11} S/cm for a-Si:H and ~ 1-2x10^{-6} S/cm for μc-Si:H was observed.

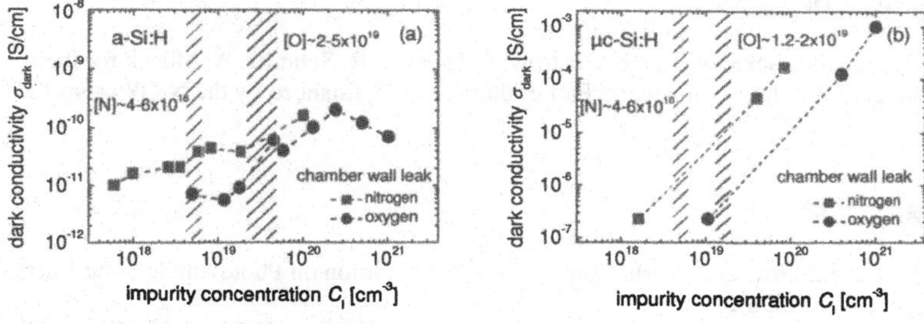

Figure 3. Dark conductivity of intrinsic (a) a-Si:H and (b) μc-Si:H films vs. the impurity (nitrogen and oxygen) concentration. The same deposition regimes as the one already shown in figure 1 were used. The impurities are inserted by controllable chamber wall leak. The critical concentrations (determined from FF_{red} of the solar cells) are indicated with dashed lines.

Figure 3 (a) shows that for a-Si:H above the critical oxygen and nitrogen concentrations the dark conductivity increases by half an order of magnitude with increasing impurity concentration up to C_I =2x10^{20} cm^{-3}. Further increase of oxygen concentration leads to a decrease of the dark conductivity attributed to Si-O alloying [3]. A corresponding effect was not observed for nitrogen within the studied range. Compared to a-Si:H intrinsic films, for μc-Si:H the dark conductivity increases stronger beyond the critical oxygen and nitrogen concentrations (see figure 3 (b)). It was demonstrated for μc-Si:H that the carrier density measured by Hall measurements also increases with the increase of the impurity concentration from 10^{13} to 10^{17} cm^{-3} [11]. Based on these conductivity data, one can conclude that at least a part of the oxygen and nitrogen impurities are incorporated in a donor-like configuration. We estimate from the respective conductivities at the critical oxygen concentration a similar Fermi level position of about 0.05 eV above midgap towards the conduction band for both a-Si:H and μc-Si:H.

CONCLUSIONS

In conclusion, we have studied for a-Si:H and μc-Si:H solar cells the incorporation of impurities (oxygen and nitrogen) introduced together with the process gases (gas pipe leak) or through a chamber leak. Similar critical concentrations in the range of $4\text{-}6 \times 10^{18}$ cm^{-3} for nitrogen and $1.2\text{-}5 \times 10^{19}$ cm^{-3} for oxygen were observed for both a-Si:H and μc-Si:H for a controllable chamber wall leak. For a-Si:H the solar cell efficiency decrease above these critical impurity concentrations due to losses in the FF under red light illumination. For μc-Si:H cells the losses in FF_{red} and in short-circuit current density deteriorate the device performance. From conductivity measurements of films a similar Fermi level position of ~ 0.05 eV above midgap is estimated at critical oxygen concentration for both materials. At impurity concentrations beyond the critical concentration, a stronger increase of dark conductivity with increasing impurity concentration was observed for μc-Si:H compared to a-Si:H. Furthermore, it is found that the critical oxygen concentration in a-Si:H solar cells depends on the contamination source.

ACKNOWLEDGMENTS

The authors acknowledge U. Zastrow, L. Niessen, R. Schmitz, A. Mueck for their important contributions to this work. Part of the work was financed by the NRW project EN/1008B "TRISO".

REFERENCES

1. B. Rech et al., Proc. 2nd World Conference and Exhibition on Photovoltaic Solar Energy Conversion, 391 (1998).
2. U. Kroll, J. Meier, H. Keppner, A. Shah, S. D. Littlewood, I. E. Kelly, and P. Giannoulès, J. Vac. Sci. Technol. A **13**, 2742 (1995).
3. J. Woerdenweber et al., J. Appl. Phys. **104**, 094507 (2008).
4. A. Morimoto, M. Matsumoto, M. Yoshita, M. Kumeda, and T. Shimizu, Appl. Phys. Lett. **57**, 2130 (1991).
5. T. Shimizu, M. Matsumoto, M. Yashita, M. Iwami, A. Morimoto, and M. Kumeda, J. Non-Cryst. Solids **137-138**, 391 (1991).
6. T. Kinoshita, M. Isamura, Y. Hishikawa, and S. Tsuda, Jpn. J. Appl. Phys. **35**, 3819 (1996).
7. P. Torres, J. Meier, R. Flückinger, U. Kroll, J.A. Selvan, H. Keppner, A. Shah, S.D. Littelwood, I.E. Kelly, and P. Giannoulès, Appl. Phys. Lett. **69**, 1373 (1996).
8. T. Ehara, Appl. Surf. Phys. **113/114**, 126 (1997).
9. T. Kamei, and T. Wada, J. Appl. Phys. **96**, 2087 (2004).
10. T. Kilper et al., J. Appl. Phys. **105**, 074509 (2009).
11. M. Stutzmann, W. B. Jackson, and C.C. Tsai, Phys. Rev. B **32**, 23 (1985).
12. F. Finger, J. Müller, C. Malten, R. Carius, and H. Wagner, J. Non-Cryst. Solids **266-269**, 511 (2000).
13. J. Woerdenweber et al., Appl. Phys. Lett. **96**, 103505 (2010).
14. Material Safety Data Sheet, Nitrogen-Silane Mixtures (Chemical Safety Associates, La Mesa, CA, USA, 2002).

Solar Cell: Light Trapping

Mater. Res. Soc. Symp. Proc. Vol. 1245 © 2010 Materials Research Society 1245-A03-02

Plasmonic Light-trapping and Quantum Efficiency Measurements on Nanocrystalline Silicon Solar Cells and Silicon-On-Insulator Devices

Hui Zhao[1], Birol Ozturk[1], E. A. Schiff[1], Baojie Yan[2], J. Yang[2] and S. Guha[2]

[1]Department of Physics, Syracuse University, Syracuse, New York 13244-1130, U.S.A.
[2]United Solar Ovonic LLC, Troy, Michigan 48084, U.S.A.

ABSTRACT

Quantum efficiency measurements in nanocrystalline silicon (nc-Si:H)solar cells deposited onto textured substrates indicate that these cells are close to the "stochastic light-trapping limit" proposed by Yablonovitch in the 1980s. An interesting alternative to texturing is "plasmonic" light-trapping based on non-textured cells and using an overlayer of metallic nanoparticles to produce light-trapping. While this type of light-trapping has not yet been demonstrated for nc-Si:H solar cells, significant photocurrent enhancements have been reported on silicon-on-insulator devices with similar optical properties to nc-Si:H. Here we report our measurements of quantum efficiencies in nc-Si:H solar cells and normalized photoconductance spectra in SOI photodetectors with and without silver nanoparticle layers. As was done previously, the silver nanoparticles were created by thermal annealing of evaporated silver thin films. We observed enhancement in the normalized photoconductance spectra of SOI photodetectors at longer wavelengths with the silver nanoparticles. For nc-Si:H solar cells, we have not yet observed significant improvement of the quantum efficiency with the addition of annealed silver films.

INTRODUCTION

Light trapping is essential in thin film solar cells due to the limited optical pathlength of photons at longer wavelengths. Their limited thickness does not allow complete light absorption at these wavelengths. Conventionally, both surfaces are textured in order to increase the scattering angles of the light from these interfaces that leads to stochastic light trapping. For a material with refractive index *n*, perfect texturing will enable as high as $4n^2$ times pathlength enhancement (about 50 for Si) as predicted by Yablonovitch [1].

Plasmonic light trapping is an emerging alternative to stochastic light trapping. A deep understanding of the plasmonic light trapping mechanism has not been presented yet, however experimental data shows that this light trapping scheme can lead to significant enhancements in semiconductor devices and solar cells. Stuart and Hall demonstrated a 20-fold photocurrent enhancement on silicon-on-insulator (SOI) photodetectors with the deposition of silver nanoparticles on the top silicon layer [2]. This experiment was reproduced by Pillai, *et al.* using SOI substrates with thicker top silicon layers [3], where they observed 18 times photocurrent enhancement at λ= 1200 nm. This group recently reported six times photocurrent enhancement on a 20 micron thick crystalline silicon solar cell [4]. Schaadt, *et al.* deposited colloidal gold nanoparticles on a silicon photodiode and observed up to 1.8 times photocurrent enhancement [5].

Here, we report our experimental results on the effects of silver nanoparticle films deposited onto the top conductive oxide layers of hydrogenated nanocrystalline silicon (nc-Si:H) solar cells. The silver films cause an overall loss in the external quantum efficiencies (QE) of solar cells. Experiments on SOI photodetectors with the same thickness of the active silicon layer and similar silver nanoparticle films resulted in photocurrent enhancements at longer wavelengths. We present an analytical calculation of the effective nanoparticle polarizability for SOI top silicon layer and ITO layer of the solar cells. In addition, we used a discrete dipole approximation (DDA) method to calculate the re-radiation efficiency of different size nanoparticles. These calculations showed that the re-radiation is enhanced with increasing diameter and particles with diameters larger than 80 nm should be used in plasmonic efficiency enhancement experiments.

EXPERIMENT

The nc-Si:H solar cells were fabricated at United Solar Ovonic. The single junction cells had *nip* configuration and were deposited on stainless-steel substrates with a nominally specular Ag/ZnO (500 nm/130 nm thick) back-reflectors [6]. The intrinsic layer thickness was chosen to be 2.0 μm in order to match the silicon layer thickness of SOI substrates that we used in this study. The top-contact/anti-reflection layer on the nc-Si:H cells was a 70 nm thick indium-tin-oxide (ITO) film

Figure 1 shows the QE of the solar cell as-received, after addition of a silver layer and annealing (to create nanoparticles), and after annealing alone. Curve (a) represents the QE of the unprocessed solar cell; note the interference fringes visible at longer wavelengths. The decline in QE at shorter wavelengths is mainly due to absorption by the *p*-layer as well as by reduced anti-reflection coating efficiency. Two solar cells with the same QE spectra as depicted by curve (a) were annealed at 200°C for two hours. For one of the cells, the ITO layer was first covered with a 17 nm thick evaporated silver film [7]. Curve (b) shows the QE of the annealed solar cell without the silver layer; the anneal resulted in a significant overall reduction in the QE, which we believe is due to the degradation of the *p* layer. Curve (c) shows results for a cell with the silver layer. After annealing, silver nanoparticles formed on the surface of the ITO layer as shown in the inset of Figure 1. The reduction in the QE of the solar cell (curve c) with the nanoparticles was larger than for the annealed sample without silver. The larger reduction at shorter wavelengths in curve (c) is attributed to the lower re-radiation efficiency of the smaller size (<50nm) nanoparticles; this will be covered in detail in the discussion section. A blue-shift of the interference fringes was observed at longer wavelengths and will be discussed elsewhere [8].

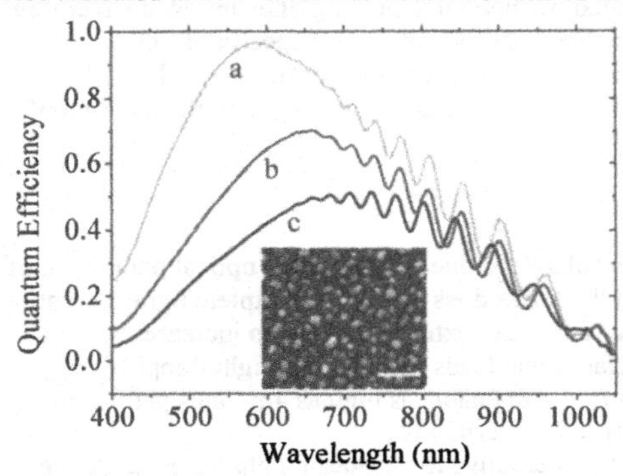

Figure 1. QEs of a nc-Si:H solar cell before and after deposition of annealed silver films; (a) as received (b) annealed without silver layer (c) with annealed silver film. Inset shows the scanning electron microscope (SEM) image of annealed silver films on top of ITO layer with 1 micron scale bar.

We have also fabricated annealed silver films on SOI photodetectors. The SOI wafers were purchased from University Wafers, Inc.. The top silicon layer was 2 μm thick with 500 nm back oxide thickness. The annealed silver films were fabricated using the same method and parameters as for the nc-Si:H solar cells. The SEM image of the resulting silver nanoparticles is shown in the inset of Figure 2. Coplanar aluminum electrodes were thermally evaporated for lateral photocurrent measurements. An external quantum efficiency setup was used to measure the wavelength-dependent normalized photoconductance spectra $G_p(\lambda)/eF(\lambda)$, where G_p is the photoconductance due to the monochromator beam, *e* is the electronic charge, and *F* is the incident photon flux from the monochromator. The spectra without the silver particles showed moderate interference fringes. The addition of silver nanoparticles caused a reduction at shorter wavelengths ($\lambda<520$ nm) and enhancement at longer wavelengths ($\lambda>670$ nm). A red-shift of the interference fringes is observed at longer wavelengths. These results are similar to our previous experiments where the Ag films were deposited onto 30 nm LiF spacer layers [8].

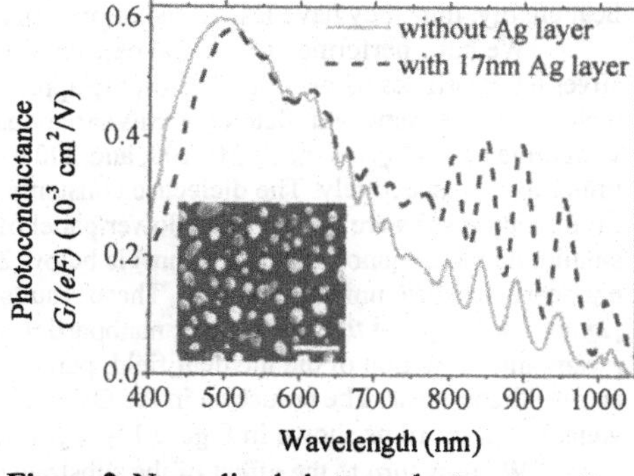

Figure 2. Normalized photoconductance spectra of the SOI photodetector with and without annealed silver film. Inset shows the SEM image of annealed silver films on the silicon layer with 1 micron scale bar.

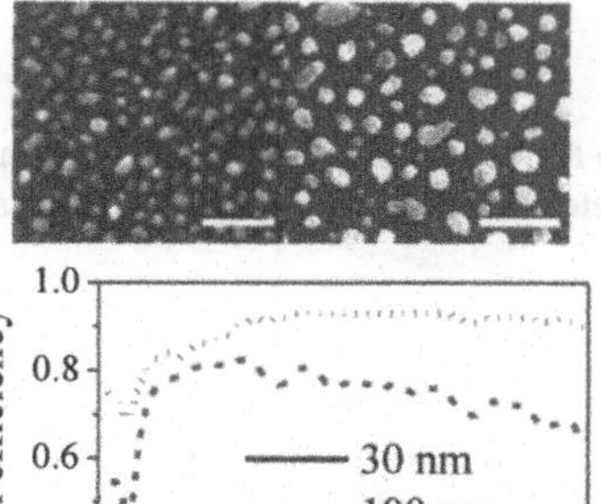

Figure 3. SEM images of the annealed silver films on ITO (top left) and on silicon (top right) surfaces. Scale bars denote 1 micron. DDA simulation results of the re-radiation efficiency dependence on the particle size (bottom).

DISCUSSION

The annealed silver films lead to different results on nc-Si:H solar cells and SOI photodetectors. The main difference between the two structures is the additional 70 nm ITO top electrode layer on the nc-Si:H solar cells. The average nanoparticle diameter in the annealed silver films varied on these substrates (Figure 3). The particles on the silicon surface have larger average diameters (~ 200 nm) compared to particles on the ITO surface (~100 nm), presumably due to different "wetting" properties of these surfaces for Ag. There is also a larger density of smaller diameter (<50nm) silver nanoparticles on the ITO surface compared to the SOI surface. These smaller nanoparticles absorb a larger fraction of the incident field and turn into heat energy, thus they have less re-radiation efficiency as will be shown subsequently.

We have performed numerical calculations of the re-radiation efficiency for different size silver nanoparticles using a discrete-dipole approximation (DDA) computer code DDSCAT 7.0 developed by Draine and Flatau [9]. Silver nanoparticles were modeled as oblate spheroids in air with dimensions $r_{maj} = 2\, r_{min} = 30$, 100, and 200 nm, where r_{maj} *and* r_{min} are radii of the major and minor axes, respectively. The dielectric constants for silver were taken from literature [10]. Simulation results are shown in the lower panel of Figure 3. The re-radiation efficiency of the smaller diameter nanoparticles (30 nm) is below 20 % and it is over 70 % for larger diameter nanoparticles (100 nm and 200 nm). These results are in good agreement with the previous work [2] and they suggest that the smaller nanoparticles on the ITO layers of the solar cells are causing plasmonic extinction of the incident field; part of the energy is lost as heat before it reaches the active silicon layer. The reduction in the QE spectra of the solar cells with the addition of annealed silver films shown in Figure 1 was partly attributed to this effect.

We now turn to the effect of the substrate on the nanoparticle polarizability α_{eff} at optical frequencies. The effects of embedding a nanoparticle into a dielectric are well established [11], and the corresponding formulae for the polarizability α_0 of a spheroidal nanoparticle are given in the appendix. In our experiments, the silver nanoparticles were located near the interface of two materials with different refractive indices instead of being surrounded by a single medium. The effective polarizability of a spherical nanoparticle that is placed near the interface of two dielectric materials is given by [12]:

$$\alpha_{\text{eff}} = \alpha_0 \frac{1+\beta}{1-\beta \dfrac{\alpha_0}{16\pi (r+d)^3}}, \qquad \beta = \frac{\varepsilon_s - \varepsilon_m}{\varepsilon_s + \varepsilon_m}, \tag{1}$$

where r is the radius of the sphere, d is the distance from the bottom edge of the particle to the interface, ε_s is the dielectric constant of the substrate material, and ε_m is the dielectric constant of the medium surrounding the nanoparticle. Following Protsenko and O'Reilly [13], we have utilized this analysis to calculate the effective polarizabilities for an oblate spheroidal silver nanoparticle near to an air/silicon and to an air/ITO interface; we identify r in eq. (1) as the semiminor axis r_{min} of the spheroid, and use the appropriately modified formulae for α_0. The wavelength dependent dielectric constant values for ITO and for Si were taken from the literature [14,15].

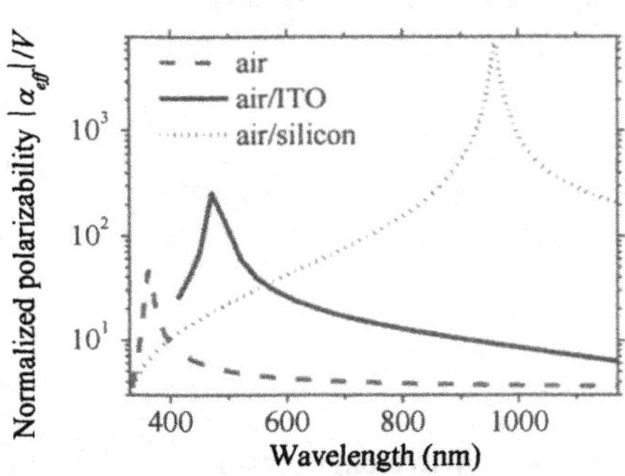

Figure 4. Calculated magnitude of the complex effective polarizability per unit volume for an oblate spheroidal silver nanoparticle (100 nm x 50 nm) in air, and 1 nm above ITO and silicon.

The magnitude of the effective polarizability of the spheroid silver nanoparticle on ITO and silicon layer is shown in Figure 4. The silver nanoparticles were modeled as oblate spheroids ($r_{maj} = 2\ r_{min} = 100$ nm and $d = 1$ nm) with the major axes parallel to the substrate surface. This calculation shows that the plasmon frequency of the nanoparticle slightly red-shifts and its polarizability magnitude increased when placed on ITO layer. A large red-shift of the plasmon frequency and more than 150-fold increase in the polarizability magnitude is apparent when the nanoparticle is placed on a silicon surface compared to air.

The particle absorption and scattering cross sections are given by [16]:

$$c_{abs} = k\,\mathrm{Im}\{\alpha\} \qquad c_{sca} = \frac{k^4}{6\pi}|\alpha|^2 \tag{2}$$

where k is the wavevector of the incident light.

Silver nanoparticles were deposited directly on top of the silicon layer in SOI photodetector experiments. The above analysis shows that the plasmon frequency of a silver nanoparticle is red-shifted and its scattering cross section has increased nearly $(150)^2$ times for a given frequency when it is placed on the silicon layer. This analysis was for a single nanoparticle and the actual experimental films have particle size distributions that broaden these resonances.

A metallic nanoparticle in a homogeneous medium scatters light uniformly in all directions. When it is placed at the interface of two dielectric materials, it will scatter most of the incident field into the material with higher refractive index [17,18]. Soller *et al.* calculated the fraction of the light that is scattered by a dipole into different types of substrates [17]; 87% of the scattered light is directed into the top silicon layer when the dipole is on a SOI substrate. Thus, the enhancement in the normalized photoconductance spectra of SOI photodetectors is due to the increased scattering efficiency of the silver nanoparticles at longer wavelengths. We expect to observe similar enhancements in nc-Si:H solar cells at longer wavelengths when the annealed silver films are deposited directly on the top silicon layer instead of ITO layer.

CONCLUSIONS

We showed the results of annealed silver film deposition on top of nc-Si:H solar cells and SOI photodetectors. We have observed an overall reduction in the quantum efficiency of the solar cells with the addition of silver nanoparticles on their top ITO layers. This reduction was attributed to the extinction of the incident field by the large density of smaller diameter silver nanoparticles on the ITO layers. The normalized photoconductance spectra of SOI photodetectors was enhanced at longer wavelengths when the silver nanoparticles are prepared directly on their top silicon layers. Our analysis suggests that the plasmon frequency of the nanoparticles shifts into red and correlates well with the enhancement at longer wavelengths. The increase in the polarizability magnitude of the nanoparticles also improves their re-radiation efficiency leading to the observed enhancement of the photoconductance signal.

ACKNOWLEDGMENTS

This research has been partially supported by the U. S. Department of Energy through the Solar America Initiative (DE-FC36-07 GO 17053). Additional support was received from the Empire State Development Corporation through the Syracuse Center of Excellence in Environmental and Energy Systems.

APPENDIX

The polarizability α_0 of a spheroid embedded in a dielectric is given in the text of Bohren and Huffman as [16]:

$$\alpha_0 = V\frac{\varepsilon - \varepsilon_m}{\varepsilon_m + L(\varepsilon - \varepsilon_m)} \tag{3}$$

where ε is the dielectric constant of the material of the spheroid, ε_m is the dielectric constant of the embedding material, and V is the particle volume. The geometrical factor L for an oblate spheroid with semimajor axis r_{maj} and semiminor axis r_{min} is given by equation 5.34 in the reference:

$$L = \frac{g(e)}{2e^2}\left[\frac{\pi}{2} - \tan^{-1} g(e)\right] - \frac{g^2(e)}{2} \qquad g(e) = \left(\frac{1-e^2}{e^2}\right)^{1/2} \qquad e^2 = 1 - \frac{(r_{min})^2}{(r_{maj})^2} \tag{4}$$

REFERENCES

1. E. Yablonovitch, *J. Opt. Soc. Am.* 72, 899 (1982).

2. H. R. Stuart and D. G. Hall, *Appl. Phys. Lett.* 69, 2327 (1996).

3. S. Pillai, K. R. Catchpole, T. Trupke and M. A. Green, *J. Appl. Phys.* 101, 093105 (2007).

4. F. J. Beck, S. Mokkapati, A. Polman, and K. R. Catchpole, *Appl. Phys. Lett.* 96, 033113 (2010).

5. D. M. Schaadt, B. Feng, and E. T. Yu, *Appl. Phys. Lett.* 86, 063106 (2005).

6. J. Yang, B. Yan, G. Yue, S. Guha, D. *Mater. Res. Soc. Symp. Proc.* 1153, A13-02, (2009).

7. In previous studies, the surface was coated with a 30 nm lithium fluoride (LiF) spacer layer before the silver deposition [2,3]. We tested the effect of LiF layer and found that this layer had little effect; we did not use the LiF spacer layer for the measurements presented here.

8. B. Ozturk, H. Zhao, E. A. Schiff, B. Yan, J. Yang, S. Guha, unpublished.

9. B. T. Draine, P. J. Flatau, *J. Opt. Soc. Am. A* 11, *1491* (1994)

10. P. B. Johnson and R. W. Christy, *Phys. Rev. B* 6, 4370 (1972)

11. H. Mertens, J. Verhoeven, A. Polman, *Appl. Phys. Lett., Vol. 85,No. 8, 23* (2004)

12. B. Knoll, F. Keilmann, *Nature 399,* 134 (1999).

13. I. P. Protsenko and E. P. O'Reilly, *Phys. Rev. A* 74, 033815 (2006).

14. Y. S. Jung, *Thin Solid Films* 467, 36 (2004)

15. E. D. Palik, *Handbook of Optical Constants of Solids Vol.2* (AP,1991) and *Vol.3* (AP, 1998).

16. C. F. Bohren and D. R. Huffman, *Absorption and scattering of light by small particles* (Wiley-Interscience, New York, 1983)

17. B. J. Soller, H. R. Stuart, and D. G. Hall, *Opt. Lett.* 26, 1421 (2001).

18. J. Mertz, *J. Opt. Soc. Am. B* 17, 1906 (2000)

Mater. Res. Soc. Symp. Proc. Vol. 1245 © 2010 Materials Research Society 1245-A03-03

Advanced Light Trapping in Thin-Film Silicon Solar Cells

M. Zeman [1], O. Isabella[1], K. Jaeger[1], R. Santbergen[1], R. Liang[1], S. Solntsev[1], J. Krc[2],
[1] Delft University of Technology, DIMES, P.O. Box 5053, 2600 GB Delft, The Netherlands
[2] University of Ljubljana, Faculty of Electrical Engineering, SI-1000 Ljubljana, Slovenia

ABSTRACT

Photon management is one of the key issues for improving the performance of thin-film silicon solar cells. An important part of the photon management is light trapping that helps to confine photons inside the thin absorber layers. At present light trapping is accomplished by the employment of the refractive-index matching layers at the front side and the high-reflective layers at the back contact of the solar cells and scattering of light at randomly surface-textured interfaces. In this article key issues and potential of light management in thin-film silicon solar cells are addressed. Approaches for light trapping are presented such as i) surface textures based on periodic diffraction gratings and modulated surface morphologies for enhanced scattering and anti-reflection, ii) metal nano-particles introducing plasmonic scattering, and iii) one-dimensional photonic-crystal-like structures for back reflectors.

INTRODUCTION

The efficiency of thin-film silicon based solar cells has to achieve a level of 20% on a laboratory scale in order to stay competitive with bulk crystalline silicon solar cells and other thin-film solar cell technologies. Photon management is one of the key issues for improving the performance of thin-film silicon solar cells and decreasing the production costs by shortening deposition times and using less material. The aim of the photon management is the effective use of the energy of the solar radiation and the maximization of absorption in desired parts of a solar cell that are called absorbers. Photon management in thin-film solar cells is accomplished by a number of techniques that are related to the following areas:

i) Effective use of the energy of the solar spectrum;
ii) Minimization of absorption outside the absorber layers;
iii) Trapping of photons inside the absorber layers.

Effective use of the energy of the solar spectrum

There are two principal optical losses that strongly reduce the energy conversion efficiency of today's solar cells. Both of these losses are related to the spectral mismatch of the energy distribution of photons in the solar spectrum and the band gap of a semiconductor material that serves as the absorber in a solar cell. The first loss is the ***non-absorption*** of photons with energy lower than the band gap energy of the absorber. These photons are in principle not absorbed in the absorber and therefore do not contribute to the energy conversion process. The second process is the ***thermalization***. In this process the electrons and holes generated by photons with energy higher than the band gap of the absorber release the extra energy as heat into the semiconductor atomic network. Multi-junction (also known as tandem) solar cells, photon up- and down-conversion are approaches to the effective utilization of the energy of the solar spectrum. The concept of a multi-junction solar cell is already widely used in thin-film silicon solar cell technology by employing a tandem solar cell structure. In this structure two [1] or more [2] solar cells are stacked on top of each other. Multi-junction solar cell approach means that the absorber layer in each component cell is tailored to a specific part of the solar spectrum. Top cells efficiently absorb short-wavelength part of the spectrum (high energy photons), whereas bottom cells absorb the remaining long-wavelength part of the spectrum (low-energy photons). In this way the thermalization losses are minimized, which is reflected in higher open-circuit voltages of the devices. Absorber layers in tandem thin-film silicon solar cells are based on hydrogenated amorphous silicon (*a*-Si:H), alloys of *a*-Si:H such as hydrogenated amorphous silicon germanium (*a*-SiGe:H) and hydrogenated microcrystalline silicon (μc-Si:H). The best laboratory stabilized efficiency of a single junction *a*-Si:H cell is 10,02% [3], of a tandem micromorph *a*-Si:H/μc-Si:H cell is 14,1% (initial) [4] and a triple junction *a*-Si:H/*a*-SiGe:H/μc-Si:H cell is 11,2% [5], *a*-Si:H/*a*-SiGe:H/*a*-SiGe:H cell is 10.4% and *a*-Si:H/μc-Si:H/μc-Si:H cell is 12,2% [6]. Recently novel absorber materials and cell concepts based on spectrum splitting on two or more laterally dislocated cells [7], up-and down-converters [8,9,10], absorbers with quantum dot superlattices [11], intermediate-band cells [12] have been investigated for a generic approach of all-silicon tandem solar cells.

Minimization of absorption outside the absorber layers

A thin-film silicon solar cell is composed of absorbers and supporting layers that are deposited on a solar cell carrier. Typical carriers are a glass plate and a metal or polymer foil. The absorption of light in the supporting layers such as doped semiconductor layers and transparent and metal electrodes leads to optical losses in the solar cell. In principle, photons that are absorbed in these supporting layers do not contribute to the energy conversion process and therefore the absorption in these layers has to be minimized. Recently a lot of effort has been dedicated to the development of transparent conductive oxides (TCO) with low absorption in a wavelength region of interest (300 nm $< \lambda <$ 1200 nm) [13,14,15]. A continuous attention is paid to the development of wide band gap doped semiconductors based on *a*-Si:H and μc-Si:H such as hydrogenated amorphous/microcrystalline silicon carbide (*a*-SiC:H/μc-SiC:H) and hydrogenated amorphous/microcrystalline silicon oxide (*a*-SiO:H/μc-SiO:H) [16,17]. Alternative solutions to metal back reflectors, i.e dielectric-based reflectors are investigated to avoid parasitic plasmonic absorption on the textured metal surfaces [18,19].

Trapping of photons inside the absorber layers

The purpose of light trapping techniques is to lead a photon through a solar cell into the absorber layer and once it enters the absorber to trap it there until it is absorbed. The most important role of light trapping is to keep the physical thickness of the absorber layer as thin as possible and to maximize its effective optical thickness. The following techniques are used to trap photons inside the absorber layer:

- In-coupling of incident photons at the front side
- Reflection at the back side
- Intermediate reflectors in tandem solar cells
- Scattering at rough interfaces
- Scattering at metal nano-particles (plasmonic effects)

Two research areas can be distinguished regarding the development of light trapping techniques. The first area deals with the manipulation of photon propagation throughout a solar cell. The techniques are related to the development and implementation of ***optically-active layers*** such as anti-reflection coatings, single or stack of layers for index matching, intermediate and back reflectors. These layers take care that photons reach the absorber layer and inside the absorber undergo multiple passes. The second area deals with the enhancement of an average photon path length inside the absorber layer. This is achieved by scattering of light at rough interfaces and/or metal nano-particles. These techniques are related to the design and fabrication of a ***surface texture*** on substrate carriers. The surface texture of a substrate introduces rough interfaces into the solar cell structure. Scattering at rough interfaces prolongs the effective path length of photons and partially leads to the total internal reflection between the back and front contacts confining the light inside the absorber. Recently, layers of ***metal nano-particles*** and composite materials with embedded metal nano-particles for efficient in-coupling and scattering of light into the absorber layer have attracted a lot of attention.

In today's thin-film solar cells standard trapping techniques are based on scattering of light at rough interfaces, the employment of high-reflective metal layers at the back contacts and refractive-index matching layers at the front side. The rough interfaces are introduced into the solar cell by using substrate carriers that are coated with a randomly surface-textured transparent conductive oxide (TCO) layer, such as fluorine-doped tin oxide (FTO) of Asahi U-type substrate with a pyramidal-like surface structure [20] and sputtered aluminum-doped zinc oxide (AZO) that after etching in a 0.5% hydrochloric acid solution shows a crater-like surface texture [21].

In this article three approaches for light trapping are presented such as i) surface textures based on periodic diffraction gratings and modulated surface morphologies for enhanced scattering, ii) plasmonic scattering using metal nano-particles, and iii) one-dimensional (1-D) photonic-crystal-like (PC) structures for back reflectors. The first approach based on periodic diffraction gratings deals with the manipulation of light scattering into (pre-) selected angles. We shall refer to this approach as the angle-selective management of scattered light at textured interfaces. The approach based on the use of modulated surface textures allows manipulation of scattering in a broad wavelength range. This approach is presented by applying substrate carriers where the surface texture is a combination of large and small surface features such as surface-textured glass substrates coated with an etched AZO layer. The second approach takes advantage of the strong scattering due to metal nano-particles embedded at the interface between two different materials, favoring light-in coupling in the higher refractive index material. The third approach deals with the manipulation of reflection and transmission at a particular interface inside a solar cell. This specially designed *optical interface* results in a wavelength-selective management of (high) reflection or transmission of light.

POTENTIAL OF LIGHT TRAPPING IN THIN-FILM SILICON SOLAR CELLS

Computer simulations using advanced programs such as the ASA program from Delft University of Technology [22] and the Sunshine program from Ljubljana University [23] are a valuable tool to investigate the potential of light trapping in thin-film silicon solar cells. Figure 1a and 1b shows calculated spectral absorption in a 300 nm thick *a*-Si:H layer and 1 μm thick *μc*-Si:H layer, respectively, when the effective optical thickness of the layers is increased 10 and 50 times. When matching the absorption in the layers to the AM1.5 spectrum one can calculate a potential photocurrent generated in the layers when used as absorbers in thin-film silicon solar

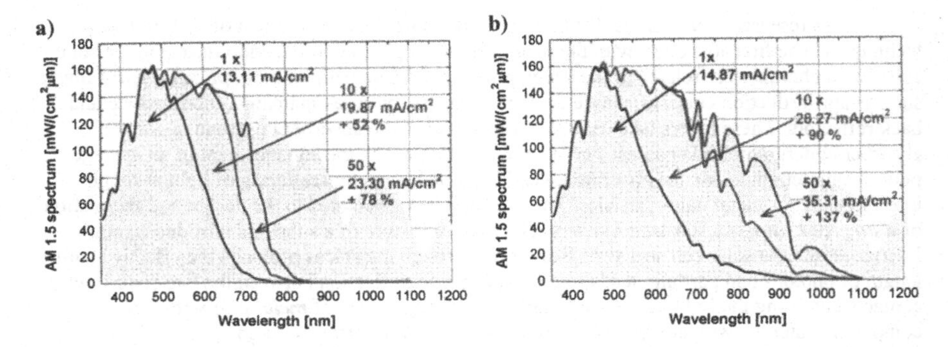

Figure 1: Spectral absorption in **a)** 300 nm thick *a*-Si:H layer and **b)** 1 µm thick *µc*-Si:H layer matched to the AM1.5 spectrum, when the effective optical thickness of the layers is increased 10 and 50 times.

cells. Increasing the optical thickness of the *a*-Si:H layer 10 and 50 times results in 52% and 78% potential enhancement of the photocurrent, respectively. The simulations demonstrate that in case of *µc*-Si:H absorber the light trapping plays even more important role. Increasing the optical thickness of 1 µm thick *µc*-Si:H layer 10 and 50 times results in 90% and 138% enhancement of the photocurrent, delivering potential photocurrent of 28 and 35 mA/cm^2, respectively.

SCATTERING MODELS

Two descriptive scattering parameters are used to evaluate the level of scattering at a nano-rough interface in the far field. The haze parameter $H(\lambda)$ describes the ratio of the scattered to the total light and can be defined for both transmitted and reflected light. The angular intensity distribution (*AID*) $I(\theta,\lambda)$, where θ is the scattering angle, describes the angular dependence of the reflected or transmitted scattered light. In order to understand scattering behavior of a rough interface a relationship between the morphology of a rough interface and the descriptive scattering parameters has been investigated. Recently, two scattering models [24,25] have been developed that calculate the scattering parameters using the root mean square (rms) roughness σ_r and the height function $\eta(x,y)$ of the surface morphology of a rough surface as input.

Both models predict the *AID* (called angular resolved scattering *ARS* in [25]) for transmitted light and are based on the *scalar scattering theory*, in which the electromagnetic fields are replaced by a complex scalar field. The intensity of the scattered light is then given by the absolute square this scalar field. The models reflect the insight that the intensity distribution of the scattered light is related to the Fourier transform of the scattering object. While the model presented in [24] is based on the first order Born approximation [26] and on Fraunhofer diffraction [27], the authors of [25] applied the Rayleigh-Sommerfeld diffraction integral [27].

Figure 2 shows measured and calculated *AID* according to [24] for pyramid-like FTO of Asahi U-Type [15], crater-like AZO that was etched after deposition in a solution of 0.5% hydrochloric acid for 20 s and 40 s, respectively [14], and pyramid-like boron doped zinc oxide

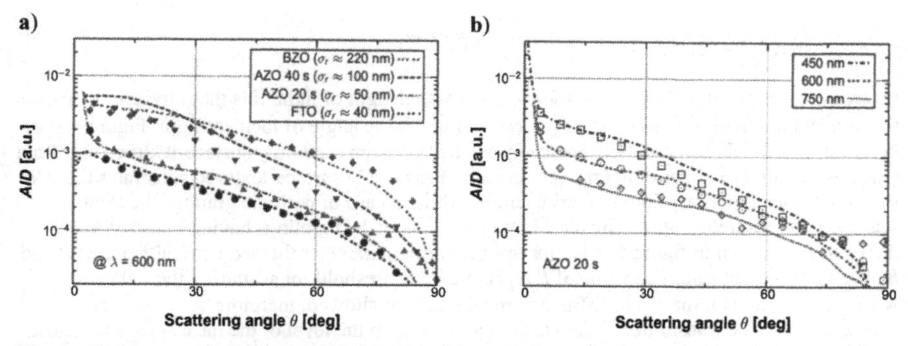

Figure 2: a) Angular intensity distributions (AID) for light scattered while traversing four different surface-textured TCO materials at 600 nm. The lines denote the calculations, the measurement results are given by the symbols ◆ (BZO), ▼ (AZO 40 s), ▲ (AZO 20 s) and ● (FTO). **b)** AID for AZO 20 s at three different wavelengths. The lines denote the calculations, the measured AID are given by the symbols ⊡ (450 nm), ⊙ (600 nm) and ◈ (750 nm).

(BZO) from the PV-LAB of the École Polytechnique Fédérale de Lausanne, Switzerland [28]. In figure 2a measured and calculated intensities for the four samples at a wavelength of λ=600 nm are shown. For all samples, the agreement between measured and calculated values is good, only for AZO etched for 40 s, the deviation is larger in the central angle range. Figure 2b shows, as an example, wavelength dependent changes of the *AID* for AZO etched for 20 s. The calculated values follow the same trend as the measured, *i.e.* the intensity of the scattered light decreases with increasing wavelength. The authors of [25] tested their model on four boron doped zinc oxide [28] samples with varying deposition and post-deposition treatment conditions at a wavelength of 543 nm and also reached good agreement between the measured and calculated *AID*. They did not show whether their model predicts wavelength dependent changes of the *AID* correctly but calculated also the haze in transmission based on their scattering model.

ADVANCED SURFACE TEXTURES

1-D periodic diffraction gratings

Diffraction gratings are introduced into thin-film solar cells in order to achieve a better control of light scattering inside the solar cells. The application of 1-D periodic gratings allows to decrease the total reflection in the wavelength region of interest and to scatter light into (pre-) selected angles by manipulating their geometrical parameters; period *P* and height *h* [29]. Figure 3 shows the surface morphology of a periodic grating patterned in a lacquer on glass substrate that was characterized by Atomic Force Microscopy (AFM). The understanding of scattering properties of 1-D gratings is important in order to make the right choice of the *P* and the *h* for the use in a particular solar cell. An extensive study based on 2-D simulations about such properties for single junction *a*Si:H based solar cells has been recently published [30]. In figure 4 the AID_T of the 1-D periodic grating and of the Asahi U-type SnO_2:F substrate, measured at 633 nm, are shown. The measurement confirmed that the 1-D periodic gratings scatter light into discrete angles according to the diffraction grating equation:

$$\varphi_{scatt} = \arcsin\left[\frac{m\lambda}{nP} - \sin(\varphi_{inc})\right], \quad (1)$$

where m represents the diffraction order, λ is the wavelength of light, n is the refractive index of the incident medium, P is the grating period, and φ_{inc} is the angle of incident light. Figure 4 also demonstrates the fundamental difference in the *AID* between randomly surface-textured substrate (such as Asahi-U) and periodic gratings. In case of randomly surface-textured substrates the *AID* is a continuous function of the scattering angle, while in case of periodic gratings the light is scattered into discrete angles. The haze in transmission of 1-D gratings having constant h and different P is shown in figure 5. Also for this scattering parameter the geometrical features P and h play an important role. The value of P represents the threshold for scattering the light. According to Eq. (1), for $\lambda > P$ diffraction modes are not allowed, therefore only specular component may be transmitted. The h is directly related to the level of the haze since for h close to zero the 1-D grating approximates a flat surface [31]. Other theoretical aspects of diffraction gratings in respect to the light scattering parameters can be found in [32].

The scattering potential of 1-D periodic surface-textured substrates in solar-cell technology was investigated by depositing a *p-i-n a*-Si:H solar cell on glass substrate with 1-D periodic grating (P = 600nm and h = 300 nm) patterned in the lacquer. A 1 µm thick AZO layer was deposited as the front contact on the patterned substrate using rf magnetron sputtering. The silicon layers were fabricated using rf-PECVD deposition technique. The solar cells structure was: glass / front AZO / p-layer (µc-Si:H) / p-layer (*a*-SiC:H) / buffer layer (*a*-SiC:H) / i-layer (*a*-Si:H) / n-layer (*a*-Si:H) /Ag/Al. After the deposition of each layer the change of the surface roughness was evaluated by AFM. The initial periodicity was preserved up till the back contact, although the height of the periodic structure decreased after the deposition of each subsequent layer. The external parameters of the solar cell are reported in Table 1 and compared with the parameters of solar cells deposited on Asahi U-type and flat AZO as references. The external quantum efficiency (*EQE*) of the solar cells is presented in figure 6.

The difference in short circuit current density (J_{SC}) between the solar cells deposited on the 1-D grating and the Asahi-U type substrate can be mainly explained by the higher absorption

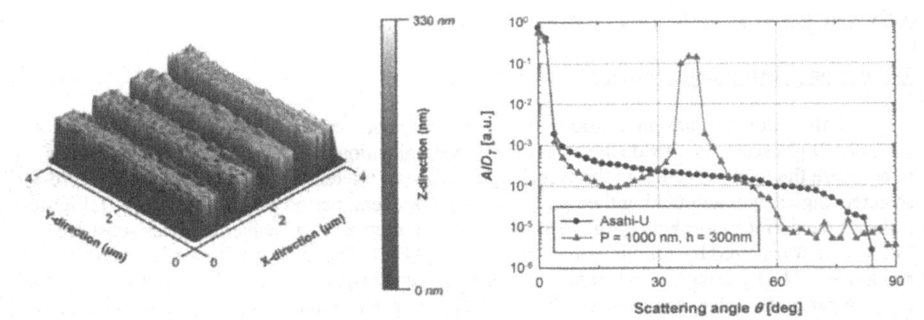

Figure 3: AFM image of a 1-D grating (patterned lacquer on glass) with P = 1000 nm and h = 300 nm.

Figure 4: AID_T of randomly surface-textured Asahi-U type substrate and of periodic grating (P = 1000 nm and h = 300 nm). The angle value of 0 deg corresponds to specular direction.

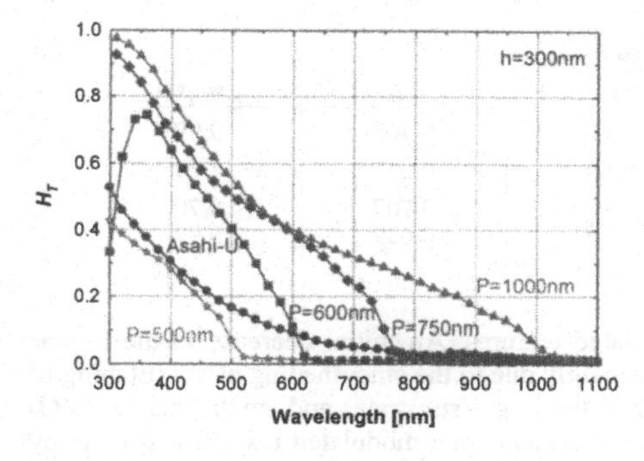

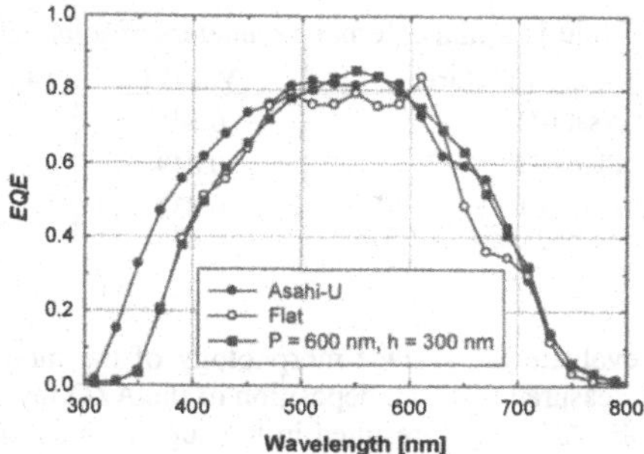

Figure 5: Haze in transmission of 1-D periodic gratings with different period *P* and constant height (h = 300 nm). Asahi-U type substrate is included as a reference for randomly surface-textured surface.

Figure 6: The *EQE* of the compared solar cells (full circles for Asahi-U type based cell, open circles for flat cell and full squares for textured cell).

in the short wavelength region in the AZO compared to the SnO_2:F front contacts. For wavelengths above 500 nm, the *EQE* of the solar cell deposited on the grating is higher than the *EQE* of the solar cell deposited on the flat substrate. Due to anti-reflective and scattering effects, the *EQE* in this wavelength region is similar to Asahi-U reference cell. This result indicates that 1-D periodic gratings are suitable substrates for improved light management in thin-film silicon solar cells. 1-D gratings were fabricated in several combinations of period and height on glass substrates on which *a*-Si:H solar cells were deposited. High performance was obtained in solar cells with diffraction grating substrates having P = 600 nm and h = 300 nm. The J_{SC} of this solar cell increased by 14.2% relative to the J_{SC} of the solar cell with flat interfaces.

Modulated surface textures

In general, the term modulated texture refers to a surface morphology that combines two or more types of textures with different statistical (vertical and lateral) surface parameters. This concept aims at enhanced light scattering in a broader wavelength range at rough interfaces where the large and small surface features are combined. In addition, anti-reflecting effects caused by small features is to be utilized at the front interfaces of the solar cell. In this article one example of the modulated surface texture is presented; a combination of random large and small textures. A random and periodic textures can be combined as well [33]. The different textures were introduced at different interfaces of the substrate, namely large random features were introduced at the glass / AZO interface and smaller random features were created at the AZO / air interface. The large random surface texture was obtained by etching the flat Corning Eagle 2000™ glass in a compound solution composed of HF and H_3PO_4 for 35 minutes. A 1 μm thick AZO film was sputtered on top of the rough glass surface. The modulated surface texture was obtained by wet-etching of the AZO layer for 40 seconds (AZO 40 s). The AFM was used to

Table 1: Initial external parameters of solar cells.

Texture	V_{OC} [V]	J_{SC} [mA/cm^2]	FF	Eff. [%]
Asahi-U	0.845	15.2	0.708	9.09
Flat AZO	0.849	13.4	0.641	7.29
P600 x h300	0.833	14.8	0.707	8.70
Etched glass/ AZO 40"	0.874	17.8	0.639	9.97

evaluate the surface morphology of the modulated textures. An initial decrease in the σ_r was measured after the deposition of the AZO layer mainly due to the smoothening effect. Etching of the AZO layer resulted in the superimposition of the large (substrate) and small (etched AZO) features on the same surface. A comprehensive analysis of the modulated texture was done by using Fourier transformation, when the vertical σ_r can be expressed as a function of spatial frequency (lateral parameter of the roughness). In this way laterally large and small roughness features can be described and classified more precisely [33]. When comparing the spatial-frequency surface representations of the etched glass and its modulated version (etched glass + etched AZO) one can observe in both cases the broad peak at low spatial-frequency components that correspond to the laterally large features. In the case of the modulated surface texture the high spatial-frequency components (holes and craters of etched AZO) are present in the representation. The haze parameter of the modulated surface textures is presented in figure 7. The initial textured surface of glass exhibited low but relatively constant haze. The surface covered with the AZO showed a similar haze behavior as the glass surface. The haze of modulated textures values was boosted because of the etching step provided additional rough features.

Single junction *a*-Si:H *p-i-n* solar cells were deposited by rf-PECVD on the modulated surface-textured substrates. The back contact consisted of AZO and silver. The initial external parameters are reported in Table 1. The *EQE* of the *a*-Si:H cells deposited on the modulated surface-textured substrate is shown in figure 8. The short wavelength *EQE* of the solar cells on

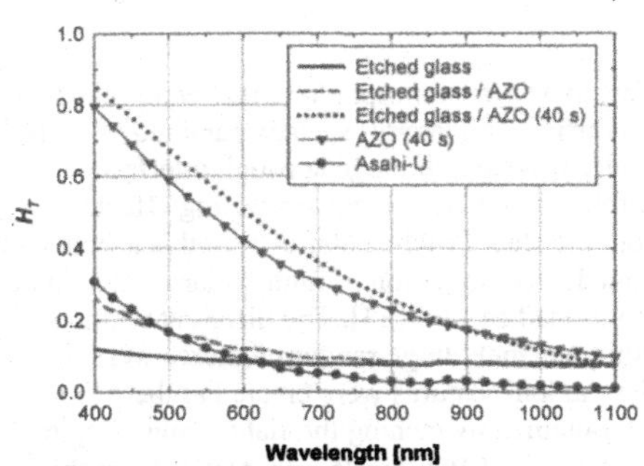

Figure 7: Haze parameter in transmission of the substrates with the regular and modulated textures.

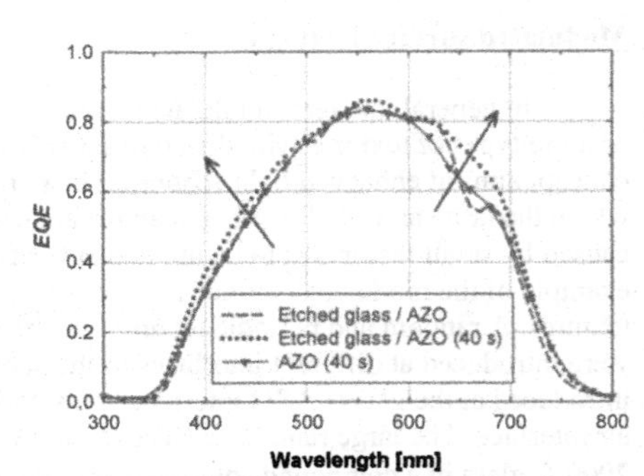

Figure 8: *EQE* of the solar cells deposited on different surface-textured substrates.

modulated surface textures was improved with respect to the solar cells deposited either on AZO 40" substrate or on etched glass covered by the AZO layer. This enhancement can be partially assigned to the different thickness of the AZO layer before and after etching but mainly to the anti-reflecting (AR) effect of the modulated surface texture. The high surface roughness of the modulated texture, which mainly affects the scattering, led to the high *EQE* in the long wavelengths region. However, the AR effect due to small features is present also at longer wavelengths. The result demonstrates the potential of modulated surface-textured substrates for enhancing the absorption of light in absorber layers in the entire wavelength region.

METAL NANO-PARTICLES

An alternative way to provide light trapping in solar cells is by means of scattering at metal nano-particles. Light incident on the particles can induce a localized surface plasmon resonance. As a result, these particles can be very efficient light scatterers in a tunable wavelength range. The size and shape of the particles and their position inside the solar cell are parameters that can be used to fine tune the scattering properties [34]. Besides scattering of light, the particles can give rise to the parasitic absorption of light. Larger nano-particles with a diameter in the order of 100 nm give rise to more scattering and less absorption [35] and are therefore desirable for solar cell applications.

The effect of a plasmonic back reflector on the performance of *a*-Si:H solar cells was investigated experimentally. AZO was deposited on glass. On this flat front TCO an *a*-Si:H *p-i-n* structure with a 150 nm thick intrinsic layer was deposited using PECVD. Finally, either a state-of-the-art 80 nm thick AZO followed by an opaque Ag layer or a plasmonic back reflector was added to finish the solar cell. The plasmonic back reflector has a similar structure, but has silver nano-particles embedded in the middle of the AZO layer. The silver nano-particles were formed by depositing a 30 nm thin Ag layer followed by an 11 hour anneal at 180 °C. Due to surface tension the Ag film breaks up into islands. A scanning electron micrograph (SEM) image of the islands is shown in figure 9. As can be seen, the Ag islands have a highly irregular shape and a size of several hundreds of nanometers. The surface coverage is about 40%. Islands with a more

Figure 9: SEM image of silver islands (nano-particles) on AZO, formed by annealing a 30 nm thick film of silver.

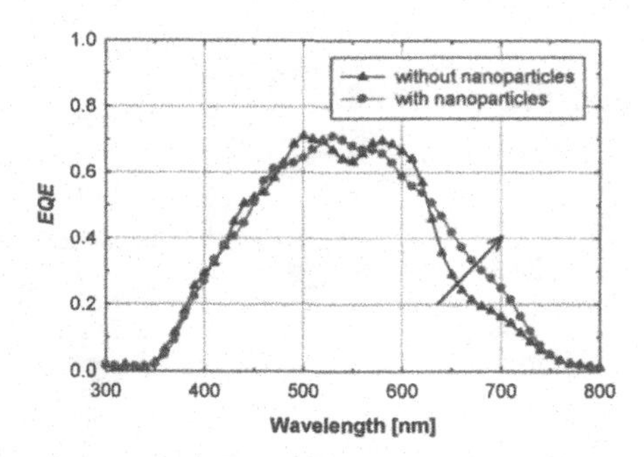

Figure 10: *EQE* of *a*-Si:H cell device with and without silver nano-particles embedded in AZO layer at the back.

regular shape could be formed by annealing a thinner layer of Ag, however this would result in smaller islands and give rise to more absorption and less scattering of light.

The *EQE* of the devices with and without nano-particles were measured and the results are shown in figure 10. It can be seen that the *EQE* of the device with nano-particles is somewhat higher in the wavelength range of 620 to740 nm. This and the less pronounced interference oscillations indicate that the particles scatter light into the absorber layer diffusely. As a result of the improved light trapping J_{sc} has increased by 5%. It is expected that larger gains in J_{sc} could be obtained if these large nano-particles could be fabricated with a more uniform size and shape.

1-D PHOTONIC CRYSTALS

Optical losses can occur at the metallic back contact of thin-film silicon solar cells because the surface-textured metal back reflectors suffer from undesired surface plasmon absorption [36]. In order to minimize these losses an optimal approach for engineering a back reflector is desired. The high-quality back reflector has a high reflectance in a broad wavelength region. These specifications match the behavior of 1-D PC in the role of distributed Bragg reflector (DBR). 1-D PC is a multilayer structure in which two layers with different optical properties (refractive indexes) are periodically alternated. When light propagates through this structure, constructive and deconstructive interferences arise, resulting in the wavelength-selective reflectance or transmittance behavior. It has been demonstrated that periodically repeated stacks of *a*-Si:H and *a*-SiN_x:H (deposited using rf-PECVD at 235 °C) on glass exhibit a highly-reflective behavior in a broad range of wavelengths [37]. 1-D PCs were fabricated on thin-film silicon compatible substrates (glass / etched AZO) and at compatible temperature < 180 °C [38]. The effects of the deposition temperature, the different substrate (from flat glass to glass/etched AZO), and the angle of incidence on the optical properties of 1-D PCs were determined. The reflectance of 80-90% is achieved in a relatively broad wavelength region of 400 nm even with a simple 6-layer PC stack. Measured results are in good agreement with the predicted ones, determined by means of simulations (not shown here). Three pairs of layers of *a*-Si:H and *a*-SiN_x:H were used to fabricate an 1-D PC operational as the DBR. The schematic

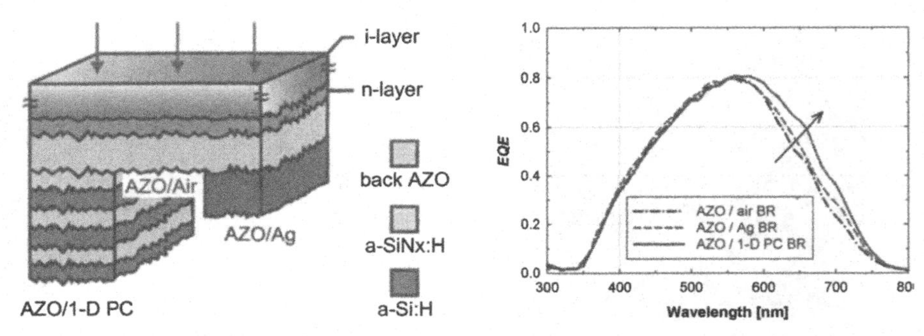

Figure 11: Back side of the solar cell with three different back reflectors (arrows indicate place where the beam of the *EQE* setup is directed).

Figure 12: *EQE* of solar cells on rough front AZO stacked with different back reflectors.

structure of the solar cell is given in figure 11. The structure consisted of glass coated with rough front AZO layer, a typical single-junction thin-film silicon *p-i-n* solar cell, a 700 nm thick AZO film used as back electrode. Three different back reflectors were applied at the back side of the solar cell, namely AZO / air (reference), AZO / Ag, and AZO / 1-D PC (see figure 11).

The measured *EQE* of solar cells with different back reflectors is presented in figure 12. The *EQE* did not show the state-of-the-art performance because of some non-optimized processing issues. Additionally, a high absorption in AZO front contact could be responsible for a low response in the short-wavelength region. The *EQE* increased in the long-wavelength region where the positive effect of the back reflectors was expected. The trend of increasing *EQE* is visualized by the arrow in the plot. In case of the textured metallic back reflector, surface plasmon absorption was responsible for a lower *EQE* in comparison with the 1-D PC back reflector, where the plasmon absorption was not expected.

CONCLUSIONS

In this article approaches for light trapping are presented such as i) surface textures based on periodic diffraction gratings and modulated surface morphologies for enhanced scattering, ii) plasmonic scattering using metal nano-particles, and iii) one-dimensional (1-D) photonic-crystal-like (PC) structures for back reflectors. A scattering model for the calculation of the *AID* of the diffused light after traversing surface-textured TCO layers is presented.

The gratings were successfully employed in single junction *a*-Si:H solar cells. High performance was obtained on solar cells using diffraction grating with P = 600 nm and h = 300 nm, where J_{SC} of this solar cell increased by 14.2% relative to the J_{SC} of the solar cell with flat interfaces. The concept of modulated surface texture was presented. Solar cells deposited on modulated surface-textured AZO/glass substrate exhibited higher *EQE* in comparison to a solar cell deposited on surface-textured ZnO or glass due to improved anti-reflective and scattering properties of rough interfaces.

Metal nano-particles are efficient light scatterers and can be used for light trapping in solar cells. Silver nano-particles were fabricated by annealing a film of silver. These particles were used to form the plasmonic back reflector of an *a*-Si:H solar cell. Compared to a similar device without nano-particles the J_{sc} increased 5%.

1-D PC can be used to obtain high reflectance at the back contact in a broad and tunable wavelength region. 1-D PCs based on *a*-Si:H and *a*-SiN_x:H layers were designed by computer simulations, fabricated and implemented in *a*-Si:H solar cells. The use of 1-D PC back reflector resulted in a higher *EQE* in the long-wavelength region in comparison to other types of back reflectors. The absence of absorption losses in the dielectric layers of 1-D PC makes the combination of AZO / PC a good candidate for back reflectors.

ACKNOWLEDGEMENTS

This work was partially carried out with a subsidy of the Dutch Ministry of Economic Affairs under EOS program (Projects No. EOSLT04029 and No. KTOT01028) and of Nuon Helianthos company. The authors gratefully acknowledge financial support from the NMP-Energy Joint Call FP7 SOLAMON Project (www.solamon.eu). Slovenian Research Agency is acknowledged for funding a part of his work (Project No. J2-0851-1538-08). The authors acknowledge OM&T company for providing the 1-D grating substrates.

REFERENCES

1. D. Fischer *et al.*, Proc. 25th IEEE PVSC, Washington, DC, 1996, p. 1053.
2. S. Guha *et al.*, Proc. PVSEC-15, 2005, Shanghai, China, p. 35.
3. S. Benagli, D. Borrello, *et al.*, Proc. 24th EUPVSEC, Hamburg, Germany (2009).
4. K. Yamamoto, M. Yoshimi, *et al.*, Sol. En. Mat. and Solar Cells, 74 (1-4) (2002) 449-455.
5. G. Yue, B. Yan, J. M. Owens, J. Yang, S. Guha, Mater. Res. Soc. Symp. Proc. **808** (2004), 808-A09-43.
6. M. A. Green, K. Emery, Y. Hishikawa, and W. Warta, Prog. Photovolt. Res. Appl. **17**, 320 (2009).
7. A. Barnett *et al.*, Prog. Photovolt: Res. Appl. **17**, 75 (2009).
8. D. Ginley, M. A. Green, R. Collins, MRS Bulletin 33/4, 2008, p. 355.
9. Ivanova *et al.*, Proc. 23rd EUPVSEC, Valencia, Spain, 2008, p. 734.
10. P. Loeper *et al.* Proc. 23rd EUPVSEC, Valencia, Spain, 2008, p.173.
11. G. Conibeer, *et al.*, Sol. Ener. Mat. Sol. C. (2010), doi:10.1016/j.solmat.2010.01.018.
12. A. Luque, A. Marti, A. J. Nozik, MRS Bulletin 32/3, 2007, p. 236.
13. S. Fay, *et al.*, Thin Solid Films **515** (2007) p. 8558.
14. M. Berginski *et al.*, Thin Solid Films **516** (2008) p. 5836.
15. M. Kambe *et al.*, Proc. 34th IEEE PVSEC, 2009.
16. C. Das, A. Lambertz, J. Huepkes, W. Reetz, and F. Finger, Appl. Phys. Lett. **92**, 053509 (2008).
17. D. Dominé, P. Buehlmann, J. Bailat, A. Billet, A. Feltrin, and Ch. Ballif, Phys. Stat. Sol. (RRL) **2**, 163–165 (2008).
18. L. Zeng *et al.*, Appl. Phys. Lett., **93** 221105 (2008).
19. O. Isabella *et al.*, Proc. 24th EUPVSEC, Hamburg, Germany, 2009, p. 2304.
20. K. Sato, Y. Gotoh, Y. Wakayama, *et al.*, Rep. Res. Lab.: Asahi Glass Co. Ltd. **42**, 129 (1992).
21. M. Berginski, J. Hüpkes, M. Schulte, et al., J. Appl. Phys. **101**, 074903 (2007).
22. M. Zeman, J.A. Willemen, L.L.A. Vosteen, G. Tao and J.W. Metselaar, Sol. Energ. Mat. Sol. C. **46**, 81 (1997).
23. J. Krc, F. Smole, and M. Topic, Prog. Photovolt. Res. Appl. **11**, 15 (2003).
24. K. Jaeger and M. Zeman, Appl. Phys. Lett. **95**, 171108 (2009).
25. D. Dominé, F.-J. Haug, C. Battaglia, and C. Ballif, J. Appl. Phys. **107**, 044504 (2010).
26. M. Born, and E. Wolf, *Principles of optics*, 7th ed. (Cambridge University Press, Cambridge, 1999), chapter 13.
27. M. Born, and E. Wolf, *Principles of optics*, 7th ed. (Cambridge University Press, Cambridge, 1999), chapter 8.
28. D. Dominé, P. Buehlmann, J. Bailat, A. Billet, A. Feltrin, and C. Ballif, Phys. Status Solidi (RRL) **2**, 163-165 (2008).
29. J. Krč, M. Zeman, *et al.*, Mater. Res. Soc. Symp. Proc. Vol. **910**, 2006, 0910-A25-01.
30. A. Campa, O. Isabella, R. van Erven, P. Peeters, H. Borg, J. Krc, M. Topic and M. Zeman, Prog. Photovolt: Res. Appl. (2010), 10.1002/pip.940.
31. O. Isabella *et al.*, in Proc. 23rd EU-PVSEC, Valencia, Spain (2008), 3AV.1.48.
32. F.-J. Haug *et al.*, in Proc. 21st EUPVSEC, Dresden, Germany (2006), .
33. O. Isabella, F. Moll, J. Krč and M. Zeman, Phys. Status Solidi A **207** (3) 642-646 (2010).
34. H.A. Atwater, A. Polman, Nat. Mater. **9**, 205-213 (2010).
35. H.R. Stuart, D.G. Hall, Appl. Phys. Lett. **73**, 3815-3817 (1998).
36. J. Springer, A. Poruba, L. Mullerova, *et al.*, J. Appl. Phys. **95**, 1427 (2004).
37. J. Krc, M. Zeman, S. Luxembourg and M. Topic, Appl. Phys. Lett. **94** (15) (2009) 153501.
38. O. Isabella, B. Lipovsek, J. Krč, M. Zeman, Mater. Res. Soc. Symp. Proc. **1153** (2009), 1153-A03-05.

Mater. Res. Soc. Symp. Proc. Vol. 1245 © 2010 Materials Research Society 1245-A03-04

A New Approach to Light Scattering from Nanotextured Interfaces for Thin-Film Silicon Solar Cells

C. Battaglia[1], J. Escarré[1], K. Söderström[1], F.-J. Haug[1], D. Dominé[1,2], A . Feltrin[1,3] and C. Ballif[1]
[1]Ecole Polytechnique Fédérale de Lausanne (EPFL), Institute of Microengineering (IMT), Photovoltaics and Thin Film Electronics Laboratory, 2000 Neuchâtel, Switzerland
[2]now at Scuola Universitaria Professionale della Svizzera Italiana (SUPSI), Institute of Applied Sustainability to the Built Environment (ISAAC), 6952 Canobbio, Switzerland
[3] now at Kaneka Corporation, Imec, 3001 Leuven, Belgium

ABSTRACT

We investigate the influence of refractive index contrast on the light scattering properties of nanotextured interfaces, which serve as front contact for p-i-n thin-film silicon solar cells. We here focus on ZnO surfaces with randomly oriented pyramidal features, known for their excellent light trapping performance. Transparent replicas, with a different refractive index, but practically identical morphology compared to their ZnO masters, were fabricated via nanoimprinting. Within the theoretical framework we recently proposed, we show how the angular and spectral dependence of light scattered by nanostructures with identical morphology but different refractive index may be related to each other allowing direct comparison of their light trapping potential within the device.

INTRODUCTION

To further improve conversion efficiencies of thin-film silicon solar cells, efficient light management schemes are crucial, as the absorption coefficient of silicon is small in the near infrared region. The most common approach to improve optical performance is by means of light scattering at randomly textured interfaces.

In the p-i-n superstrate configuration, light trapping is traditionally achieved by scattering at the interface between the transparent front electrode and the absorbing silicon layers by exploiting either the natural, randomly-oriented pyramidal texture of ZnO grown via low-pressure chemical vapor deposition (LP-CVD) [1] or the crater-like texture of sputtered ZnO obtained by wet-etching in HCl [2].

We recently validated the use of replicated nanostructures fabricated via ultraviolet nanoimprinting as front contacts for the p-i-n configuration [3] by demonstrating short-circuit current densities as high as for state-of-the-art nanotextured ZnO master structures, known for their excellent light trapping performance [4, 5].

In this contribution we investigate, both experimentally and theoretically, the influence of refractive index contrast on the scattering behavior of nanotextured interfaces. As the replicated structures exhibit a refractive index n=1.5, which is lower than that of the corresponding ZnO master with n=2, a direct comparison of the measured optical scattering properties is not possible.

We here use a new methodology, we recently presented, to simulate the angular and spectral dependence of light diffusely scattered across nanotextured interfaces [6]. Our method allows us to relate the measured optical scattering data of the master and its corresponding replica via the experimentally determined surface profile and the refractive indices.

EXPERIMENT AND THEORY

Pyramidally-textured boron-doped ZnO layers of thickness 2 μm (called Z2 used for the development of amorphous silicon solar cells) and 5 μm (called Z5 used for the development of microcrystalline silicon solar cells) serving as a master for the replication process were grown via LP-CVD on 0.5 mm thick borosilicate glass. The 5 μm thick sample was subjected to a 20 min plasma treatment to adapt the surface morphology to the growth of microcrystalline silicon layers [4]. The UV nanoimprinting process is described in Ref. [7, 8].

Surface morphologies were measured using atomic force microscopy (AFM) probing an image size of 10x10 μm^2 with a resolution of 512x512 pixels. Angle-resolved scattering (ARS) of light was measured with a detector on a home-built goniometer under normal incidence onto the glass side using a laser at a wavelength of 543 nm. The spectral dependence of the ratio between diffuse to total optical transmittance, called haze, was measured with a photospectrometer equipped with an integrating sphere.

Our theoretical model makes use of a slightly modified Rayleigh-Sommerfeld diffraction integral proposed by Harvey [9] and requires only measured profile data and the refractive indices as input. In this model, light passing across the nanotextured interface, with a peak-to-valley roughness z_0, is assumed to acquire a phaseshift proportional to $z \cdot n_1 + (z_0 - z) \cdot n_2$, where z is the distance from the maximum peak height to the interface travelled in the first medium with refractive index n_1, and $(z_0 - z)$ the distance travelled after the interface in the second medium with refractive index n_2. Since the interface is textured, z depends on the morphology of the surface and is therefore a function of the lateral coordinates x and y. Thus the local phaseshift acquired by a plane wave after crossing the interface is $(n_1 - n_2) \cdot z(x, y) + n_2 \cdot z_0$. Summing up all plane waves exiting the roughness zone, taking into account this phaseshift, we obtain the radiance L in direction cosines space for a given wavelength λ.

$$L(\alpha_k, \beta_l, \lambda) \propto \left| \sum_{i=1}^{N} \sum_{j=1}^{M} e^{2 \cdot \pi \cdot i \cdot \frac{n_2}{\lambda} \cdot (\alpha_k \cdot x_i + \beta_l \cdot y_j)} \cdot e^{2 \cdot \pi \cdot i \cdot \frac{|n_1 - n_2|}{\lambda} z(x_i, y_j)} \right|^2$$

Note that the term $n_2 \cdot z_0$ in the phaseshift leads to a position independent phaseshift and drops out when taking the absolute value for the radiance. The double sum can be handled efficiently using a fast Fourier transform approach. Here α_k and β_l are direction cosines in reciprocal space, related to the wavevector $\vec{k}$ via $\vec{k} = \frac{2 \cdot \pi}{\lambda} (\alpha, \beta, \sqrt{1 - \alpha^2 - \beta^2})$.

For comparison of the angular dependence with experiment it is convenient to transform the radiance from cosine direction space (α, β) into spherical coordinates (ϑ, φ) using

$\alpha = \sin \vartheta \cdot \cos \varphi$

$\beta = \sin \vartheta \cdot \sin \varphi$

For light scattered at random interfaces, which does not show any azimuthal dependence, the experimental ARS curve for a given wavelength is obtained by integration over all azimuthal angles φ and by taking into account the projection factor $\cos\vartheta = \sqrt{1-\alpha^2-\beta^2}$ [9].

$$ARS(\vartheta,\lambda) = I(\lambda)\int_0^{2\pi} d\varphi \cdot L(\vartheta,\varphi,\lambda)\cdot\cos(\vartheta)$$

where $I(\lambda)$ is an angle independent factor taking into account the intensity of the incoming light. The spectral dependence of the scattered light as measured by the haze may directly be obtained from the radiance

$$Haze(\lambda) = \left(\sum_{k,l}^{\sqrt{\alpha_k^2+\beta_l^2}\le 1} L(\alpha_k,\beta_l,\lambda) - L(0,0,\lambda)\right) / \sum_{k,l}^{\sqrt{\alpha_k^2+\beta_l^2}\le 1} L(\alpha_k,\beta_l,\lambda)$$

As proposed by Harvey [9], the sum should only be carried out over real propagating modes characterized by $\sqrt{\alpha_k^{\ 2}+\beta_l^{\ 2}} \le 1$, as modes outside the unit circle are evanescent. The specular beam is represented by $L(0,0,\lambda)$, whereas experiment includes polar angles up to approximately 7° (for a discussion of this approximation see Ref [6]).

DISCUSSION

Fig. 1 shows AFM images of the master and replicated Z2 and Z5 ZnO surfaces with their pyramidal structure. In order to quantify the quality of the replication process, we compare the local height and angle histogram extracted from the AFM images. Both histograms are practically identical for master and replica indicating the high fidelity of the replication process.

We now focus on the angular dependence of the scattered light. In Fig. 2 the experimental and calculated ARS curves of the master and replicated structures are presented for scattering into air. As input for all calculations we used the AFM images of the masters in Fig. 1, adapting only the refractive index. The calculated curves nicely reproduce the experimental ARS curves. We note however that they deviate slightly at large scattering angles, presumably because we model the optical interface by a random phase-screen with zero thickness which neglects diffraction within the peak-to-valley depth of the rough interface [10]. The oscillations in the calculation are due to the finite AFM image size and may be reduced by either taking larger images or averaging over several images.

As can be seen the replicas generally scatter significantly less light into large angles compared to their masters. From the results of our calculations, we conclude that this is due to the reduced refractive index contrast when scattering from the replica (n=1.5) into air (n=1) compared to scattering from ZnO (n=2) into air. For sample Z2 the change in refractive index leads to a global reduction of the diffuse scattering. The shape of the ARS curve remains approximately the same. This is different for the Z5 sample. In this case, we observe enhanced scattering into smaller angles. This non-trivial behavior, observed in experiment, is well reproduced by the calculation.

In Fig. 3 we show the spectral dependence of the scattered light. The theoretical results for the Haze for scattering into air agree nicely with the experimental data. The reduction in Haze observed in experiment, when switching from the master to the replica, is well reproduced in the calculation and can therefore again be attributed to the change in refractive index.

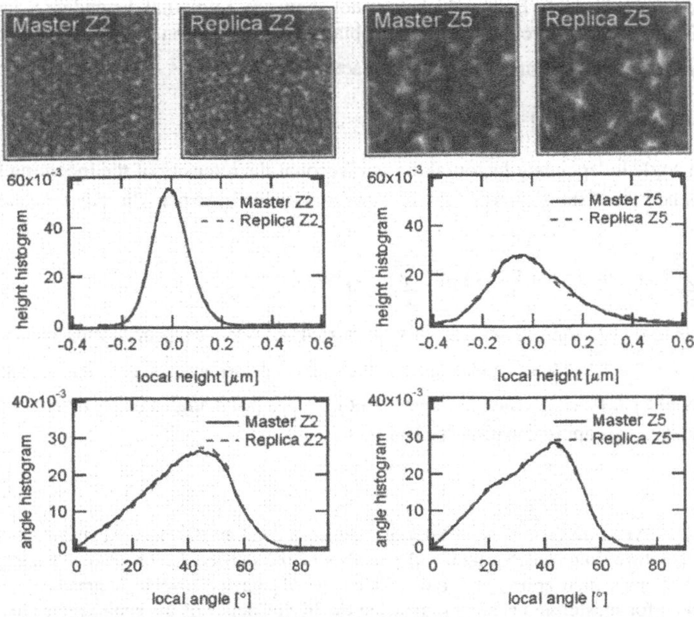

Figure 1: AFM topographies of ZnO masters and their transparent replicas, image size 10x10 μm^2, with the extracted local height and angle histograms.

However, note that once integrated into a solar cell device, scattering takes place into silicon (n=4), resulting in a higher refractive index contrast for scattering from the replica into silicon than from the master into silicon. Although not experimentally accessible, our theory allows determining the scattering behavior into silicon. As can be seen from Fig. 3, the higher refractive index contrast indeed leads to a higher haze value for the replica than the master.

It is important to note that the reflection behavior differs for scattering into air and into silicon. For the former case, total internal reflection may occur, which is not the case for scattering into silicon. In both situations however, the additional randomization produced by those reflections are not taken into account by our model.

At this point, it is interesting to come back to the theoretical expression for the radiance L. As the difference between replica and master only enters into the expression for the radiance through the terms $\frac{n_2}{\lambda}$ and $\frac{|n_1 - n_2|}{\lambda}$, we find that scattering from the replica ($n_1 = 1.5$) into air ($n_2 = 1$) at a certain wavelength λ, yields exactly the same value for the radiance for scattering from the ZnO master ($n_1 = 2$) into silicon ($n_2 = 4$) at the wavelength 4λ (exact for wavelength independent refractive indices). This scaling property of our model is useful in the sense that it allows to determine directly the scattering characteristics of the transmitted fraction of the light, such as ARS and haze at the interface between ZnO and silicon, by measuring the scattering of the replica into air at a quarter of the considered wavelength.

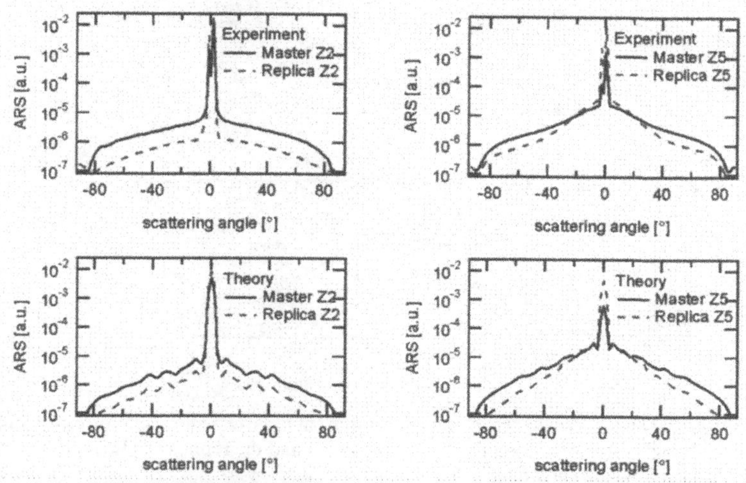

Figure 2: Experimental and calculated ARS curves for the ZnO masters and their corresponding replicas for a wavelength of 543 nm. For all calculations, only the AFM images of the masters in Fig. 1 were used, but the refractive index was adapted to account for scattering from either the ZnO (n=2) or the replica (n=1.5) into air (n=1).

CONCLUSIONS

We investigated the influence of refractive index contrast on the angular and spectral dependence of light diffusely scattered across nanotextured interfaces, which serve as front contact for p-i-n thin-film silicon solar cells. For this we compared the measured and simulated angle-resolved scattering intensity and its spectral dependence (haze) of pyramidally-textured ZnO surfaces and their replicas, exhibiting different refractive indices, but practically identical morphologies. We find a good match between theory and experiment for scattering into air. Our theoretical model also nicely reproduces the non-trivial change in scattering behavior when switching from the master to the replica, which may be attributed to the change in refractive index. Our theory furthermore allows to calculate scattering into silicon and offers us a scaling law, which allows to directly relate the scattering of transmitted light at the replica-air interface to the scattering of transmitted light at the master-silicon interface.

ACKNOWLEDGMENTS

Stimulating discussions with P. Cuony are gratefully acknowledged. The authors acknowledge support by the Swiss Federal Office for Energy (OFEN) under contract 101191 and the Swiss National Science Foundation under grant 200021-12577/1.

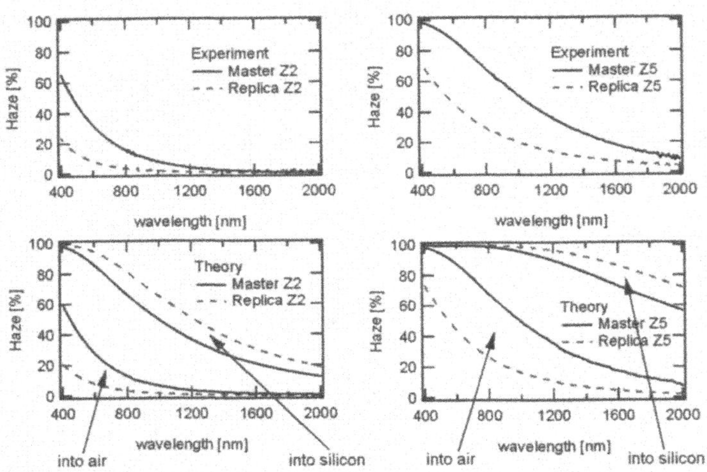

Figure 3: Experimental and calculated Haze curves for the ZnO masters and their corresponding replicas. Theoretical curves for scattering from the ZnO (n=2) and the replica (n=1.5) into air (n=1) and into silicon (n=4) are shown.

REFERENCES

1. S. Faÿ, J. Steinhauser, N. Oliveira, E. Vallat-Sauvain, and C. Ballif, Thin Solid Films 515, 8558 (2007)
2. M. Berginski, J. Hüpkes, W. Reetz, B. Rech, and M. Wuttig, Thin Solid Films 516, 5836 (2008)
3. C. Battaglia, K. Söderström, J. Escarré, F.-J. Haug, D. Dominé, P. Cuony, M. Boccard, G. Bugnon, C. Denizot, M. Despeisse, A. Feltrin, and C. Ballif, Appl. Phys. Lett. 96, 213504 (2010)
4. J. Bailat, D. Dominé, R. Schlüchter, J. Steinhauser, S. Faÿ, F. Freitas, C. Bucher, L. Feitknecht, X. Niquille, T. Tscharner, A. Shah, and C. Ballif, Proc. 4th Conf. Photovoltaic Energy Conversion, 1533 (2006)
5. S. Benagli, D. Borello, E. Vallat-Sauvain, J. Meier, U. Kroll, J. Hoetzel, J. Bailat, J. Steinhauser, M. Marmelo, G. Monteduro, and L. Castens, Proc. 24th European Photovoltaic Solar Energy Conference, 3BO.9.3 (2009)
6. D. Dominé, F.-J. Haug, C. Battaglia, and C. Ballif, J. Appl. Phys. 107, 044504 (2010)
7. J. Escarré, K. Söderström, O. Cubero, F.-J. Haug, and C. Ballif, in *Amorphous and Polycrystalline Thin-Film Silicon Science and Technology – 2010* (Mater. Res. Soc. Symp. Proc. **1245**, Warrendale, PA, 2010), A7.4
8. K. Söderström, J. Escarré, O. Cubero, F.-J. Haug, S. Perregaux, and C. Ballif, accepted for publication in Prog. Photovolt: Res. Appl.
9. J. E. Harvey, C. L. Vernold, A. Krywonos, and P. L. Thomson, Appl. Opt. 38, 6469 (1999)
10. D. Dominé, Ph.D. thesis, Université de Neuchâtel (2009)

Low Gap Materials

Mater. Res. Soc. Symp. Proc. Vol. 1245 © 2010 Materials Research Society 1245-A04-02

Influence of Hydrogen on the Germanium Incorporation in a-$Si_{1-x}Ge_x$:H for Thin-Film Solar Cell Application

C.M. Wang, Y.T. Huang, K.H. Yen, H.J. Hsu, C.H. Hsu, H.W. Zan and C.C. Tsai,
Dept. of Photonics, National Chiao Tung University, 1001 Ta Hsueh Road, Hsinchu, Taiwan

ABSTRACT

In this work, we examined the Ge incorporation and the accompanied defect formation during PECVD deposition of hydrogenated amorphous silicon-germanium alloys (a-$Si_{1-x}Ge_x$:H). In particular, we studied the effect of hydrogen on film growth, defect formation, Ge and Si incorporation efficiencies, and the H-bonding configuration. Our results indicate that hydrogen has a strong effect on improving the a-$Si_{1-x}Ge_x$:H film quality and the Ge incorporation in a-$Si_{1-x}Ge_x$:H. With adequate hydrogen dilution, the a-$Si_{1-x}Ge_x$:H thin-film quality significantly improved. However, excessive hydrogen dilution degraded the film properties. A number of analytical tools were employed, including FTIR, XPS, UV-Visible spectroscopy, photoconductivity, etc. The a-$Si_{1-x}Ge_x$:H material having 24% Ge content and a bangap of 1.61ev produced the solar cell with a conversion efficiency of 7.07%.

INTRODUCTION

Improving the quality of amorphous silicon-germanium (a-$Si_{1-x}Ge_x$:H) film is one of the key topics for achieving high efficiency multi-junction solar cells. This is because a-$Si_{1-x}Ge_x$:H, with its bandgap of 1.1-1.7ev [1], can enhance the optical absorption in the near-infrared spectral range. However, the Ge incorporation always introduces defects in the material [2], which affects the cell performance. Previous studies had shown that micro-structural defects can be removed by hydrogen dilution [3]. In this work, we examined the influence of hydrogen on the Ge incorporation in a-$Si_{1-x}Ge_x$:H alloys.

EXPERIMENT

The a-$Si_{1-x}Ge_x$:H alloys were simultaneously deposited on Corning Eagle 2000 glass substrates and crystalline silicon wafers using a mixture of SiH_4, GeH_4, and H_2 in the plasma enhanced chemical vapor deposition (PECVD) at 27.12 MHz. The influences of hydrogen on a-$Si_{1-x}Ge_x$:H thin-film quality with different germane to silane ratios were investigated by changing gas flow rates. The germane concentration in the gas, X_g, is defined as a ratio of germane gas flow rate to the reactant gas flow rate: $X_g = GeH_4 / (GeH_4+SiH_4)$. The hydrogen dilution ratio, R_H, is defined as a ratio of hydrogen gas flow rate to the reactant gas flow rate: $R_H = H_2 / (GeH_4+SiH_4)$. We increased the X_g from 8.3% to 16.7% and varied the R_H from 0 to 8, while other deposition parameters were listed in Table 1.

The photoconductivity and cell performances were measured under the AM1.5 illumination. The transmittance of a-$Si_{1-x}Ge_x$:H alloys was obtained through the UV-Visible spectroscopy, and the band gap was evaluated from the Tauc's plots [4]. The information on a-$Si_{1-x}Ge_x$:H bonding configuration was obtained by Fourier Transform Infrared Spectroscopy (FTIR) [5]. We calculated the bonding intensity ratio of the Si-dihydride to silicon-hydrogen stretching modes, which is also called the microstructure factor: $F = I_{SiH_2} / (I_{SiH_2} + I_{SiH})$. Bonded hydrogen contents were calculated using FTIR absorption peaks of Si–H and Ge–H wagging modes and the calibration factors [6]. The Ge contents in the a-$Si_{1-x}Ge_x$:H alloy were measured by X-ray photoelectron spectroscopy (XPS). We have also fabricated single-junction solar cells with different Ge contents and hydrogen dilution ratios.

Table 1 Deposition parameters of a-$Si_{1-x}Ge_x$:H alloys

Substrate temperature (T_S)	190-210 °C
Substrate to plasma electrode distance (E/S)	13-25 mm
Growth pressure (P_g)	300 mTorr
RF power density (P_{rf})	30-60 mW/cm^2
SiH_4 flow rate	25-55 sccm
GeH_4 flow rate	2-10 sccm
H_2 flow rate	0-480 sccm

RESULTS AND DISCUSSION

Shown in the Fig.1(a) is the dependence of deposition rate on the R_H. Shown in the Fig.1(b) is the dependence of microstructure factor on the R_H. Exhibited in Fig.1(a), we found that the growth rates of the a-$Si_{1-x}Ge_x$:H films reduced with increasing H-dilution for various X_g of 8.3, 11.1, 16.7%. Due to a hydrogen etching effect, the deposition rate obviously decreased from 0.55 to 0.2 nm/s. But the germane concentration did not affect the deposition rates.

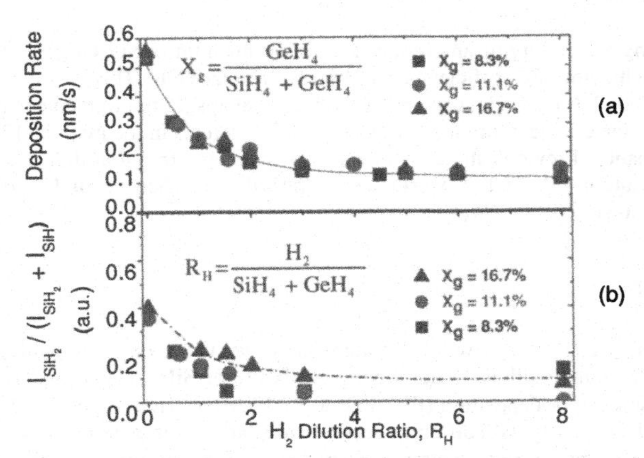

Fig.1 (a) The deposition rate (b) The microstructure factor, as a function of R_H for three different X_g at 8.3, 11.1, 16.7%

The FTIR spectra with different hydrogen dilution ratio at a fixed germane concentration of 8.3% were shown in Fig.2. The absorption peaks at 2090 and 2000 cm^{-1} are related to the Si-dihydride (SiH_2) bonding and the Si mono-hydride (SiH) bonding, respectively. Similarly, the absorption peaks at 1980 and 1880 cm^{-1} are related to the Ge-dihydride (GeH_2) bonding and the Ge mono-hydride (GeH) bonding. The features between 755 and 890cm^{-1} are related to the SiH_2 and the GeH_2 bending modes. When the H_2 flow was increased from 0 to 2 in the reactant gas flow, the SiH_2 stretching mode at 2090 cm^{-1} was significantly reduced, while the SiH stretching mode at 2000 cm^{-1} dominated the film quality. Shown in the Fig.1(b) is the dependence of the microstructure factor on the R_H. The SiH_2 modes decreased when H-dilution increased, which means that SiH_2 was preferentially eliminated with increasing H-dilution. Also, we estimated the hydrogen content from the wagging modes in the FTIR spectra. Our results indicated that the hydrogen content varied from 17 to 11 at.% with increasing hydrogen dilution, independent of X_g.

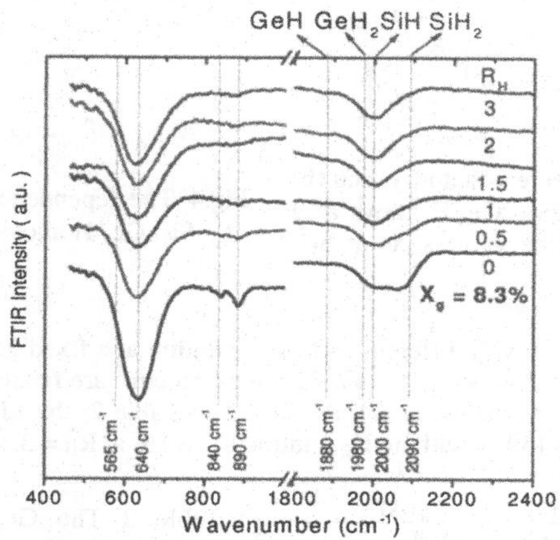

Fig.2 The Fourier transform infrared spectra of a-$Si_{1-x}Ge_x$:H alloys with different R_H = 0-3 at X_g = 8.3%

We also found that the hydrogen dilution influenced on the Ge incorporation in a-$Si_{1-x}Ge_x$:H films. X_f is defined as the film Ge contents of a-$Si_{1-x}Ge_x$:H alloys. The film Ge contents as a function of the hydrogen dilution for three different germane concentrations at 8.3, 11.1, 16.7% were presented in Fig.3(b). Increasing hydrogen dilution resulted in more Ge incorporation in the films. The correlations of the optical band gap with varying R_H for three different X_g are provided in the Fig.4. With increasing Ge incorporation in the films, the optical band gap decreases from 1.7eV to 1.48 eV. Germanium was found to be preferentially incorporated into the growing film from the gas phase compared to Si. The enhancement factor, y, is the ratio of film germanium content to germane gas concentration. The enhancement factor versus hydrogen dilution for three different X_g is shown in Fig.3(a). The enhancement factor rises more rapidly at lower hydrogen dilution, and slows down at hydrogen dilution ratio higher than 2. In addition, increasing germane concentration decreases the enhancement factor.

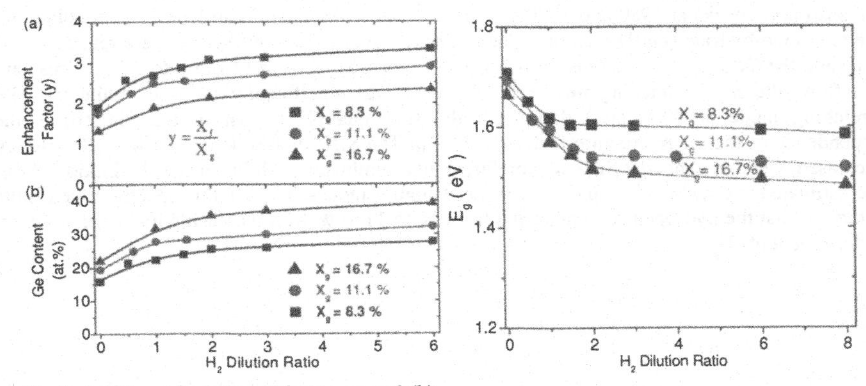

Fig.3 (a) The enhancement factor, y, and (b) The Ge contents in a-$Si_{1-x}Ge_x$:H alloys as a function of R_H for three various X_g at 8.3, 11.1, 16.7%

Fig.4 The dependence of optical band gap of a-$Si_{1-x}Ge_x$:H alloys on R_H at various X_g

The Raman Spectra with different hydrogen dilution at a fixed germane concentration of 8.3% are shown in Fig.5. The peaks at 480, 370 and 255 cm^{-1} are related to Si-Si bonds, Si-Ge bonds and Ge-Ge bonds, respectively [7]. As shown in Table 2, the film Ge content increases with the H_2 dilution from 15.8% without H_2 dilution to 26.1% at $R_H = 3$.

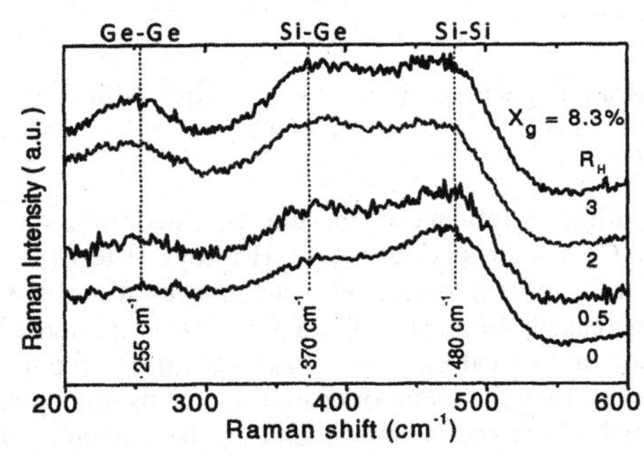

Fig.5 The Raman Spectra of a-$Si_{1-x}Ge_x$:H alloys with different R_H at $X_g = 8.3\%$

Table 2 The Ge contents and the enhancement factor, y, of $Si_{1-x}Ge_x$:H alloys with different hydrogen dilution ratios R_H at $X_g = 8.3\%$

R_H	X_f (%)	y
0	15.8	1.9
0.5	21.5	2.6
1.5	24.0	2.9
2	25.6	3.1
3	26.1	3.1

For a fixed X_g, both the film Ge content and the enhancement factor increased with increasing H-dilution. In PECVD, Ge is more readily incorporated into a-$Si_{1-x}Ge_x$:H alloys than Si with an enhancement factor about 2-3. But not all the additional Ge contributed to the Si-Ge bonding. In fact we found that some of the Ge ended up in Ge-Ge clusters, as shown in Fig.5, which are probably related to some defect states. The Ge related precursor was believed to have a higher sticking coefficient and a shorter lifetime on the deposited surface compared to the

Si-species [8]. Therefore, Ge is probably more readily incorporated into the network before finding a minimum energy configuration, which results in more defects.

The photoconductivity is an indication of the a-$Si_{1-x}Ge_x$:H film quality. The photo- and dark- conductivity versus hydrogen dilution for three different germane concentrations were shown in Fig.6. The photo- and dark- conductivity seems insensitive to the film Ge contents. As the hydrogen dilution ratio was increased from 0 to 2, the photoconductivity increased, indicating the improved material quality. This is consistent with the reduction of the Si-dihydride bonding by the hydrogen dilution as discussed earlier. With further hydrogen dilution beyond R_H = 2, the photoconductivity stayed unchanged. And photoconductivity properties are all similar in three different germane gas concentrations. The level of the photoconductivity indicates the films are of reasonable quality.

We have fabricated the a-$Si_{1-x}Ge_x$:H single-junction solar cells with 300nm active layer thickness on textured SnO_2:F substrates with different germane concentrations in the reactant gas. The quantum efficiency of two a-$Si_{1-x}Ge_x$:H cells are shown in Fig.7. The result of an a-Si:H solar cell is also included for comparison. The J-V curve of solar cell with a germane concentration of 8.3% is shown in the Fig.8.

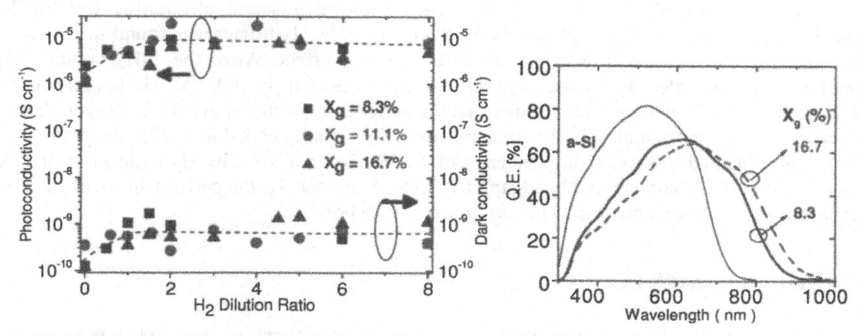

Fig.6 The dependence of photoconductivity and dark conductivity of a-$Si_{1-x}Ge_x$:H alloys on R_H at various X_g

Fig.7 The quantum efficiency of a-Si:H and a-$Si_{1-x}Ge_x$:H solar cells

It is interesting to note that the deposition rate, Ge content, E_g and photoconductivity changed within H_2 dilution ratio of 2. It is understandable that increasing the hydrogen concentration initially increases the H-etching effect, leading to a slower growth rate, fewer defects related to SiH_2 bonding structure, thus higher photoconductivity. But once the H-etching and the Si-Ge film formation reach an equilibrium near R_H = 2, increasing the supply of etch species does not seem to increase of the H-etch rate any more. Shown in the Table 3 is the comparison of the performances of the a-$Si_{1-x}Ge_x$:H single-junction solar cells with different R_H at X_g = 8.3 and 16.7%. At higher X_g, the cell performance degraded, which is probably due to increased Ge-Ge bonds or the Ge clusters. The Ge clusters induced voids or other defects, which increased the series resistance of the cell. Both the open-circuit voltage (V_{oc}) and the fill factor (FF) worsened at higher X_g, as a result of smaller optical band gap (from 1.61 ev to 1.51 ev) and increased Ge-Ge bonds. We obtained a conversion efficiency of 7.07% for an a-Si_{1-x} Ge_x:H single-junction solar cell with 24 at.% Ge content and bandgap of 1.61ev.

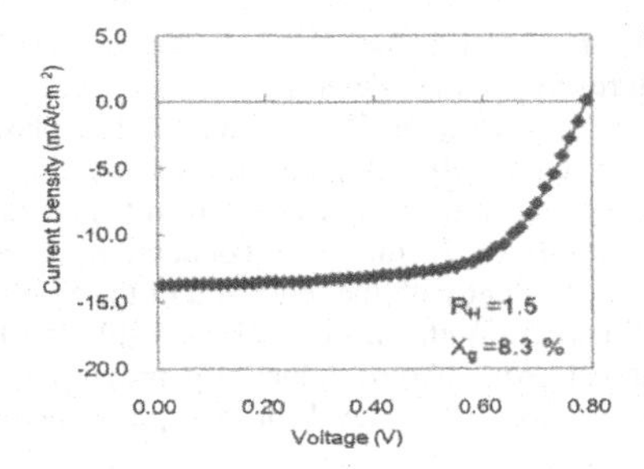

Fig.8 The performances of the a-$Si_{1-x}Ge_x$:H single-junction solar cells at X_g = 8.3%.

Table 3 The performances of the a-$Si_{1-x}Ge_x$:H single-junction solar cells at various X_g.

X_g (%)	8.3	16.7
X_f (%)	24	37.2
R_H	1.5	3
E_g (ev)	1.61	1.51
V_{OC} (v)	0.79	0.68
J_{sc} (mA/cm^2)	13.73	14.70
FF (%)	65.12	48.71
Efficiency (%)	7.07	4.85

CONCLUSION

We have found that the hydrogen dilution of the reactant gas during the PECVD deposition of a- $Si_{1-x}Ge_x$:H alloys has the following effects. Hydrogen was found to reduce the film growth rate, consistent with an increased etching effect. Also, the SiH_2 bonding was preferentially eliminated with increasing the hydrogen dilution. In PECVD, Ge is more readily incorporated than Si with an enhancement factor about 2-3. Furthermore, the hydrogen dilution increased the Ge incorporation in the film, and some Ge actually ended up in Ge clusters.

We obtained a conversion efficiency of 7.07% for an a-$Si_{1-x}Ge_x$:H single-junction solar cell with 24 at.% Ge content and bandgap of 1.61ev. At higher X_g, the performances of solar cell degraded, which is probably due to the increased Ge-Ge bonds.

ACKNOWLEDGMENTS

The study was supported by the Center for Green Energy Technology (CGET) of the National Chiao Tung University.

REFERENCES

1. P. Wickboldt, D. Pang, W. Paul, J.H. Chen, F. Zhong, C.C. Chen, J.D. Cohen and D.L. Williamson, *J. Appl. Phys.*, **81**, 6252 (1997).
2. D.L. Williamson, A.H. Mahan, B.P. Nelson and R.S. Crandall, *J. Non-Cryst. Solids*, **114**, 226 (1989).
3. S. J. Jones, Y. Chen, D. L. Williamson, R. Zedlitz, and G. Bauer, *Appl. Phys. Lett.*, **62**, 3267 (1993).
4. J. Tauc, R. Grigorovici, A. Vancu, *Phys. Stat. Sol.* , **15**, 627 (1966).
5. B.P. Nelson, Y. Xu, J.D. Webb, A. Mason, R.C. Reedy, L.M. Gedvilas and W.A. Langford. *J. Non-Cryst. Solids*, **266–269**, 680 (2000).
6. M. Cardona, *Phys. Stat. Sol. (b)*, **118**, 463 (1983).
7. J. Xu, S. Miyazaki and M. Hirose, *Jpn. J. Appl. Phys.*, **34**, L203 (1995)
8. S. Hazra, A.R. Middya and S. Ray, *J. Phys. D: Appl. Phy.*, **29**, 1666 (1996)

Mater. Res. Soc. Symp. Proc. Vol. 1245 © 2010 Materials Research Society 1245-A04-03

High Quality, Low Bandgap A-Si Films And Devices Produced Using Chemical Annealing

Vikram Dalal, Ashutosh Shyam, Dan Congreve and Max Noack
Iowa State University, Dept. of Electrical and Computer Engr. and Microelectronics Research Center, Ames, Iowa 50011

ABSTRACT

We report on the growth and properties of novel amorphous Silicon (a-Si:H) p-i-n devices prepared using chemical annealing with argon gas. The i layer in the p-i-n devices was grown using a layer by layer approach, where the growth of a very thin a-Si:H layer (7-30 angstroms) grown using a silane:argon mixture was followed by chemical anneal by argon ions. Repeated cycling of such growth/anneal cycles was used to produce the desired total thickness of the i layer. The thickness of the a-Si layer, and duration of the anneal time, were varied systematically. Pressure and power of the plasma discharge were also systematically varied. It was found that a thin a-Si layer, <10 angstroms, and low pressures which led to relatively high ion flux on the surface, gave rise to a significantly smaller bandgap in the device, as indicated by a significant lateral shift in the quantum efficiency vs. photon energy curve to lower energies. The smallest Tauc gap observed was in the range of 1.62 eV. Corresponding to this smaller bandgap, the current in the solar cell increased, and the voltage decreased. The Urbach energies of the valence band tail were also measured in the device, using the quantum efficiency vs. energy curve, and found to be in the range of45 meV, indicating high quality devices. Too much ion bombardment led to an increase in Urbach energy, and an increase in defect density in the material. Raman spectra of the device i layer indicated an amorphous structure. When hydrogen was added to argon during the annealing cycle, some materials turned microcrystalline, as indicated by the Raman spectrum, and confirmed using x-ray diffraction.

INTRODUCTION

Amorphous Silicon is an important electronic material. It is well known that its optical properties, i.e. optical bandgap (Tauc gap) is governed by the deposition temperature, with higher temperatures resulting in a lower bandgap. Most device-quality a-Si:H films have Tauc gaps in the range of 1.75-1.8 eV. While a lower gap in the range of 1.52-1.6 eV is possible by increasing the temperature during growth, and may be very desirable for devices such as tandem junction solar cells, such a material is not generally of high quality, and therefore is avoided for solar cell devices. In previous work from the Shimizu group [1-4], it has been shown that chemical anneal, namely layer by layer growth where the growth of a thin layer is followed by annealing by argon ions, leads to a smaller bandgap. We speculate that this result arises from the fact that energetic argon ions, which have about the same mass as Si ions, can be expected to help surface diffusion of silyl radicals produced during growth by a PECVD process, and thereby reduce the associated high hydrogen concentration (10%) typical of a-Si:H. However, the quality of devices produced was rather poor, with low fill factors (<60%) and high Urbach energies for valence band tails(>50 meV).
In this work, we have done a systematic study of the influence of ion bombardment by argon on the material and device properties of a-Si:H. The systematic study includes deliberate variations in pressure during growth, time of annealing, power during deposition, and the time of post-

growth annealing cycle. We have concentrated on fabricating p-i-n devices, and studying material properties using the data from devices.

EXPERIMENTS

The growth of devices was done using a standard PECVD growth technique. The frequency used was in the VHF range, ~45 MHz. The growth was done using a mixture of silane and argon, and the annealing was done using just argon, except for a few cases where hydrogen was used. For comparison, we also used non-annealed (continuous growth) samples deposited using either silane and argon mixture, or silane and hydrogen mixtures. For annealed devices, the pressure was varied between 15 mT and 100 mT, and the power between 3W and 6W. The pressure variation was done so as to change the ion energy and flux impinging on the surface. A lower pressure results in both higher ion energy and higher flux of ions impinging on the surface, and thus we could vary the amount and energy of ion bombardment. The objective was to study how this factor affected the bandgap and quality of the resulting material.

The growth/anneal cycle is represented schematically in Figure 1. The durations of growth (T1) and anneal cycles (T2) were individually varied. By varying the growth time, we could control the thickness of the layer grown per individual cycle. This was done because Argon cannot penetrate too deeply into the surface unless very high energy is used, and one does not want to use very high energies because they may lead to lattice distortion during growth. In general, energies greater than ~15 eV lead to distortion in the lattice and damage during growth [5]. Therefore, we need to keep argon energies small, and this fact dictates that the layer thickness/cycle should be kept small so that the ions can influence the structure of the entire thickness. The smallest thickness per cycle was ~7 angstroms, and the largest, ~30 angstroms.

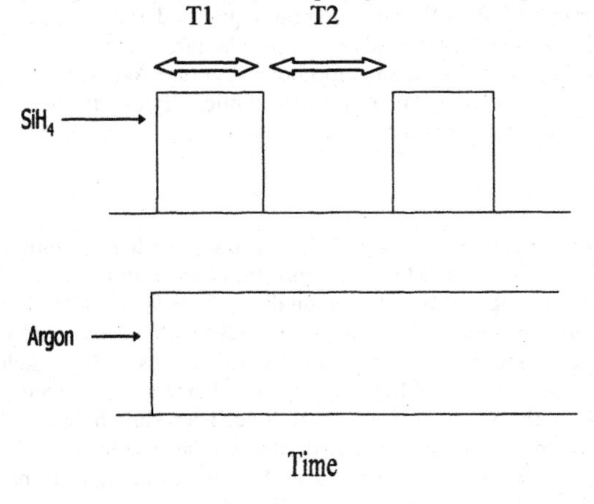

Fig. 1 Schematic diagram of the chemical annealing cycle for growing i layers in p-i-n devices

The devices were of the standard p-i-n type, with the middle i layer being deposited using chemical annealing. The outer p and n layers were deposited using standard, continuous deposition conditions. All devices were deposited on stainless steel substrates without an enhanced reflector. The devices were provided with a top ITO contact. The devices were measured for their illuminated I-V curves, and were also probed for quantum efficiency vs. photon energy under both zero bias and forward bias conditions so as to study hole transport properties[6]. We also measured quantum efficiency vs. photon energy in the subgap region so as to determine Urbach energy of valence band tails [7]. A comparison of such subgap QE data for devices made using annealed and unannealed i layers allows one to estimate the Tauc bandgap of the i layers. This is a more accurate way than measuring Tauc gaps in films related to devices, since the devices are deposited on a steel substrate and therefore may be subjected to a different ion bombardment condition than films which are deposited on an insulating substrate. A few of the devices were measured for their crystallinity using a Raman apparatus.

RESULTS

1. Results on I-V curves

In Fig. 2, we show a typical I-V curves for a devices fabricated using 30 mT pressure and chemical annealing. The growth per cycle was layer was 11.5 angstroms. The thickness of the i layer was 0.25 micrometer. The good fill factor (65%) indicates a high quality device. The voltage is only 0.82V, a result of chemical annealing, as shown in the next figure. The device efficiency was ~6.3%. In Fig. 3, we plot the open circuit voltage vs. four different deposition conditions, namely a device fabricated with hydrogen dilution, one with Argon dilution, both of these without any chemical annealing, and for two devices produced using chemical annealing under different conditions. It appears from Fig. 3 that using argon vs. hydrogen significantly lowers the voltage, and indication of a lower bandgap, and chemical annealing reduces voltage.

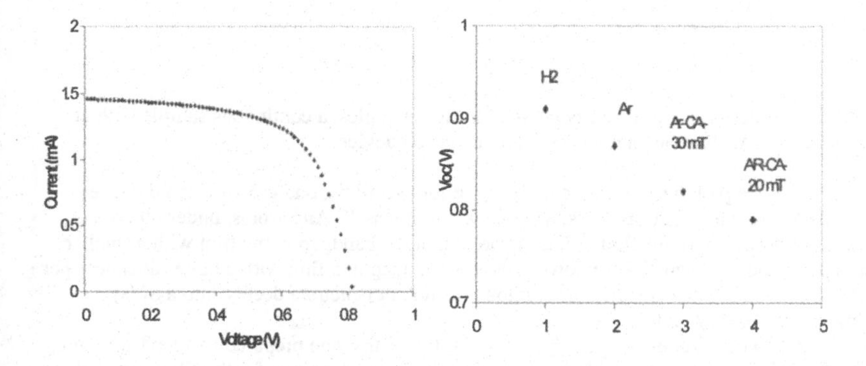

Fig. 2 I-V curve of a chemical annealed cell

Fig. 3 Open circuit voltage vs. preparation Condition of the cell

2. Results on absorption

A convincing proof for the assertion at the end of the preceding paragraph is provided by studying subgap absorption data under a strong reverse bias, and then deducing absorption coefficient α from the equation:

$$QE = (1-R)\exp(-\alpha t_1)[1-\exp(-\alpha t_2)] \quad [1]$$

where R is the reflection from the cell, and t_1 and t_2 are the thicknesses of the p and i layers respectively. The absorption coefficient is plotted in Fig. 4 for three i layers, one grown with hydrogen dilution, one grown with argon dilution, and one grown using chemical annealing at 20 mT. Quite clearly, the figure shows that the film grown using chemical annealing at 20 mT has a smaller bandgap than a film grown using either hydrogen or argon dilution. From thelateral shift of the alpha vs. photon energy curve, we estimate that the annealed film has a bandgap in the range of ~1.63 eV.

The slope of the alpha vs energy curves gives an estimate of the Urbach energy of valence band tail states. We estimate that the Urbach energy is ~45 meV.

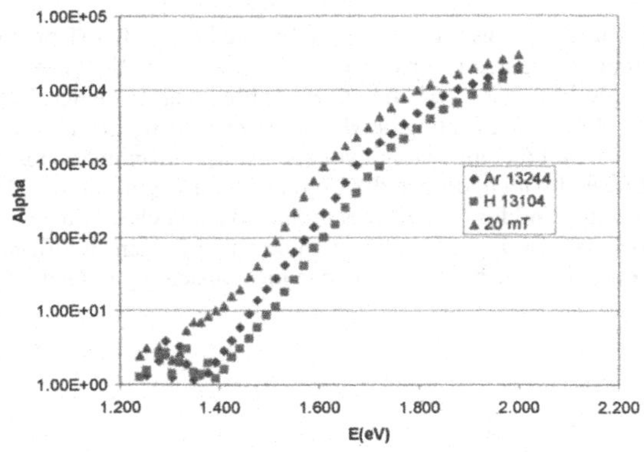

Fig. 4 Subgap alpha vs. photon energy curves for three samples, a continuous sample with H dilution, one with Ar dilution, and a chemically annealed device.

In Fig. 5, we plot the alpha vs. energy curve for two films, one grown when the layer thickness per cycle was 9.8 Angstroms, and one where it was 17 Angstroms, under otherwise identical conditions. From this figure, it is apparent that the bandgap in the film with a smaller thickness per cycle was significantly lower than the bandgap in a film with a larger thickness per cycle. This result is understandable since Argon ions do not penetrate deeply into a Si layer unless the energy is very high.

In Fig. 6, we plot alpha vs. energy curves for two films, one prepared at 30 mT pressure, and one at 15 mT under otherwise identical conditions. While the curve for the film prepared at 15 mT indicates that it has a smaller bandgap, that film also has a significantly higher absorption shoulder than the film grown at 30 mT. This result indicates that very low pressure growth, with its attendant higher ion energies, induces additional defects in the material. Indeed, when we studied the device data in such devices, the fill factors were rather poor, in the range of 55%, as opposed to ~65% for films prepared under lower ion energy conditions.

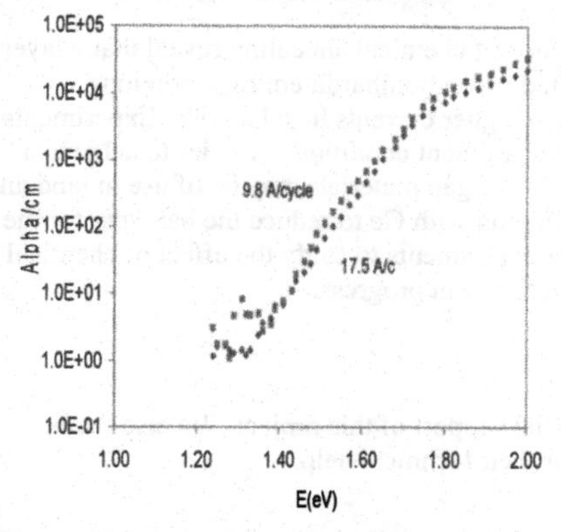

Fig. 5 Alpha vs. energy for two films, one deposited at 9 A/cycle and one at 17.5 A/cycle

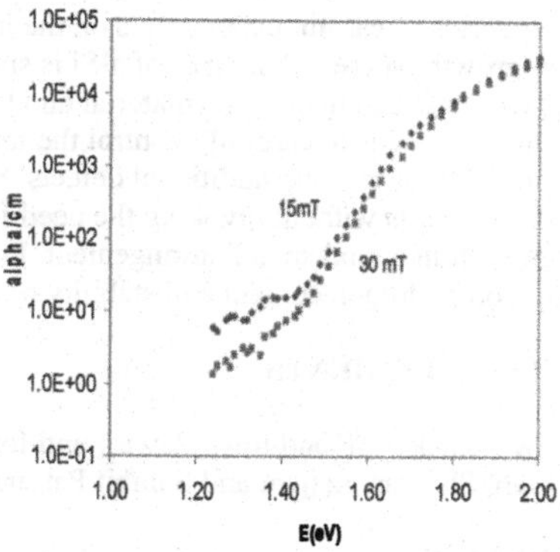

Fig. 6 Alpha vs. energy for two films, one deposited at 15 MT and one at 30 mT

3. Results on structure

In Fig. 7, we show the results of Raman spectroscopy measurement on a typical chemically annealed film prepared at 20 mT. Quite clearly, the absence of a 520 cm-1 peak and a broad shoulder at 485 cm^{-1} indicates that the film is amorphous.

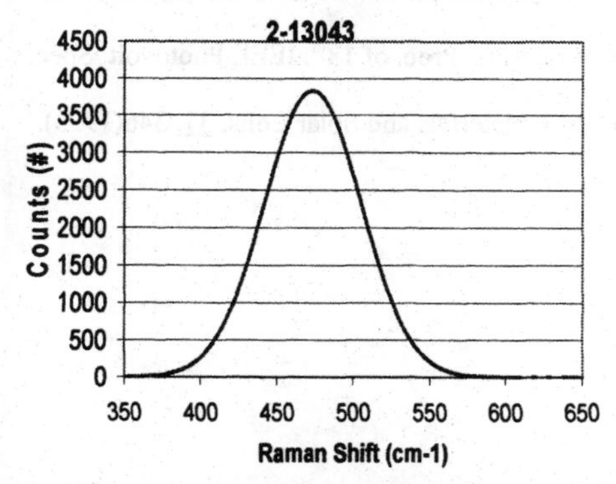

Fig. 7 Raman spectrum for a chemically annealed device

CONCLUSIONS

A series of careful experiments on the influence of chemical annealing reveal that a layer-by-layer growth where a thin layer of a-Si is subjected to ion bombardment by argon ions reduces the Tauc bandgap of the material and leads to higher currents in solar cells. Experiments reveal that one needs to carefully control the ion bombardment conditions in order to achieve a lower gap without creating additional defects. Such lower gap materials may be of use in tandem junction solar cells without invoking the need for alloying with Ge to reduce the bandgap for the lower gap cell in a tandem cell arrangement. Further experiments to study the effect of chemical annealing on hydrogen content and stability of the cells are in progress.

ACKNOWLEDGMENTS

We thank NSF and Iowa PowerFund for partial support of this project. We also thank Keqin Han, Shantan Kajjam and Sambit Pattnaik for their technical help.

REFERENCES

1. H. Sato, K. Fukutani K, W. Futako and I. Shimizu, Solar Energy Mater. and Solar Cells,66,321(2001).
2. W. Futako, T. Kamiya , C. Fortmann and I. Shimizu, J. Non-Cryst. Solids, 266,630(2000).
3. M. Kambe, Y. Yamamoto , K. Fukutani, C. Fortmann and I. Shimizu, Proc. Of MRS, Vol. 507, 205(1999).
4. K. Fukutani , T. Sugawara T, W. Futako, C. Fortmann and I. Shimizu, Proc. Of MRS, Vol. 507,211(1999).
5. A. Shah , J. Meier, E. Vallat-Sauvain, Solar Energy Materials and Solar Cells, 78 , 469-491 (2003).
6. V. L. Dalal, M. Leonard, J. F. Booker and A. Vaseashta, Proc. of 18th. IEEE Photovolt. Spec. Conf., 847(1985).
7. V. L. Dalal, R. Knox and B. Moradi, Solar Energy Materials and Solar Cells, 31, 346(1993).

Mater. Res. Soc. Symp. Proc. Vol. 1245 © 2010 Materials Research Society 1245-A04-04

Microstructure Effects in Amorphous and Microcrystalline Ge:H Films

W. Beyer[1,2], F. Einsele[1], M. Kondo[3], T. Matsui[3], F. Pennartz[1]
1. IEF5-Photovoltaik, Forschungszentrum Jülich GmbH, D-52425 Jülich, Germany
2. Malibu GmbH & Co. KG, Böttcherstrasse 7, D-33609, Bielefeld, Germany
3. National Institute of Advanced Industrial Science and Technology (AIST), Tsukuba, Ibaraki 305-8568, Japan

ABSTRACT

The characterization of void-related microstructure in amorphous and microcrystalline Ge:H films is reported. Various methods are applied including effusion measurements of hydrogen and of implanted helium and neon, measurements of the infrared absorption of C-H bonds due to in-diffusion of contaminants and of the stretching modes of bonded hydrogen. Several microstructure effects like interconnected voids and isolated voids and a quite different material homogeneity are detected and are found to depend on the preparation conditions. Amorphous Ge:H can be prepared with a (largely) homogeneous structure while microcrystalline Ge:H tends to consist of compact grains surrounded by more or less open voids. Enhanced substrate temperatures ($T_S \approx 250°C$) favour the growth of more compact material.

INTRODUCTION

One major defect in hydrogenated silicon and related materials is a void-related microstructure involving voids of various shapes and sizes [1]. While in microcrystalline materials these voids are expected to be related to grain boundaries, in amorphous materials growth effects like shadowing [2] and incorporation of hydrogen clusters at positions interrupting the amorphous network [3] are considered to be the main reasons. Here we focus on void related microstructure effects in hydrogenated amorphous and microcrystalline germanium (Ge:H) films which are of interest for application in thin film silicon solar cells. These microstructure effects are studied by effusion measurements of hydrogen and of implanted helium and neon, by measurements of hydrogen-related infrared absorption and by the measurement of the post-deposition incorporation of contaminants.

EXPERIMENT

Several series of undoped Ge:H films were grown by plasma enhanced chemical vapour deposition (PECVD) using gas mixtures of GeH_4 and H_2 as process gases. For comparison, amorphous and microcrystalline Si:H films were also investigated. Deposition conditions are listed in Table 1. Crystalline Si and Ge wafers were used as substrates. The deposition rate ranged from 0.2 nm/s to about 0.5 nm/s. The crystallinity of μc-Ge:H was studied by Raman scattering by evaluating the crystalline Ge peak near 290 cm^{-1} with reference to the total Raman scattering signal. The infrared absorption of bonded hydrogen and the effusion of implanted helium (He) [4,5] and neon (Ne) [6] were used for void-related microstructure characterization. In order to avoid major implantation-related microstructure changes [4,5], He^+ and Ne^+ ions (40 and 100 keV, respectively) were implanted at doses $\leq 10^{16}$ cm^{-2} for the microstructure analysis. Gas effusion was performed as described elsewhere using a heating rate of 20K/min [7].

Table 1 Deposition conditions

Material type	GeH_4 flow (sccm)	SiH_4 flow (sccm)	H_2 flow (sccm)	pressure (mbar)	RF Power (W)	RF Power (W/cm^2)
a-Ge:H Jülich	3	-	10-30	1	10	0.15
μc-Ge:H Tsukuba	1-3	-	280-300	2	30	0.23
a-Si: H Jülich	-	3	0	0.5	10	0.2
μc-Si:H Jülich	-	2	300	8	40	0.5

RESULTS AND DISCUSSION

In Fig. 1a the Raman crystallinity X of the samples prepared at Tsukuba is plotted as a function of the GeH_4 flow, while the hydrogen flow was fixed between 280 and 300 sccm. It is seen that X is near unity at low GeH_4 flow (high hydrogen dilution) and decreases as the GeH_4 flow is rising. A decrease in substrate temperature T_S leads to a lower crystallinity, in particular at the high GeH_4 flow, i.e. to a mixed amorphous-microcrystalline phase. As seen in Fig. 1b, the H content (measured by H effusion) decreases with increasing crystallinity X and increasing T_S. Like for Si:H growth, the latter effect is attributed to an increasing desorption of H_2 during growth and a more complete polymerization. The low solubility of hydrogen in crystalline Si and Ge [8] can qualitatively explain the decrease in H content with increasing crystallinity since, accordingly, within crystalline grains little H is expected.

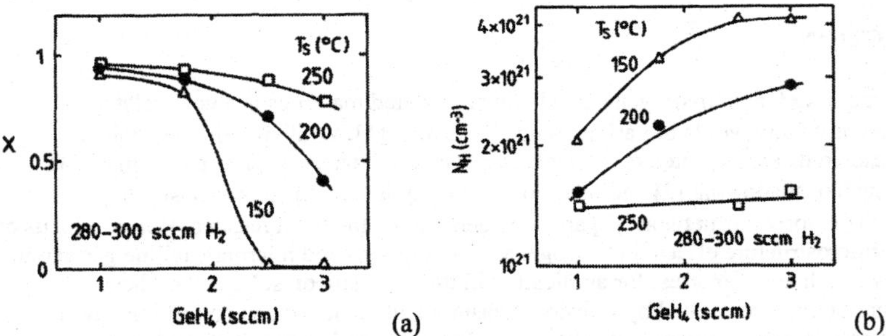

Figure 1 (a) Raman crystallinity X and (b) hydrogen density N_H of Ge:H films as a function of GeH_4 gas flow for different substrate temperatures T_S (hydrogen flow at 280 – 300 sccm).

In the following, we compare the properties of the highly crystalline Ge:H films prepared at Tsukuba at a GeH_4 flow of 1 sccm (series I) with the fully amorphous Ge:H films prepared at Jülich (series II). Figs. 2a and b show the H effusion spectra for films prepared at T_S = 150, 200 and 250 °C. The H effusion shows for the amorphous films (Fig. 2a) a clear shift of the major H effusion peak with rising substrate temperature from a position near 300°C to temperatures exceeding the crystallization temperature of 500°C. For the μc-Ge:H films (Fig. 2b) the maximum H effusion rate lies for all T_S near 600°C. From previous work on H effusion in a-Ge:H [7], H effusion maxima near 300°C indicate the presence of interconnected voids whereas effusion maxima near or above the crystallization temperature of 500°C (for films of typically 1 μm in thickness) suggest the presence of a rather compact material. In order to analyze the

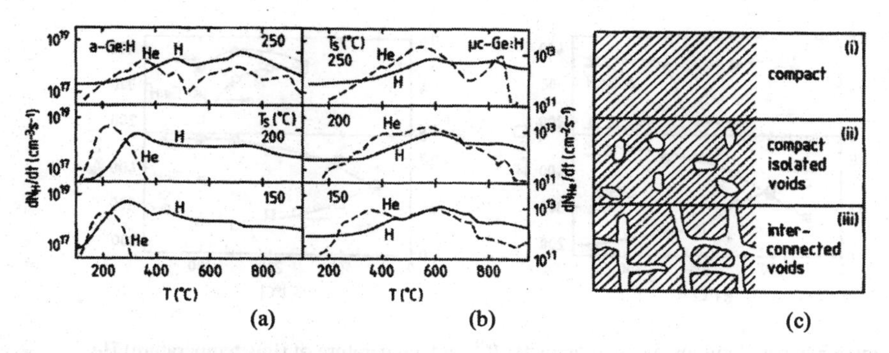

Figure 2 Effusion rate dN_H/dt of hydrogen and dN_{He}/dt of helium (10^{16} cm^{-2} implantation dose) as a function of temperature T for (a) a-Ge:H films (series II) and (b) μc-Ge:H films (series I) prepared at three different substrate temperatures T_S. Schematic view of microstructure (c).

microstructure of the films in more detail, effusion measurements of implanted He were performed and results for an implantation dose of 10^{16} cm^{-2} are also shown in Fig. 2a,b. Since He does not bind to the Ge atoms, the effusion of He is strongly affected by the presence of void related microstructure. Basically, three structure types can be distinguished: compact material with negligible void content (i) and with isolated voids (ii), and material with interconnected voids (iii), as depicted schematically in Fig. 2c. The a-Ge:H films show a low temperature (LT) He effusion peak which shifts with increasing T_S to higher T. This peak shift indicates (in agreement with the H effusion data) the growth of material with interconnected voids at lower T_S and of a more dense material (with little interconnected voids) at higher T_S [4,5]. In addition, for T_S = 250°C material, a He effusion peak at high temperatures (HT) near 700°C appears which has been attributed [4,5] to the presence of isolated voids in the as-deposited material. As discussed elsewhere [5], in the presence of isolated voids, diffusing He either reaches the film surface contributing to the lower temperature He effusion peak or gets trapped in isolated voids. The only way for He atoms to leave an isolated void (if this void survives the crystallization process) is a permeation process, which is expected to take place once the He pressure within the void gets high enough. Note that for material with a high concentration of interconnected voids only few isolated voids are expected, except for a highly inhomogeneous material.

In case of the microcrystalline Ge:H (Fig. 2b), the He effusion spectra show for T_S = 150°C a LT effusion peak near 350°C which shifts to around 400°C for the higher substrate temperatures of 200 and 250°C. However, at T_S = 250°C, only a weak (but reproducible) structure remains. In addition, a broad He effusion maximum appears near 600°C which must be attributed to the presence of isolated voids and which becomes more dominant with rising T_S. These results together with the H effusion data (as discussed above) suggest that the present μc-Ge:H films are all quite compact and that they are only slightly getting more interconnected voids and less isolated voids with decreasing T_S. However, the rather broad He effusion peaks for all microcrystalline films and for the amorphous one of T_S = 250°C suggest the presence of rather inhomogeneous materials.

In Fig. 3, the He effusion results are compared with infrared absorption data, both for

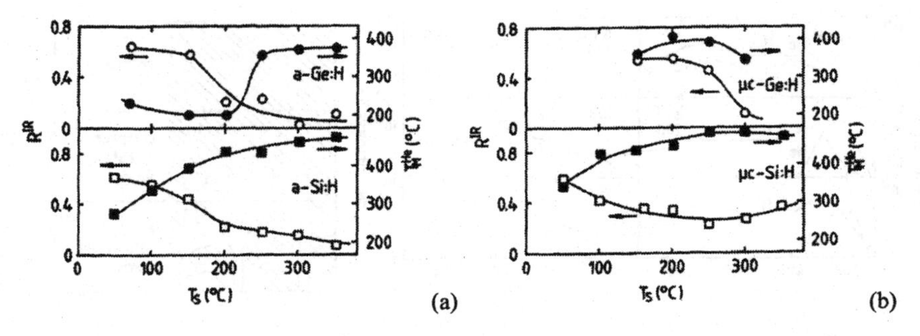

Figure 3 Infrared microstructure parameter R^{IR} and temperature of (low temperature) He effusion T_M^{He} versus substrate temperature T_S (a) for amorphous Ge:H and Si:H, (b) for microcrystalline Ge:H and Si:H.

Ge:H and for Si:H samples. Plotted is as a function of T_S the infrared microstructure parameter R^{IR}, defined for a-Ge:H as the ratio $R^{IR} = I(2000)/[I(2000) + I(1900)]$, where I(1900) and I(2000) are the integrated absorptions of the Ge-H stretching peaks near 1900 and 2000 cm^{-1}, respectively. For Si:H, the corresponding parameter is given by $R^{IR} = I(2100)/[I(2100) + I(2000)]$. Plotted in Fig. 3 is, furthermore, the temperature T_M^{He} of the maximum (low temperature) He effusion rate. A largely mirror-like behaviour for R^{IR} and T_M^{He} is observed as a function of substrate temperature, i.e. when T_M^{He} is low, R^{IR} is high and vice versa. This result demonstrates, as discussed previously [6], that for the present samples both microstructure parameters give roughly similar information. Both parameters indicate that at decreasing substrate temperatures a less compact material forms for amorphous as well as microcrystalline germanium. However, when comparing the substrate temperature dependences of R^{IR} and T_M^{He} in detail, clear differences are apparent, namely that for microcrystalline Ge, T_M^{He} is rather high for all substrate temperatures indicating a rather compact material, whereas R^{IR} shows for the same samples fairly high values suggesting a void rich material. These findings can be reconciled, if for µc-Ge at all investigated substrate temperatures the presence of a granular material with fairly dense crystalline Ge grains is assumed (which, however, may contain isolated voids). These grains then would give rise to the rather high T_M^{He}, while hydrogen, primarily incorporated at grain boundaries, would give rise to a fairly high IR microstructure parameter.

This picture is also supported by the effusion results of implanted neon, as depicted in Fig. 4. When neon is implanted in crystalline Ge wafers, the effusion spectrum as shown in Fig. 4a (bottom) is observed. Ne effusion sets in near 800°C, showing a peak near 950°C. Due to the size of neon of about 2.5 Å in diameter (diameter of He: 2 Å) which roughly equals that of H_2 [9], apparently neon cannot diffuse in crystalline Ge except at highest temperature. Note that the material melts at 937°C. For the µc-Ge:H films of series I (upper part of Fig. 4 a) neon effusion is observed to occur at $T \leq 800$°C down to temperatures below 400°C. Thus channels for fast diffusion of small neon concentrations must exist. The effect is quantified in Fig. 4b which shows, as a function of substrate temperature, the fraction F^{Ne} of integrated Ne effusion at $T <$ 800°C, compared to the total neon effusion (which equals the implanted neon dose). As seen, F^{Ne} decreases with rising T_S, i.e. the (microscopic) film density increases.

While so far results of fully amorphous and highly crystalline Ge:H films were discussed,

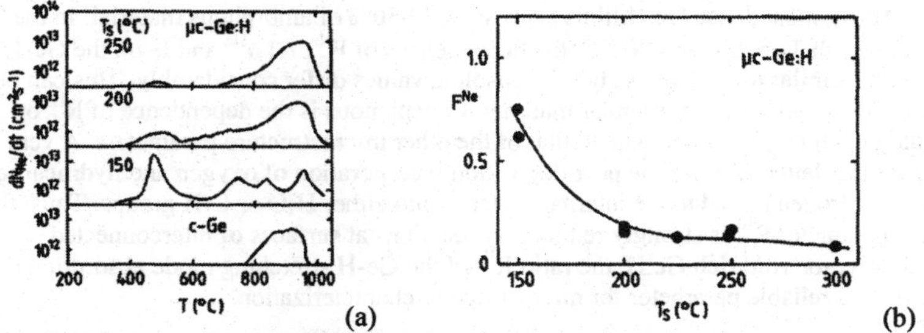

Figure 4 (a) Neon effusion spectra for µc-Ge:H films of series I (crystallinity $X \geq 0.9$) and crystalline Ge. (b) Quantity F^{Ne} as a function of substrate temperature T_S.

some results for material with a mixed amorphous and microcrystalline phase are shown in Fig. 5. Plotted are for the Ge:H films prepared at Tsukuba the microstructure parameters F^{Ne}, ΔT_M^{He} (defined as $\Delta T_M^{He} = T_M^{Max} - T_M^{He}$, with T_M^{Max} the highest value of T_M^{He} observed in these series of Ge:H films) versus GeH_4 flow in Fig. 5a and the microstructure parameters R^{IR} and I_C in Fig. 5b. Here, I_C is the integrated absorption of C-H groups detected by infrared absorption in the range of 2800-3000 $cm^{-1}$and attributed to post-deposition in-diffusion of contaminants like CO or CO_2. While the detailed nature of this effect is unclarified, it seems clear that molecular species like CO diffuse through interconnected voids into the material and presumably cause reactions of the type: Ge-H + CO $\Rightarrow$ C-H + GeO. We find hydrocarbon contamination to be always accompanied by Ge oxidation. Note that Raman crystallinity X and hydrogen density N_H of these series of samples are shown in Fig. 1. As it is seen in Fig. 5, the microstructure parameters differ in some cases considerably. Best agreement is obtained for films of $T_S = 250°C$. Here, in particular the material deposited with 3 sccm GeH_4 shows the lowest microstructure according to all four microstructure parameters. Note that this latter material has according to Fig.1a the lowest Raman crystallinity ($X \approx 0.75$) for this substrate temperature. This result suggests that,

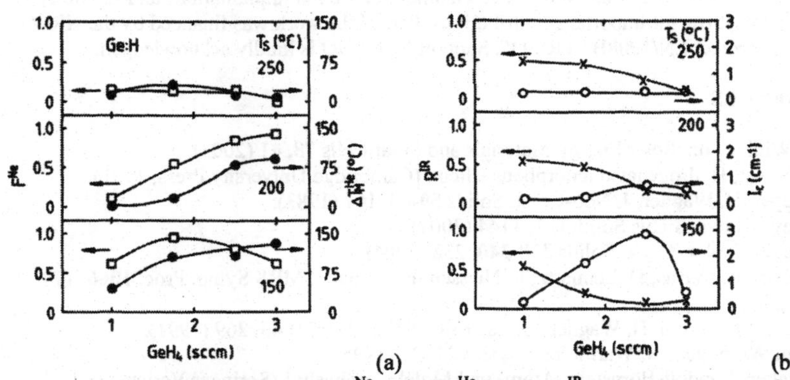

Figure 5 Microstructure parameters (a) F^{Ne} and ΔT_M^{He} and (b) R^{IR} and I_C versus GeH_4 flow (H_2 flow 280-300 sccm) for Ge:H films prepared at Tsukuba.

similar to a-Si:H [6], most dense Ge:H films grow with 20-30% of amorphous material. In the series of samples with T_S = 150 and 200°C, the dependences of F^{Ne}, ΔT_M^{He} and I_C on the GeH_4 flow rate are quite similar to each other, but the absolute values differ considerably. This can be explained, e.g., by the presence of granular material. Conspicuous is the dependence of R^{IR} on GeH_4 flow rate which is largely opposite to that of the other microstructure parameters. A very likely reason for this latter effect is the post-deposition incorporation of oxygen and hydrocarbon which converts hydrogen bound to Ge internal surfaces into either H_2O or C-H groups. Thus, the microstructure parameter R^{IR} is strongly reduced by reactions at surfaces of interconnected voids. Accordingly, for void rich Ge:H, the ratio R^{IR} of the Ge-H stretching mode absorptions turns out not to be a reliable parameter for microstructure characterization.

CONCLUSIONS

It is shown that various methods, like effusion of hydrogen and of implanted helium and neon can be successfully applied for microstructural characterization of Ge:H films. Thus, in addition to the standard method of microstructure characterization by analysis of infrared absorption of stretching modes of bonded hydrogen, several microstructure parameters can be defined, and a comparison of these parameters leads to meaningful results. While for amorphous and highly crystalline Ge:H the various microstructure parameters give a consistent picture, partly controversial results are obtained for mixed phase materials. For both a-Ge:H and μc-Ge:H a rather dense material is obtained at high substrate temperatures ($T_S \geq 250$°C). Microcrystalline Ge:H is found to be most dense when the Raman crystallinity is near 70 %. Under non-optimized conditions, Ge:H films appear to consist of compact grains with fairly wide voids in between, open for out-diffusion of H_2 and for in-diffusion of gases from the ambient. In this latter case, the microstructure characterization based on the analysis of infrared Ge-H stretching modes can fail.

ACKNOWLEDGEMENTS

The authors would like to thank A. Dahmen for the ion implantations and L. Niessen, P. Prunici, U. Zastrow for important contributions. Part of the work was financed by the NRW (Germany) project EN/1008B "TRISO". Support by NEDO is kindly acknowledged.

REFERENCES

1. D.L. Williamson, Solar Energy Materials and Solar Cells **78**, 41 (2003).
2. R.A. Street, Hydrogenated amorphous silicon (Cambridge University Press, 1991).
3. W. Beyer, H. Wagner, J. Non-Cryst. Solids **59-60**, 161 (1983).
4. W. Beyer, Phys. Status Solidi C **1**, 1144 (2004).
5. W. Beyer, J. Non-Cryst. Solids **338-340**, 232 (2004).
6. W. Beyer, R. Carius, D. Lennartz, L. Niessen, F. Pennartz, MRS Symp. Proc. **1066**, 179 (2008).
7. W. Beyer, J. Herion, H. Wagner, U. Zastrow, Philos. Mag. B **63**, 269 (1991).
8. A. Van Wieringen, N. Warmoltz, Physica **22**, 849 (1956).
9. A. Eucken, Landolt-Börnstein, Atom- und Molekularphysik I (Springer Verlag, Berlin, Germany, 1950) pp. 325 and 369.

Poster Session: Crystallization

Mater. Res. Soc. Symp. Proc. Vol. 1245 © 2010 Materials Research Society 1245-A05-05

Rapid thermal annealing of amorphous silicon thin films grown by electron cyclotron resonance chemical vapor deposition

Pei-Yi Lin, Ping-Jung Wu and I-Chen Chen*
Institute of Materials Science and Engineering, National Central University, Jhongli 32001, Taiwan

ABSTRACT

Hydrogenated amorphous silicon (a-Si:H) thin films were deposited on pre-oxidized Si wafers by electron cyclotron resonance chemical vapor deposition (ECRCVD). The rapid thermal annealing (RTA) treatments were applied to the as-grown samples in nitrogen atmosphere, and the temperature range for the RTA process is from 450 to 950 °C. The crystallization and grain growth behaviors of the annealed films were investigated by Raman spectroscopy and X-ray diffraction (XRD). The onset temperature for the crystallization and grain growth is around 625 ~ 650 °C. The crystalline fraction of annealed a-Si:H films can reach ~ 80% , and a grain size up to 17 nm could be obtained from the RTA treatment at 700 °C. We found that the crystallization continues when the grain growth has stopped.

INTRODUCTION

Due to its extensive applications on solar cells and thin film transistors, the hydrogenated amorphous silicon (a-Si:H) film has been fabricated by various techniques, such as by glow discharge [1], plasma enhanced chemical vapor deposition (PECVD)[2], hot wire chemical vapor deposition (HWCVD)[3] and electron cyclotron resonance chemical vapor deposition (ECRCVD) [4]. Among various growth methods of silicon thin films, ECR-CVD is a promising, low temperature process to achieve high deposition rate and expected to be a potential approach for the next generation of Si thin film solar cell production since ECR plasma can generate high electron density, low energy ions and highly active species.

Recrystallization of a-Si:H has been studied as an approach for the production of inexpensive and lower defect-density nanocrystalline and polycrystalline thin films. Nanocrystalline Si has better optical and electrical properties than a-Si:H in terms of the device stability under light soaking in solar cells and gate bias stress in thin film transistors. The thermal annealing treatment is a way to enhance the grain growth and crystallinity, and reduce the defects in the as-grown a-Si:H films, thus they could highly improve the performance of Si thin film devices. Three conventional thermal treatments are used to anneal the material: excimer laser annealing (ELA), furnace annealing (FA), rapid thermal annealing (RTA). Among them, RTA process is most desirable because of its simplicity, fast processing and low capital cost. However, there has been little work done on investigation of annealing effects on ECRCVD-grown films.

In this study, we investigated the post-growth annealing effects on ECRCVD-grown a-Si:H films by RTA processes. Raman spectroscopy, X-ray diffraction (XRD) and scanning electron microscopy (SEM) were used for characterization and structure examination.

EXPERIMENT

The undoped a-Si:H films were deposited on thermally oxidized Si wafers to a thickness of 500 nm in a homemade ECRCVD system. The mixed Ar, H_2 and SiH_4 gases were introduced into the reaction chamber at 5 mtorr for the growth of a-Si:H films.A circular polarized microwave source was used to increase the film uniformity. The frequency and power of the microwave generator are 2.45 GHz and 800 W respectively. The growth temperature is about 200°C, and the growth rate is 1 nm/s. The as-grown films were annealed using RTA treatments method at an annealing temperature (T_A) between 450 and 950°C in nitrogen ambient. The range of the process time (t_A) is from 5s to 10 min, depending on the annealing temperature. A thermocouple behind the sample holder was used to monitor T_A. In the annealing process, the temperature ramp rate and cooling rate is ~ 50°C/s and ~5°C/s separately.

The Raman spectra were measured using Jobin-Yvon HR800 micro-Raman spectrometer at room temperature over a range from 200 to 2400 cm^{-1} with a spectral resolution of 0.54 cm^{-1}. The power of the laser was kept below 0.5 mW to eliminate the possibility of laser induced crystallization, the spectra were decomposed from the integrated Raman intensity ratio of three peaks, crystalline silicon peak at 520 cm^{-1}, the intermediate component peak at 494~507 cm^{-1}, and amorphous silicon peak at 480 cm^{-1} [5]. The crystalline fraction X_c was calculated from $X_c = (I_c+I_{gb})/[I_c+I_{gb}+0.8xI_a]$, where I_c, I_a and I_{gb} were integrated intensities of crystalline, amorphous and intermediate peak, and 0.8 is the ratio of the cross section for the amorphous to crystalline phase [6]. The XRD spectra were measured by a Bruker D8 x-ray system operated at 40kV and 40 mA. The grain size and orientation of crystallites were calculated respectively from the (111) diffraction peak full width at half maximum (FWHM) using Sherrer's formula and the ratio of diffraction peak intensity from (220) to (111).

DISCUSSION

Figure 1 shows the relationship between crystalline fraction/ grain size and annealing temperature. The crystallization and grain growth don't occur until ~ 650 °C, and reach a steady value above ~ 700 °C. The onset temperature of crystallization of a-Si:H in our experiment is ~ 650°C, which is similar to the previous study [7]. This specific temperature is thus chosen as initial RTA temperature for analyzing the structural transformation of a-Si:H films. In general, infrared (IR) absorption spectra were utilized to inspect for SiHx configurations and H concentration [8]. The high stretching mode (2070-2100 cm-1) indicated silicon dihydrides (SiH2) which generally existed near the boundary of microvoids embedded in a-Si:H matrix, and the low stretching mode (1980-2010 cm-1) indicated silicon monohydrides (SiH) which dispersed in a-Si:H matrix [9]. Since the stretching modes range shown in the IR spectra are also presented in the Raman scattering spectra, we utilized Raman spectroscopy to observe our samples grown on non-IR transparent oxidized substrates [10].

Figure 2 shows the relative intensity of Raman spectra of a-Si:H films between 1800 to 2300 cm^{-1} versus the RTA temperature to monitor the changes of SiH_x. The main peak located at 1995 cm^{-1} confirms that the as-grown film was a compact amorphous tissue with dispersed monohydrides since only few amounts of dihydrides related to microvoids are found. The decrease in the intensity of stretching modes shows that the hydrogen concentration of a-Si:H films reduces significantly in short annealing time and vanishes as temperature up to 650°C. This phenomenon is in agreement with previous study that H atoms can be release from Si-H_2 bonds to transport level leading to diffusion at temperatures higher than 300°C. Furthermore, Si-H bonds can also be broken at temperatures higher than 600°C. It must be noted that the microstructures of these films annealed under temperatures lower than 650°C are still in amorphous phase. The transverse optical (TO) modes of Raman shown in the Figure 3 can reveal these films are amorphous. We can thus demonstrate that the most H atoms are almost diffused out of a-Si:H films in the incubation period of crystallization during thermal annealing and do not play an important role at the onset of crystallization.

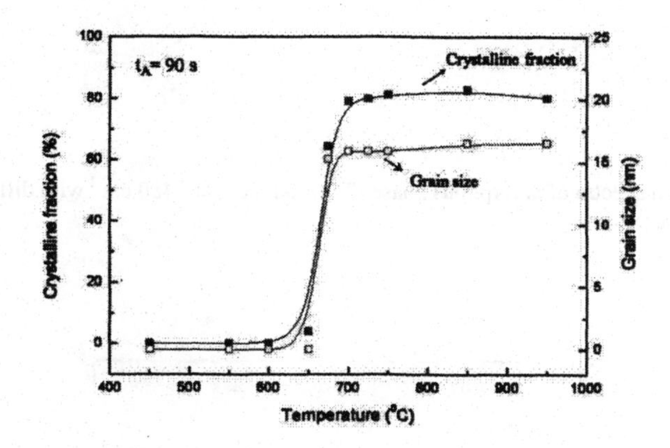

Figure 1. Crystalline fraction and grain size vs. annealing temperature.

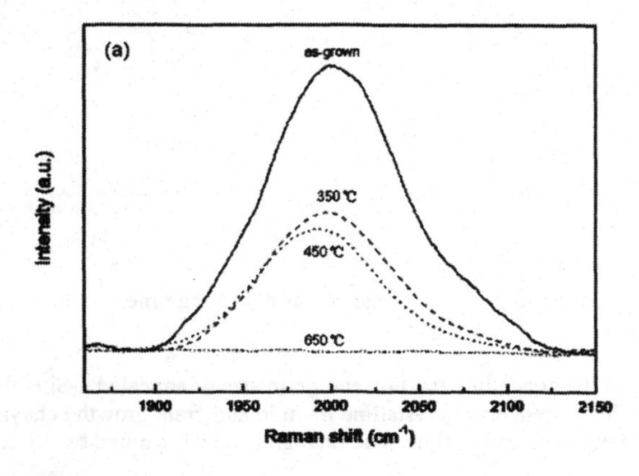

Figure 2. Raman spectra of Si-H stretching mode with different annealing temperatures.

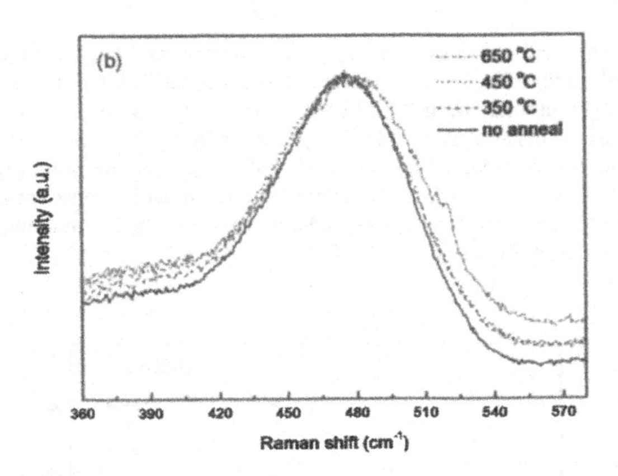

Figure 3. Raman spectra of amorphous phase (TO mode) are at ~480 cm^{-1} with different annealing temperatures.

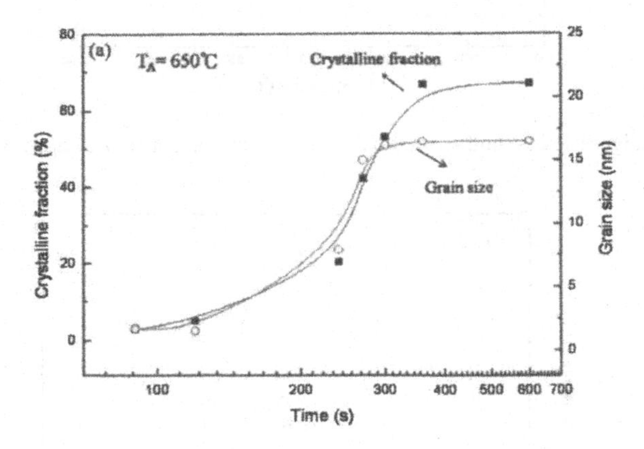

Figure 4. Crystalline fraction and grain size versus annealing time.

Figure 4 shows the crystalline fraction and grain size of annealed a-Si:H thin films versus annealing time. The tendencies of crystalline fraction and grain growth behaviors are nearly the same until up to longer annealing time. The final grain size is limited by the columnar structures, as shown in Figure 5(a) and (b). The annealed film surface is smoother, but the size of columnar structure does not changed. In the low-temperature solid phase crystallization, the grain size is restricted by the microstructure of the films. The final grain size is ~ 17 nm, which is similar to the size of the columnar structure. The crystallinity continually increases from 52 % to ~ 80 % when the grain growth has stopped. This may be due to some new crystallites formed during longer annealing time. More research work is undergoing and the optimized conditions for different annealing treatments are under developing.

Figure 5 (a) The SEM image of the as-grown a-Si:H film. (b) The SEM image of the annealed a-Si:H film.

ACKNOWLEDGMENTS

The research was supported by the National Science Council through a grant No. NSC98-2120-M-008-005-, NSC98-2218-E-008-002- .

REFERENCES

1. M. H. Brodsky, M. Cardona, and J. J. Cuomo, Phys. Rev. B **16**, 3556 (1977).
2. S. Guha, J. Yang, D. L. Williamson, Y. Lubianiker, J. D. Cohen, and A. H. Mahan, Appl. Phys. Lett. **74**, 1860 (1999).
3. A. H. Mahan, L. M. Gedvilas, and J. D. Webb, J. Appl. Phys. **87**, 1650 (2000).
4. M. Kitagawa, K. Setsune, Y. Makabe and T. Hirao. Jap. J. App. Phys. **27** (1988).
5. G. Yue, J. D. Lorentzen, J. Lin, D. Han, and Q. Wang, Appl. Phys. Lett. **75**, 492 (1999).
6. C. Smit, R. A. C. M. M. van Swaaij, H. Donker, A. M. H. N. Petit, W. M. M. Kessels, and M. C. M. van de Sanden, J. Appl. Phys. **94**, 3582 (2003).
7. A. H. Mahan, T. Su, D. L. Williamson, L. M. Gedvilas, S. P. Ahrenkiel, P. A. Parilla, Y. Xu, and D. A. Ginley, Adv. Funct. Mater. 19, 1 (2009).
8. A. A. Langford, M. L. Fleet, B. P. Nelson, W. A. Lanford, and N. Maley, Phy. Rev. B 45, 13367 (1992).
9. A. H. M. Smets, T. Matsui, and M. Kondo, Appl. Phys. Lett. 92, 033506 (2008).
10. E. V. Johnson, L. Kroely, and P. Roca I Cabarrocas, Solar Energy Materials & Solar Cells 93, 1904 (2009).

Poster Session: Solar Cells

Mater. Res. Soc. Symp. Proc. Vol. 1245 © 2010 Materials Research Society 1245-A07-01

Effects of Grain Boundaries on Performance of Hydrogenated Nanocrystalline Silicon Solar Cells

Tining Su[1], David Bobela[2], Xixiang Xu[1], Scott Ehlert[1], Dave Beglau[1], Guozhen Yue[1], Baojie Yan[1], Arindam Banerjee[1], Jeff Yang[1], and Subhendu Guha[1]
[1]United Solar Ovonic LLC, 1100 West Maple Road, Troy, MI, 48084, U.S.A.
[2]National Renewable Energy Laboratory, 1617 Cole Blvd, Golden, CO, 80401, U.S.A.

ABSTRACT

We investigate the effect of hydrogenation of grain boundaries on the performance of solar cells for hydrogenated nanocrystalline silicon (nc-Si:H) thin films. Using hydrogen effusion, we found that the amplitude of the lower temperature peak in the H-effusion spectra is strongly correlated to the open-circuit voltage in solar cells. This is attributed to the hydrogenation of grain boundaries in the nc-Si:H films.

INTRODUCTION

Hydrogenated nanocrystalline silicon (nc-Si:H) thin films are getting a great deal of attention for use in the bottom cell in multi-junction solar cells. The nc-Si:H based cell has excellent long wavelength response, and is less vulnerable to light-induced degradation than hydrogenated amorphous silicon (a-Si:H) based cells [1-4]. One disadvantage of the nc-Si:H solar cell is its relatively low open-circuit voltage (V_{oc}). A typical value of V_{oc} for nc-Si:H solar cells is about 0.5 V, and is often much lower. This compares poorly to about 1 V for a-Si:H cells. The lower than optimal V_{oc} for the nc-Si:H solar cells can be attributed to many reasons, such as crystalline volume fraction and the defect density in *i*-layer, as well as ambient degradation. Ambient degradation refers to degradation of V_{oc} and short-circuit current density (J_{sc}) observed for certain cells when exposed to the ambient conditions for an extended period of time. Electron-spin-resonance (ESR) in nc-Si:H films shows that the neutral defects mostly reside at the grain boundaries [5]. These defects can act as recombination centers, resulting in lower V_{oc}. In addition, it has been suggested that oxygen related defects at the grain boundaries causes increased dark current, and hence also reduced V_{oc} [6].

It has been suggested that improved hydrogenation of grain boundaries can improve V_{oc} [7]. Also, depositing the *i*-layer at lower temperatures has been shown to improve V_{oc} [6]. This was attributed to increased hydrogen concentration in the film resulting in better hydrogenation of the grain boundaries. Although it is well known that lower deposition temperature tends to reduce the crystalline volume fraction, and increase hydrogen incorporation in the films, it does not necessarily increase the degree of hydrogenation of the grain boundaries. To our best knowledge, no direct evidence has been reported that demonstrates a positive correlation between increased hydrogen concentration and increased hydrogenation of the grain boundaries. To distinguish hydrogen at the grain boundaries from that in the amorphous phase, sophisticated techniques are required for typical local probes, such as ^{1}H NMR; and interpretation of the results is not trivial. In addition, these techniques usually cannot be directly applied to solar cells, due to the presence of doped layers in the solar cells.

We use hydrogen effusion to detect different hydrogen sites in the nc-Si:H films. Hydrogen effusion has been extensively used to study hydrogenated micro-crystalline (μc-Si:H) and a-Si:H films made near the transition between amorphous and nanocrystalline phase. In μc-

Si:H made with plasma enhanced chemical vapor deposition (PECVD), the effusion peak near 400 °C is attributed to the desorption of hydrogen at the grain boundaries. These hydrogen molecules can quickly diffuse out along the paths at the grain boundaries [8,9]. In a-Si:H made with very high hydrogen dilution, an effusion peak near 400 °C was also observed [10-12]. This peak was attributed to hydrogen at the "grain boundaries" of those very small crystallites that are assumed to exist in the film, probably with a size less than 10 Å [9]. In both cases, the effusion peaks at about 400 °C are attributed to the hydrogen on the grain boundaries. Typical H-effusion spectra in nc-Si:H also exhibit this sharp peak at about 400 °C. This peak is attributed to the hydrogen sites that can find a "fast" path to the film surface, such as through the grain boundaries and columnar shaped voids, similar to the case in a-Si:H with very high hydrogen dilution and μc-Si:H. Careful comparison of this peak in solar cells made with different deposition conditions and cell performance may provide insight into the hydrogenation of the grain boundaries and the effect of such hydrogenation on cell performance.

EXPERIMENTAL

Large-area solar cells were deposited at United Solar, using modified very high frequency (MVHF). Details of MVHF technique can be found elsewhere [3]. Four small-area samples were cut from large area films deposited over a 15"x14" area for various characterizations. These cells have different V_{oc} and J_{sc} in their as-deposited states. Cell A is the reference cell, cells B and C have similar J_{sc} but different V_{oc} as sample A. Cell D shows low V_{oc} and J_{sc}, and exhibits significant ambient degradation. All four cells have similar thickness of around 2.5 μm. Table I summarizes the J-V characteristics of these four cells.

Hydrogen effusion, Raman spectroscopy, and X-Ray diffraction (XRD) spectroscopy were carried out on the whole solar cells at the National Renewable Energy Laboratory (NREL).

Table I. J-V characteristics of the samples described in the text.

Sample	V_{oc} (V)	J_{sc} (mA/cm^2)	FF (AM1.5)	FF (blue)	FF (red)	η (%)
A	0.49	24.56	0.61	0.65	0.68	7.38
B	0.42	24.09	0.50	0.57	0.58	5.03
C	0.49	24.50	0.51	0.62	0.64	6.17
D	0.41	17.05	0.52	0.61	0.60	3.62

RESULTS AND DISCUSSION

Hydrogen effusion results

Figure 1 shows the H-effusion spectra for the four cells listed in Table I. The thin black line represents the spectrum from sample A, the reference sample. The thick line, the gray dashed line, and the gray solid line represent the spectra from samples B, C, and D, respectively. For sample B, no hydrogen effusion data was obtained between 260 °C and 330 °C. The spectra are normalized to their sample volume. Therefore the amplitude of the spectra reflects the hydrogen content in these cells. In Fig. 1, samples A, B, and C all show a sharp effusion peak near 390 °C, and also increasing effusion near 600 °C. On the other hand, sample D shows only a small increase of effused H_2 near 390 °C, and almost all the hydrogen is effused after the temperature reaches 600 °C. The hydrogen concentration in these samples is also significantly different. Samples A and C have the highest hydrogen concentration, while sample D has the lowest. In all cases, samples having a higher peak near 390 °C also have a higher peak near 600 °C.

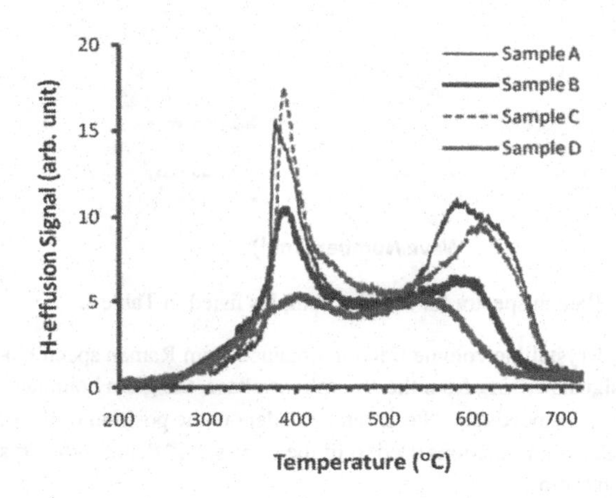

Figure 1. Hydrogen effusion spectra for the fours samples listed in Table I.

Raman results

Figure 2 shows the Raman spectra for the four samples, using 633 nm light. The black dashed line is from sample A. The gray solid line, the black solid line, and the gray dotted line are from samples B, C, and D, respectively. The crystalline volume fraction for each sample is calculated by fitting the curve with three components that are attributed to the amorphous phase, the crystalline phase, and the grain boundaries. A detailed discussion can be found elsewhere [13]. Table II lists the crystalline volume fraction in the intrinsic layer for each sample. From Table II, one can see that these four cells have very different crystalline volume fractions. The two samples with high V_{oc} and high J_{sc} both have much lower crystalline volume fractions, 25% and 36% respectively, while sample D has the highest crystalline volume fraction, 82%.

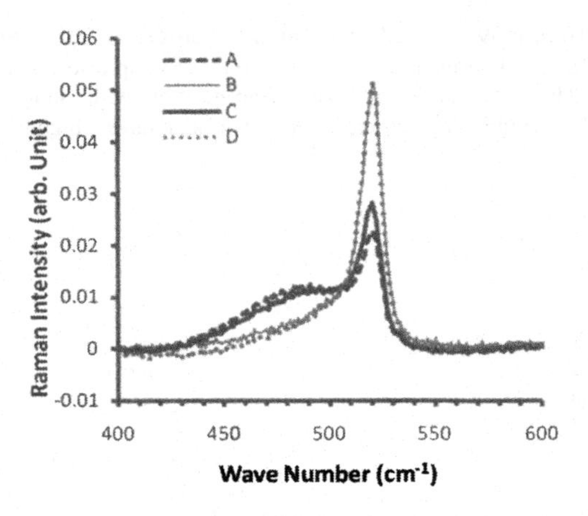

Figure 2. Raman spectra for the four samples listed in Table I.

Table II. Crystalline volume fraction obtained from Raman spectra. a-Si, GB, and c-Si indicate signals arising from the amorphous phase, the grain boundaries, and the crystalline phase, respectively. "Peak" and "σ" denote the position of the peaks and the witdth of the Gaussian functions used to fit the curves. "f_c" denotes the crystalline volume fraction in each sample.

Sample	Components	Peak (cm^{-1})	2σ (cm^{-1})	f_c
A	a-Si	485	52	0.25
	GB	514	20	
	c-Si	520	8	
B	a-Si	485	56	0.73
	GB	514	28	
	c-Si	520	8	
C	a-Si	485	50	0.36
	GB	512	26	
	c-Si	520	8	
D	a-Si	495	56	0.82
	GB	508	28	
	c-Si	520	10	

Discussion

The Raman results show that the two high performance cells, samples A and C, both have very low crystalline volume fractions. Sample A shows no ambient degradation, while sample C shows a slight ambient degradation. On the other hand, samples B and D show severe ambient degradation after exposure to air for three months. The degree of ambient degradation were measured by the J-V characteristics and the external quantum efficiency (QE). The correlation between cell performance and crystalline volume fraction is consistent with a previous report [13]. On the other hand, the XRD results did not show any significant difference in terms of crystalline orientation in these samples; all samples have dominant orientations along (111) and (220), and similar volume fractions along each orientation.

The H-effusion results provide interesting information about the structure in the intrinsic nc-Si:H layer. A general trend is that cells with better performance have higher hydrogen content in the *i*-layer. For all four samples, the spectra intensities between 400 °C and 600 °C are rather similar, and probably represent the hydrogen in the "bulk" amorphous region. The most obvious differences are the intensities of the peak near 390 °C and 600 °C. The lower temperature peak at 390 °C is very similar for the two high performance cells (A and C). A sharp reduction of V_{oc} occurs when the intensity of this peak is reduced by about 50% (sample B), and further reduction of the peak intensity results in only a slight reduction of V_{oc} (sample D). This suggests that for sample A and C, the grain boundaries are highly hydrogenated. Since the hydrogen concentration is much higher than that of the defects, a slight reduction of hydrogenation creates enough defects on the grain boundaries to reduce V_{oc} significantly. It is also possible that when these samples are exposed to moisture and oxygen, the "exposed" sites on the grain boundaries allow for the oxygen related defects to more readily form, thus causing ambient degradation and further reducing V_{oc}.

The peak near 600 °C has been attributed to the presence of "compact" amorphous or crystalline silicon, which may form during the effusion [14], and is limited by the diffusion of atomic hydrogen in the amorphous phase. However, it is not easy to explain the occurrence of the peak unless there is a large difference in hydrogen concentration along the growth direction. It should be noted that this temperature is similar to that for the solid-state crystallization of amorphous silicon thin films. In fact, amorphous silicon films can be crystallized on a time scale of 30 minutes at 600 °C [15]. It is conceivable that crystallization can be faster in nc-Si:H films since no incubation is needed. Therefore it is possible that the peak at 600 °C is partially caused by the crystallization in these films, resulting in a rapid displacement of a large quantity of hydrogen as the crystalline volume fraction grows rapidly.

SUMMARY

In summary, we studied the correlation between the hydrogenation of grain boundaries and the V_{oc} in nc-Si:H based solar cells. Hydrogen effusion spectra suggests a strong correlation between hydrogenation of grain boundaries and V_{oc}. Higher hydrogenation of grain boundaries results in higher V_{oc}, and the dependence of V_{oc} on the hydrogenation is rather strong. This provides direct evidence of how microscopic hydrogen bonding at the grain boundaries affects the nc-Si:H cell performance.

ACKNOWLEDGEMENTS

The authors thank K. Younan, D. Wolf, T. Palmer, N. Jackett, L. Sivec, B. Hang, R. Capangpangan, J. Piner, G. St. John, A. Webster, J. Wrobel, B. Seiler, G. Pietka, C. Worrel, D. Tran, Y. Zhou, S. Liu, and E. Chen for sample preparation and measurements. We also thank H.

Fritzsche for the in-depth discussions. The work was supported by US DOE under the Solar America Initiative Program Contract No. DE-FC36-07 GO 17053.

REFERENCES

1. J. Meier, R. Flückiger, H. Keppner, and A. Shah, Appl. Phys. Lett. **65**, 860 (1994).
2. B. Yan G. Yue, and S. Guha, in *Amorphous and Polycrystalline Thin-Film Silicon Science and Technology*, edited by Virginia Chu, Seiichi Miyazaki, Arokia Nathan, Jeffrey Yang, and Hsiao-Wen Zan (Mat. Res. Soc. Symp. Proc. **989**, Pittsburgh, PA, 2007) pp. 335.
3. X. Xu, B. Yan, D. Beglau, Y. Li, G. DeMaggio, G. Yue, A. Banerjee, J. Yang, S. Guha, P. Hugger, and D. Cohen, in *Amorphous and Polycrystalline Thin-Film Silicon Science and Technology*, edited by Arokia Nathan, Andrew Flewitt, Jack Hou, Seiichi Miyazaki, and Jeffrey Yang (Mater. Res. Soc. Symp. Proc. **1066**, Pittsburgh, PA, 2008) pp. 325
4. G. Ganguly, G. Yue, B. Yan, J. Yang, S. Guha, in *Conf. Record of the 2006 IEEE 4th World Conf. on Photovoltaic Energy Conversion*, Hawaii, USA, May 7-12, 2006, p.1712.
5. T. Su, T. Ju, B. Yan, J. Yang, S. Guha, and P. C. Taylor, J. Non-Cryst. Solids, **354**, 2231 (2008)
6. Y. Nasuno, M. Kondo, and A. Matsuda, Appl. Phys. Lett. **78**, 2330 (2001).
7. M. Kondo, *et al.*, in *Proceedings of 31st IEEE PVSC* (IEEE, New York, 2005) pp. 1377 (and references therein).
8. F. Finger, K. Prasad, S. Dubail, A. Shah, X.-M Tang, J. Weber, and W. Beyer, in *Amorphous Silicon Technology*, edited by A. Madan, Y. Hamakawa, M. Thompson, P. C. Taylor, and P. G. LeComber (Mat. Res. Soc. Symp. Proc. **219**, Pittsburgh, PA, 1991) pp. 383.
9. W. Beyer, P. Harpke, and U. Zasreow, in *Amorphous and Microcrystalline Silicon Technology*, edited by E.A.Schiff, M.Hack, S.Wagner, R.Schropp, I.Shimizu (Mater. Res. Soc. Symp. Proc. **467**, Pittsburgh, PA , 1997) pp.343.
10. X. Xu, J. Yang, and S. Guha, J. Non-Cryst. Solids, **198-200**, 60 (1996).
11. S. Guha, J. Yang, A. Banerjee, B. Yan, K. Lord, Sol. Ener. Mater. Sol Cells, **78**, 329 (2003) (and references therein).
12. A. H. Mahan, J. Yang, S. Guha, and D. L. Williamson, Phys. Rev. B **61**, 1677 (2000).
13. G. Yue, B. Yan,G. Ganguly, J. Yang, S. Guha, C. W. Teplin,, and D. Williamson, in *Conf. Record of the 2006 IEEE 4th World Conf. on Photovoltaic Energy Conversion*, Hawaii, USA, May 7-12, pp. 1588 (and references therein).
14. W. Beyer, in *Tetrahedrally-Bonded Amorphous Semiconductors*, edited by S. D. Adler and H.H. Fritzshe (Plenum, New York, 1985) pp.129.
15. P. Stradins, D. Young, Y, Yan, E. Iwaniczko, Y. Xu, R. Reedy, H. Branz, and Q. Wang, Appl. Phys. Lett. **89**, 121921 (2006).

Mater. Res. Soc. Symp. Proc. Vol. 1245 © 2010 Materials Research Society 1245-A07-03

Material Properties of a-SiGe:H Solar Cells as a Function of Growth Rate

Peter G. Hugger[1], Jinwoo Lee[1], J. David Cohen[1], Guozhen Yue[2], Xixiang Xu[2], Baojie Yan[2], Jeff Yang[2] and Subhendu Guha[2]
[1]University of Oregon, Department of Physics. Eugene, Oregon, United States
[2]United Solar Ovonic LLC. Troy, Michigan, United States

ABSTRACT

We have examined a series of a-SiGe:H alloy devices deposited using both RF and VHF glow discharge in two configurations: *SS/n⁺/i (a-SiGe:H)/p⁺/ITO nip* devices and *SS/n⁺/i (a-SiGe:H)/Pd* Schottky contact devices, over a range of deposition rates. We employed drive-level capacitance profiling (DLCP), modulated photocurrent (MPC), and transient junction photocurrent (TPI) measurement methods to characterize the electronic properties in these materials. The DLCP profiles indicated quite low defect densities (mid 10^{15} cm^{-3} to low 10^{16} cm^{-3} depending on the Ge alloy fraction) for the low rate RF (~1Å/s) deposited a-SiGe:H materials. In contrast to the RF process, the VHF deposited a-SiGe:H materials did not exhibit nearly as rapid an increase of defect density with the deposition rate, remaining well below 10^{17} cm^{-3}. up to rates as high as 10Å/s. Simple examination of the TPI spectra on theses devices allowed us to determine valence band-tail widths.. Modulated photocurrent (MPC) obtained for several of these a-SiGe:H devices allowed us to deduce the *conduction* band-tail widths. In general, the a-SiGe:H materials exhibiting narrower valence band-tail widths and lower defect densities correlated with the best device performance.

INTRODUCTION

Hydrogenated amorphous silicon germanium alloys (a-$Si_{1-x}Ge_x$:H) have been critical materials in the development of multi-junction thin film photovoltaics. However, despite this success and the easily tunable nature of its optical bandgap to energies from 1.7 eV to 1.3 eV, a substantial challenge remains in the deposition of high quality a-SiGe:H films at rates above ~3 Å/s using PECVD glow discharge techniques. Typically, as deposition rates increase cell performance parameters and carrier mobility-lifetimes drop while, simultaneously, deep defect densities increase.

To maintain high-quality photovoltaic material and at the same time increase deposition rates, United Solar has developed a modified VHF glow discharge deposition technique that has been shown to allow deposition rates 3 to 5 times greater than typical RF methods while still maintaining low densities of deep defects [1,2]. Our previous studies for series of a-SiGe:H alloys with similar bandgaps indicate that the Urbach energies determined from transient photocapacitance (TPC) and transient photocurrent spectra (TPI) correlate very well with the measured defect densities determined by our drive-level capacitance profiling (DLCP) in the manner predicted by certain models of defect creation. [3,4]

In light of these previous results, here we evaluate the application of a complementary technique, the "modulated photocurrent" (MPC) method. This technique has been shown capable of providing good estimates of the distribution of the conduction bandtail states. We then compare these results with the deep defect distributions in these samples as well as the valence bandtail distributions determined via TPI measurements.

EXPERIMENTAL DETAILS

Description of Samples

A set of a-SiGe:H samples were deposited at United Solar Ovonic LLC for this study using both RF and VHF glow discharge techniques. Two device configurations were examined: the *n-i-p* structure: *SS/n+(a-Si:H)/i(a-SiGe:H)/p+nc-Si:H/ITO,* and the Schottky geometry: *SS/n+ (a-Si:H)/i (a-SiGe:H)/Pd.* Intrinsic layers were between 0.8 and 1.3 microns thick. All devices were classified as "low gap" with optical bandgaps, E_{04}, all within the range 1.4±.05 eV. One special sample, BMW12200, was prepared in a bifacial geometry: *Glass/TCO/n+(a-Si:H)/i(a-SiGe:H)/p+nc-Si:H/TCO* allowing for illumination both from the "top" (*p*-layer side) and "bottom" (*n*-layer side).

All measurements presented here have been characterized the fully lightsoaked metastable state of these materials, produced by exposing devices to 200 mW/cm² of 610 nm long-pass-filtered white light for 100h .

Measurement Methods

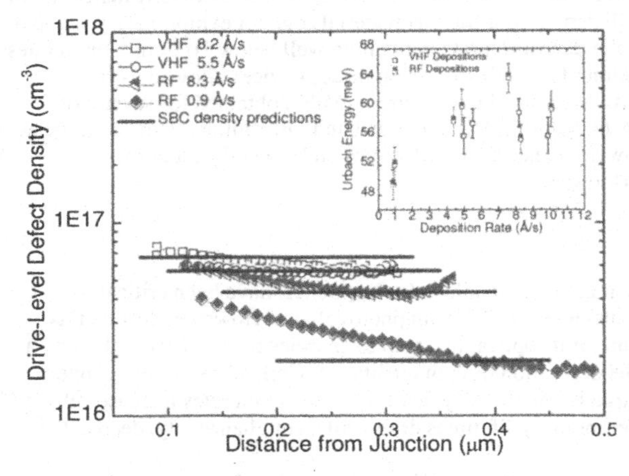

Figure 1. Many samples showed a correlation between deep state densities and valence bandtail Urbach energies. This correlation seems consistent with defect creation models such as the spontaneous bond conversion (SBC) model [4].

Deep defect densities in a set of RF and modified VHF deposited a-SiGe:H n/i/p devices were studied using drive level capacitance profiling. This measurement method has previously been described fully but can be easily understood as a density resulting from fitting the measured depletion capacitance *vs.* the amplitude of the oscillating voltage amplitude, δV, to the quadratic form: $C=C_0+C_1\delta V+C_2(\delta V)^2+\ldots$ The drive-level defect density is obtained directly from the coefficients of this fit. The resulting density gives a direct estimate of the density of mid-gap states below the Fermi level to a maximum energy depth given by the measurement temperature and frequency . When this density is then plotted against the spatial position variable $\langle x\rangle=\varepsilon A/C_o$, where ε is the

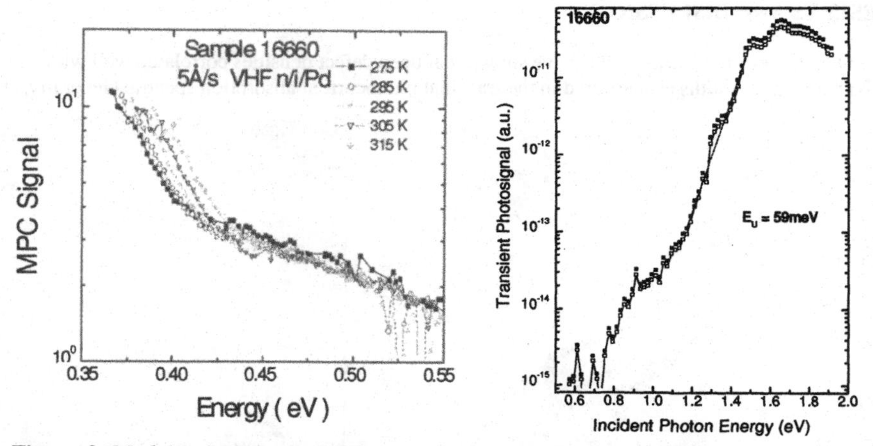

Figure 2. Modulated photocurrent (left) and transient photocurrent (right) measurements of bandtails in a 5 A/s a-SiGe:H device of optical bandgap 1.4±.05 eV. Transient photocurrent yields a valence band Urbach energy of 59 ± 3 meV, MPC yields a conduction band Urbach energy of 35 ± 2 meV.

dielectric constant, and A is the area of the junction, one can then obtain a deep defect profile across the i-layer in the device.

In addition to DLCP, transient photocurrent (TPI) was used to measure the band-to-band absorption edge on these materials. An exponential fit to the absorption edge in these spectra was used to identify the Urbach energies, E_U. Although The E_U obtained from TPI measurements represents a convolution of Urbach energies of both the conduction and valence bands, it is generally assumed to disclose the characteristic width energy of the *valence* bandtail, which is the broader bandtail.

A third measurement technique was used to determine the conduction bandtail independently of the valence bandtail. That is, to measure an Urbach energy that is not a convolution of both bandtails. To accomplish this, we used the modulated photocurrent (MPC) method. This method involves observing the majority carrier drift current as it traverses the intrinsic region of a *p-i-n* structure. As carriers become trapped and are subsequently thermally ejected to the mobility edge during this drifting process an affect is seen in the resultant photocurrent amplitude and phase. This information can be analyzed to deduce a density of states within the bandgap [5,6]. Here, we have utilized the analysis of Hattori *et. al.* [6] and it is important to recognize that the energetic resolution of this analysis method is $k_BT/2$. Therefore Urbach energies can only be accurately determined when their values exceed this limit

RESULTS AND DISCUSSION

Our studies on these materials have shown that deep defect densities correlated well with Urbach energies widths as measured in the transient photocurrent absorption spectra. Generally,

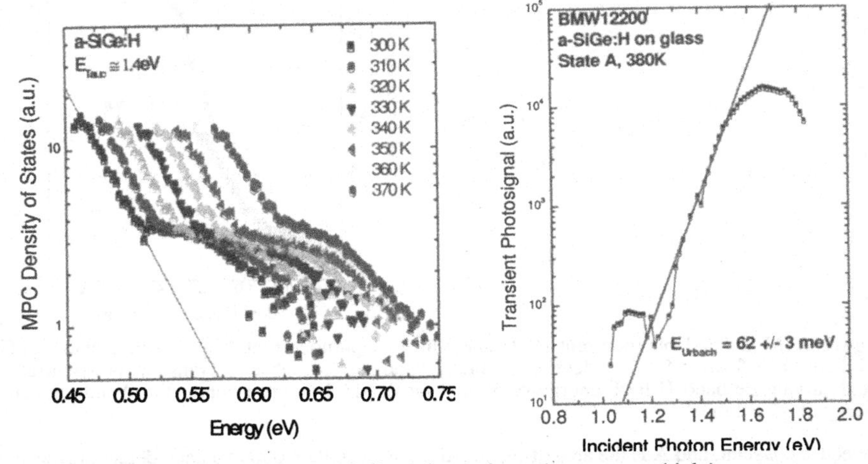

Figure 3. Modulated photocurrent (left) and transient photocurrent (right) measurements of bandtail distributions in a bifacial a-SiGe:H device of optical bandgap 1.4±.05 eV. Transient photocurrent yields a wide valence band Urbach energy of 62 ± 3 meV, while MPC yields a conduction band Urbach energy of 32 ± 2 meV. Poor device performance qualities of this film are attributed to the wide valence bandtail. Attempts to measure the valence bandtail distribution using MPC via backside illumination were not successful (see below).

as deposition rates increased, Urbach energies became broader and deep defect densities rose. Moreover, this correlation occurred in a way consistent with the spontaneous bond-breaking model [4]. The defect density profiles obtained by DLCP for several a-SiGe:H samples are displayed in Fig. 1 along with the values predicted by the spontaneous bond conversion model.

Given the range of electronic properties for these a-SiGe:H alloys that we are able to determine with our methods we wanted to identify which might be the best indicators of device quality. Time of flight-based measurements for these materials have shown that carrier mobilities depend strongly on the bandtail state distributions and these are in turn also strong indicators of device quality [7].

An important first step in our study was been to verify that the MPC method is able to extract individual bandtails for these materials at all. Figure 2(a) shows MPC spectra for an a-SiGe:H Schottky device deposited at 5 A/s which has a TPI determined valence bandtail Urbach energy of 59 ± 2meV. The Urbach energy of the conduction bandtail, as given by MPC, was found to be 35 ± 2meV, a value consistent with bandtail energies measured previously using time-of-flight methods [8].

This result may be compared to data from a sample specially prepared in the bifacial configuration (BMW12200),. These data, shown in Figure 3, indicate a similar, or even slightly

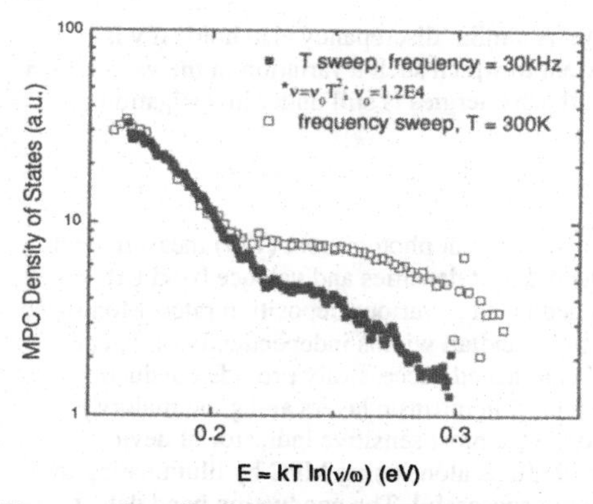

Figure 4. Similarity of MPC-deduced densities of states when frequency (open symbols) or temperature (closed symbols) is varied. The magnitude discrepancy between the deepest states is attributed to nonlinearities of the MPC signal depending on measurement temperature and optical intensity [10].

sharper, conduction band Urbach energy of 32 ± 2meV, but a significantly broader valence bandtail width of 62 ± 2meV. We attribute this wider bandtail, indicating increased long to mid-range structural disorder, to the non-optimized growth geometry of this particular sample. Particularly, the lower substrate temperature associated with deposition on glass. Indeed, considering previous results, this broad valence bandtail suggests poor electronic properties for this sample, a result which time of flight measurements carried out in another laboratory have confirmed.[9] Therefore, the significance of these data is twofold, 1) that it is possible to obtain plausible conduction and valence bandtail distributions in both Schottky and heterojunction configurations by applying MPC and the TPI technique, and 2) that it appears likely that valence bandtail width is the most appropriate figure of merit when investigating these a-SiGe:H materials.

We also attempted to reproduce the valence bandtail widths obtained by our TPI spectra using the MPC method by employing back-side illumination to the bifacial device to examine the drift of hole carriers across the device. However, this experiment revealed a bandtail distribution that appeared nearly identical to the front-side illumination result displayed in Fig. 3. This indicates that MPC is not able to observe hole transport properites, most likely due to the inherently slow drift velocity of these minority carrier across the intrinsic region, thus allowing recombination events with the majority electrons to dominate the overall time-response.

Finally, although our results of the conduction bandtail distributions obtained using MPC seemed reasonable, we performed one further experiment to help verify their validity. One way to do this is to perform the MPC experiment by using temperature as the experimental variable and holding frequency constant. The good qualitative agreement of these 'temperature sweep' with the 'frequency sweep' spectra, shown in Figure 4, verifies that the MPC experiment is behaving with the expected thermally activated behavior. However matching these spectra in the manner shown in Fig. 4 also required addressing an important issue in such MPC results; namely, the proper determination of the thermal energy axis.

The thermal energy depth at which unoccupied mid-gap states respond in an MPC measurement is given by the usual expression: $E = k_B T \ln(\nu/\omega)$, where ν is the attempt frequency and ω is the experimental (angular) frequency. Therefore the energy axes in these experiments are obtained by assuming a value for $\nu \equiv \nu_o T^2$. However, our analysis indicates that no single value of ν_o can produce an MPC derived density of states with both bandtail and deep defect states at their expected energy depths. Figures 3 and 4 demonstrate this clearly. That is, we have found that deep state densities appear at energy values near E_C - 0.7 eV by assuming $\nu_o \sim 10^8$ s^{-1}/K but that bandtail features in these spectra require a value of $\nu_o \sim 10^4$ s^{-1}/K to fall within the expected energy range of E_C - 0.1 to E_C - 0.2 eV. A similar discrepancy also holds if ν is assumed to be temperature independent. The extent to which such a variation in the value of if ν affects our MPC determined conduction-band Urbach energies is still under investigation.

CONCLUSIONS

Drive level capacitance profiling (DLCP) and transient photocurrent (TPI) measurements were used to determine a correlation between deep defect densities and valence band Urbach energies in a series of low-gap a-SiGe:H alloys deposited at various deposition rates. Modulated photocurrent (MPC) was used to obtain conduction bandtail widths independently of valence bandtails. Preliminary results suggest that MPC can indeed successfully provide conduction bandtail distributions in these materials despite their thin intrinsic layers and good majority carrier mobilities, but that valence bandtails may be the most sensitive indicator of device quality. Attempts to obtain valence bandtail distributions alone using MPC by illuminating the sample from the back (substrate) direction were not successful. The conduction band defect distributions given by MPC were confirmed by repeating the MPC experiment with temperature as an experimental variable instead of frequency. To within nonlinearities which have been previously reported [10] these distributions agreed qualitatively well. All MPC data, whether taken in the frequency or temperature energy domain, appear to exhibit attempt frequencies (i.e. thermal prefactors) that are orders of magnitude different for deep and shallow mid-gap states. While interesting in its own right, this result may affect the accuracy with which MPC can determine Urbach energies in these particular devices.

ACKNOWLEDGMENTS

This work was partially supported by NREL under the Thin Film Partnership Program No. ZXL-5-44205-11 at the University of Oregon, and by US DOE under the Solar America Initiative Program Contract No. DE-FC36-07 GO 17053 at both United Solar and the Univ. of Oregon, and the NSF IGERT fellowship.

REFERENCES

1. G. Yue, B. Yan, J. Yang and S. Guha, *Mat. Res. Soc. Symp. Proc.* **989**, 359 (2007)
2. S. Guha, J. Yang, S. Jones, Y. Chen and D. Williamson, *Appl. Phys. Lett.* **61**, 1444 (1992)
3. P. Hugger, J. Lee, David J. Cohen, G. Yue, X. Xu, B. Yan, J. Yang, S. Guha, . *Mat. Res. Soc. Symp. Proc.* **1183**, A07-12 (2009)
4. M. Stutzmann, Philo. Mag. B**60**, 531 (1989)
5. R. Brüggemann, C. Main, J. Berkin and S. Reynolds (1990). Phil. Mag. B **62** p.29-4
6. K. Hattori, Y. Niwano, H. Okamoto, Y. Hamakawa,. J. Non-Cryst. Solids **137-138**, 363 (1991).
7. E.A. Schiff, Solar Energy Materials and Solar Cells **78**, 567 (2003).
8. Q. Wang, H. Antoniadis, E.A. Schiff. Phys. Rev. B**47**, p. 9435 (1993)
9. E.A. Schiff, private communication.
10. F. Zhong and J.D. Cohen, *Mat. Res. Soc. Symp. Proc.* **258**, 813 (1992).

UV-embossed textured back reflector structures for thin film silicon solar cells

Jordi Escarré[1], Karin Söderström[1], Oscar Cubero[1], Franz-Josef Haug[1], and Christophe Ballif[1]
[1]Ecole Polytechnique Fédérale de Lausanne (EPFL), Institut of Microengineering (IMT), Photovoltaics and Thin Film Electronics Laboratory, 2000 Neuchâtel, Switzerland

ABSTRACT

In this work, we study the replication of nanotextures used in thin film silicon solar cells to enhance light trapping onto inexpensive substrates such as glass or polyethylene naphtalate (PEN). Morphological analysis was carried out to asses the quality of these replicas. Moreover, single and tandem a-Si:H solar cells were deposited on top of the master and replica structures to verify their suitability to be used as substrates for solar cells in n-i-p configuration. We find stabilized efficiencies around 8 % which are similar for tandem cells on masters and PEN replicas.

INTRODUCTION

The use of plastic foils as substrates for thin film amorphous silicon solar cells is an interesting approach to reduce manufacturing costs by means of roll to roll deposition [1]. In these devices, optical confinement techniques play a crucial role in order to reduce the thickness of the intrinsic layer, leading to solar cells with better carrier collection and higher stable efficiencies. The light trapping is achieved by texturing some of the layers or the substrate itself. In n-i-p configuration, the texture is usually obtained on the back reflector by growing rough metallic [2] or transparent conducting oxides layers [3] at high temperatures. These temperatures make plastics as PET or PEN not suitable to be used as substrates. In order to avoid substrate limitations, in this contribution we study the controlled transfer of a roughness on to another substrate by means of an "in house" ultra violet nano imprinting lithography (UV-NIL) system, a technology that can easily be scaled up to large areas and roll to roll processing [4]. As grown textured boron doped zinc oxide deposited by low-pressure chemical vapor deposition (LPCVD-ZnO) and as grown silver deposited at high temperature (~ 400 °C) by sputtering were used as masters in the replication process. The replicas were obtained on glass and PEN. The quality of the replication was evaluated by a morphological comparision of both, the master and the replica surfaces. Moreover, single and tandem a-Si:H solar cells were deposited on masters and replicas, showing comparable performance.

EXPERIMENT

Glass (0.5 mm thick, Schott AF45) and polyethylene-naphtalate (50 μm thick PEN) with a size of 4×4 cm^2 were used as substrates in the replication process. An "in house" [5] UV-NIL system was utilized in this work to transfer nanometric random textures onto the substrates. The original texture (Master) is molded in polydimethylsiloxane (PDMS, Sylgard 184, Dow Corning). The PDMS is prepared with a ratio 10:1 between the base and the curing agent, degassed in vacuum, dispensed onto the master, and cured at 46°C for 12 hours. After separation, a negative of the master texture is transferred to the surface of the PDMS. For replication of the master surface, a 28 μm thick sol-gel UV sensitive lacquer layer is deposited by spin coating onto a glass or PEN substrate. Then, this substrate and the PDMS are put into an evacuated chamber, followed by the application of a pressure of 1 bar. This pressure enables the UV lacquer to adopt the mold surface while it is cured under a UV light exposure of 1.4 mW/cm^2 for 20 minutes ($\lambda \sim 365$ nm). After separation from the PDMS mold, the master texture is reproduced in the surface of the cured UV lacquer.

Three different master textures were used; type A is as-grown textured silver (hot-Ag) deposited by sputtering at high temperature on glass, whereas Type B and C masters consist of textured boron doped ZnO grown by low-pressure chemical vapor deposition (LP-CVD ZnO).

Atomic force microscopy (AFM) was used to perform morphological analysis using the software presented in ref. [6]. The root mean square roughness (σ_{RMS}), the correlation length (L), the mean angle and the ratio between the scanning area and its projected flat surface (A_{flat}/A_{scan}) were extracted to compare master and replica structures in a quantitative way. σ_{RMS} gives information about the vertical size of the roughness whereas L represents the lateral size distribution of the surface texture. Note that σ_{RMS} and L represent the commonly used approximation of the surface height distribution and the autocorrelation function, respectively, by Gaussian distributions [7]. L was calculated as the radius where the auto-correlation peak drops to 1/e of its maximum value, assuming circular shape. The mean angle is the average angle between the surface normal and the local inclination evaluated for every point of the AFM image. As A_{flat}/A_{scan}, the mean angle gives both vertical and lateral information.

Single and tandem a-Si:H solar cells were grown on masters and replicas by plasma enhanced chemical vapor deposition (PECVD) at a frequency of 70 MHz and a substrate temperature of 190 °C from a silane/hydrogen mixture with dilution $[H_2]/[SiH_4]$ equal to 2. Each substrate contains around 20 independent devices (5x5 mm^2) which allowed us to compare statistically the quality of the solar cells obtained on masters and on replicas.

The solar cells were characterized by measuring the external quantum efficiency (EQE) and the current-voltage characteristics; the latter were measured under a dual lamp spectrum simulator (Wacom WXS-140S-10) in standard test conditions (25 °C, AM 1.5 spectrum and 1000 W/m^2) to extract the open circuit voltage (V_{oc}) and the fill factor (FF) of the solar cells. The short circuit current density (J_{sc}) of the devices was calculated integrating over the EQE curve by weighing with the AM 1.5g solar spectrum.

RESULTS AND DISCUSSION

Morphological analysis

The morphology of the three used masters and their corresponding UV replicas obtained on glass are presented in figure 1. A similar appearance is observed between the AFM images of masters and replicas; parameters with vertical (σ_{RMS}), lateral (L), and both vertical and lateral (mean angle, σ_{RMS}/L, A_{flat}/A_{scan}) roughness information were calculated to compare these randomly textured surfaces (table I). A good correlation is observed between the mean angle, σ_{RMS}/L and A_{flat}/A_{scan} as shown in figure 2 (A). Consequently, only σ_{RMS}/L was used to compare the aspect ratio of masters and replicas. Figure 2 (B) shows the relative aspect ratio loss of the replicas compared with the original values (losses of 15, 37 and 32 % for type A, B and C replicas, respectively).

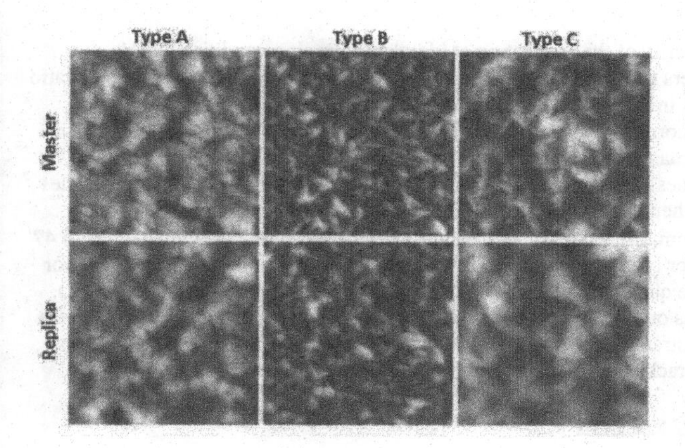

Figure 1. AFM images of the three master textures (above). Corresponding replicas obtained on glass substrates (below). The scanned area is 5 × 5 μm^2.

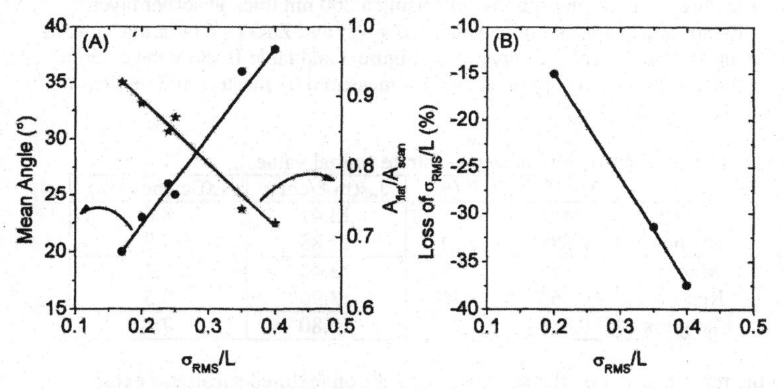

Figure 2. Morphological parameters of A, B and C textures: (A) Mean angle and A_{flat}/A_{scan} as a function of the aspect ratio parameter σ/L for both masters and replicas. (B) Loss of aspect ratio σ/L after replication as a function of the aspect ratio of the master texture.

Table I. Summary of the parameters extracted from the AFM analysis.

Texture		σ_{RMS} (nm)	L (nm)	σ_{RMS}/L	Mean Angle (°)	A_{flat}/A_{scan}
Type A	Master	46	228	0.20	23	0.89
	Replica	47	280	0.17	20	0.92
Type B	Master	67	167	0.40	38	0.72
	Replica	51	202	0.25	25	0.87
Type C	Master	124	356	0.35	36	0.74
	Replica	99	413	0.24	26	0.85

We observe that the quality of the replication at identical stamping conditions (UV exposure, pressure …) differs for our master textures. Figure 2 (B) shows that higher aspect ratio of the master texture results in more severe loss of texture in replication.

Table I shows that the correlation lengths estimated for the replicas are always higher than the values obtained for the masters. Interpreting the correlation length as a mean size of the surface features, higher values of L could be explained by the loss of smaller details during the replication. For example, when the morphologies of the master substrates are altered with a square averaging over 300 nm, the values of σ_{RMS} and L obtained from the smoothed data are 47 and 281 nm, respectively, for type A master, 57 and 203 nm for type B and 180 and 378 nm for the type C. These values are quite similar to the ones calculated from the replicas (see table 1). This simple treatment points out the losses induced by the replication process. However, texture smoothing has been demonstrated to be beneficial from the electrical properties of the devices reducing the formation of cracks during the growth of the layers [8].

Single junction amorphous solar cell

We compared the light scattering properties of the replicated textures with their respective masters by incorporating them into thin film silicon solar cells. Here we concentrate on the use of the type A and B textures in n-i-p amorphous cells using a 200 nm thick absorber layer. Identical back reflector structures consisting on sputtered silver and ZnO (~ 65 nm thick) were deposited on master and replicas at room temperature. Figure 3 and table II show the external quantum efficiency and the initial cell parameters values measured by masters and replicas on to these two types of texture.

Table II. Cell parameters of the best cell on each substrate (initial values).

Texture		V_{oc} (V)	FF (%)	J_{sc} (mA/cm^2)	Efficiency (%)
Type A	Master	0.884	72.7	13.41	8.6
	Replica	0.881	70.1	12.85	7.9
Type B	Master	0.848	64.8	13.42	7.4
	Replica	0.865	70.0	13.96	8.5
Reference cell	Flat glass	0.888	74.5	10.80	7.1

Compared to the reference cell on flat substrate, all cells on textured substrates exhibit increased light confinement. At long wavelengths, the four cells differ considerably, but trends between the two types of substrate texture can be indentified; for the type A texture, a comparison of the EQEs shows that the cell on the replica does not reach the level of light trapping that is obtained on the type A master. This behavior might be attributed to the loss of the smaller features during the replication process. For the type B texture, the slightly more pronounced interference fringes due to the cell thickness suggest that the replica is flatter than the master. We observe that this loss of aspect ratio is not necessarily detrimental for the cell quality. Higher V_{oc} and FF on the type B replica with respect to the cell on the master suggest that in this case the replica has a more favorable surface for the growth of cells.

The cell on the type B replica exhibits better response between 550 and 800 nm. This is surprising because in case of the type A texture, the loss of small features resulted in a severe loss in this region. As the total absorption in the cell on the replica is lower than the one observed on the master (not shown) while the photocurrent on the replica is higher, parasitic absorption in the master appears plausible. Eventually, these results show that no direct and intuitive explanation can be used to understand the role of the small features in light trapping, as appearing on our samples.

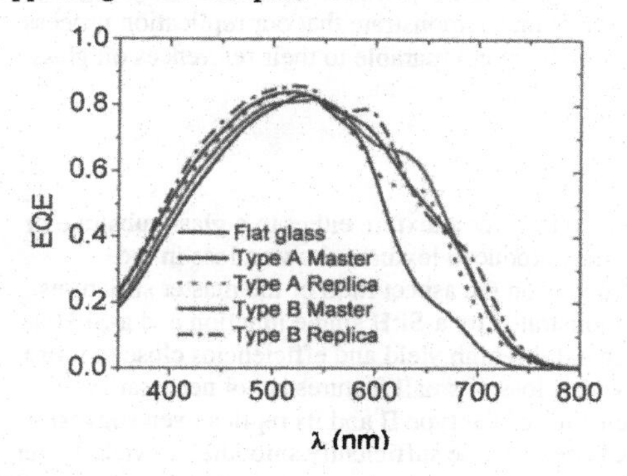

Figure 3. External quantum efficiencies of a flat reference and cells on the various textured substrates.

Tandem solar cell

In this section, we present tandem cells deposited on type A and B samples. The original type A structure deposited on glass is our best home-made textured substrate for tandem cells. The other substrate is a replica on PEN obtained from the type B master. Type B replication was found to be the best replicated texture for single amorphous cells on glass. In order to obtain a good current matching, the tandems cells on glass (PEN) have top and bottom cells thicknesses of 70 nm (60 nm) and 300 nm (300 nm), respectively.

We use I-V measurement at low illumination to assess the suitability of different substrates for solar cell processing; problems in substrate preparation often result in shunt paths which decrease the parallel resistance. We decided to consider cells with a $V_{oc} < 1.2$ V or a FF below 30% under low illumination as non functional. Out of the 20 cells deposited on each substrate, this measurement showed a yield of 90% on the PEN substrate, and of 85 % on the glass substrate. The V_{oc} and FF values shown in table III are obtained by statistical average on the working cells under AM 1.5. The J_{sc} and the efficiency are given for the best cell of each substrate. Table III presents a summary of these parameters. Values in bold represent stabilized values (1000h of light soaking at 65°C and light intensity of 100 mW/cm^2).

Table III. Tandem cell parameters in initial and stabilized state; stabilized values are in bold.

Substrate	Mean on cells		Best cell	
	V_{oc} (V)	FF (%)	J_{sc} (mA/cm^2)	Efficiency (%)
Type A on glass	1.73 ± 0.01	71.70 ± 1.84	7.54	9.8
	1.68 ± 0.02	**59.59 ± 3.45**	**7.30**	**8.3**
Type B replica on PEN	1.70 ± 0.01	72.80 ± 1.05	7.62	9.5
	1.68 ± 0.01	**62.33 ± 1.83**	**7.31**	**8.1**

The equivalent yields between the cells on both substrates prove the ability of the replication process to produce good substrates for photovoltaic applications. The degradation of the parameters is equivalent between the two tandems and it proves that the lacquer has no influence on the degradation. The best cell efficiency achieved on the replication on PEN is only slightly lower than for the best cell on glass. These three points demonstrate that our replication process on PEN permits to grow cells of high quality which are comparable to their references on glass substrate.

CONCLUSIONS

We have successfully developed a process to transfer a texture either to a glass substrate or to a flexible polymer sheet. Detailed analysis of reproduced textures shows a loss in the replication fidelity of small features which depends on the aspect ratio of the master structures. The potential of our texturing approach is demonstrated by a-Si:H single junction and a-Si:H/a-Si:H tandem cells in n-i-p configuration which exhibit high yield and efficiencies close to those of reference cells on glass. These results show that loss of small features is not necessarily detrimental for solar cell performance; the behavior of the type B and its replica even suggests that some detrimental aspects of pyramidal textures may be sufficiently smoothed to yield better cell performance.

ACKNOWLEDGMENTS

The authors gratefully acknowledge support by the Swiss Federal Office for Energy (OFEN) under contract 101191, the Swiss Commission for Technology and Innovation (CTI) within project No. 8809.2 and the Swiss National Science Foundation under grant 200021-125177.

REFERENCES

1. C. Ballif, V. Terrazzoni-Daudrix, F.-J. Haug, D. Fischer, W. Soppe, J. Loffler, J. Andreu, M. Fahland, H. Schlemm, M. Topic, M. Wurz. Proceedings of the 22nd European PVSEC (Milano, 2007).
2. S. Guha, J. Yang. *Journal of Non-Crystalline Solids* **352**, 1917-1921 (2006)
3. M. Kambe, M. Fukawa, N. Taneda, Y. Yoshikawa, K. Sato, K. Ohki, S. Hiza, A. Yamada, and M. Konagai. *Proceedings of the 3rd World Conference Photovoltaic Energy Conversion*, 1812-1815 (2003)
4. S.H. Ahn, J.-S. Kim and L. J. Guo. *J. Vac. Sci. Technol. B* **25(6)** 2388-2391 (2007)
5. K. Soderstrom, J. Escarré, O. Cubero, F.-J. Haug, S. Perregaux and C. Ballif, accepted for publication in *Progress in Photovoltaics: Research and Applications*
6. I. Horcas, R. Fernandez, J. M. Gomez-Rodriguez, J. Colchero, J. Gomez-Herrero, and A. M. Baro. *Review of Scientific Instruments* **78**, 013705 (2007)
7. H. Davies, *The reflection of electromagnetic waves from a rough surface.* Proceedings – American Academy for Jewish research (1954)
8. M. Python, E. Vallat-Sauvain, J. Bailat, D. Domine, L. Fesquet, A. Shah and C. Ballif. *Journal of Non-Crystalline Solids* **354**, 2258-2262 (2008)

Mater. Res. Soc. Symp. Proc. Vol. 1245 © 2010 Materials Research Society 1245-A07-05

Improving Performance of Amorphous Silicon Solar Cells Using Tungsten Oxide as a Novel Buffer Layer between the SnO_2/p-a-SiC Interface

Liang Fang, Seung Jae Baik, Koeng Su Lim, Seung Hyup Yoo, Myung Soo Seo and Sang Jung Kang

Department of Electrical Engineering, KAIST, 335 Gwahak-ro, Yuseong-gu, Daejeon 305-701, Republic of Korea

ABSTRACT

A thermally evaporated *p*-type amorphous tungsten oxide (*p-a*-WO_3) film was introduced as a novel buffer layer between SnO_2 and *p*-type amorphous silicon carbide (*p-a*-SiC) of *pin*-type amorphous silicon (*a*-Si) based solar cells. By using this film, a-Si solar cells with a *p-a*-WO_3/*p-a*-SiC double p-layer structure were fabricated and the cell photovoltaic characteristics were investigated as a function of *p-a*-WO_3 layer thickness. By inserting a 2 nm-thick *p-a*-WO_3 layer between SnO_2 and a 6 nm-thick *p-a*-SiC layer, the short circuit current density increased from 9.73 to 10.57 mA/cm^2, and the conversion efficiency was enhanced from 5.17 % to 5.98 %.

INTRODUCTION

One of the strategies to improve the efficiency of *pin*-type amorphous silicon (*a*-Si) based solar cells is to insert a buffer layer between the *p*-type layer and transparent conducting oxide (TCO) electrode, or between the *p*-type and active layers. With a suitable material and configuration, the buffer layer can benefit *a*-Si:H solar cells by preventing recombination loss at the *p*/*i* interface [1-3], the carbon-alloyed amorphous silicon carbide (*a*-SiC:H) buffer layer has been developed to solve the heterojunction interface problem between the *p*-type a-SiC:H (*p-a*-SiC:H) window and intrinsic amorphous silicon (*i-a*-Si:H) active layers in a-Si:H based pin type thin film solar cell. Because the structure misfit at the *p-a*-SiC:H/*i-a*-Si:H interface could be reduced effectively by using this buffer layer, the open circuit voltage (V_{oc}) and short wavelength response could be dramatically enhanced, and thus a high efficiency of 11.2 % was obtained (by using an Ag back electrode) [3]; or by enhancing hole collection via lowering of the interface potential barrier [4]. However, the high defective *p-a*-SiC:H window and a-DLC:H buffer layers with a short carrier life time, and newly formed interface between the window and buffer layers still generate the considerable recombination loss of photo-generated carriers. In addition, the serious resistance of the cell, originated from the low dark conductivity (σ_d) value of the buffer layer, severely limits the overall cell performance. Recently, an amorphous tungsten oxide (*a*-WO_3) has been successfully used as a buffer layer for an organic solar cell [5] and organic light emitting diode (OLED) [6]. In this letter, we investigate the effects of inserting an *a*-WO_3 buffer layer between SnO_2 and *p*-type *a*-SiC (*p*-*a*-SiC) on the *a*-Si solar cell performance. Furthermore, in order to gain insight into the buffering effect of the *a*-WO_3 layer, a transport analysis based on the Schottky barrier model was performed.

EXPERIMENTAL

WO_3 films were thermally deposited on bare glass (Corning 7059) and surface textured SnO_2/glass (Asahi U-type) substrates without heating the substrates. The total transmittance and the haze spectra of the SnO_2/glass and WO_3 deposited SnO_2/glass (WO_3/SnO_2/glass) substrates were measured by the UV/vis transmission spectrum. The *a*-Si solar cells were fabricated using a two-chamber system, which consists of a photo-assisted chemical vapor deposition chamber for the *p* and *n* layers, and an rf-plasma enhanced chemical vapor deposition chamber for the intrinsic (*i*) layer. The fabricated solar cells have a simple structure of glass/SnO_2/*p*-*a*-WO_3/*p*-*a*-SiC/*i*-*a*-Si/*n*-*a*-Si/Al without a back reflector. The cell area is 0.092 cm^2. The deposition conditions of each layer are summarized in Table I. Photocurrent density versus voltage measurements were performed under AM 1.5, 88 mW/cm^2 irradiation. A constant energy spectrophotometer was used for quantum efficiency (QE) measurements.

Table I. The deposition conditions of each layer in the fabricated pin-type a-Si based solar cells.

	buffer layer (*p*-*a*-WO_3)	*p* layer (*p*-*a*-SiC)	*i* layer (*i*-*a*-Si)	*n* layer (*n*-*a*-Si)
Deposition method	Thermal evaporator	photo-CVD	RF-PECVD	photo-CVD
Substrate temperature (℃)	—	250	250	250
Pressure (Torr)	2×10^{-6}	0.5	0.7	0.5
Flow rate (sccm)	WO_3 source (Alfa Aesar 99.99%)	SiH_4:C_2H_4:B_2H_6 =5:0.7:1.7	SiH_4:H_2=5:10	SiH_4:PH_3=5:3
Thickness (nm)	2, 5	4, 6, 8, 10, 12	500	40

RESULTS AND DISCUSSION

The *a*-WO_3 used in this study, which was deposited by the same evaporation system [5, 6], is a *p*-type material with a high dark conductivity of 6×10^{-6} S/cm. It possesses a wide optical energy band gap of up to 3.5 eV [5, 6]. Figure 1 shows the total transmittance and haze spectra of a 10 nm-thick *p*-type *a*-WO_3 (*p*-*a*-WO_3) layer deposited on SnO_2/glass (*p*-*a*-WO_3(10nm)/SnO_2/ glass), and those of the SnO_2/ glass are provided as references. The fluctuation of the curves is due to the scattering effect of the textured surface of the SnO_2/glass. The total transmittance of the *p*-*a*-WO_3(10nm)/SnO_2/ glass is almost identical with that of the reference. This is consistent with the wide optical band gap characteristic of the *p*-*a*-WO_3 film. Furthermore, the good conformal deposition property of the *p-a*-WO_3 is verified by the haze spectra. It is very important to maintain the surface texture of SnO_2, as it is critical to attain large short circuit current density (J_{sc}) [7].

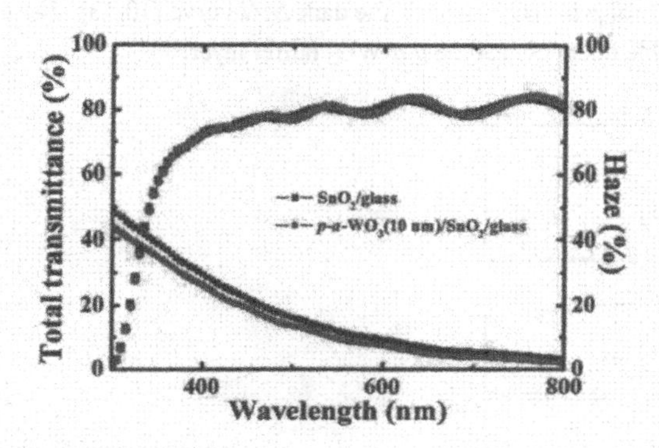

Figure 1. Total transmittance and haze spectra of the SnO_2/glass and WO_3/SnO_2/glass substrates.

When a *p-a*-WO_3 layer replaces part of the *p-a*-SiC window layer, its photo absorption can be neglected, because the absorption coefficient of the *p-a*-WO_3 layer is 7.3×10^3 cm^{-1} at 400 nm, which is almost two orders smaller than that of the *p-a*-SiC (3.75×10^5 cm^{-1}) layer. Figure 2 shows the effect of *p-a*-WO_3 thickness variation on the photovoltaic parameters.

The V_{oc} of the reference cell with a 6 nm-thick *p-a*-SiC layer was as low as 0.774V. This is due to the Schottky barrier formed between the SnO_2 and *p-a*-SiC [8]. When a *p-a*-WO_3 layer with a work function of around 4.85 eV is inserted between the SnO_2/*p-a*-SiC interface, the built-in potential (V_{bi}) can be enhanced by ΔV_{bi}, because the energy band of the *p-a*-SiC is raised by the Fermi level E_F of the *p-a*-WO_3. Therefore, the interface Schottky barrier (ϕ_{s1}) between the SnO_2 and *p-a*-SiC, as shown in Fig. 3, is replaced by an interface barrier (ϕ_{s2}) formed between the *p-a*-WO_3 and *p-a*-SiC. Thus, ΔV_{bi} can be expressed as

$$\Delta V_{bi} = \phi_{s1} - \phi_{s2}, \quad (1)$$

As shown in Fig. 2, the V_{oc} was enhanced from 0.770 V to 0.821 V when an ultrathin (2 nm-thick) *p-a*-WO_3 film was inserted between the SnO_2 and 6 nm-thick *p-a*-SiC. At the same time, the photo-generated holes flowing toward the SnO_2 experience a reduced interface Schottky barrier as a result of inserting the *p-a*-WO_3 layer. The electron back diffusion can thus be effectively suppressed. As a result, photo-generated holes and electrons can be collected more effectively [4]. Consequently, the J_{sc} was increased from 9.73 to 10.57 mA/cm^2, and the conversion efficiency from 5.17 % to 5.98 %. The fill factor was increased from 0.687 to 0.690.

When the thickness of the p-a-WO_3 increases from 2 nm to 5 nm, the cell exhibits an efficiency of 5.90% (V_{oc} =0.823 V, J_{sc} =10.52 mA/cm^2, FF=0.682). The deceasing of J_{sc} and FF is because the thicker WO_3 has relatively high defect density [9], which was confirmed by calculating the density of the hopping states $N(E_F)$ of p-a-WO_3 based on the dark J_D-V curve [10-13]. Therefore, we concluded that 2 nm is a suitable thickness for the p-a-WO_3 buffer layer.

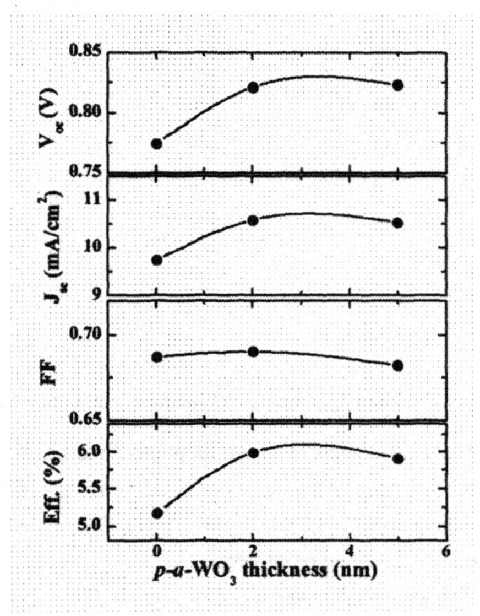

Figure 2. Effect of p-a-WO_3 thickness on solar cell performance.

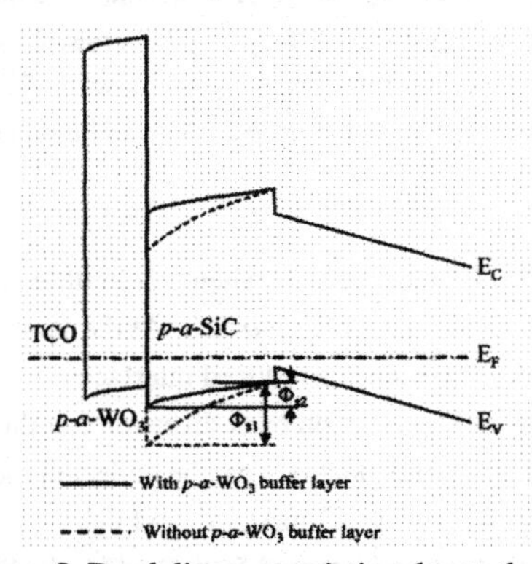

Figure 3. Band diagram variation due to the p-a-WO_3 buffer layer at the SnO_2/p-a-SiC interface.

As shown in Fig. 3, the photo-generated holes flowing toward the SnO_2 experience a reduced interface Schottky barrier as a result of inserting the *p-a*-WO_3 layer. At the same time, the electron back diffusion can be effectively suppressed. As a result, photo-generated holes and electrons can be collected more effectively. Therefore, as shown in Fig. 4, the QE of the solar cell with the *p-a*-WO_3(2nm)/*p-a*-SiC(6nm) buffer layer was remarkably improved in the visible wavelength region, as compared to that of the cell with a 6 nm-thick *p-a*-SiC buffer layer. Consequently, the J_{sc} was increased from 9.73 to 10.57 mA/cm^2, and the conversion efficiency from 5.17 % to 5.98 %. By optimization of the thicknesses of the *p-a*-WO_3 and *p-a*-SiC layers, the cell efficiency can be enhanced further to 6.33% (V_{oc} =0.864 V, J_{sc} =10.46 mA/cm^2, FF=0.701) with a *p-a*-WO_3(2 nm)/*p-a*-SiC(8 nm) window layer, the conversion efficiency was increased by about 7.3 % compared to the optimized bufferless cell, which has a conversion efficiency of 5.90 % (V_{oc}= 0.857 V, J_{sc}= 9.76 mA/cm^2, FF= 0.706) with a 10 nm-thick p-a-SiC window layer. Moreover, there is still considerable room to enhance the cell performance by variety of post-deposition processes such as the ion and laser irradiation, which have been used to enhance the electrical conductivity (σ_D) of *a*-WO_3 [14,15].

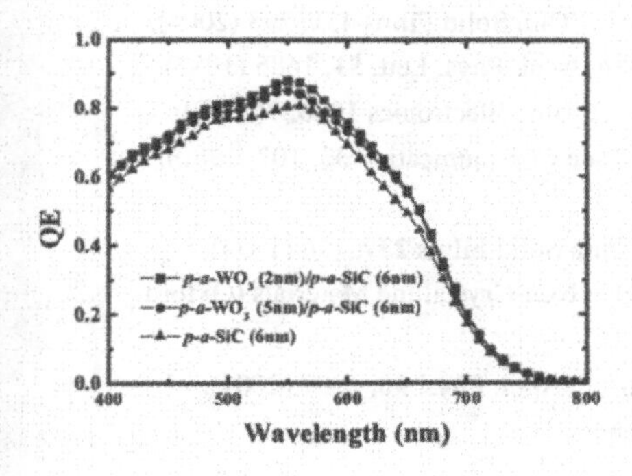

Figure 4. Spectral response of *pin*-type a-Si based solar cells with a *p-a*-SiC layer and *p-a*-WO_3/*p-a*-SiC double window layer, respectively.

CONCLUSIONS

In summary, we have applied a *p-a*-WO_3 layer at a SnO_2/*p-a*-SiC interface as a buffer layer and found that the V_{oc} and short wavelength response increased with lowering of the Schottky barrier height between the SnO_2 and *p-a*-SiC, and the photovoltaic performance of the cell was increased by about 7.3 % compared to the optimized bufferless cell with 10 nm p-a-SiC. Therefore, it is concluded that thermally evaporated ultrathin (~2 nm) *p-a*-WO_3 is a very promising candidate material for a buffer layer to enhance the performance of *pin*-type *a*-Si based solar cells.

ACKNOWLEGMENT

The authors gratefully acknowledge the Korea Science and Engineering Foundation (KOSEF) grant funded by the Korea government (MEST) (No. 2008-0062241) supported this research.

REFERENCES

1. K. S. Lim, M. Konagai and K. Takahashi, J. Appl. Phys. **56**, 538 (1984).
2. C. H. Lee, J. W. Jeon and K. S. Lim, J. Appl. Phys. **87**, 8778 (2000).
3. W. Y. Kim, H. Tasaki, M. Konagai, and K. Takahashi, J. Appl. Phys. **61**, 3071 (1987).
4. C. H. Lee and K. S. Lim, Appl. Phys. Lett. **72**, 106 (1998).
5. S. C. Han, W. S. Shin, M. S. Seo, D. Gupta, S. J. Moon and S. Yoo, Organic Electronics **10**, 791(2009).
6. H. S. Cho, C. H. Yun, J. W. Park and S. Yoo, Organic Electronics **10**, 1163(2009).
7. J. Krc, M. Zeman, O. Kluth, F. Smole and M. Topic, Thin Solid Films **451**, 298 (2004).
8. K. Itoh, H. Matsumoto, T. Kobata and A. Fujishima, Appl. Phys. Lett. **51**, 1685 (1987).
9. M. J. Son, S. H. Kim, S. N. Kwon and J. W. Kim, Organic Electronics **10**, 637(2009).
10. S. Benci, M. Manfredi and G. C. Salviati, Solid State Communication **33**, 107 (1980).
11. N. F. Mott, J. Non-Cryst. Solids. **1**, 1(1968).
12. C. Bechinger, S. Herminghaus and P. Leiderer, Thin Solid Films **239**, 156 (1994).
13. N. F. Mott and E. A. Davis, Electronic Processes in Non-Crystalline Materials (Oxford University Press, London, 1979).
14. B. Heinz, M. Merz, P. Widmayer and P. Ziemann, J. Appl. Phys. **90**, 4007 (2001).
15. Y. F. Lu and H. Qiu, J. Appl. Phys. **88**, 1082 (2000).

Mater. Res. Soc. Symp. Proc. Vol. 1245 © 2010 Materials Research Society 1245-A07-07

Preparation of narrow-gap a-Si:H solar cells by VHF-PECVD technique

Do Yun Kim[1], Ihsanul Afdi Yunaz[1], Shunsuke Kasashima[1], Shinsuke Miyajima[1], and Makoto Konagai[1,2]
[1] Department of Physical Electronics, Tokyo Institute of Technology, Japan
[2] Photovoltaics Research Center (PVREC), Tokyo Institute of Technology, Japan

ABSTRACT

Optical, electrical and structural properties of silicon films depending on hydrogen flow rate (R_H), substrate temperature (T_S), and deposition pressure (P_D) were investigated. By decreasing R_H and increasing T_S and P_D, the optical band gap (E_{opt}) of silicon thin films drastically declined from 1.8 to 1.63 eV without a big deterioration in electrical properties. We employed all the investigated Si thin films for p-i-n structured solar cells as absorbers with i-layer thickness of 300 nm. From the measurement of solar cell performances, it was clearly observed that spectral response in long wavelength was enhanced as E_{opt} of absorber layers decreased. Using the solar cell whose E_{opt} of i-layer was 1.65 eV, the highest QE at long wavelength with the short circuit current density (J_{sc}) of 16.34 mA/cm^2 was achieved, and open circuit voltage (V_{oc}), fill factor (FF), and conversion efficiency (η) were 0.66 V, 0.57, and 6.13%, respectively.

INTRODUCTION

As the world has come under pressure to stop burning fossil fuel, a solar cell, which is an environment-friendly and unlimited energy source, has received a great deal of attention. For now, wafer-based crystalline silicon solar cells have occupied by over 90% of PV markets [1], however, Si-based thin film solar cells have emerged as promising candidates to replace bulk crystalline silicon solar cells. Si-based thin film solar cells have notable advantages of low production cost, large-scale deposition, low-temperature production, and so on. Especially, the most noticeable advantage is that it is possible to use wide range of solar radiation spectra by stacking a few number of unit cells on the top of the others, so called a multi-junction solar cell [2]. In this manner, it is an important issue to control the optical band gap of an absorber layer of the each unit cell. The optical band gap of silicon films can be easily tuned by incorporating various elements such as carbon [3] and oxygen [4] for a wide-gap solar cell, and germanium [5] for a narrow gap solar cell. Also, the optical band gap of silicon thin films depends largely on hydrogen contents (C_H) of the films [6].

In this study, we have fabricated narrow gap silicon films and solar cells without an introduction of any germanium sources. We have systematically investigated a dependence of optical, electrical and structural properties of Si thin films and solar cells on hydrogen flow rate (R_H), substrate temperature (T_S) and deposition pressure (P_D).

EXPERIMENTAL DETAILS

Silicon films were prepared with very high frequency plasma enhanced chemical vapor deposition (VHF-PECVD, 60 MHz) from a gas mixture of silane (SiH_4) and hydrogen (H_2) on the Corning 7059 glass and Si (100) substrates. During the deposition, R_H, T_S, and P_D were systematically varied while VHF power density and distance between substrate and electrode were kept constant at 6.5 mW/cm^2 and 2 cm, respectively. The p-i-n structured solar cells with an area of 0.086 cm^2 were fabricated on Asahi U-type (textured SnO_2:F) glass substrates. The solar cell has the structure of glass/SnO_2:F/p-a-SiC:H (15 nm)/i-a-SiC:H buffer (10 nm)/i-a-Si:H (300 nm)/n-a-Si:H (15 nm)/ZnO:B/Ag/Al. The absorption coefficient and thickness of the films were obtained by spectroscopic ellipsometry (SE) and, then, the optical band gap (E_{opt}) of the films was calculated by the method established by Tauc *et al* [7]. For structural analysis, FT-IR spectrometer and Raman spectrometer were used. The dark (σ_d) and photo conductivities (σ_{ph}) of the films were characterized by *J-V* measurement using Al coplanar electrodes. The photo conductivity of films and photo *J-V* characteristics of solar cells were measured under standard 1-sun illumination (AM 1.5, 100 mW/cm^2). The quantum efficiency (QE) measurement was also carried out to examine the spectral response of the solar cells.

RESULT AND DISCUSSION

Effects of hydrogen flow rate (R_H)

First of all, effects of R_H on silicon films were investigated. In figure 1, Raman spectra for the films deposited with different R_H are presented. The peaks positioned near 480, 510 and 520 cm^{-1} correspond to the transverse optical (TO) mode of a-Si:H, intermediate fraction due to small crystallites and a defective part of the crystalline phase, and TO mode of crystalline silicon, respectively [1, 8]. The degree of crystallinity is indicated by X_C given by $X_C = (I_{510} + I_{520})/(I_{480} + I_{510} + I_{520})$, where I_i denotes an integrated intensity at i cm^{-1} [1].

Under the high R_H of over 70 sccm, it is clearly seen that sharp and narrow peak centered near 518 cm^{-1} breaks through the broad peak centered near 480 cm^{-1} which means that the degree of crystallinity becomes enhanced. It is well known that atomic hydrogen can etch the growing surface and break the weakly bonded atoms. Then, structural ordering onto stable nucleation sites at the growing surface is accelerated [9, 10].

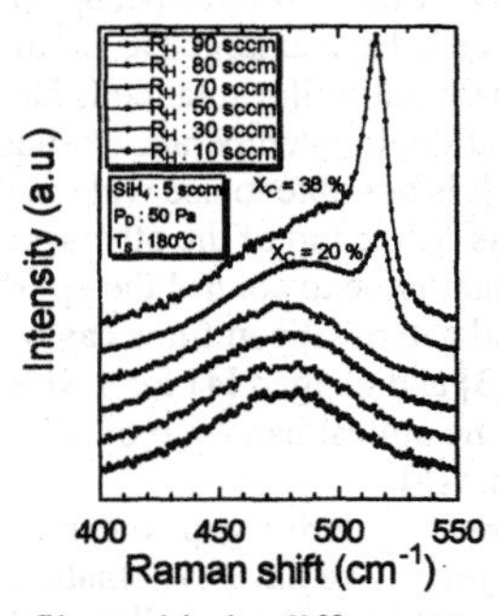

Figure 1. Raman spectra of silicon films with the different R_H.

Figure 2 (a) shows hydrogen concentration (C_H) and deposition rates (R_D) of the films as a function of R_H. Since all of the H bonded to Si contributes to the Si-H wagging mode near 640 cm^{-1} of IR spectra, the peak is chosen to calculate C_H of the films [1, 11]. C_H gradually increases with an increase in R_H up to 70 sccm. Here, it is noteworthy that C_H of the samples even with the low R_H of 10 sccm is over 20 at.%. On the other hands, further increment of R_H over 70 sccm induces the abrupt reduction in C_H because of the crystallization. Figure 2 (b) represents the variation of E_{opt} as a function of R_H. Unfortunately, it is very difficult to use Tauc plot for μc-Si:H, therefore, the E_{opt} of μc-Si:H is not included in the figure 2 (b). As R_H increases up to 70 sccm, E_{opt} gradually increases due to the increase in C_H. The presence of hydrogen causes the valence-band edge to move towards the lower energy side by removing the Si-Si bonding orbitals in the neighborhood of the valence band edge. Consequently, the E_{opt} is expanded as the content of bonded hydrogen increases.

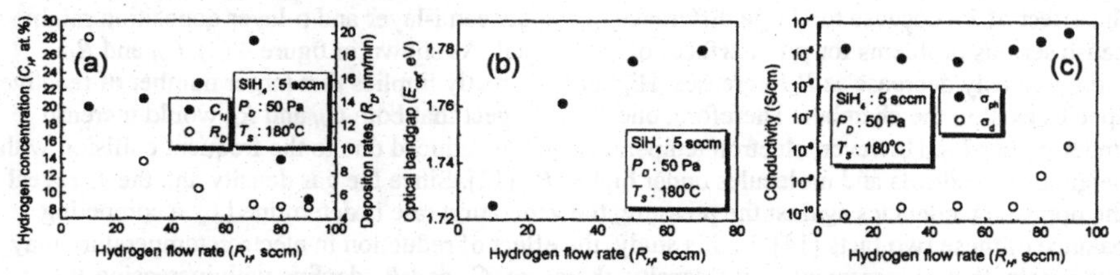

Figure 2. Variation of (a) hydrogen concentration, deposition rates, (b) optical band gap, (c) dark (σ_d) and photo conductivities (σ_{ph}) of silicon films as a function of R_H.

In figure 2 (c), σ_d and σ_{ph} of the samples are shown. Interestingly, σ_d decreases and σ_{ph} increases as R_H decreases from 50 to 10 sccm, although C_H decreases monotonously. Considering the fact that the a-Si:H prepared at R_H of 10 sccm contained C_H of over 20 at.%, it seems like too high C_H in the films deteriorate the electrical properties of the a-Si:H. On the other hands, under high R_H over 70 sccm, which is the phase boundary, electrical properties begin to show typical behavior of μc-Si:H.

Effects of substrate temperature (T_S)

Secondly, we examined effects of T_S on a-Si:H films while R_H and P_D were fixed at 10 sccm and 50 Pa, respectively. In figure 3 (a), it is shown that C_H and R_D linearly decrease as T_S increases. This is a result of an enhanced hydrogen effusion which is due to the increase in kinetic energy of the species on the film surface during the growth [12].

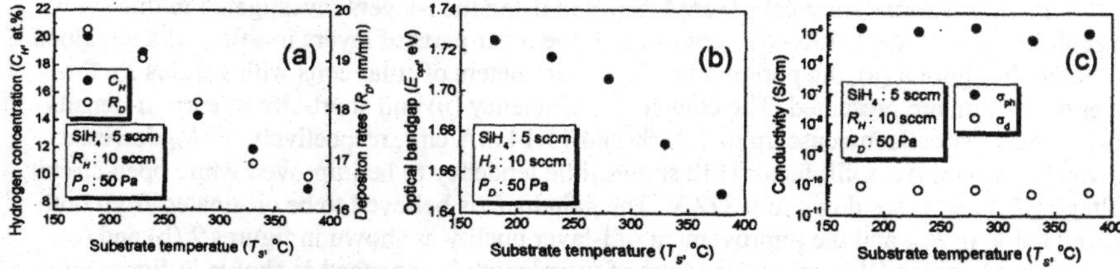

Figure 3. Variation of (a) hydrogen concentration, deposition rates, (b) optical band gap, (c) dark (σ_d) and photo conductivities (σ_{ph}) of a-Si:H films as a function of T_S.

An effect of T_S on E_{opt} is shown in figure 3 (b). The decrease in E_{opt} with increasing T_S is simply originated from the reduction in C_H. Figure 3 (c) exhibits the dependence of σ_d and σ_{ph} on

T_S. Although there is a drop in C_H as T_S increases, σ_d and σ_{ph} are almost not changed. It is believed that the structural relaxation due to the high T_S compensates the reduction of C_H, which could induce a formation of dangling bonds.

Effects of deposition pressure (P_D)

Next, influence of P_D on a-Si:H under R_H of 10 sccm and T_S of 330°C were studied. Although higher T_S was benefit for band gap narrowing, T_S of 380°C was not used to investigate the effect of P_D because too large difference in T_S between i-layer and p-layer deposition might cause serious problems for p/i interfaces of a solar cell. As shown in figure 4 (a), C_H and R_D monotonously decrease as P_D increases. Higher P_D directly implies that larger number of reactive species exist in the chamber. Therefore, one could expect that both C_H and R_D would increase under higher P_D. However, electron temperature (T_e) is reduced due to the frequent collision with neighboring radicals and molecules under higher P_D [13]. Since the gas density and the T_e reveal the opposite tendencies against the pressure, the generation rate is determined by a competing balance of these two facts [14]. In this study, the effect of reduction in electron temperature may overwhelm that of increment in gas density, therefore, C_H and R_D decline with increasing P_D.

As P_D increases, E_{opt} decreases as shown in figure 4 (b). Also, both σ_d and σ_{ph} show little tendency to increase although the variation is so small, as shown in figure 4 (c).

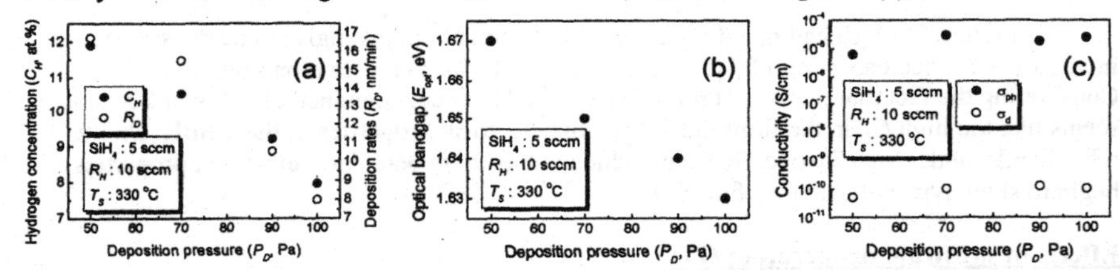

Figure 4. Variation of (a) hydrogen concentration, deposition rates, (b) optical band gap, (c) dark (σ_d) and photo conductivities (σ_{ph}) of a-Si:H films as a function of P_D.

Performances of a-Si:H solar cells

Finally, various solar cells were fabricated using the i-layers investigated in this study. Except the i-layers, deposition conditions of all the other parts of layers in solar cells remained the same. In figure 5 (a), the photovoltaic (PV) parameters of solar cells with various R_H for i-layer deposition are presented. The conversion efficiency (η) and short-circuit current-density (J_{sc}) of the solar cells increase up to 9.23 % and 15.14 mA/cm^2, respectively, as R_H decreases down to 10 sccm. Also, fill factor (FF) shows little tendency to be improved while open circuit voltage (V_{oc}) decreases down to 0.872 V. The gain in J_{sc} is believed to be originated from both the reduction in E_{opt} and the improvement of i-layer quality as shown in figures 2 (b) and (c), respectively because QE over whole range of wavelength is enhanced as shown in figure 6 (a).

Figure 5 (b) exhibits the PV parameters of the solar cells as a function of T_S for i-layer deposition. With increasing T_S up to 330°C, all the parameters decrease whereas J_{sc} increase up to 15.64 mA/cm^2. On the other hands, further increment of T_S induces a drop in J_{sc}. In figure 6 (b), it is shown that QE at long wavelength is enhanced as T_S increases which reveals the band gap narrowing. However, QE at short wavelength is deteriorated, which indicates the increase in recombination loss at p/i interfaces. Therefore, it is believed that the band gap difference between p/i interfaces become significant by increasing T_S. However, since the further drop in J_{sc} with increasing T_S of over 330°C is correlated to the decrement in QE at middle wavelength, the drop could be originated from using worse TCO [15] or electric field weakening of the i-layer [16]. Since we used the same TCO, the drops is believed to be caused by the later one. Furthermore, the electrical weakening of the i-layer is likely due to impurity diffusion from a p-layer to an i-layer caused by high T_S because as previously shown in figure 3 (c) the electrical properties of the film themselves were almost same with various T_S.

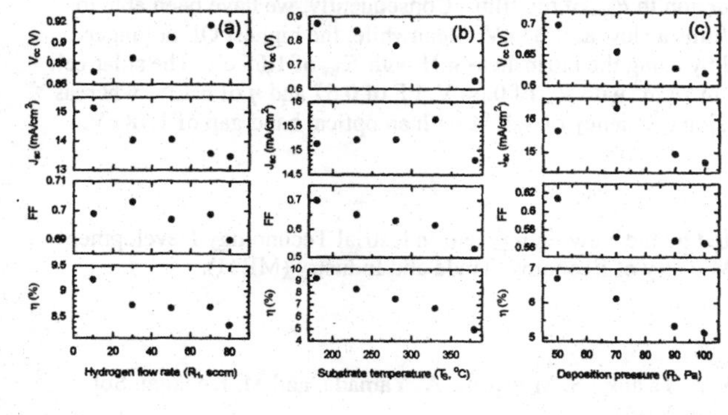

Figure 5. Photovoltaic parameters of the solar cells employing different i-layers fabricated under various (a) R_H, (b) T_S, and (c) P_D.

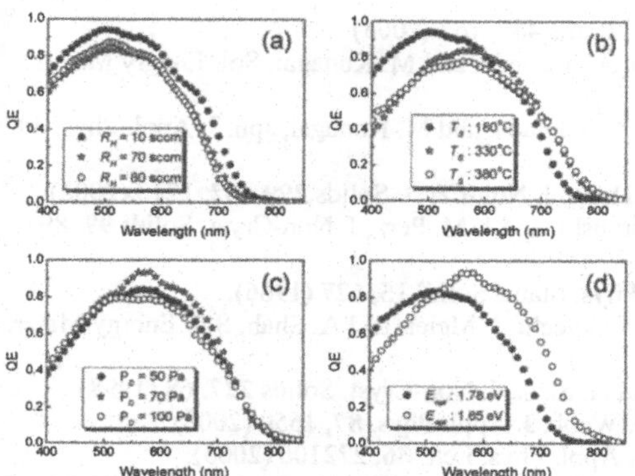

Figure 6. Changes in spectral response for the solar cells employing different i-layers fabricated under various (a) R_H, (b) T_S, (c) P_D, and (d) comparison of spectral response between two a-Si:H solar cells fabricated in this study, whose E_{opt} of i-layers is 1.78 eV and 1.65 eV, respectively.

Figure 5 (c) represents the dependence of PV parameters of the solar cells on P_D of i-layers. As P_D goes up to 70 Pa, J_{sc} increases up to 16.34 mA/cm^2 and QE at long and, especially, middle wavelength is enhanced as the result of the band gap narrowing. However, further increment in P_D leads to the abrupt drops in J_{sc} and the reduction in QE is observed at the long and middle wavelength as shown in figure 6 (c).

In figure 6 (d), the difference in QE between two solar cells fabricated in this study is represented in order to give more clear comparison. One is the a-Si:H solar cell, whose i-layer has the highest E_{opt} of 1.78 eV, and the other is the narrow-gap a-Si:H solar cell, whose i-layer shows the highest QE at longer wavelength with the E_{opt} of 1.65 eV. Here, it is clearly seen that how much QE at long wavelength is enhanced over the systematically varied deposition conditions in this study.

CONCLUSION

In this study, we have investigated the influence of R_H, T_S and P_D on optical, structural, electrical properties of silicon films and the solar cell performance. It was found that low R_H, high T_S and P_D induce the reduction in E_{opt} of the films. Consequently, we have been able to make narrow gap a-Si:H thin films as low as 1.63 eV. Meanwhile, the highest QE at longer wavelength region is achieved by using the intrinsic a-Si:H with E_{opt} of 1.65 eV. The solar cell shows the highest J_{sc} of 16.34 mA/cm^2 with V_{oc} of 0.66 V, FF of 0.57 and η of 6.13% whereas the best cell presents a conversion efficiency of 9.23% with an optical band gap of 1.78 eV.

ACKNOWLEDGEMENT

This work was supported by the New Energy and Industrial Technology Development Organization (NEDO) under Ministry of Economy, Trade and Industry (METI).

REFERENCES

1. S. Y. Myong, K. Sriprapha, Y. Yashiki, S. Miyajima, A. Yamada, and M. Konagai, Sol. Energy Mater. Sol. Cells **92**, 639 (2008).
2. J. Yang, B. Yan, and S. Guha, Thin Solid Films **487**, 162 (2005)
3. I. A. Yunaz, K. Hashizume, S. Miyajima, A. Yamada, and M. Konagai, Sol. Enegry Mater. Sol. Cells **93**, 1056 (2009).
4. S. Inthisang, K. Sriprapha, S. Miyajima, A. Yamada, and M. Konagai, Jpn. J. Appl. Phys. **49**, 122402-1 (2009).
5. P. Agarwal, H. Povolny, S. Han, and X. Deng, J. Non-Cryst. Solids **299-302**, 1213 (2002).
6. R. V. Kruzelecky, D. Racansky, S. Zukotynski, and J. M. Perz, J. Non-Cryst. Solids **99**, 89 (1988)
7. J. Tauc, R. Grigorovici, and A. Vancu, Phys. Status Solidi **15**, 627 (1966).
8. C. Droz, E. Vallat-Sauvain, J. Bailat, L. Feiknecht, J. Meier, and A. Shah, Sol. Energy Mater. Sol. Cells, **81**, 61 (2004)
9. U. Kroll, J. Meier, P. Torres, J. Pohl, and A. Shah, J. Non-Cryst. Solids **227**, 68 (1998).
10. A. H. Maha, L. M. Gedvillas, and J. D. Webb, J. Appl. Phys. **87**, 1650 (2000).
11. C. H. Lee, A. Sazonov, and A. Nathan, Appl. Phys. Lett. **86**, 222106 (2005).
12. M. T. Gutirrez, J. Carabe, J. J. Gandia and A. Solonko, Sol. Energy Mater. Sol. Cells **26**, 259 (1992).
13. L. Guo, M. Kondo, M. Fukawa, K. Saitoh, and A. Matsuda, Jpn. J. Appl. Phys. **37**, L1116 (1998).
14. M. Kondo, M. Fukawa, L. Guo, and A. Matsuda, J. Non-Cryst. Solids **266**, 84 (2000).
15. J. Sutterluti, S. Janki, A. Hugli, J. Meier, and F. P. Baumgartner, Proc. 21st EU PVSEC, Dresden, Germany, (2006) pp. 1749-1752.
16. K. Dairiki, A. Yamada, and M. Konagai, Proc. 26th IEEE PVSC, Anaheim, CA, (1997) pp. 779-782.

Mater. Res. Soc. Symp. Proc. Vol. 1245 © 2010 Materials Research Society 1245-A07-11

The study of Optical and Electrical Properties of a-SiC:H for Multi-junction Si Thin Film Solar Cell.

J.H. Shim, W.K. Yoon, S.T. Hwang, S.W. Ahn, and H.M. Lee
LG Electronics Inc., 16 Woomyeon-dong, Seocho-gu, Seoul, Korea

ABSTRACT

Studies have shown that wide band gap material is required to investigate for multi-junction applications. Here, we address proper deposition condition for a-SiC:H film adopting various deposition conditions. Those conditions were realized to single layers and cells to analyze their electrical and optical properties. In high power high pressure regime, we observed that the defect density get much lowered to the similar level of a-Si:H film with high H_2 dilution. Solar cells fabricated with the optimized condition show high efficiency and lower LID effect with only 13 % reduction indicating that a-SiC:H is promising materials for multi-junction solar cells.

INTRODUCTION

For the Si based thin film solar cell, tandem structured solar cell has been studied to utilize solar spectrum more effectively for the higher efficiency solar cells. Many theoretical simulations show high efficiencies with tandem solar cells, there are still discrepancy between simulations and realized cells due to the material limitations, such as low optical absorption and high defect density [1,2] Yunaz et al.[3] had simulated multi-junction cells and showed that wide bandgap (Eg) materials are needed as a top cell to make solar cell with efficiency higher than 20%. Then also mentioned that one of the promising candidate materials with wide Eg is hydrogenated amorphous silicon carbide (a-SiC:H) [3]. Whereas, a good quality a-SiC:H film with low defect densitty has not been reported till now due to its microstructural disorder reducing photoelectronic quality significantly with incorporated carbon content [4,5].

The researchers have tried to make a-SiC:H film with increase of an excitation frequency or high hydrogen dilution ratio of the reactive gases (R ratio), for the better incorporation of the carbon atom into a-Si:H matrix to obtain higher Eg suppressing the defect states. Whereas, the increase of an excitation frequency from RF(13.65MHz) to VHF(60, 70 MHz) or microwave there were no clear relations observed between deposition condition and film quality [5,6,7,8] The a-SiC:H film quality was enhanced with the increase of hydrogen dilution ratio to silane (SiH_4) gas.

In Fourier transformed infrared spectroscopy (FT-IR), although there are still controversial to understand the origin of the distribution of the peaks near at 2000 cm^{-1}[9,10,11], many experimental results supported that the peaks near 2000 cm^{-1} possesses two stretching modes with the monohydride configuration of Si-H bonding at 1980-2030 cm^{-1} and the dihydride configuration of Si-H_2 at 2060-2160 cm^{-1} [12,13]. The latter mode is regarded as mostly related to the defect states which will affect the light induced degradation (LID) known as the Staebler-Wronski effect.

In this work, we focused our research on the reduction of defect states in the a-SiC:H films by changing the deposition conditions to control film properties. The photoelectronic and electrical properties of a-SiC:H films were analyzed by adopting Spectroscopic

Ellipsometry (SE), the silicon-hydrogen bonding and the defect lever of the deposited film were measured with FT-IR and constant photocurrent measurements (CPM). The defect level of silicon carbide film can be lowered as intrinsic amorphous silicon. The fabricated single junction cells with the low defect states as an silicon showed higher efficiency and lower LID effect than more defective cells.

EXPERIMENTAL DETAILS

The intrinsic a-SiC:H films were made by using a capacitively coupled parallel plate PECVD system with RF excitation frequency power source. The substrate temperature was 180°C during film growth. Silane (SiH_4), and methane (CH_4) were used as reactant gases and supplied to have various gas ratios, CH_4/SiH_4 (0.5 ~ 2.0). The reactant gases were also diluted with H_2 with various gas ratios, H_2/SiH_4 (R ratio, 50~ 300). The total pressure and plasma power density were controlled from 2 to 4 Torr and from 52 to 260 mW/cm^2, respectively. The deposited film thicknesses were adjusted to be 200 nm. Samples for CPM, SE, and conductivity measurements were prepared on Corning glasses and for FTIR on crystalline silicon wafers. The cell fabrications with a-SiC:H as an intrinsic layers were prepared on Asahi VU type glass as a p-i-n structure. The solar cells were structured to be glass/p-a-SiC:H(12nm)/i-a-SiC:H(200nm)/n-a-Si:H(20nm) with the cell area $1cm^2$. The efficiencies of the cells were measured under AM1.5 illumination (100 mW/cm^2) at room temperature and LID was tested at 50°C for 200 hours.

DISCUSSION

First, the effects of plasma power and CH_4/SiH_4 ratio on the properties of a-SiC:H films were investigated separately at 2torr. The optical bandgaps of deposited films were calculated from the fitting of Tauc plots [14] by the SE measurement. In Fig. 1, the calculated bandgaps of grown films are plotted with a function of plasma power and CH_4/SiH_4 gas ratio. The increase of the bandgap is more with the change of power than that of gas ratio. With the changing of power, the bandgap increases sharply with power density upto 175 mW/cm^2 and then the increments slowdowns afterwards.

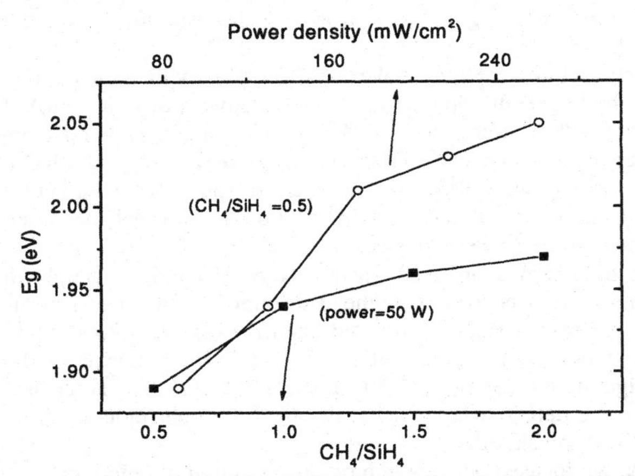

Figure 1. Optical band gap intrinsic a-SiC:H films plotted against CH_4/SiH_4 gas ratio and power density indicating that E_g has stronger dependence on the applied power.

The deposited films were characterized with FT-IR measurements to identify bonding in the films with different power densities. The results are shown in Fig. 2 representing peaks near 650, 790, 860~890, 950~1100, 1250, and 2000~2200 cm^{-1} which correspond to the rocking mode of Si-H, stretching mode of Si-C, bending mode of $(Si-H_2)_n$, rocking (wagging) mode of CH_n, stretching mode of $Si-CH_3$ and stretching mode of $Si-H_n$ (n=1,2), respectively [15,16] The film grown at low power density shows non-dissociated CH_n and $Si-CH_3$ peaks. Whereas, with the increase of power, methane is completely dissociated so the CH_n and $Si-CH_3$ are not found. The changes of FT-IR spectra are not observed with different CH_4/SiH_4 gas ratio, that are not shown here. The result of SE and FT-IR analysis indicate that high power condition is needed for the incorporation of carbon atoms effectively into a-Si:H matrix and for the dissociation of CH_4 fully to make high bandgap film.

To investigate the effect of the non-dissociated CH_n and $Si-CH_3$ in the a-Si:H films, we systematically explored the regime with different power and pressure ranges. The high quality amorphous films are observed at low power density (< 87 mW/cm^2), low pressure (< 2 torr) (LPLP) and high power density (> 174 mW/cm^2), high pressure (> 4 torr) (HPHP) regimes showing the lower dark conductivities and the higher photo conductivities with sensitivities around 10,000 as shown in Fig. 3. Hydrogenated a-SiC films were prepared in both regimes with various hydrogen dilution ratios keeping the CH_4/SiH_4 ratio = 1.0. In Table 1., the measured bandgap, defect density, and silicon hydrogen bonds for the low power low pressure (LPLP) regime and the high power high pressure (HPHP) regime, are summarized. For the LPLP regime, the optical bandgap increases with H_2 dilution from 1.85 eV to 1.93 eV.

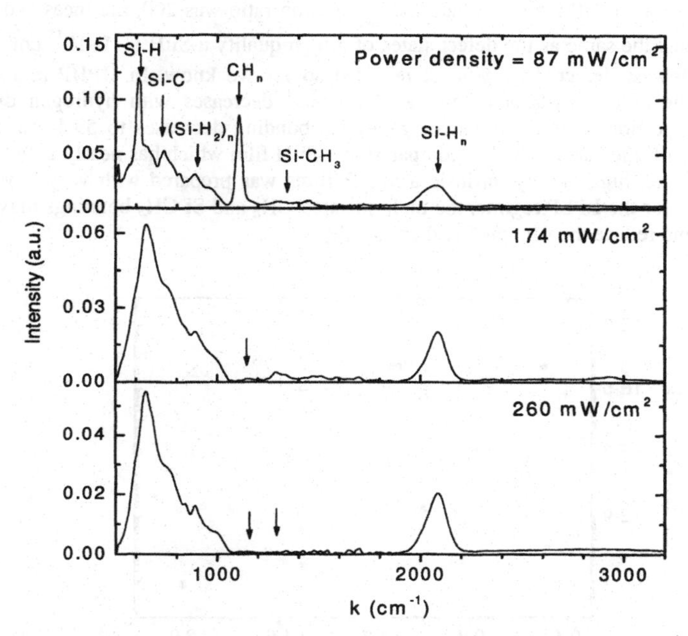

Figure 2. FT-IR spectra of a-SiC:H films plotted with power density showing that as power increases, CH_n peaks are dissociated.

Table 1. Energy bandgap, defect density, and Si-H bonds measurements for two growth regimes of LPLP and HPHP with various R ratios.

Deposition condition		Eg (eV)	Defect density (cm^{-2})	Si-H_n (n=1,2) bonds	
				I_{2000} / (I_{2000}+I_{2100})	I_{2100} / (I_{2000}+I_{2100})
87 mW/cm^2, 2 torr (LPLP)	R = 50	1.85	7.17E+15	24.9	75.1
	R = 100	1.89	2.93E+15	35.1	64.9
	R = 150	1.92	6.04E+15	33.6	66.4
	R = 200	1.93	-	30.1	69.9
	R = 250	1.93	-	26.3	73.7
176 mW/cm^2, 4 torr (HPHP)	R = 50	1.87	3.27E+16	13.9	86.1
	R = 100	1.86	5.71E+15	25.7	74.3
	R = 150	1.86	5.17E+15	34.0	66.0
	R = 200	1.86	1.87E+15	39.4	60.6
	R = 250	1.86	-	40.6	59.4

On the other hand, there is no such an increase of bandgap for the HPHP regime with increase of the H_2 dilution ratio. The increase of bandgap in the LPLP regime may be explained by the deposition rate which is three times slower than the HPHP regime, resulting incorporation of impurities such as oxygen into the films [7]. The defect density decreased at H_2 dilution ratio in HPHP regime. When the H_2 dilution ratio was 200, the measured defect level was almost the same as the defect states of a high quality a-Si:H (~ 1×10^{15} cm^{-3}). This result is the lowest defect density level reported so far we know. In HPHP regime, the intensity of the silicon hydrogen bond at 2100 cm^{-1} decreases with hydrogen dilution. Although, the portion of Si-H_2 in silicon hydrogen bonding decreases to 59.4% at 250 of dilution ratio, still the value is higher compared to a-Si:H film which has nearly 20% [17]. In the HPHP regime, high quality intrinsic a-SiC:H films was prepared with very low defect state, whereas, for the LPLP regime, the undissociated CH_n and Si-CH_3 bondings may act as defect states and reduce the film quality dramatically.

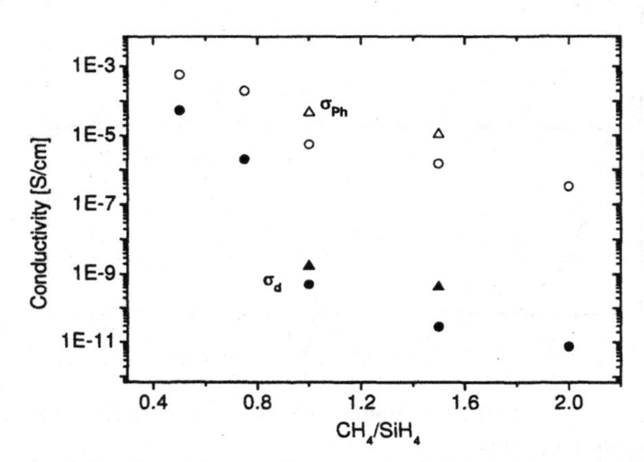

Figure 3. Conductivity plot of the samples grown on LPLP (circle) and HPHP (triangle). Dark conductivity (σ_d) and photo conductivity (σ_{ph}) are compared with CH_4/SiH_4 gas ratio.

Single-junction cell fabrications were carried out using a-SiC:H as an intrinsic layer, and they were deposited in HPHP regimes with various H_2 dilution ratios. The cell efficiencies were measured after LID test by the designated interval for 200 hours. The normalized efficiency changes with LID test were plotted in Fig. 4. The graphs show that the efficiency change was slowdowned after 100 hours in most samples as the degradation was saturated. The intrinsic layer degradation was prevented at higher hydrogen dilution ratio. In the HPHP regime, the lowest efficiency drop was only 13% with dilution ratio of 250. Whereas, the efficiency drops was more than 60% with low hydrogen dilution. This tendency of the efficiency degradation is very similar to the detected Si-H_2 peaks in the FT-IR spectra as the defect states consisting of the dangling bond are related to the degradation of the solar cell. The highest initial cell efficiency was about 7% which was a little bit lower than expected, although, defect level was almost same as a-Si:H films. The measured open circuit voltage (V_{oc}) was as high as 0.97 V, even the interface between p layer and i layer is not optimized (any buffer layer is not used),. This result shows that a-SiC:H is very promising material as atop cell in the high efficiency multi-junction solar cell.

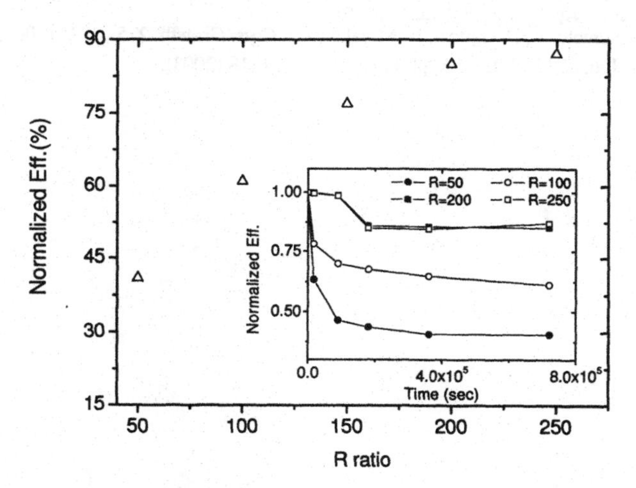

Figure 4. For various R ratio, normalized efficiency vs. time plotted for the samples grown on HPHP regime (inset). The LID effect of a-SiC:H cell in the HPHP regime.

CONCLUSION

In conclusion, high quality a-SiC:H films was prepared by the optimization of the deposition conditions at the high power high pressure regime. In this regime, the methane gas is well dissociated and the defect densities get lowered upto the level of an a-Si:H film with hydrogen dilution. The fabricated cells show high open circuit voltage and low degradation.

ACKNOWLEDGEMENTS

This work was partly supported by the Ministry of Knowledge Economy under the Contract No 2008NPV12J032100.

REFERENCES

1. J. Muller, B. Rech, J. Springer, and M. Vanecek, Sol. Energy 77, 917 (2004).
2. B.J. Simonds, F. Zhu, J. Hu, A. Madan, P.C. Taylor, Proc. Of SPIE Vol. 7409 (2009).
3. I.A. Yunaz, A. Yamada, M. Konagai, Jpn. J. Appl. Phys. 46, L1152 (2007).
4. I.A. Yunaz et al.Sol. Energy Matt. Sol. Cells 93, 1056 (2009).
5. B. von Roedern, A.H. Mahan, D.L. Williamson, and A. Madan, Symp. Mater. New Process. Tech. Photovoltaics, 5th, New Orleans (1984).
6. S. Zhang et al., Sol. Energy Matt. Sol. Cells, 87, 343 (2005).
7. R. Platz, D. Fischer, A. Shah, MRS Symp., 377, 645 (1995).
8. N. Andoh, H. Nagayoshi, T. Kanbashi, K. Kamisako, Sol. Energy Matt. Sol. Cells, 49, 89 (1997).
9. W. Beyer and M.S. Abo Ghazala, Mater. Res. Soc. Symp. Proc. 507, 601 (1998).
10. J.D. Ouwens and R.E.I. Schropp, Phys. Rev. B 54, 17759 (1996).
11. G. Lucovsky, R.J. Nemanich, and J.C. Knights, Phys. Rev. B 19, 2964 (1979).
12. M.H. Brodsky, M. Gardona, and J.J. Cuomo, Phys. Rev. B 16, 3556 (1977).
13. A.H.M. Smets, W.M.M. Kessels, and M.C.M. Van de Sanden, Appl. Phys. Lett. 82, 1547 (2003).
14. J. Tauc, in Amorphous and Liquid Semiconductors, Chap.6, Plenum, London, (1976).
15. J.P. Conde, et al., J. Appl. Phys. 85, 3327 (1999).
16. T. Kaneko, D. Nemoto, A. Horiguchi, N. Miyakawa, J. Crys. Growth, 275 e1097 (2005).
17. J.Y. Ahn, K.H. Jun, and K.S. Lim, Appl. Phys. Lett. 82, 1718 (2003).

Mater. Res. Soc. Symp. Proc. Vol. 1245 © 2010 Materials Research Society 1245-A07-15

P-LAYER OPTIMIZATION IN HIGH-PERFORMANCE A-SI:H SOLAR CELLS

Yueqin Xu[1], Bill Nemeth[1], Falah Hasoon[1], Lusheng Hong[2], Anna Duda[1], and Qi Wang[1]
[1] National Renewable Energy Laboratory (NREL), Golden, CO 80401, USA
[2] National Taiwan University of Science and Technology, Taipei, Taiwan

ABSTRACT

We report our progress toward high-performance hydrogenated amorphous silicon (a-Si:H) solar cells fabricated in NREL's newly installed multi-chamber film Si deposition system. The a-Si:H layers are made by standard radio frequency plasma-enhanced chemical vapor deposition. This system produces a-Si:H *p-i-n* single-junction devices on Asahi U-type transparent conducting oxide glass with >10% initial efficiency. The importance of the *p*-layer to the cell is identified: it plays a critical role in further improving cell performance. Our optimization process involves changing *p*-layer parameters such as dopant levels, bandgap, and thickness in cells as well as applying a double *p*-layer. With the optimized *p*-layer, we are able to increase the fill factor of our cells to as high as 72% while maintaining high open-circuit voltage.

INTRODUCTION

Hydrogenated amorphous silicon (a-Si:H) has long been studied as a unique thin-film material for a variety of applications, especially photovoltaics. Amorphous Si-based solar cells have evolved from single junction to tandem to advanced triple-junction structures. Despite the well-known Staebler-Wronski deterioration of cell performance [1], improvements have been made in the development of cells, and the best laboratory stable cell efficiency ranges up to 12.5% for multijunction configurations [2]. In addition, a-Si:H-based solar cell production has increased significantly in recent years along with rising global photovoltaic production. Despite the increasing interest in a-Si:H-based solar cells as a source for power, these devices still have many facets requiring further research and understanding to improve cell efficiency. These include minimizing light-induced degradation, managing light trapping [3], engineering bandgaps and interfaces, and improving *p*-layers.

The *p*-layer in an a-Si:H solar cell serves to establish the built-in potential of a device electrically as well as act as a window layer for optical transmission of incident light into the device. In a superstrate configuration, the *p*-layer requires a transparent conducting substrate, which can be accomplished in many ways. Ideally, the *p*-layer will allow all usable light to pass into the absorbing intrinsic layer of the device and interact with the transparent conducting substrate to promote efficient hole extraction. Typical *p*-layers employ methods to minimize optical absorption by either incorporating carbon or hydrogen to widen the bandgap or create proto- or microcrystalline phases. Boron, a typical *p*-type dopant, can be found in several precursor gases used in plasma-enhanced chemical vapor deposition (PECVD) such as diborane (B_2H_6) and trimethylboron (TMB). Additionally, the deposition of the *p*-layer should not damage nor dissociate the underlying substrate. These criteria prove to be very elusive, hence the longevity and depth of research involved in studying *p*-layers in a-Si:H devices.

In this paper, we address *p*-layer issues in a-Si:H *p-i-n* single-junction solar cells in a newly installed multi-chamber system at NREL. With the improved control the system allows, we were able to observe the subtle effects that the *p*-layer has on both the transparent conducting oxide (TCO) and *i*-layer. With optimization using double *p*-layers, we were able to achieve initial cell efficiencies >10% and stable efficiencies at 8% after a 1000-hr, 1-sun light soaking at 50 °C.

EXPERIMENTAL

The schematic of the newly installed system (MVSystems, Inc.) is shown in Figure 1. This system has ten ports for connecting chambers in a cluster configuration. Two ports are designated for the load-lock and outside sample transport. Five chambers employ radio frequency (RF) PECVD for *p*-layer, *n*-layer, *i*-layer, and SiN_x deposition as well as plasma etching. One chamber utilizes hot-wire chemical vapor deposition (CVD) for *i*-layer growth, while another very high frequency (VHF) CVD chamber is used for μc-Si deposition. The final sputter chamber is used to grow ITO and ZnO:Al layers. The substrate is transported into each chamber via a robot arm located in the central chamber. The largest substrate that can be accommodated is 157 mm x 157 mm. All chambers have independent power supplies and gas delivery systems, and they can be operated simultaneously. The nature of the system permits maintaining vacuum conditions while transferring substrates among the various processing chambers. This makes it possible to conduct solar cell research without exposure to air and other contaminants, which provides a unique and advanced environment to study and develop thin-film Si technologies for photovoltaic applications.

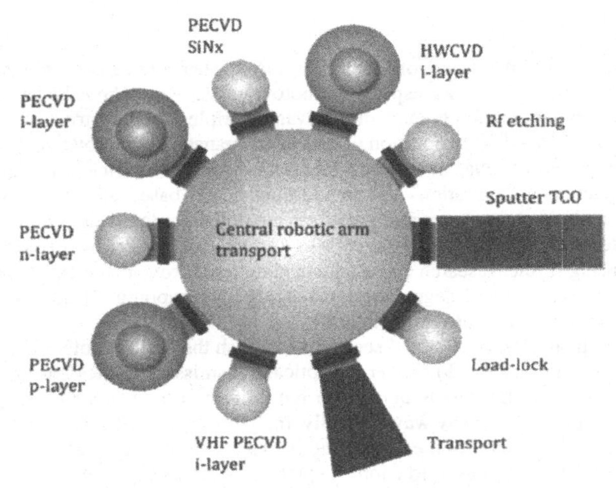

Figure 1. Schematic of the Si cluster tool

A series of solar cells with a structure of *p-i-n* were grown on Asahi U-type SnO:F substrates. Substrates were ultrasonically cleaned in DI water, rinsed with acetone and isopropanol, and dried with N_2. All a-Si:H layers were grown in the PECVD chambers.

P-layers were studied in the cell by varying flow rates of SiH_4, CH_4, and 3% TMB in He source gases at a substrate temperature of 250 °C. We selected two *p*-layers, as shown in Table I, and varied the thicknesses for optimization.

Table I. *P*-layer Growth Conditions and Characteristics

p-layer	SiH_4 (sccm)	CH_4 (sccm)	TMB/He (sccm)	P (mTorr)	E_{Tauc} (eV)	σ_{dark} (S/cm)
p^+	10	5	5.0	500	1.86	1.10×10^{-5}
p^-	26	50	6.6	600	2.04	1.98×10^{-7}

We maintained a constant *i*-layer and *n*-layer for all solar cells in this study. The intrinsic layer was grown using 16 sccm SiH_4 without hydrogen dilution at 400 mTorr and 250 °C. The layer thickness is 4000 Å with a deposition rate of 2 Å/s. This process results in $E_{Tauc} = 1.78$ eV and $\sigma_{dark} \sim 2 \times 10^{-10}$ S/cm. The 300 Å *n*-layer used SiH_4 (20 sccm) and 2% PH_3 in H_2 (4 sccm) source gases at 500 mTorr and 250 °C. It results in $E_{Tauc} = 1.75$ eV and $\sigma_{dark} \sim 2 \times 10^{-2}$ S/cm. All layers were deposited with 9.5 mW/cm^2 13.56 MHz RF power. Finally, 3000 Å Ag back contacts (with areas of 0.05 cm^2, 0.25 cm^2, 0.5 cm^2, and 1 cm^2) were deposited by electron beam evaporation. A schematic of the device can be seen in Figure 2.

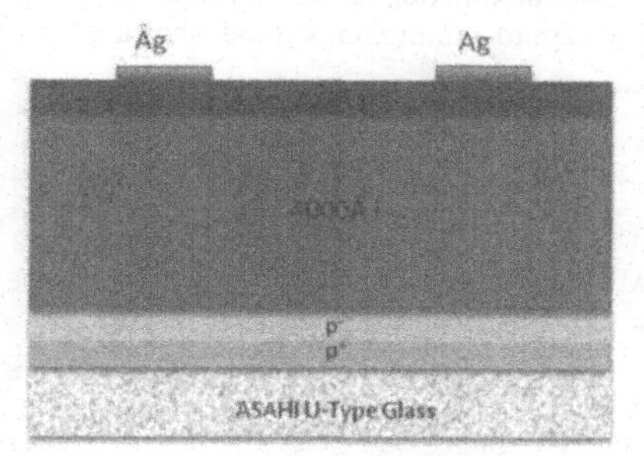

Figure 2. Structure of a *p-i-n* solar cell and a picture of sample

An indium grid was soldered to the front TCO and used as the front contact for current density voltage (J-V) measurements. The J-V measurements via a Keithley model 2400 were made using an ELH projector lamp light source calibrated with an AM 1.5 standard reference solar cell. The resulting measurements reported are averaged over all device areas. Layer thicknesses, reflection, and transmission measurements were measured using an n&k 1700 R-T analyzer from n&k Technology, Inc. Photoconductivity measurements were made using standard current-voltage (I-V) measurement under illumination utilizing 1 cm bars spaced 1 mm apart under a 100 V bias.

RESULTS AND DISCUSSIONS

For our initial study, we identified the *p*-layer optimization procedure based on a dual purpose of assisting in future device development as well as debugging and improving operation of the newly installed cluster tool. Initial a-Si:H solar cells were grown with a single *p*-layer with the same *i*- (4000 Å) and *n*- (300 Å) layers. Table II shows typical cell performance with the layers from Table I with mixed results. It can be seen that the higher-bandgap p^- layer resulted in better J_{sc} and V_{oc} values, whereas the more conductive p^+ layer contributed a better fill factor.

Table II. Cell Performance of a *P-I-N* Solar Cell with a Single *P*-Layer

p-layer	Thickness (Å)	J_{sc} (mA/cm^2)	V_{oc} (V)	FF	Eff (%)
p^+	40	16.6	0.800	0.70	9.23
p^-	90	18.2	0.874	0.59	9.28

Two fundamental issues governing *p-i-n* performance concern the front-contact TCO/*p* interface and the *p/i* interface [4]. It has been shown in the past that an undoped SiC buffer layer inserted between the *p*- and *i*- layers improves device performance by reflecting electrons, which is due to high conduction-band offset without impeding hole extraction [5]. More recently, dual *p*-layers employing a proto-crystalline buffer layer have shown improvements in device performance [6]. Furthermore, the wide bandgap of the buffer layer reduces the absorption losses of incident light. This study sought to understand the function of dual *p*-layers. It is apparent that optimized *p*-layer thicknesses should be kept to a minimum without shorting the device.

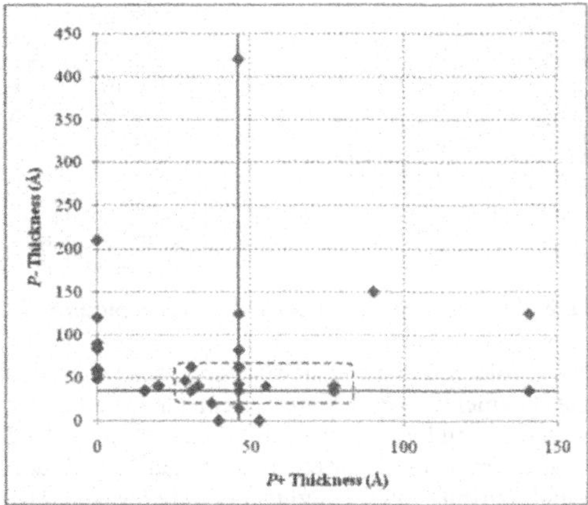

Figure 3. Optimization map of *p*-layer via various p^+ and p^- thicknesses. The red lines are the guide for a cross cut of *p*-layer optimization, and the dotted outlined area indicates enhanced device performance.

We applied dual *p*-layers to our a-Si:H solar cells, which were deposited using various thicknesses of two candidate layers (p^+ and p^-) in an effort to optimize the device performance with high *FF*, V_{oc}, and J_{sc}. The thickness variation of the double *p*-layer optimization can be seen in Figure 3. The more conductive p^+ layer was grown first on the TCO followed by the less conductive but wider gap p^- layer. It can be seen from Figures 4a and 4b that p^+ should be kept between 40 and 100 Å thick, and p^- should range between 30 and 60 Å, where electrical characteristics will balance the absorption losses. By marrying the FF benefits from the p^+ layer to the V_{oc} and J_{sc} characteristics of the p^- layer, initial efficiencies of about 10% were measured.

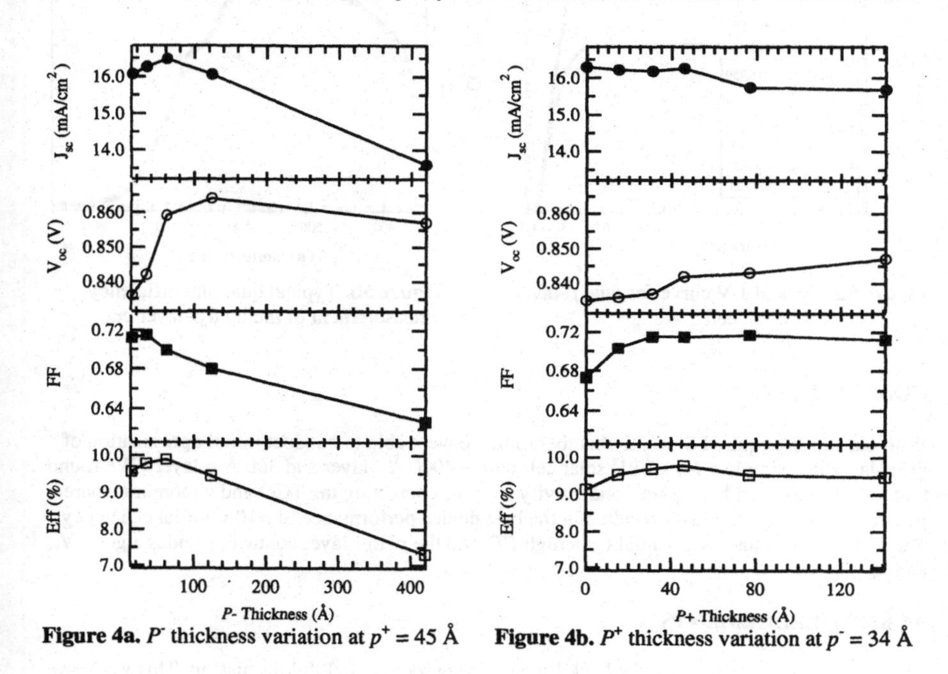

Figure 4a. P^- thickness variation at p^+ = 45 Å **Figure 4b.** P^+ thickness variation at p^- = 34 Å

Figure 5a shows the typical cell performance of a 50 Å p^+ / 40 Å p^- cell, and Figure 5b shows a typical quantum efficiency measurement of a 30 Å p^+ / 50 Å p^- cell. We also performed standard light soaking of 1000 hr and 50 °C on selected a-Si:H solar cells. The cell performance greatly decreased, by as much as 20%, which is normal for a standard PECVD *i*-layer.

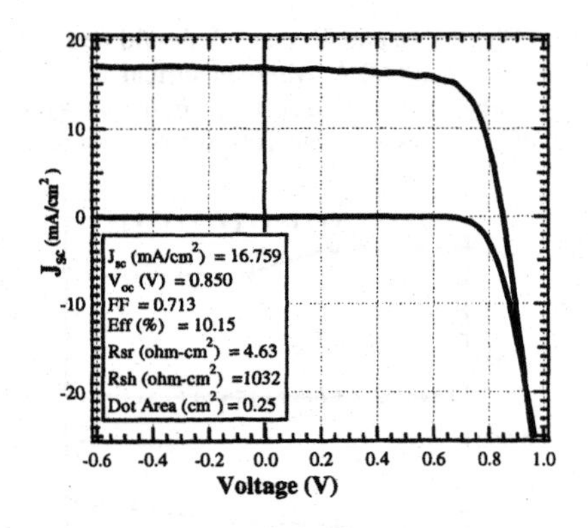

Figure 5a. Typical J-V curve for dual *p*-layer cell

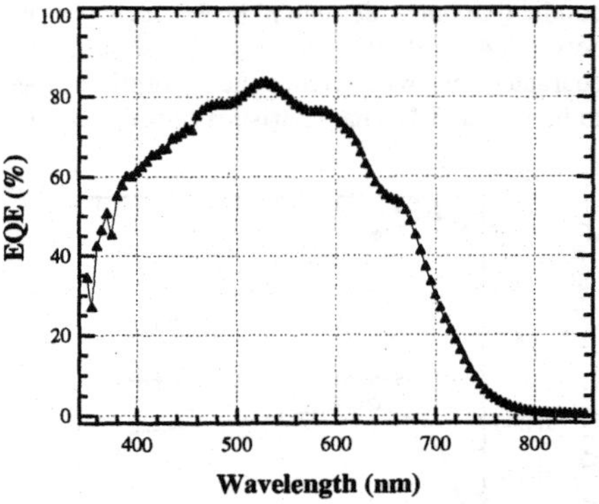

EQE (%)

Wavelength (nm)

Figure 5b. Typical quantum efficiency measurement of the dual *p*-layer cell

CONCLUSIONS

With our newly acquired Si multi-chamber tool, we were able to conduct initial optimization of the *p*-layer in a simple *p-i-n* a-Si:H solar cell with a 4000 Å *i*-layer and 300 Å *n*-layer. We found that a dual *p*-layer with a higher conductivity p^+ layer contacting the TCO and a more transparent p^- layer contacting the *i*-layer resulted in the best device performance of >10% initial efficiency. The p^+ to TCO contact is essential for a high *FF*, and the p^- to *i*-layer contact provides a good V_{oc} and J_{sc}.

ACKNOWLEDGEMENTS

The authors would like to thank Eugene Iwaniczko for helpful discussion. This work was supported by the U.S. Department of Energy under Contract DE-AC36-08-GO28308 to NREL.

REFERENCES

1. D. L. Staebler and C.R. Wronski, *Appl. Phys. Lett.* **31**, 292, 1977.
2. M.A. Green, K. Emery, Y. Hishikawa, and W. Warta, *Prog. Photovolt: Res. Appl.* **17**, 320, 2009.
3. E. Yablonovitch and G. Cody, *IEEE Trans. Elec. Dev.* **29**, 300, 1982
4. H. Stiebig, F. Siebke, W. Beyer, C. Beneking, B. Rech, and H. Wagner, *Solar En. Matls. and Solar Cells* **48**, 351, 1997.
5. Y. Tawada. K. Tsuge, M. Kondo, H. Okamoto, and Y. Hamakawa, *J. Appl. Phys.* **53**, 5273, 1982.
6. S. Myong, K. Lim, and J. Pears, *Appl. Phys. Letts.* **87**, 193509, 2005.

Mater. Res. Soc. Symp. Proc. Vol. 1245 © 2010 Materials Research Society 1245-A07-17

Equivalent-circuit Modeling of Microcrystalline Silicon pin Solar Cells prepared over a Wide Range of Absorber-layer Compositions

Steve Reynolds [1] and Vladimir Smirnov [2]

[1] Carnegie Laboratory of Physics, University of Dundee, Dundee DD1 4HN, UK.

[2] IEF-5 Photovoltaik, Forschungszentrum Jülich, D-52425 Jülich, Germany.

ABSTRACT

An equivalent-circuit electrical model is used to simulate the photovoltaic properties of mixed-phase thin-film silicon solar cells. Microcrystalline and amorphous phases are represented as separate parallel-connected photodiode equivalent circuits, scaled by assuming that the photodiode area is directly proportional to the volume fraction of each phase. A reasonable correspondence between experiment and simulation is obtained for short-circuit current and open-circuit voltage *vs.* volume fraction. However the large dip in fill-factor and reduced PV efficiency measured for cells prepared in the low-crystalline region is inadequately reproduced. It is concluded that poor PV performance in this region is not due solely to shunting by more highly-crystalline filaments, which supports the view that the low-crystalline material has transport properties inferior to either microcrystalline or amorphous silicon.

INTRODUCTION

The optoelectronic properties of thin-film silicon solar cells may be controlled and enhanced by the use of a process gas consisting of silane diluted in hydrogen. At low silane concentrations $SC = [SiH_4]/([H_2] + [SiH_4])$, of the order of 5% under typical VHF PECVD deposition conditions, mixed-phase microcrystalline silicon consisting of roughly equal volumes of crystalline and amorphous material is obtained. When this material forms the absorber layer, optimized PV conversion efficiencies in excess of 8% can be achieved, with open-circuit voltages V_{OC} in the region of 500-550 mV [1]. As SC is increased towards 10%, both the optical band-gap of the absorber and V_{OC} increase as the crystalline fraction continues to fall, but ultimately a point is reached where both the fill-factor FF and short-circuit current I_{SC} decrease rapidly and PV efficiency drops to as low as 2%. Further increase in SC yields material containing no detectable crystalline volume fraction when analyzed by Raman spectroscopy and the efficiency begins gradually to improve. By re-optimization of the deposition conditions in this regime good-quality solar cells may again be obtained, with V_{OC} now in the region 850-1000 mV [2]. Thus there are two regimes, that when approached from the microcrystalline and amorphous 'ends' of the dilution spectrum, may yield good quality solar cells.

Why, then, do solar cells with what will be termed here 'low-crystallinity' absorbers lying between these two regimes, exhibit such poor PV performance? ESR studies [3] have shown that paramagnetic defects remain at a low level in this region and thus there is no reason to suspect an increase in defect-mediated recombination. However, time-of-flight transport measurements [4] indicate that both electron and hole mobilities are significantly reduced relative to their values in optimized microcrystalline silicon films [5]. Indeed, the electron mobility appears even lower than in good-quality *amorphous* films. It has also been suggested that the *p-i* interface under these growth conditions is particularly defective, which results in a space-charge of trapped holes [6] and consequent reduced efficiency. These variations in transport properties with composition were recently investigated by the present authors and their co-workers by means of two-beam photogating measurements [7].

In this paper we consider an alternative approach to interpreting such behavior. The United Solar group [8,9] has previously proposed a model where microcrystalline and amorphous phases are represented by two parallel-connected photodiodes, in which the photocurrent generation and diode transport parameters are scaled in proportion to the volume fraction of the constituents. This model was previously shown to predict *qualitatively* the observed increase in V_{OC} with increasing amorphous content, and with light-soaking, and is consistent with conductive AFM measurement of local current flows in mixed-phase material down to quite low volume fractions. Here we develop this model in terms of a PSPICE computer simulation [10] which enables a more detailed evaluation in the low-crystallinity regime than hitherto.

EXPERIMENTAL

Sample preparation

Samples were prepared in p-i-n sequence using a cluster-tool system, as described previously [11]. The i-layer, typically 4 μm thick, was deposited by PECVD at 20 W VHF power, at a rate of typically 0.5 nm/s onto a 10 cm × 10 cm glass substrate held at 200 °C. Two types of back contact were deposited. One half of the substrate was prepared with millimeter dots of sputtered ZnO:Al, enabling illumination from either the p- or the n-side for carrier transport measurements such as time-of-flight [4,5], which is the reason why these cells are substantially thicker than optimal, and the other half with evaporated Ag back contacts in a standard test pattern to enable solar cell statistics to be measured. Raman spectroscopy was used as an estimate of crystallinity. The integrated intensities of the phonon bands at 480 cm^{-1}, associated with disordered material, and at 505 and 518 cm^{-1}, associated with crystalline grains or columns, yield the Raman intensity ratio, $I_{CRS} = (I_{505} + I_{518})/(I_{480} + I_{505} + I_{518})$. Because the crystalline bands in low-crystallinity materials are small, to improve accuracy a background spectrum of amorphous silicon normalized to the 480 cm^{-1} band was subtracted prior to integration.

Modeling

The electrical model used to represent the mixed-phase solar cell is shown schematically in figure 1. The sources I_L represent the photocurrents generated by the microcrystalline and amorphous phases, denoted by the additional subscripts M and A respectively. Similarly, the components D, R_S and R_P embody charge transport through these phases between the top and bottom contacts. Lateral transport between phases is not included, and carrier transport paths through both phases are considered uniform. In this model therefore, the cross-sectional area is proportional to the volume of each phase, and is thus a scaling factor for each circuit element.

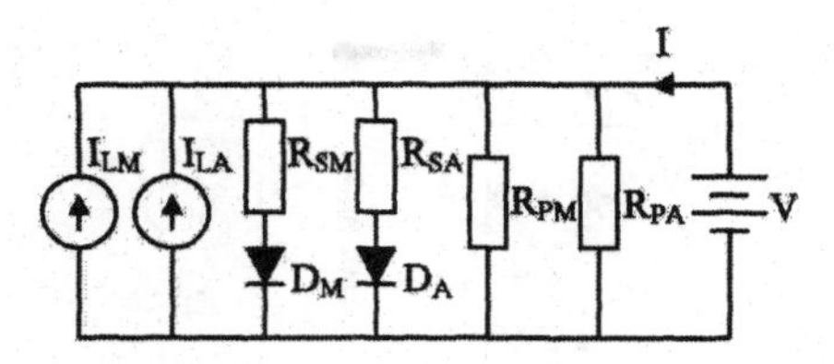

Figure 1. Equivalent-circuit model of mixed-phase solar cell.

The starting point for the simulation requires the equivalent-circuit parameters for the limiting cases of '100% microcrystalline' and '100% amorphous' cells to be determined. However there is a problem if a '100% microcrystalline' cell is used to define the electrical parameters for the 'M' phase. It is well-known that very highly-crystalline cells ($I_{CRS} \geq 0.8$) have poor efficiencies (2-3%) and a low V_{OC} around 400 mV, which correlates with a high dangling-bond spin-density [12]. Increasing *SC* towards 6% yields progressively more efficient cells (~ 8%) and V_{OC} above 500 mV, prior to the onset of the rapid fall in efficiency described above. These improvements are thought to result from passivation of unsatisfied bonds located at crystalline column boundaries by the amorphous material, as evidenced by reduced spin-density [12]. This cannot be accounted for by the two-diode model since the optoelectronic 'quality' of the two components is inter-dependent. Therefore, it is more appropriate to define the '100% M' limit as the optimum phase mixture of $I_{CRS} \sim 50\%$. Best fits to the limiting *I-V* curves yield the parameters summarized in Table 1.

The volume fraction used in model scaling will be referred to simply as *F*. In scaling the 'M' parameters, the diode saturation current I_S is multiplied by *F*, the diode ideality factor *N* is unchanged, and the series and parallel resistances R_S and R_P are divided by *F*. The 'A' parameters are similarly scaled, but by the corresponding factor (1-*F*).

RESULTS AND DISCUSSION

Representative experimental (AM1.5) and simulated *I-V* characteristics over a range of absorber layer compositions are shown in figures 2(a) and 2(b) respectively. It can be seen that the experimental variations in V_{OC} and I_{SC} are broadly reproduced by the model. However, it is not possible to make a detailed appraisal from these characteristics alone. Accordingly, parameters from a wider range of solar cells are shown in figures 3, 4, 5 and 6, along with the model predictions. Data from four depositions, at $SC = 5.5$, 6.25, 6.5 and 10%, are presented.

It is evident that considerable variation in parameters occurs within each deposition. Under the present growth conditions, material deposited on the central region of the substrate is more crystalline than at the edges, and this typically yields cells with lower V_{OC} [4]. Yan *et al* [9]

Table 1. Equivalent-circuit model parameters.

'M' phase				'A' phase			
I_S (A)	*N*	R_S (Ω)	R_P (Ω)	I_S (A)	*N*	R_S (Ω)	R_P (Ω)
4.5E-07	1.7	1.4	450	8.5E-07	3.0	5.0	350

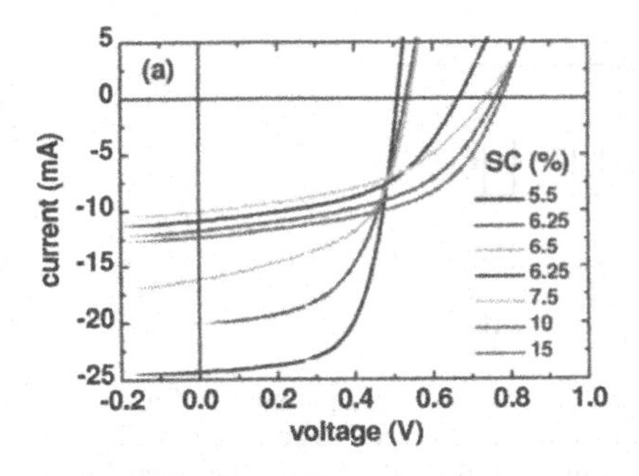

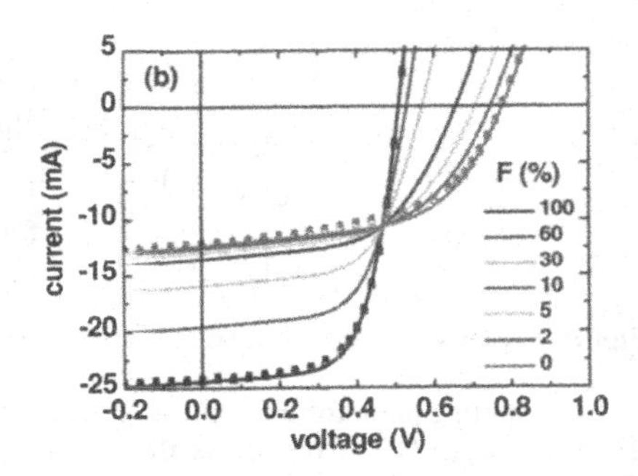

Figure 2. (a) Experimental *I-V* curves for cells deposited over a range of *SC* between 5.5 and 15%; (b) Simulated *I-V* curves computed between $F = 0$ and 100%. Symbols are the experimental data to which the $F = 0$ (circle) and $F = 100\%$ (square) curves were fitted.

report a similar effect, but in their case the central region appears more amorphous.

Figure 3 demonstrates the correlation between I_{CRS} and V_{OC} for a variety of cells deposited by VHF PECVD under similar conditions. This is useful because it enables other PV parameters to be plotted versus V_{OC}, which is available for every cell, rather than I_{CRS}, which would require extensive Raman measurements. Also, an absorber may show no discernable crystalline Raman peaks, yet systematic variations in I_{SC} and *FF* with V_{OC} may continue to occur [4]. In figures 4, 5 and 6 we adopt the same approach for both experiment and simulation, to enable comparison.

The two-diode simulation correctly reproduces the trend of I_{CRS} *vs.* V_{OC}. Small quantities of 'M' phase significantly reduce V_{OC} because D_M is strongly forward-biased and the $D_M + R_M$ combination acts as a shunt while generating little additional photocurrent. As the proportion of 'M' phase increases, the *relative* change in V_{OC} is reduced because it begins to dominate current

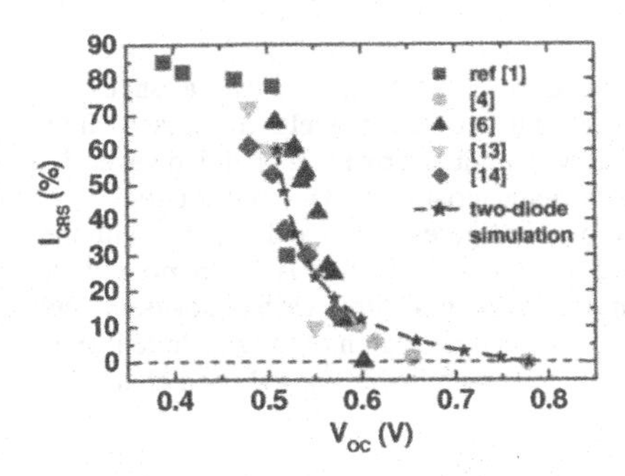

Figure 3. Correlation between I_{CRS} and V_{OC} for a number of experimental studies, and for the two-diode simulation. $F = 100\%$ has been aligned with $I_{CRS} = 60\%$.

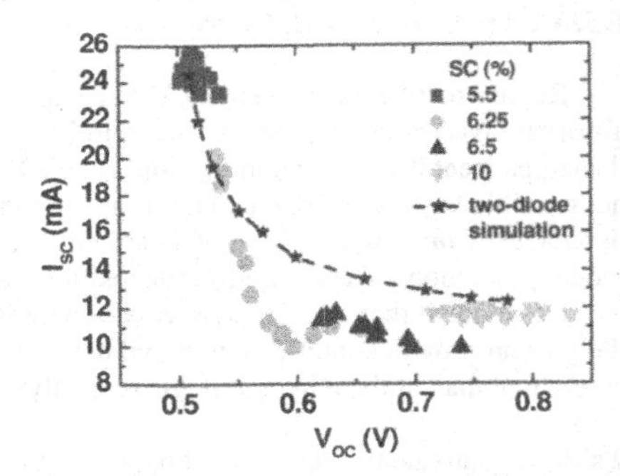

Figure 4. Variation in I_{SC} for a range of cells plotted vs. V_{OC}, and corresponding simulation.

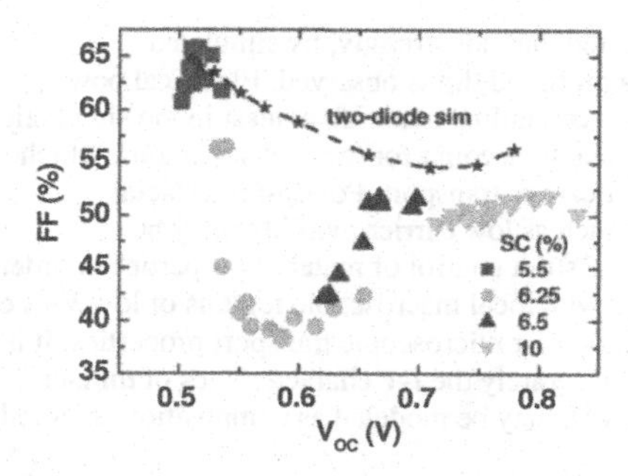

Figure 5. Variation in *FF* for a range of cells plotted vs. V_{OC}, and corresponding simulation.

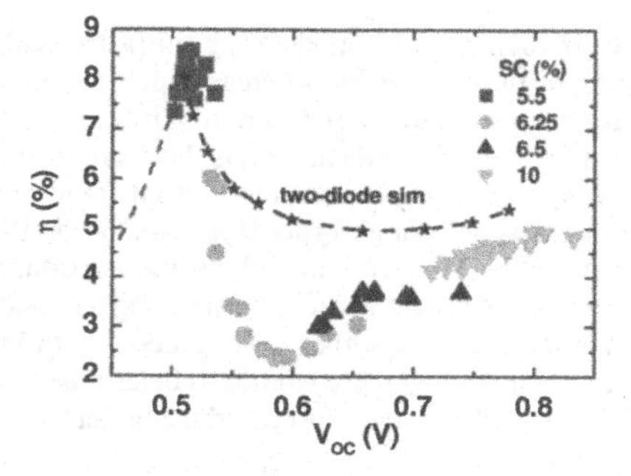

Figure 6. Variation in PV efficiency for a range of cells plotted vs. V_{OC}, and corresponding simulation.

generation and progressively less current flows through D_A. The model somewhat over-estimates the crystallinity of the experimentally 'Raman amorphous' material for $V_{OC} > 650$ mV.

Figure 4 shows that the simulation reproduces to some extent the observed fall in current in the low-crystalline regime, suggesting that a simple linear superposition of the currents generated by the microcrystalline and amorphous fractions is a realistic assumption. However, the experimental relationship falls somewhat more steeply, and this may be an indication of increased carrier loss due to stronger recombination in the low-crystallinity material.

The greatest discrepancy between experiment and model is evident in Figure 5. The experimental fill-factor falls from 65% to below 40% when V_{OC} reaches 600 mV. Referring to figure 3 it can be seen that this corresponds to I_{CRS} of 5 to 10%. At higher *SC* the absorber becomes Raman amorphous and *FF* recovers somewhat, to around 55%. In contrast, the simulation predicts a much more gradual fall in *FF*, reaching a shallow minimum of 55% at around 700 mV, with $F = 5\%$. As discussed above, the reason for the minimum in *FF* in the two-diode model is the loss of current available to an external load due to the internal shunting effect of the microcrystalline phase. This is quite insufficient to explain the experimental results, and it is again necessary to propose that the low-crystallinity material has substantially poorer PV properties than either of the optimized phase mixtures.

Finally, figure 6 compares the experimental and simulated PV efficiencies. While the simulation bears the correct general shape, it is not in good quantitative agreement for the reasons discussed above.

CONCLUSIONS

A two-diode equivalent-circuit model of a mixed-phase microcrystalline silicon *pin* solar cell has been evaluated with regard to its ability to simulate the PV parameters of experimentally-measured test structures. Open-circuit voltage and short-circuit current trends are reasonably accurately reproduced. However, the region of very low PV efficiency when the absorber layer lies between optimal 'microcrystalline' and optimal 'amorphous' compositions, is not predicted by this model. The fill-factor is affected, but not strongly, by simulated compositional changes, whereas experimentally a profound dip is observed. Electrical power dissipation due to internal shunting by more highly-crystalline material, at least in the simplistic form embodied in this model, is therefore insufficient to account for these observations, which thus appear to be a consequence of inherently poor carrier transport. Poor *p-i* interfacial properties, or poor absorber layer bulk properties such as low carrier mobility, may be responsible. This work underlines the importance of strict control of material properties in order to minimize power loss due to shunting associated with local macroscopic regions of low V_{OC}, or incorporation of absorber-layer material with locally poor microscopic transport properties. It is planned to extend these studies to determine how accurately the *I-V* characteristics of thinner devices, and also those prepared using hot-wire CVD, may be modeled as composition is varied.

ACKNOWLEDGMENTS

The authors are grateful for the assistance of A. Gordijn and Y. Mai in depositing the cells investigated in this study, to W. Reetz for *I-V* measurements, and to M. Hülsbeck for Raman characterizations.

REFERENCES

1. O. Vetterl, F. Finger, R. Carius, P. Hapke, L. Houben, O. Kluth, A. Lambertz, A Mück, B. Rech and H. Wagner, *Sol. Energy Mater. Sol. Cells* **62**, 97 (2000).
2. S. Myong, K. Sriprapha, Y. Yashiki, S. Miyajima, A. Yamada and M. Konagai, *Sol. Energy Mater. Sol. Cells* **92**, 639 (2008).
3. O. Astakhov, R. Carius, F. Finger, Y. Petrusenko, V. Borysenko and D. Barankov, *Phys. Rev. B* **79**, 104205 (2009).
4. S. Reynolds, R. Carius, F. Finger and V. Smirnov, *Thin Solid Films* **517**, 6392 (2009).
5. T. Dylla, S. Reynolds, R. Carius and F. Finger, *J. Non-Cryst. Solids* **352**, 1093 (2006).
6. Y. Mai, S. Klein, R. Carius, H. Stiebig, X. Geng and F. Finger, *Appl. Phys. Lett.* **87**, 073503 (2005).
7. S. Reynolds, C. Main, V. Smirnov and A. Meftah, *Phys. Stat. Sol. (c)* **7**, 505 (2010).
8. B. Yan, J. Yang, G. Yue, K. Lord and S. Guha, Proceedings of 3rd World Conference on Photovoltaic Energy Conversion (Osaka, May 11-18 2003) , Vols. A-C, p. 1627.
9. B. Yan, C. Jiang, C. Teplin, H. Moutinho, M. Al-Jassim, J. Yang and S. Guha, *J. Appl. Phys* **101**, 033712 (2007).
10. *Modelling Photovoltaic Systems using PSpice*, L. Castaner and S. Silvestre (Wiley, 2002).
11. Y. Mai, S. Klein, R. Carius, J. Wolff, A. Lambertz, F. Finger and X. Geng, *J. Appl. Phys.* **97**, 114913 (2005).
12. A. Baia Neto, A. Lambertz, R. Carius and F. Finger, *J. Non-Cryst. Solids* **299-302**, 274 (2002).
13. A. Lambertz, F. Finger and R. Carius, Proceedings of 3rd World Conference on Photovoltaic Energy Conversion (Osaka, May 11-18 2003), Vols. A-C, p. 2738.
14. J. Bailat *et al*, Proceedings of the 19th EU-PVSEC (Paris, June 2004), p. 1548.

Mater. Res. Soc. Symp. Proc. Vol. 1245 © 2010 Materials Research Society 1245-A07-20

Amorphous Silicon Solar Cells with Silver Nanoparticles Embedded Inside the Absorber Layer

Rudi Santbergen, Renrong Liang and Miro Zeman
Delft University of Technology, Photovoltaic Materials and Devices Laboratory/DIMES, P.O. Box 5053, 2600 GB Delft, Netherlands, Tel: +31-15-2784425, Email: r.santbergen@tudelft.nl

ABSTRACT

A novel light trapping technique for solar cells is based on light scattering by metal nanoparticles through excitation of localized surface plasmons. We investigated the effect of metal nanoparticles embedded inside the absorber layer of amorphous silicon solar cells on the cell performance. The position of the particles inside the absorber layer was varied. Transmission electron microscopy images of the cell devices showed well defined silver nanoparticles, indicating that they survive the embedding procedure. The optical absorption of samples where the silver nanoparticles were embedded in thin amorphous silicon layer showed an enhancement peak around the plasmon resonance of 800 nm. The embedded particles significantly reduce the performance of the fabricated devices. We attribute this to the recombination of photogenerated charge carriers in the absorber layer induced by the presence of the silver nanoparticles. Finally we demonstrate that the fabricated solar cells exhibit tandem-like behavior where the silver nanoparticles separate the absorber layer into a top and bottom part.

INTRODUCTION

Thin-film hydrogenated amorphous silicon (a-Si:H) solar cells have the potential to provide large-scale solar electricity at low cost [1]. To keep the inherent light-induced degradation and the production costs of the a-Si:H solar cell low, the thickness of its absorber layer has to be kept as thin as possible. Reducing the thickness of the absorber layer will reduce the absorption especially in the near infrared part of the solar spectrum. To compensate for this reduction, light-trapping techniques are implemented to significantly increase the path length of weakly absorbed light [2]. The standard techniques are based on light scattering at textured interfaces and reflection of light by the back reflector. A novel technique for light scattering in solar cells, not relying on surface roughness is light scattering through excitation of localized surface plasmons in metal nanoparticles. Metal nanoparticles scatter visible and near infrared light extremely efficiently and their optical properties are highly tunable by manipulating the particles' size, shape and mutual distance. Recent progress in nanotechnology has made it possible to produce suitable metal nanoparticles. This has opened up the possibility to realize very efficient light trapping schemes in solar cells [3].

However, metal nanoparticles can also introduce optical losses by absorbing light or by scattering it away from the absorber layer. Therefore, there is still room to improve the design of a nanoparticle enhanced solar cell. Thus far, mainly solar cell designs with nanoparticles in front of or behind the absorber layer have been investigated [4,5]. The objective of this work is to improve solar cell performance by embedding the nanoparticles inside the absorber layer. It is anticipated that in this design most of the forward and the backward scattered light is trapped.

The most suitable position of the metal nanoparticles and their effect on the transport and recombination of electrons and holes are investigated. In this study we focus on a-Si:H thin-film solar cells because their fabrication process allows the embedding of nanoparticles inside the absorber layer. Silver is the metal of choice for the nanoparticles because it has a relatively low parasitic absorption.

EXPERIMENT

Figure 1 shows the schematic diagram of the solar cell structure. The intrinsic a-Si:H layer (i-layer) is the absorber layer. The silver nanoparticles divided the absorber layer into a top part and a bottom part with the thicknesses of t_{top} and t_{bot}, respectively. In order to study the effect of the position of the silver nanoparticles they were embedded at a depth of 50, 100, 150, 200 or 250 nm inside the absorber layer. The total thickness of the absorber layer was kept fixed at 300 nm.

The following procedure was used to fabricate the a-Si:H solar cells. The devices were prepared on Asahi U-type superstrates. The silicon layers were deposited using plasma enhanced chemical vapor deposition (PECVD). After depositing the p-layer (a-SiC:H) and the buffer layer (a-Si:H), the top absorber layer with thickness t_{top} was deposited. The sample was taken out of the PECVD deposition chamber and loaded into the metal evaporator where a 3 nm thin silver film was deposited using thermal evaporation. A deposition rate of 0.1 nm/s was used. Silver nanoparticles were formed by annealing the samples in nitrogen ambient at 180°C for 90 minutes. Subsequently, the sample was reloaded into the PECVD deposition chamber and the bottom absorber layer with the thickness t_{bot} was deposited. After finishing the n-layer (a-Si:H) deposition, an 80 nm thick layer of ZnO:Al was deposited by RF magnetron sputtering. Finally, a 100 nm thick silver and a 200 nm thick aluminum film were applied as back contact. The device area was defined to be 0.16 cm^2 by the shadow mask during the evaporation of the back contact. At the same time reference devices were made without particles. These reference devices underwent the same process steps except for the deposition of silver.

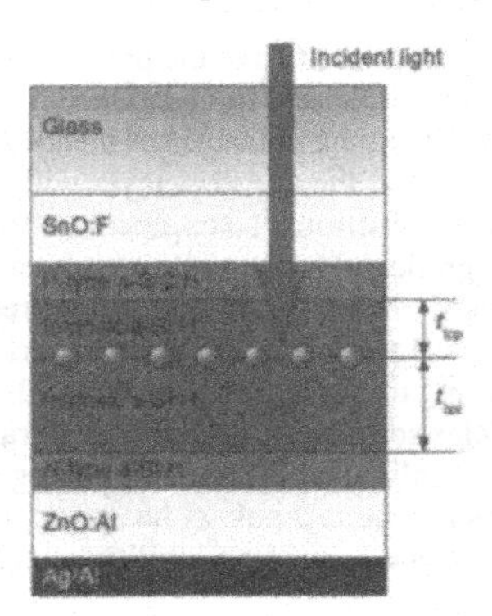

Figure 1: Schematic cross-section of a-Si:H solar cell with silver nanoparticles embedded in the absorber layer.

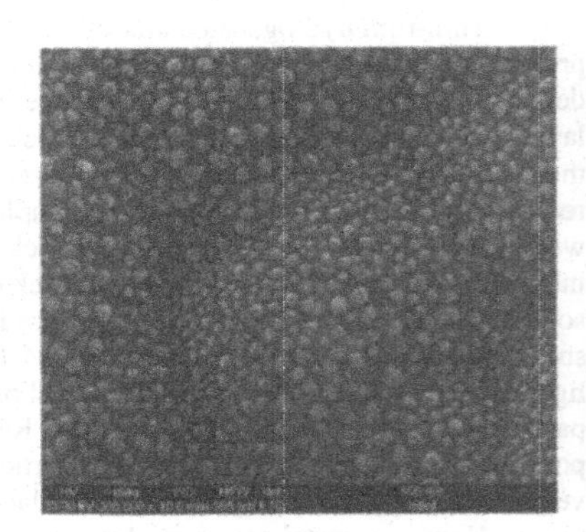

Figure 2: SEM image of silver nanoparticles fabricated on an a-Si:H overcoated Asahi U-type substrate.

RESULTS

Optical properties of silver nanoparticles embedded in a-Si:H

Prior to device fabrication, semi-transparent test structures were deposited to characterize the particles' optical properties. Particles were formed on a 20 nm thick a-Si:H layer deposited on a Asahi U-type substrate. The procedure of metal particle formation was described above. The scanning electron micrograph (SEM) image of this test structure is shown in figure 2. The SEM picture confirms that well separated silver nanoparticles with a diameter of 10 - 40 nm were formed. Note that the large scale roughness, visible as hills and valleys, is caused by the roughness of the underlying Asahi U-type substrate.

This test structure was overcoated with another 20 nm of a-Si:H, thereby embedding the silver nanoparticles. The reflectance (R) and transmittance (T) of this test structure with embedded nanoparticles were measured. From this the absorptance (A) was calculated using $A=1-R-T$. In figure 3 the results are shown and compared to the absorptance of a similar structure that has no embedded nanoparticles. A-Si:H is highly absorbing for wavelengths below 400 nm, but with increasing wavelength it becomes less absorbing (i.e. more transparent). Note that in both curves some weak interference fringes are present. For the reference sample without embedded particles, the absorptance drops to 30% for wavelengths longer than 700 nm. The sample with embedded particles, however, has a much enhanced absorptance at these wavelengths. The enhancement is largest around a wavelength of 800 nm, where the particles have their plasmon resonance wavelength. Note that similar particles formed on a glass substrate have a resonance wavelength of 500 nm (not shown), but for the samples considered here the resonance wavelength is shifted to 800 nm due to the high refractive index of the embedding a-Si:H. This red-shift is in agreement with Mie-theory [6,7]. Also note that the enhanced absorption can be caused by enhanced absorption in a-Si:H, which is the desired effect that is beneficial for solar cell performance, or by parasitic absorption in the silver nanoparticles, which is an undesirable loss mechanism. To distinguish between the two effects, the electrical performance of the solar cell devices will be analyzed as discussed in the next section.

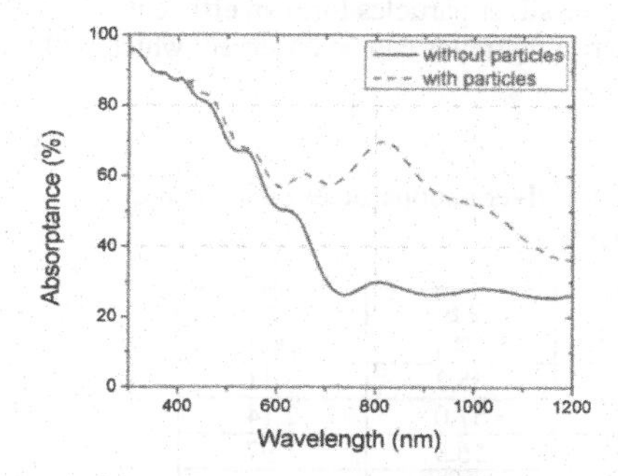

Figure 3: Absorptance of the glass/SnO:F/a-Si:H sample with and without silver nanoparticles embedded in the a-Si:H layer.

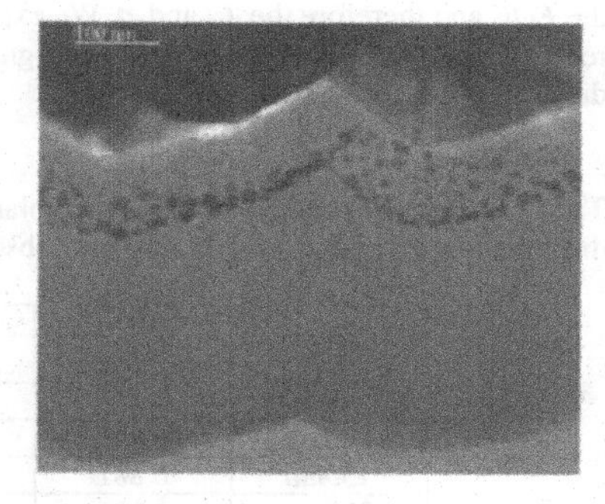

Figure 4: TEM cross-section of a-Si:H solar cell with silver nanoparticles embedded at a depth of 50 nm in the absorber layer.

Electrical performance of a-Si:H solar cells with embedded nanoparticles

The complete a-Si:H devices with embedded silver nanoparticles (shown in figure 1) were fabricated. The particles were embedded at depths of 50, 100, 150, 200 or 250 nm inside the 300 nm thick absorber layer. To check whether the particles survived the embedding procedure and that they were at the intended position, the cross-sectional transmission electron micrographs (TEM) of the solar cell structures were taken. In figure 4 the TEM image is shown for the device with the particles embedded at a depth of 50 nm. In the TEM image the silver nanoparticles do not appear to form an interface separating the top and bottom cell, but the particles seem to be scattered throughout the absorber layer. However, this is an artifact caused by the fact that a TEM image is a projected image through a cross-sectional slice of a finite thickness. Due to the highly textured surface of the substrate, the particles located deeper into this slice appear at a different vertical position although they are still located at the interface. In addition it can be seen from this image that the silver particles are on average somewhat compressed in the vertical direction, resembling oblate spheroids.

The current voltage characteristics of the solar cell devices were measured and an overview of the external parameters is given in table 1. The external parameters of a solar cell without particles are given as reference. Note that this reference solar cell was treated similar to the other cells, i.e. the growth of the absorber layer was interrupted, the sample was taken out of the PECVD system (exposing it to air) to be annealed in a nitrogen environment. All devices with embedded nanoparticles have a much lower short circuit current (J_{SC}) as compared to the reference cell. As a result the efficiency (η) dropped from 6.7% for the reference device to 1.2 to 3.1% for the devices with embedded metal nanoparticles. The J_{SC} was lowest for the device with the particles embedded at the maximum depth of 250 nm. The shallower the particles were embedded, the higher J_{SC} was observed. To investigate this in more detail, the wavelength dependent external quantum efficiency (*EQE*) was measured in the range between 300-800 nm. The results are shown in figure 5. The *EQE* of the reference solar cell without particles exceeded 0.80 (not shown). However, in devices where the particles were embedded, the maximum *EQE* varied from 0.15 (for a particle depth of 250 nm) to 0.35 (for a particle depth of 50 nm). Clearly, the embedding of silver particles in the absorber layer of a-Si:H solar cell significantly reduced the *EQE* and therefore the J_{sc} and η. We expect that the silver particles form an efficient recombination channel. Nonetheless, in figure 5 interesting trends can be observed, which will be discussed in the next section.

Table 1: External parameters of a-Si:H solar cells with silver nanoparticles embedded at different depths inside the 300 nm thick absorber layer.

t_{top}	V_{OC} [V]	J_{SC} [mA/cm^2]	FF [%]	η [%]
50nm	0.882	-6.38	55.3	3.11
100nm	0.899	-4.18	57.0	2.14
150nm	0.881	-3.96	56.5	1.97
200nm	0.876	-3.14	53.7	1.48
250nm	0.889	-2.47	52.2	1.15
no particles	0.827	-17.2	47.5	6.75

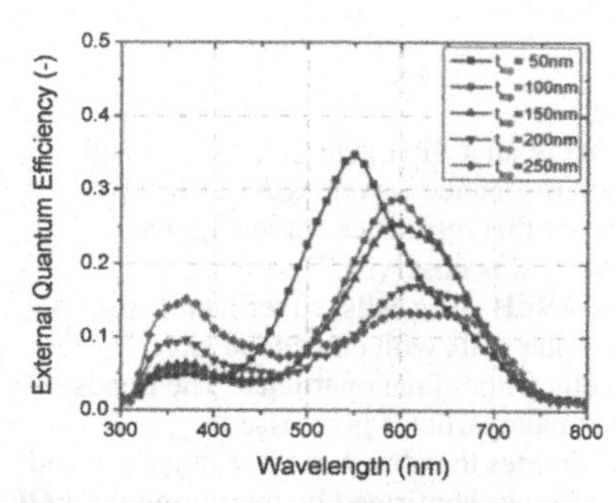

Figure 5: EQE of the a-Si:H solar cells with silver nanoparticles embedded at different depths in the 300 nm thick absorber layer.

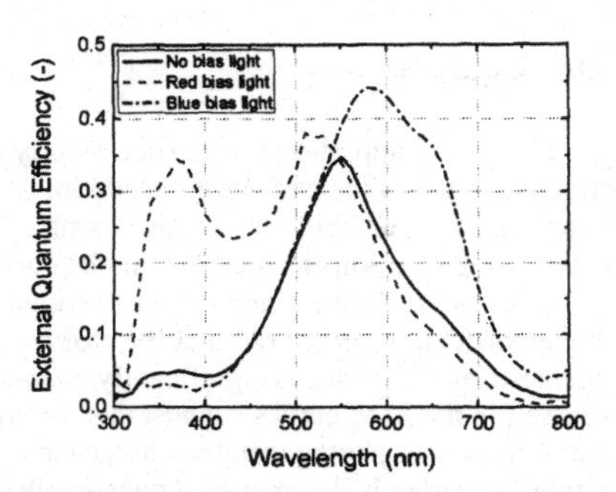

Figure 6: EQE of a-Si:H solar cells with silver nanoparticles embedded at a depth of 50 nm under red, blue and without bias illumination.

DISCUSSION

The *EQE* curves shown in figure 5 have a dip in the *EQE* in the wavelength region of 400 to 500 nm. One can observe two local maxima in the *EQE* curves, the first maximum at the shorter and the second maximum at the longer wavelengths. Interestingly, when the embedding depth of the particles increased, the short wavelength maximum increased while the long wavelength maximum decreased. These opposite trends are attributed to a quasi-tandem behavior of the solar cell. It will be demonstrated that the layer of metal nanoparticles functions like an interlayer between the top and bottom cell in tandem devices. In this way, the position of the interlayer affects current matching between the top and bottom cell.

Figure 6 shows the *EQE* of the solar cell devices with nanoparticles embedded at a depth of 50 nm. At a wavelength of 350 nm the *EQE* is 0.05 in the standard measurement, without bias illumination (solid line). However, light of this wavelength is absorbed in the top part of the cell only. In the standard *EQE* measurement the bottom part of the cell does not generate current to match the current generated in the top part of the cell. By applying red bias illumination during the *EQE* measurement, a matching current is generated in the bottom cell. This demonstrates that the *EQE* of the top cell at a wavelength of 350 nm is 0.35 (see figure 6). Similarly, more current is generated in the bottom cell than in the top cell for wavelengths longer than 550 nm. Using a blue bias illumination to generate a matching current in the top cell reveals that the *EQE* of the bottom cell exceeds 0.40 at a wavelength of 600 nm.

The poor performance of a-Si:H solar cells with silver nanoparticles embedded in the absorber layer is attributed to recombination of photogenerated charge carriers. This outweighs any positive effect due to improved light trapping. Recombination might be prevented by electrically passivating silver nanoparticles before they are embedded in the absorber layer. In case recombination is caused by diffusion of silver into a-Si:H, it might be beneficial to deposit the silver nanoparticles using a room temperature deposition method.

CONCLUSIONS

Silver nanoparticles were successfully embedded in an a-Si:H layer. Due to the high refractive index of the embedding medium, the plasmon resonance wavelength of the silver nanoparticles is red-shifted to a wavelength of 800 nm. At this resonance wavelength a significant absorption enhancement in the composite layer was observed.

At five different depths in the absorber layer of a-Si:H solar cells, silver nanoparticles were embedded. It turned out that the efficiency of the solar cells with embedded silver nanoparticles was reduced significantly compared to cells without nanoparticles. The trends observed in the *EQE* curves suggest that a film of silver nanoparticles gives rise to recombination of photogenerated charge carriers. This divides the absorber layer into a top and bottom cell, effectively creating a quasi-tandem cell. This was confirmed by measuring the *EQE* under red and blue bias illumination.

ACKNOWLEDGMENTS

The authors would like to acknowledge Martijn Tijssen, Kasper Zwetsloot and Frans Tichelaar for their technical assistance and Janez Krč and Tristan Temple for their helpful discussions. Financial support from the NMP-Energy Joint Call FP7 SOLAMON Project (www.solamon.eu) is acknowledged.

REFERENCES

1. Miro Zeman, *Advanced Amorphous Silicon Solar Cell Technology*, in Thin Film Solar Cells: Fabrication, Characterization and Applications, p. 173-236, eds. J. Poortmans and V. Archipov, Wiley 2006.
2. J. Krč, M. Zeman, F. Smole and M. Topič, in Amorphous and Polycrystalline Thin-Film Silicon Science and Technology – 2006, edited by H.A. Atwater, Jr., V. Chu, S. Wagner, K. Yamamoto, H-W. Zan, (Mater. Res. Symp. Soc. Proc. **910**, Warrendale, PA 2006) A25-01.
3. H.A. Atwater and A. Polman, Nature Materials, **9**, 205-213 (2010)
4. E. Moulin et al., Journal of Non-Crystaline Solids, **354**, 2488-2491 (2008)
5. T.L. Temple et al., Solar Energy Materials and Solar Cells, **93**, 1978-1985 (2009)
6. G. Mie, Annalen der Physik, 25, 377-445 (1908)
7. C.F. Bohren and D.R. Huffman, *Absorption and scattering of light by small particles*, Wiley-VCH Verlag GmbH & Co. KGaA, Weinheim, 2004

Mater. Res. Soc. Symp. Proc. Vol. 1245 © 2010 Materials Research Society 1245-A07-21

Photonic Crystal Back Reflectors for Enhanced Absorption in Amorphous Silicon Solar Cells

Benjamin Curtin[1], Rana Biswas[1,2] and Vikram Dalal[1]
[1]Microelectronics Research Center; Dept. of Electrical and Computer Engineering, Iowa State University, Ames, Iowa 50011, U.S.A.
[2]Physics & Astronomy; Ames Lab, Iowa State University, Ames, Iowa 50011, U.S.A.

ABSTRACT

Photonic crystal back reflectors offer enhanced optical absorption in thin-film solar cells, without undesirable losses. Rigorous simulations of photonic crystal back reflectors predicted maximized light absorption in amorphous silicon solar cells for a pitch of 700-800 nm. Simulations also predict that for typical 250 nm i-layer cells, the periodic photonic crystal back reflector can improve absorption over the ideal randomly roughened back reflector (or the '$4n^2$ classical limit') at wavelengths near the band edge. The PC back reflector provides even higher enhancement than roughened back reflectors for cells with even thinner i-layers. Using these simulated designs, we fabricated metallic photonic crystal back reflectors with different etch depths and i-layer thicknesses. The photonic crystals had a pitch of 760 nm and triangular lattice symmetry. The average light absorption increased with the PC back reflectors, but the greatest improvement (7-8%) in short circuit current was found for thinner i-layers. We have studied the dependence of cell performance on the etch depth of the photonic crystal. The photonic crystal back reflector strongly diffracts light and increases optical path lengths of solar photons.

INTRODUCTION

Light trapping and advanced photon management are key techniques for efficiently harvesting solar photons and improving solar cell efficiency. Commonly used amorphous silicon and crystalline silicon solar cells suffer from the large absorption length of long-wavelength photons with energies just above the band edge.

For hydrogenated amorphous silicon (a-Si:H) the hole diffusion length (l_d) is ~300-400 nm, limiting the solar cell absorber layer thickness to less than l_d. For a-Si:H midgap cells, blue and green solar photons have absorption lengths (l_a) less than 250 nm[1,2], and are effectively absorbed in the thin absorber layer. l_a grows rapidly for $\lambda>600$ nm and exceeds 7 μm for photons near the band edge (750-800 nm). Thus red/near-IR photons are very difficult to absorb in thin a-Si:H layers. Light-trapping is necessary to harvest these long-wavelength photons. Similarly in c-Si solar cells, $l_a>10$ μm (for $\lambda>900$ nm) and harvesting of near-IR photons ($\lambda=900$-1100 nm) above the band edge is very difficult [3,4].

Conventional light trapping schemes utilize textured silver/aluminum-doped zinc oxide (Ag/ZnO:Al) back reflectors that scatter light within the a-Si:H absorber layer and increase the optical path length [5]. Although textured metallic back reflectors are widely used, they suffer intrinsic losses from surface plasmon modes generated at the granular metal-dielectric interface [6,7]. The plasmonic excitations in textured Ag/ZnO back reflectors have been extensively analyzed with spectroscopic ellipsometry [7], and it is proposed that a large fraction of light coupled into localized plasmonic modes at the Ag/ZnO interface can be re-radiated as scattered light in optimized textured back reflectors [7]. The idealized classical limit [2] of loss-less scattering (with an enhancement of $4n^2$[8]) is difficult to achieve, and it is estimated that optical path length enhancements of ~10 can be achieved in practice [9]. We develop an alternative light trapping technique with a photonic/plasmonic crystal back reflector, following advances in manipulation of light with diffractive scattering [10] and 2-d periodic crystals [4,11]. The solar cell architecture (Fig. 1a) consists of a triangular lattice photonic crystal (PC) on a silicon substrate with a conformal layer of Ag to generate the metallic PC. A thin e-transporting ZnO:Al layer is coated on Ag, followed by the a-Si:H n-i-p solar cell, and the top anti-reflecting indium tin oxide (ITO) transparent contact.

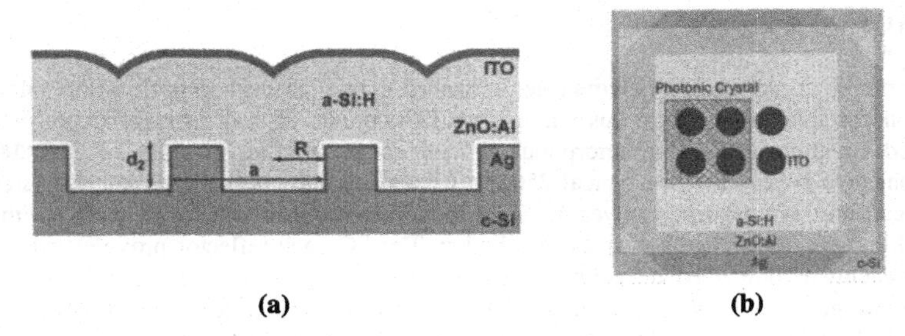

Figure 1. Schematics of the a) cross-sectional view of PC back reflector solar cell and the b) layout of the solar cell used to measure and compare flat and back reflector devices. The ITO top contacts define each of the a-Si:H devices.

APPROACH

To fabricate this solar cell, we first determined optimum parameters with rigorous scattering matrix simulations, where Maxwell's equations are solved in Fourier space for both polarizations [2,12]. From the scattering matrix S, we obtain the reflection and absorption [2,12]. Optimal simulated absorption occurs for a PC depth (d_2) of 250 nm, radius R/a ~ 0.30, and pitch a ~740 nm, coupled with a standard 65 nm ITO layer. The PC back reflector is assumed to be atomically flat in the simulation. An a-Si:H i-layer thickness (d) of 250 nm was used. There is a large simulated absorption of long wavelength photons (λ=650-780 nm) for the PC compared to the reference flat Ag reflector. Absorption maxima between 650-780 nm are caused by diffraction resonances (standing waves) in the absorber layer[4,10,11], where path length and dwell time are enhanced. The electromagnetic fields are concentrated within the PC. Below 600 nm, the crystal has little effect, since photons with $l_a<d$ are effectively absorbed without reaching the back surface. This metallic PC is much easier to fabricate than the previously proposed distributed Bragg reflector [2].

The layout of the a-Si:H devices on a single substrate is shown in Figure 1b. The optimized PC back reflector was fabricated adjacent to a flat region. After the a-Si:H n-i-p device was deposited on the substrate, ITO top contacts electrically isolated back reflector and flat reference devices for subsequent characterization. The device closest to the reference is referred to as PC1 and the outside device is PC2. The flat references established a baseline for the a-Si:H solar cell performance and allowed for an accurate measurement of the absorption enhancement introduced by the PC back reflector.

PROCESS DEVELOPMENT

We fabricated this optimized PC solar cell architecture on a c-Si substrate using photolithography and etching. The PC was patterned into 480 nm of photoresist and 80 nm of bottom antireflective coating. We used a PlasmaTherm 700 reactive-ion etching (RIE) system with a CF_4 and O_2 plasma to etch the PC into the c-Si wafer. The RIE etching parameters were optimized for photoresist to c-Si etch selectivity and sidewall etch anisotropy. After RIE, the remaining photoresist and antireflective coating were removed with an O_2 plasma.

A thin reflective Ag layer with a thickness of 50 nm was thermally evaporated on the PC at room temperature to avoid agglomeration from surface diffusion. A thin (~100 nm) ZnO:Al film was then sputtered at low temperature (150°C) to encapsulate the Ag and prevent surface roughening from Ag agglomeration.

An a-Si:H n-i-p solar cell was deposited using plasma enhanced chemical vapor deposition (PECVD) with device parameters similar to those on flat substrates. The wide bandgap n-layer was a-SiC:H with a thickness of roughly 200 nm. Although absorption is optimal with thinner n-layers, a minimum thickness of 100 nm was necessary to avoid electrical shorts. ITO top contacts were sputtered with a 65 nm thickness that was optimized for transparency, conductivity, and antireflective properties. Finally, the devices were annealed in atmosphere at 200°C for 30 minutes.

RESULTS

A scanning-electron microscope (SEM) characterized the back reflector structures and surface roughness between each processing step. The lattice pitch (Fig. 2a) was ~760 nm with excellent long range order across the 12 x 12 mm solar cell.

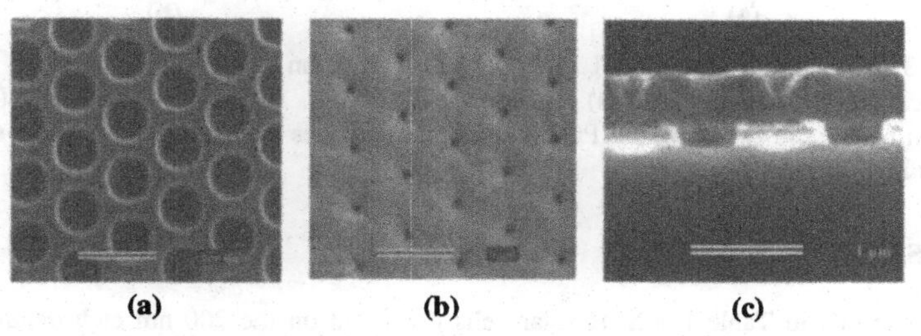

(a) (b) (c)

Figure 2. SEM images of the photonic crystal back reflector taken after: a) RIE etching and plasma cleaning and b) a-Si:H deposition and ITO sputtering. c) Cross-sectional SEM of the a-Si:H solar cell on the PC back reflector. All images have a 1 μm scale bar.

After ZnO:Al sputtering, the substrate was relatively uniform and showed only weak signs of Ag agglomeration and roughness. Conformal a-Si:H growth on the PC (Fig. 2b) resulted in a slightly non-uniform top surface with the spatial period of the underlying PC. Cross-sectional SEM images (Fig. 2c) showed the patterned top surface and a-Si:H filled cavities for two periods of the PC.

We deposited a-Si:H devices (E_G~1.75 eV) with identical process parameters on two separate PC back reflectors, with each adjacent to a flat reference on the same substrate. The PC back reflectors had etch depths of 200 and 250 nm and i-layer thicknesses of 220 nm and 290 nm, respectively. The i-layer thickness was verified with reverse bias C-V.External quantum efficiency (EQE) was measured to determine wavelength-dependent absorption and is shown in Fig 3. The ratio of photo-generated current within the device and a c-Si reference cell with known EQE is used to determine relative EQE of the device. EQE is then normalized to 90%. The short circuit current J_{SC} was estimated by summing the product of electron charge (q), EQE (A(λ)), and AM1.5 solar flux (Φ(λ)) at each measured wavelength:

$$J_{SC,EQE} = \sum_{\lambda=400nm}^{800nm} qA(\lambda)\Phi(\lambda) \quad (1)$$

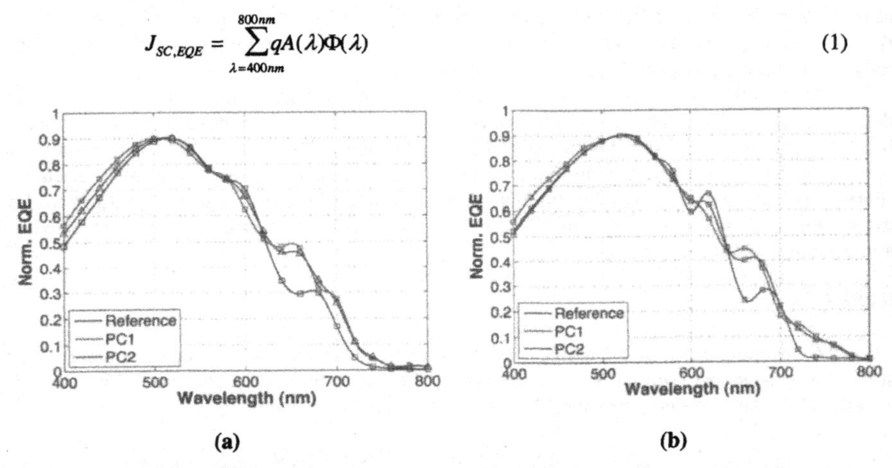

(a) (b)

Figure 3. EQE for similar a-Si:H solar cells deposited on photonic crystal back reflector substrates with PC etch depths of a) 200 nm and b) 250 nm. A flat reference device (squares) is shown beside the two different PC back reflector devices measured on the same substrate (PC1 and PC2).

DISCUSSION

As shown in Table I, a-Si:H solar cells deposited on the 200 nm etch depth PC back reflectors had an EQE J_{SC} of 13.22 mA/cm^2 and 13.17 mA/cm^2. The reference device was found to have an EQE J_{SC} of 12.27 mA/cm^2, which corresponded to a 7.7% and 7.3% increase in EQE J_{SC} due to the back reflector. The 250 nm etch depth PC back reflectors had a slightly larger EQE J_{SC} for the two photonic crystal back reflector devices and the reference, which were 13.30 mA/cm^2, 13.28 mA/cm^2 and 12.59 mA/cm^2 respectively. Despite the larger short circuit current, the overall improvement in absorption over the reference (~5.6%) was lower than the 200 nm PC back reflector. Although the optimal etch depth is near 250 nm, greater enhancement is found for thinner i-layers where the absorption edge is shifted towards shorter wavelengths. These gains are less than those observed with textured Ag/ZnO back-reflectors in a-SiGe:H devices, which showed ~30% improvement in photocurrent for optimal designs [13]. Similar a-Si:H devices deposited by our group on optimized etched Ag/ZnO showed a slightly larger, but comparable, increase in EQE J_{SC}. Further studies are necessary to determine the reason for this discrepancy.

Table I. Measured EQE J_{SC} data for flat and photonic crystal back reflector a-Si:H solar cells substrates with different etch depths and i-layer thicknesses.

	Etch Depth (nm)	i-layer Thickness (nm)	EQE J_{SC} (mA/cm^2)
PC1	200	220	13.22
PC2	200	220	13.17
Reference	-	220	12.27
PC1	250	290	13.30
PC2	250	290	13.28
Reference	-	290	12.59

EQEs are similar below 600 nm where photons are efficiently absorbed in the i-layer (Fig. 3). The PC devices showed enhanced collection for $\lambda>600$ nm compared to the flat reference devices. The ratio of the EQE for the PC back reflector to the reference device (Fig. 4) show considerable enhancement at near-infrared wavelengths. Enhancement from the PC increased with wavelength, but each etch depth had characteristic maxima and minima. At longer λ, the PC back reflector devices showed an enhancement of ~5-6 and the 250 nm etch depth device had the greatest enhancement of ~6.8 at 740 nm. Significantly, there is increase of absorption below 500 nm, which results from light trapping in the conformal top surface of the a-Si:H cell.

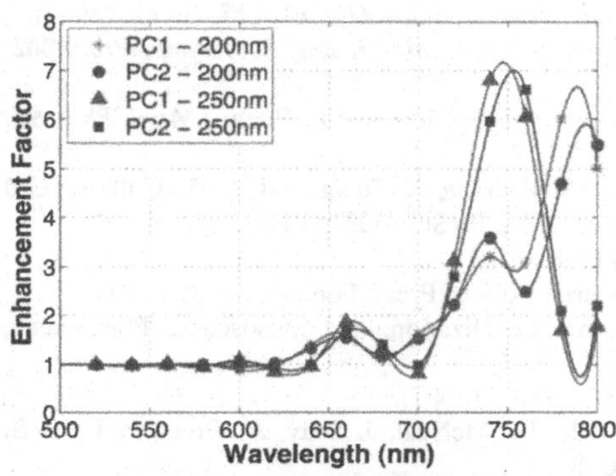

Figure 4. Enhancement factor for similar devices deposited on a photonic crystal back reflector with different etch depths (200 nm and 250 nm). PC1 and PC2 refer to the two PC back reflector devices.

In terms of device parameters, the open circuit voltage V_{OC} was nearly identical between the back reflector devices (~0.78 V) and less than the reference devices (~0.84 V). The fill factors decreased for greater etch depths, with the 200 nm etch depth back reflector devices having a slightly larger fill factor than the 250 nm etch depth devices. This reduction in V_{OC} and dependence of fill factor on etch depth might be due to defects and voids forming during the a-

Si:H deposition, resulting in parasitic resistive paths between the n and p-doped layers. A decrease in shunt resistance was observed in the PC back reflector devices compared to the references during I-V characterization. Recent efforts within our group [14] and elsewhere [15] have reduced V_{OC} and fill factor losses on patterned substrates with a decreased a-Si:H growth rate, which prevents the formation of voids around the textured back reflector.

CONCLUSIONS

We provide proof-of-concept for enhanced photon harvesting in thin film amorphous silicon solar cells with photonic crystals. Photonic crystal back reflectors are expected to have larger enhancements in nano-crystalline silicon and in multi-junction solar cells.

ACKNOWLEDGEMENTS

We thank K. Han, N. Chakravarty, S. Pattnaik, and M. Noack for assistance with samples. We thank D. Vellenga and the North Carolina State University Nanofabrication Center for photolithography. We acknowledge support from the NSF under grants ECCS-0824091, ECS0601377 and the Iowa Powerfund. The Ames Laboratory is operated for the Department of Energy by Iowa State University under contract No. DE-AC0207CH11385.

REFERENCES

1. A.S. Ferlauto, G. M. Ferreira, J. M. Pearce, C. R. Wronski, R. W. Collins, X. Deng, G. Ganguly, *J. Appl. Phys.* **92**, 2424 (2002).
2. D. Zhou and R. Biswas, *J. Appl. Phys.* **103**, 093102 (2008).
3. L. Zeng et al, *Appl. Phys. Lett.* **93**, 221105 (2008).
4. P. Bermel, C. Luo, L. Zeng, L.C. Kimerling, J. Joannopoulos, *Opt. Exp*, **15**, 16986, (2007).
5. B. Yan, J. M. Owens, C. Jiang, J. Yang and S. Guha, *Mater. Res. Soc. Symp. Proc.* **862**, A23.3.1 (2005).
6. J. Springer, A. Poruba, L. Mullerova, M. Vanecek, O. Kluth and B. Rech, *J. Appl. Phys.* **95**, 1427 (2004).
7. L. R. Dahal, D. Sainju, J. Li, J. A. Stoke, N. Podraza, X. Deng, and R. W. Collins, 33rd IEEE Photovoltaic Specialists Conference, 2008. PVSC 2008, pg.1-6.
8. E. Yablonovitch, *J. Opt. Soc. Am.* **72**, 899 (1982).
9. J. Nelson, The Physics of Solar Cells, (Imperial College Press, London, 2003), p. 279.
10. P. Sheng, Introduction to Wave Scattering, Localization, and Mesoscopic Phenomena, (Academic Press, Boston, 1995).
11. K. Catchpole, M. Green, *J. Appl. Phys.* **101**, 063105 (2007).
12. R. Biswas, C.G. Ding, I. Puscasu, M. Pralle, M. McNeal, J. Daly, A. Greenwald and E. Johnson, *Phys. Rev. B.* **74**, 045107 (2006).
13. J. Yang, B. Yan, G. Yue, and S. Guha, *Mater. Res. Soc. Symp. Proc.* **1153**, A13.02 (2009).
14. S. Pattnaik, R. Biswas, N. Chakravarty, J. Jin, L. Kaufman, V. Dalal, *IEEE PVSC* (2010).
15. V. E. Ferry, M. A. Verschuuren, H. B. T. Li, R. E. I. Schropp, H. A. Atwater, and A. Polman, *Appl. Phys. Lett.* **95**, 183503 (2009).

Novel Devices

Mater. Res. Soc. Symp. Proc. Vol. 1245 © 2010 Materials Research Society 1245-A08-01

A Highly Sensitive Integrated a-Si:H Fluorescence Detector for Microfluidic Devices

Toshihiro Kamei[1] and Amane Shikanai[1,2]
[1]National Institute of Advanced Industrial Science and Technology,
Tsukuba Central 2nd bldg. 1-1-1 Umezono Tsukuba, Ibaraki, 305-8568, Japan
[2]Present Address: Fujikura Ltd., 1440 Mutsuzaki, Sakura, Chiba, 285-8550, Japan

ABSTRACT

Most of micromachined and/or integrated fluorescence detectors suffer from high limit of detection (LOD) compared to conventional optical system that consists of discrete optical components, which is mainly due to higher laser light scattering of integrated optics rather than detector sensitivity. In this work, we have reduced background (BG) photocurrent of an integrated hydrogenated amorphous Si (a-Si:H) fluorescence detector due to laser light scattering by nearly one order magnitude, significantly improving a LOD. The detection platform comprises a microlens and the annular fluorescence detector where a thick SiO_2/Ta_2O_5 multilayer optical interference filter (>6 μm) is monolithically integrated on an a-Si:H pin photodiode. With a microfluidic capillary electrophoresis (CE) device mounted on the platform, the system is demonstrated to separate DNA restriction fragment digests (LOD: 58 pg/μL) as well as 2 nM of fluorescein-labeled single strand DNA (LOD: 240 pM) with high speed, high sensitivity and high separation efficiency. The integrated a-Si:H fluorescence detector exhibits high sensitivity for practical fluorescent labeling dyes as well as feasibility of monolithic integration with a laser diode, making it ideal for application to point-of-care microfluidic biochemical analysis.

INTRODUCTION

Microfluidics is the science and technology of systems that process or manipulate small (10^{-9} to 10^{-18} liters) amount of fluid, using channels with dimensions of tens to hundreds of micrometers[1]. Similar to the scaling law of a MOSFET that, as it gets smaller, it can switch faster and use less power, microfluidic approach can speed up biochemical analysis and require less reagent consumption. Representative examples include polymerase chain reaction (PCR) and capillary electrophoresis (CE). PCR can double DNA with a specific region in each thermal cycle. Twenty thermal cycles result in million ($=2^{20}$) fold amplification of DNA, but the total time is typically limited by heating and cooling speed. Therefore, a reduction of reagent volume from microliter to nanoliter has dramatically reduced time required for PCR due to the smaller thermal capacity. Similarly, sample plug in CE has dramatically reduced to a length of 100 μm by electrokinetic manipulation of fluid in a microchannel, achieving unprecedented high speed and high separation efficiency[2]. Microfluidic plumbing technology based on a silicone elastomeric membrane has been used to perform cell sorting and combinatorial screening of protein crystallization conditions[3,4]. Despite the rapid progress in microfluidic biochemical assays, a microfluidic biochemical assays require a high sensitivity bulky detection system such as confocal laser induced fluorescence detection system in order to detect small amount of sample fluid.

Fluorescence detection system has to be miniaturized and integrated in order to exploit potential point-of-care benefits of microfluidic biochemical analysis devices. In this regard, a-Si:H is an ideal choice of material for various reasons (a) it exhibits high sensitivity at the emission wavelength (500-600 nm) of most practical labeling dyes such as green fluorescence protein, DNA intercalators, ethidium bromide and fluorescein; (b) it exhibits low dark current due to a wide bandgap of a-Si:H, which is advantageous for low-noise measurement; (c) it can be monolithically integrated on a blue-green light-emitting InGaN laser diode due to its random structure and low temperature fabrication process; and (d) its manufacture is mature and inexpensive[5,6]. With this in mind, we have proposed an annular fluorescence detector in which an optical interference filter is monolithically integrated on an a-Si:H pin photodiode[5]. This allows laser light to pass through the detector and to irradiate a microchannel, enabling coaxial configuration of excitation source and detector. Such a configuration is advantageous for miniaturization and scalable to a detector array when combined with a vertical cavity surface emitting laser (VCSEL) diode. A high BG photocurrent due to laser light scattering, however, is observed in the integrated a-Si:H fluorescence detector, the noise level of which limits the LOD of 7 nM for fluorescein solution[7].

In this work, we have reduced the BG photocurrent by nearly one order magnitude and then performed a microfluidic separation of DNA restriction fragment digests and fluorescein-labeled single strand DNA (ssDNA) with high speed, high sensitivity and high separation efficiency.

EXPERIMENT

The microfluidic CE device was fabricated according to the previously published procedure[5]. Briefly, Borofloat glass wafers (76 mm dia., 1.1 mm thick) were micromachined by using HF wet chemical etching to produce separation channels 50 μm deep × 100 μm wide with cross injection arms. The etched plate was then thermally bonded to a blank glass plate. Prior to all electrophoretic operations, the channel walls were charge-neutralized with linear polyacrylamide, following a modified procedure of Hjerten coating[8]. Hydroxyethylcellulose (HEC) dissolved in a 1× Tris/acetate/EDTA buffer (TAE; 40 mM Tris base, 40 mM acetate, 1 mM EDTA) at 1.4 w/v % was used as the sieving matrix with 1 μM oxyazole yellow (YO) for separation of DNA restriction fragment digests, while a linear polyacrylamide sieving matrix was used for separation of fluorescein-labeled ssDNA. A *Hae*III digest of ϕX174 bacteriophage DNA was diluted with 1×TAE buffer to 25 ng/μL, while fluorescein-labeled ssDNA (20-mer) was diluted with water to 2 nM. Each DNA sample was placed in the sample reservoir, and an appropriate buffer was loaded into the waste, cathode, and anode reservoirs to make electrical contact with the platinum electrodes. The DNA restriction fragment digests were injected 30-60 s from the sample to waste reservoirs at 750 V/cm while applying a floating potential at the cathode and anode reservoirs. Separation was then performed at 170 V/cm while applying a back-bias electrical field of 110-180 V/cm between the injection cross point and both the sample and waste reservoirs. Separation of the ssDNA was also performed with similar conditions. The effective separation length was approximately 2.7 cm.

The cross-sectional structure of an a-Si:H integrated fluorescence detector is shown in Figure 1A. An a-Si:H pin photodiode was deposited at low temperature (250°C) by plasma enhanced chemical vapor deposition (PECVD) and patterned to form an annular shape using photolithography. After silicon oxide film was deposited on the a-Si:H pin photodiode by PECVD, a SiO_2/Ta_2O_5 multilayer optical interference filter was coated by ion-assisted deposition and patterned by a lift-off process using a polyimide/Si bilayer as a sacrificial layer. An optical micrograph of the top view of the fluorescence detector is shown in Figure 1B. Dark current of the fluorescence detector was approximately 1 pA at a reverse bias of 1V.

As is shown in Figure 1A, the integrated fluorescence detector and a microlens with a through hole were assembled by black anodized Al, forming a compact detection platform. A microfluidic CE device was mounted on the platform. In addition, a prism was optically coupled with the microfluidic CE devices with index-matching fluid. Incident laser light from an optically pumped frequency doubled VCSEL (488 nm, 110 μW) was introduced normal to the microfluidic CE through the detector and loosely focused on a CE channel. Fluorescence was collected and approximately collimated by the microlens and spectrally filtered with the integrated band-pass filter (511-615 nm, simulated Optical Density (OD) > 6 at 488 nm). The photocurrent generated in the a-Si:H photodiode with a reverse bias of 1 V was synchronized to the chopped laser light at 27 Hz and detected by a lock-in amplifier. The filtered LIA output (300 ms) was digitized at 20 Hz.

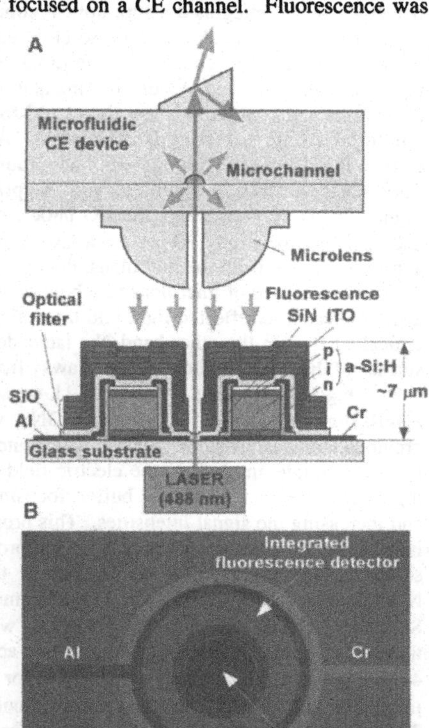

Figure 1. A Schematic cross-sectional view of a microlens and an integrated a-Si:H fluorescence detector, forming a compact platform where microfluidic CE device is mounted. **B** Optical micrograph of the top view of the integrated fluorescence detector.

RESULTS AND DISCUSSION

As is shown in Figure 1A, the design of the integrated a-Si:H fluorescence detector exploits the optical transparency of the glass substrate to combine a pinhole in the detector and vertical laser excitation through the detector, facilitating the construction of a coaxial excitation and detection module. This design would not be possible with a crystalline Si (c-Si) photodiode without an expensive deep-etcher to make a pinhole through the c-Si wafer, because the c-Si wafer is opaque to visible light. Furthermore, an annular Cr bottom electrode acts as an aperture for incident laser light to prevent scattered light from getting into the detector. Laser light, however, contains by at least ten orders magnitude more photons than fluorescence, so that it is fundamental challenge to reduce LOD in the current optical configuration, because even a tiny fraction of laser light, when scattered, could generate high BG photocurrent, limiting the LOD. Namely, the LOD of an integrated fluoresce detector is determined by the efficacy of an integrated optics rather than by detector sensitivity. In a previous work (with neither a through hole of microlens nor a prism on microfluidic CE device), the BG level measured from a buffer-filled channel was around 260 pA for a laser power of 110 μW. The noise level of the BG photocurrent limited LOD value (S/N = 3) of the fluorescence detector to be 7 nM for fluorescein solution in the 50-μm-deep channel at a flow rate of 800 μm/s[7]. This flow rate is comparable to the velocity of small DNA fragments (approximately 100 base pairs (bp)) under typical CE conditions, rendering a determined LOD value relevant to microfluidic CE analysis.

The high BG photocurrent due to laser light scattering is a limiting factor in the LOD value not only for our integrated fluorescence detector, but also for other integrated fluorescence detector. In the integrated setup, optical components such as the excitation source, the optical filter, the lens and the detector are so close each other that it is difficult to get rid of laser light scattering. A further reduction of the LOD value is essential and possible just by reducing BG photocurrent, evidenced by our experimental results that the LOD value of the a-Si:H photodiode placed in a confocal setup is less than 1 nM for a fluorescein solution where no BG photocurrent due to laser light scattering is detected[5]. Although it is virtually impossible to predict what causes laser light scattering, we have found that the through hole in the center of the fluorescence-collecting microlens and the prism on the microfluidic CE device, as shown in Figure 1A, is very effective: the BG photocurrent was reduced by nearly one order magnitude and now approximately 30 pA for a laser power of 110 μW. In addition to an elimination of autofluorescence from the microlens, the former shifts laser light scattering/reflection point from the convex surface of microlens to a bottom surface of microfluidic CE device, making incident angle of scattered/reflected light into the optical filter higher and increasing the efficiency of the optical filter. On the other hand, the latter deflects laser light scattering/reflection at the upper surface of the microfluidic CE device away from the detector.

Figure 2 presents a series of 11 consecutive microfluidic separations and detection of *Hae*III-digested ϕX174 bacteriophage DNA with the integrated fluorescence detector. In our previous work, DNA stock solution was diluted with water. Due to a difference of conductivity between sample and buffer, the electric field across sample plug inside a separation channel is higher than that in background buffer, focusing negatively charged DNA around zone boundary and increasing the signal intensities. This process is called stacking. With stacking, however, it is difficult to repeat separation with high reproducibility. Therefore, in this case, the DNA stock solution was diluted with 1×TAE buffer, the same as a run buffer, causing no stacking. Nevertheless, all 11 peaks of the DNA fragments could be successfully detected with sufficient S/N ratios. While 250 pg/μL in a previous work, an average LOD was reduced to 58 pg/μL in this work without stacking, meaning that an actual improvement of LOD is more than a factor of 4. The present LOD corresponds to 29 fg for a volume of sample plug (100 μm × 100 μm × 50 μm = 500 pL). Since average weight of double strand DNA is 660 Da/bp, the LOD would be 3×10^5 in the case of 100 bp DNA. Therefore, it would be possible to detect single molecular DNA when combined with PCR. The number of theoretical plates for the longest fragment was 28,000 while 95,000 on average for other fragments. These results demonstrate that the integrated fluorescence detector coupled with microfluidic CE devices provides high speed, high sensitivity, high separation efficiency and high reproducibility for point-of-care clinical, genetic and pathogen diagnosis.

Figure 3 presents the microfluidic separation and detection of 2 nM fluorescein-labeled ssDNA. With stacking, the ssDNA has been detected with LOD of 240 pM. Sanger DNA sequencing, a workhorse of a human genome project, comprises many steps, but the final step is electrophoretic separation of DNA sequencing products, that is, ssDNA, with stacking. Each DNA sequencing product is labeled with an energy transfer fluorescent reagent that involves

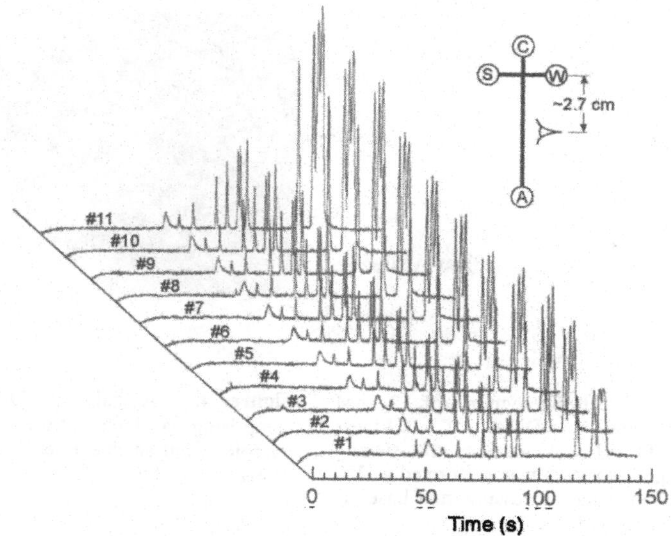

Figure 2. A series of 11 consecutive separations of the HaeIII digest of ϕX174 bacteriophage DNA (25 ng/μL) with the integrated fluorescence detector. 1.4 w/v% of HEC dissolved in the 1× TAE buffer was used as the sieving matrix and the separation was performed at 170 V/cm. On-column fluorescent labeling was accomplished with the DNA intercalating dye, oxyazole yellow (YO, 1 μM). S, W, C and A stand for sample, waste, cathode and anode reservoirs, respectively. Neither data processing nor baseline subtraction has been done.

fluorescein as a donor molecule and its concentration is around 1 nM. Therefore, this result demonstrates feasibility of DNA sequencing with the integrated fluorescence detector.

Finally, a palm-top laser-induced fluorescence (LIF) detection module (120×100×25 mm) comprising a InGaN laser diode (473 nm), the microlens and the integrated fluorescence detector has also been constructed, as shown in Figure 4. Neither the through hole of the microlens nor the prism on the microfluidic CE device is adopted in the palmtop LIF module so that the BG photocurrent is still high. Nevertheless DNA restriction fragment digests (100 ng/μL) has been successfully detected.

CONCLUSIONS

BG photocurrent of the integrated fluorescence detector due to laser light scattering has been reduced by nearly one order magnitude by making a through hole in the center of the microlens and placing the prism on the microfuidic CE device. With the microfluidic CE device mounted on the integrated detection platform, the system is demonstrated to separate and detect DNA restriction fragment digests with LOD of 58 pg/μL and 2 nM of fluorescein-labeled ssDNA with LOD of 240 pM, demonstrating feasibility of single molecular DNA detection combined with PCR and DNA sequencing, respectively. Low dark current of the integrated fluorescence detector has the potential for a further reduction of LOD. Moreover, a palm-top

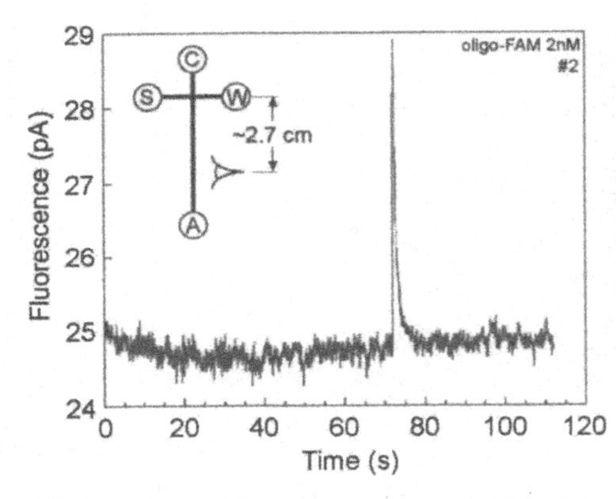

Figure 3. Electropherogram of 2 nM fluorescein-labeled ssDNA. Linear polyacrylamide gel was used as the sieving matrix and the separation was performed at 220 V/cm. Neither data processing nor baseline subtraction has been done.

Figure 4. A palmtop LIF module comprising a InGaN laser diode, a microlens and an integrated fluorescence detector.

LIF detection module comprising an InGaN laser diode, the microlens and the integrated fluorescence detector has been constructed.

ACKNOWLEDGMENTS

This work is partly supported by the Japan Society for the Promotion of Science (JSPS) through its "Funding Program for World-Leading Innovative R&D on Science and Technology (FIRST Program)."

REFERENCES

1. G. M. Whitesides, Nature **442**, 368 (2006).
2. D. J. Harrison, K. Fluri, K. Seiler, Z. H. Fan, C. S. Effenhauser, and A. Manz, Science **261**, 895 (1993).
3. C. L. Hansen, E. Skordalakes, J. M. Berger, and S. R. Quake, Proceedings of the National Academy of Sciences of the United States of America **99**, 16531 (2002).
4. A. Y. Fu, H. P. Chou, C. Spence, F. H. Arnold, and S. R. Quake, Anal. Chem. **74**, 2451 (2002).
5. T. Kamei, B. M. Paegel, J. R. Scherer, A. M. Skelley, R. A. Street, and R. A. Mathies, Anal. Chem. **75**, 5300 (2003).
6. A. C. Pimentel, A. T. Pereira, D. M. F. Prazeres, V. Chu, and J. P. Conde, Appl. Phys. Lett. **94**, 164106 (2009).
7. T. Kamei and T. Wada, Appl. Phys. Lett. **89** (2006).
8. S. Hjerten, J. Chromatogr. **347**, 191 (1985).

Reviewing photo-sensing devices using a-SiC based materials

M. Vieira[1,2,3], M. Fernandes[1,2], P. Louro[1,2], A. Fantoni[1,2], M. A Vieira[1,2], J. Costa[1,2], M. Barata[1,2]
[1]Electronics Telecommunication and Computer Dept. ISEL, R.Conselheiro Emídio Navarro, 1949-014 Lisboa, Portugal Tel: +351 21 8317290, Fax: +351 21 8317114, mv@isel.ipl.pt
[2]CTS-UNINOVA, Quinta da Torre, 2829-516, Caparica, Portugal.
[3]DEE-FCT-UNL, Quinta da Torre, 2829-516, Caparica, Portugal.

ABSTRACT

In this paper a double pi'n/pin a-SiC:H voltage and optical bias controlled device is presented and it behavior as image and color sensor, optical amplifier and multiplex/demultiplex device discussed. The sensing element structure (single or tandem) and the light source properties (wavelength, intensity and frequency) are correlated with the sensor output characteristics (light-to-dark sensitivity, resolution, linearity, bit rate and S/N ratio). Depending on the application, different readout techniques are used. When a low power monochromatic scanner readout the generated carriers the transducer recognize a color pattern projected on it acting as a color and image sensor. Scan speeds up to 10^4 lines per second are achieved without degradation in the resolution. If the photocurrent generated by different monochromatic pulsed channels is readout directly, the information is multiplexed or demultiplexed. It is possible to decode the information from three simultaneous color channels without bit errors at bit rates per channel higher than 4000bps. Finally, when triggered by appropriated light, it can amplify or suppress the generated photocurrent working as an optical amplifier. An electrical model is presented to support the sensing methodologies. Experimental and simulated results show that the tandem devices act as charge transfer systems. They filter, store, amplify and transport the photogenerated carriers, keeping its memory (color, intensity and frequency) without adding any optical pre-amplifier or optical filter as in the standard p-i-n cells.

INTRODUCTION

Whether we hope to view images or count photons, we use devices that work by absorbing photons and turning them into information. Traditional optics has relied heavily on the human eye for the evaluation of light distributions, using photographic films as an intermediate storage medium, when necessary. With the advent of computers and digital image processing, imaging electronic sensors have become indispensable tools. The current need for communication also demands the transmission of huge amounts of information. To increase the capacity of transmission and allow bidirectional communication over one strand fiber, wavelength-division multipexing (WDM) is used [1]. WDM systems have to accomplish the transient color recognition of two or more input channels in addition to their capacity of combining them onto one output signal without losing any specificity (wavelength and bit rate). Only the visible spectrum can be applied when using polymer optical fiber for communication. So, the demand of new optical processing devices is a request.

Three forces drive the evolution: Advances in technology determine what is possible; algorithms determine what is practical, and applications determine what is desirable. If we wanted to generalize about areas in which detectors can improve, we could talk about collecting information more quickly and efficiently, at more wavelengths, with more compact and user-friendly devices or interfaces. All of these devices use essentially the same light sensing mechanism. Photons penetrating a depletion region generate electron-hole pairs. These carriers are swept away by the electric field across the depletion region and generate a small transverse photocurrent. Except under very bright conditions, it is not possible to use this photocurrent directly. Thus, in order to achieve a reasonable signal-to-noise ratio, these currents are usually integrated to produce an accumulated charge output.

Amorphous silicon-carbon (a-SiC:H) is a material that exhibits excellent photosensitive properties. This feature together with the strong dependence of the maximum spectral response with the applied bias has been intensively used for the development of color devices. Various structures and sequences have been suggested [2, 3, 4, 5, 6]. In our group efforts have been devoted towards the development of a new kind of color sensor. Large area hydrogenated amorphous silicon single and stacked p-i-n structures with low conductivity doped layers were proposed as color Laser Scanned Photodiode (LSP) image sensors [7, 8, 9]. These sensors are different from the other electrically scanned image sensors as they are based on a single sensing element with an opto-mechanical readout system. No pixel architecture is needed. The advantages of this approach are quite obvious like the feasibility of large area deposition and on different substrate materials (e.g. glass, polymer foil, etc.), the simplicity of the device and associated electronics, high resolution, uniformity of measurement along the sensor and the cost/simplicity of the detector. The design allows a continuous sensor without the need for pixel-level patterning, and so can take advantage of the amorphous silicon technology. It can also be integrated vertically, *i. e.* on top of a read-out electronic, which facilitates low cost large area detection systems where the signal processing can be performed by an ASIC chip underneath.

In this paper a double pi'n/pin a-SiC:H heterostructure with two optical connections for light triggering in different spectral regions is presented and its behavior as color image sensor and multiplex/demultiplex device is discussed. Electrical models are present to support the sensing methodologies. Experimental and simulated results show that the tandem devices act as charge transfer systems. They filter, store and transport the photogenerated carriers, keeping their memory (color, intensity and frequency) without adding any optical pre-amplifier or optical filter as in the standard p-i-n cells.

EXPERIMENTAL DETAILS

Device configuration and sample preparation

Voltage controlled devices, were produced by PECVD in different architectures, as displayed in Figure 1, and tested for a proper fine tuning in the visible spectrum.

The simplest configuration is a p-i-n photodiode (#1) where the active intrinsic layer is a based on an a- Si:H thin film. In the other two (#2), the active device consists of a p-i'(a-SiC:H)-n / p-i(a-Si:H)-n heterostructures (#2a). To test the efficiency of the internal n-p junction, a third transparent contact was deposited in-between (#2b). The thickness (200nm) and the optical gap (2.1 eV) of all a-SiC:H intrinsic layers (i'-) are optimized for blue collection and red transmittance. The thickness (1000 nm) of the a-Si:H i-layers was adjusted to achieve full absorption in the green and high collection in the red spectral ranges. As a result, both front and back diodes act as optical filters confining, respectively, the blue and the red optical carriers, while the green ones are absorbed across both [10]. The deposition conditions of the i- and i'- intrinsic layers were kept constant in all the devices. Both the intrinsic layers present good optoelectronic properties with conductivities between 10^{-11} and $10^{-9}\Omega^{-1}$ cm^{-1} and photosensitivity higher than 10^4 under AM1.5 illumination (100 mW/ cm^2) for the i-layer. To decrease the lateral currents which are crucial for device operation [11, 12], low doping levels were used and methane was added during the deposition process. The doped layers (20 nm thick) have high resistivity (>$10^7 \Omega$cm) and optical gaps around 2.1 eV. Transparent contacts have been deposited on front and back surfaces to allow the light to enter and leave from both sides.The back contact defines the active area of the sensor (4×4 cm^2). The front and back contacts are based on ZnO:Al or ITO and have an average transmission around 80% from 425 nm to 700 nm and a resistivity around 9×10^{-4} Ωcm.

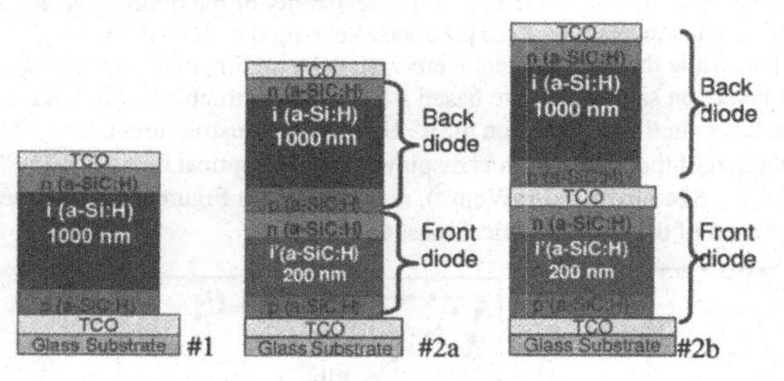

Figure 1. Sensor element configuration.

Spectral photocurrent

The characterization of the devices was performed through the analysis of the photocurrent dependence on the applied voltage and spectral response measurements, under different optical and electrical bias conditions. The responsivity was obtained by normalizing the photocurrent to the incident flux. To suppress the *dc* components and achieved good signal to noise ratio, all the measurements were performed using the lock-in technique. Figure 2a displays the spectral photocurrent for different applied bias, the internal transparent contact was kept floating in all measurements. In Figure 2b the spectral photocurrent, is displayed for the front and the back diodes, the internal transparent contact was used in all the measurements.

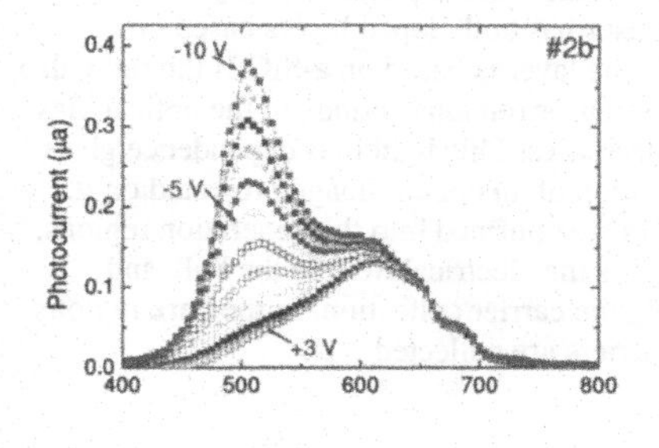

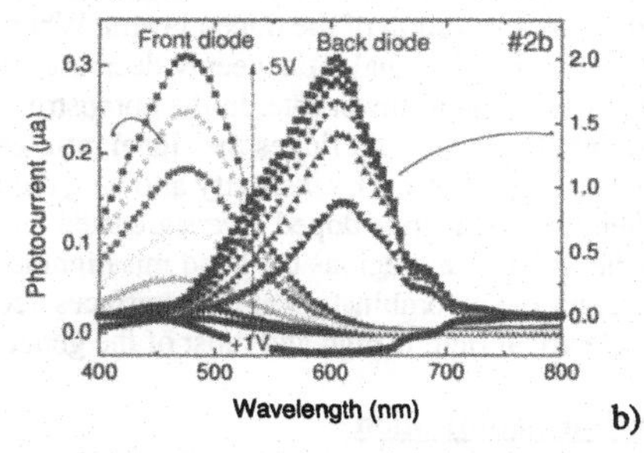

Figure 2. a) p-i'-n-p-i-n spectral photocurrent under different applied voltages, b) Front, p-i' (a-SiC:H)-n, and back, p-i (a-Si:H)-n spectral photocurrents under different applied bias.

Results confirm that the front and back photodiodes act, separately, as optical filters. The front diode cuts the wavelengths higher than 550nm while the back one cuts wavelengths lower than 500 nm. Each diode, separately, presents the typical responses of single p-i-n cells with intrinsic layers based, respectively, on a-SiC:H or a-Si:H materials. Since the current across the device has to remain the same it is clearly observed (Figure 2a) the influence of both front and back diodes modulated by its series connection through the internal n-p junction.

Light-to-dark sensitivity

To improve the light-to-dark sensitivity, in sensor #1, the resistivity of the doped layers and the optical gap were optimized. Three samples were produced keeping the deposition conditions of the i-layers constant while the doped layers were varied, by adding methane during the deposition process. All the layers on sample #1a are based a-Si:H (homostructure), while the p-layer in #1b and the p- and n-layers in #1 are based on a-SiC:H alloy (heterostructures). In Figure 3 the sensitivity as a function of the wavelength is displayed without optical bias (Φ_L=0) and under uniform illumination (λ_L=524 nm, Φ_L=2 mWcm^{-2}), respectively. In Figure 4 we plot the light to dark sensitivity as a function of the applied optical bias, Φ_L.

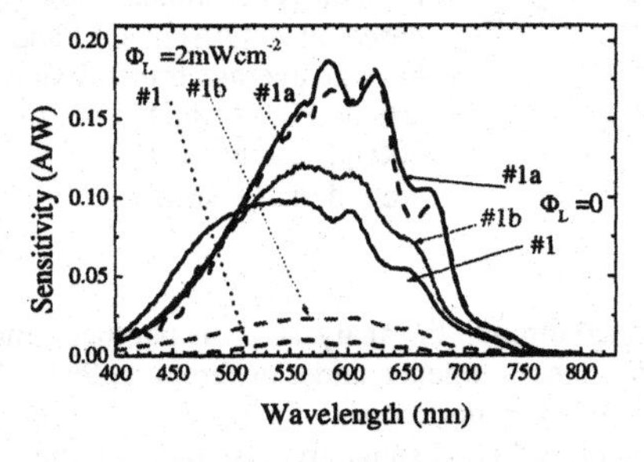

Figure 3. Spectral sensitivity with (dash) and without (solid) applied optical bias.

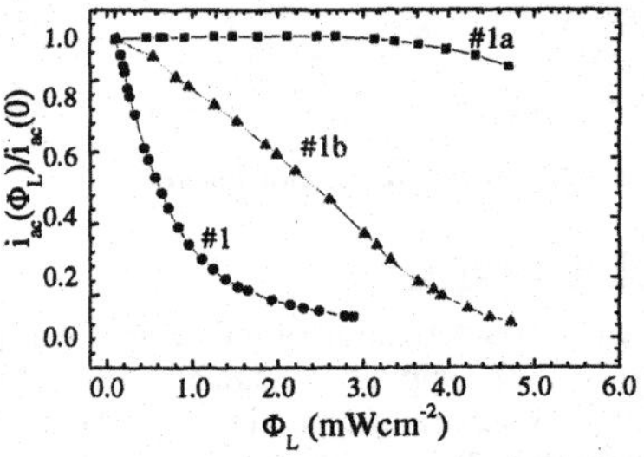

Figure 4. $i_{ac}(\Phi_L)/i_{ac}(0)$ ratio dependence with Φ_L.

Data reveal that when wide band gap doped layers are used the sensitivity is lower and decreases significantly with the optical bias. The light-to-dark ratio depends strongly on the carbon concentration of the doped layers. When the sensor has both doped layers based on a-SiC:H layers the signal ratio steeply decreases. If only one layer is based on a-SiC:H the ratio also decreases but at a slower rate. In the homostructure the sensor remains "blind" to the optical bias and only at higher light fluxes the signal ratio gently decreases. This light bias dependence gives the sensor light-to-dark sensitivity allowing the recognition of an optical image projected on it. When low conductive doped layers are used the carriers are confined into the generation regions. At the illuminated regions the band misalignment reduces the electrical field in the bulk and increases the recombination at the interfaces decreasing the carrier collection. In the dark regions the electrical field is high and most of the generated carriers are collected.

Optical amplification

We have shown that in a #2-like configuration a self biasing effect occurs under an imbalanced photogeneration [13]. When an external electrical bias (forward or reverse) is applied to the structure, it mainly influences the field distribution within the less photo excited sub-cell. If compared with the electric field profile under thermo-dynamical equilibrium conditions, the field under illumination is lowered in the most absorbing cell (self forward bias effect) while the less absorbing reacts by assuming a reverse bias configuration (self reverse bias effect). Thus, opposite behaviours are observed under red and blue background lights while under green irradiation the redistribution of the field profile is balanced between the two sub-cells.

To analyze the self bias effect under transient conditions and uniform irradiation, three monochromatic pulsed lights (input channels): *red* (R: 626 nm; 51μW/cm^2), *green* (G: 524 nm; 73μW/cm^2) and *blue* (B: 470nm; 115μW/cm^2) illuminated separately the device (#2a). Steady state *red* (λ_R=626 nm;Φ_R=102 μW/cm^2), *green* (λ_G=524 nm; Φ_G=71 μW/cm^2) and *blue* (λ_B=470 nm; Φ_B=293 μW/cm^2) optical bias were superimposed separately and the photocurrent generated measured at -8V and +1 V. In Figure 5 the signal is displayed for each monochromatic channel. The photocurrent without optical bias was normalized to the unit.

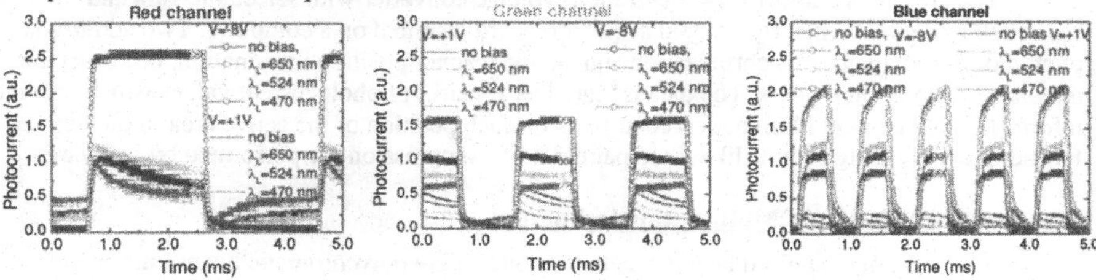

Figure 5. Input red, green and blue channels under negative and positive bias, with and without red, green and blue steady state optical bias (λ_L).

Results show that under negative bias the blue optical bias enhances the red (α_R=2.25) and the green (α_G=1.5) channels and quench the blue one (α_B=0.8). The red bias has an opposite behavior, it reduces the red and green channels and amplifies de blue (α_R=0.9, α_G=0.5, α_B=2.0). The green optical bias reduces the green channel (α_G=0.75) keeping the other two almost unchangeable.

The reinforcement of the electric field in the back diode, under blue irradiation, increases the collection of the carriers generated by the red channel and decreases the blue one while, under red bias, the electric field is enhanced in the front diode and an opposite collection behavior is observed. The green bias absorption is balanced in both front and back diodes; consequently, the green channel collection is reduced while the red and blue collections are almost insensitive to the green irradiation.

This optical nonlinearity makes the transducer attractive for optical communications. As it will be explain in the next sections, this nonlinear effect under transient conditions will be used to distinguish a wavelength, to read a color image, to amplify or to suppress a color channel or to multiplex or demultiplex an optical signal.

READOUT TECHNIQUES

Laser scanned photodiode (LSP) technique

An optical image is projected onto the active surface from the glass side and is scanned by sequentially detecting scene information at discrete XY coordinates. A 633 nm low power solid-state laser was used as scanner. The beam deflection is controlled by a two axis deflection system capable of high speed scan. The read-out of the injected carriers is achieved by measuring, under appropriated applied voltages, the *ac* component of the generated current, i_{ac}, which depends on the intensities of both the image and the scanner [11]. This component can be analytically described by the equation 1, where $R(\Phi_L)$ is the small signal responsivity at a given illumination

Φ_L, $\Phi_S(t)$ is the average power density of the moving scanner beam and d its diameter. The signal-to-noise power ratio (S/N) depends on both components of the current ($I=i_{ac}+I_{DC}$) and is given by equation 2, where Δf is the bandwidth and R_0 the sensor resistance at zero voltage bias.

$$i_{ac} = R(\Phi_L)\cdot\Phi_S(t)\cdot\frac{\pi d^2}{4} \quad (1) \qquad\qquad S/N = \frac{i_{ac}^2}{(4kT/R_0 + 2q.I_{DC})\Delta f} \quad (2)$$

The current is amplified by a current to voltage converter with selectable gain and converted to digital format by a signal acquisition card installed on a computer. Two additional photodiodes provide the synchronization signals for scanner position information, necessary for the image restoration process. The data is stored as a matrix of photocurrent values which provide information about local illumination conditions on each position of the active area of the device. Further processing algorithms like fixed pattern noise suppression are performed by software.

Wavelength-Division (de)Multiplexing techniques

Monochromatic pulsed beams together or one single polychromatic beam (mixture of different wavelength) impinge on the device and are absorbed, according to their wavelength (Figures 2 and 5). By reading out, under appropriated electrical bias, the photocurrent generated by the incoming photons, the input information is electrically multiplexed or demultiplexed. In the multiplexing mode the device faces the modulated light incoming together (monochromatic input channels). The combined effect of the input signals is converted to an electrical signal, via the device, keeping the input information (wavelength, intensity and modulation frequency). In the demultiplexing mode a polychromatic modulated light beam is projected onto the device and the readout performed by shifting between forward and reverse bias in order to tune the different color channels.

THREE OPTICAL TRANSDUCERS INTO ONE PHOTODETECTOR

LSP image and color transducer

In order to analyze the behavior of the single configuration under different voltages, a pattern composed by dark and illuminated regions was projected on device #1.

In Figure 6 the scans of the image under forward, reverse and zero bias voltage are represented. The image brightness (2μWcm^{-2}), scanner intensity (two order of magnitude lower) and spot diameter (50 μm) were kept constant. Results show that under reverse bias the image sensitivity and the dynamic range are higher than in short circuit or forward bias. In reverse mode a dynamic rate two orders of magnitude higher with a sensitivity of 6 mA/W and a responsivity of 17 μWcm^{-2} was achieved. As a possible application in Figure 6b we also display a grayscale photo and a fingerprint representation acquired under short circuit and Φ_L =10 μWcm^{-2}. No image processing algorithms were used.

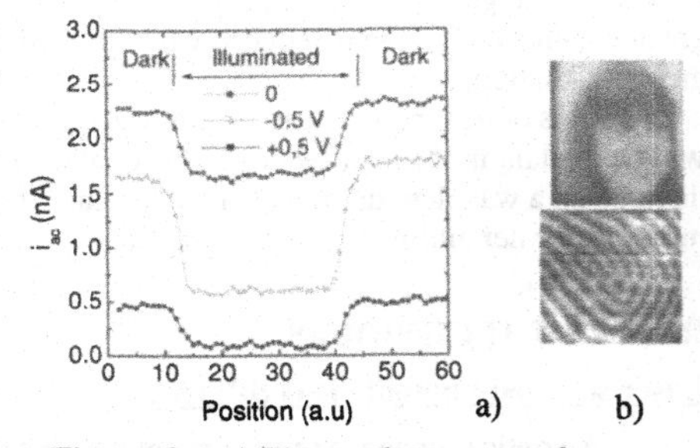

Figure 6. a) The *ac* photocurrent for one dimension scans under forward, reverse and zero bias. b) Grayscale photo and fingerprint representations.

This image presents a good contrast and a resolution around 30 μm showing the potential of these devices for biometric applications.

In Figure 7 the image output signal [14] defined as the difference between the *ac* component of the photocurrent in dark (Φ_L=0) and under illumination (λ_L=524 nm, 626 nm and Φ_L=10 μWcm^{-2}) is displayed as a function of the electrical bias.

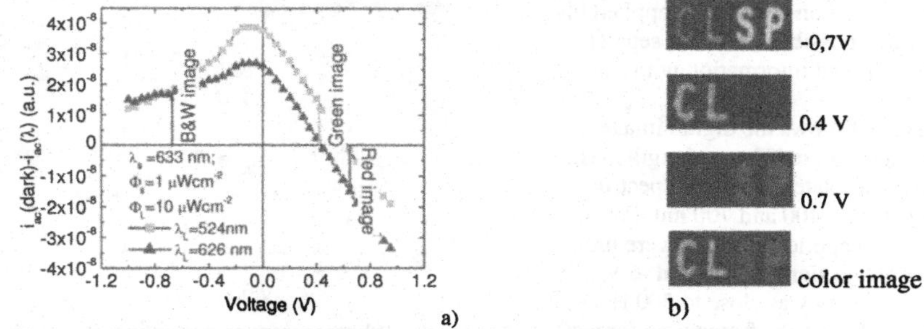

Figure 7. a) Image intensity as a function of the electrical bias under different optical bias. b) A picture image of the sensor acronym CLSP with green "CL" and red "SP".

Data show that at -0.7 V, the image intensity for the red or the green image presents the same magnitude and signal. No color information can be extracted at this voltage, which leads to a black and white image. In this mode the brightness of the image is proportional to the output signal (i_{ac}), which gives to the sensor the ability of acquiring monochrome gray level images (see Figure 6b). Color information can only be obtained under forward bias. By tuning the voltage to 0.4 V the red signal is the same as in dark, the red image is suppressed allowing the green image recognition. The red image information is obtained at 0.7 V, where the green image signal goes down to zero. Combining the signal information at these voltages (see arrows in figure) enables the reconstruction of the color image without the need of the usual color filters or pixel architecture. A picture image of the sensor acronym *CLSP* (Color Laser Scanned Photodiode) with green *"CL"* and red *"SP"* was projected onto the sensor at the appropriated applied voltages. The full color image was obtained by combining the information. The result is shown in Figure 7b.

In #2a configuration full color detection is attempted based on spatially separated absorption of the red, green and blue photons (Figure 2). In Figure 8 the photocurrents (i_{ac}) generated by the scanner (λ_S) under different steady-state illumination (λ_L) conditions are displayed. At the acquired voltages the image output signals are shown as inserts. Here the same *green*, *red* and *blue* pictures (**5**) were projected, one by one, onto the front diode and acquired through the back one with a moving red scanner. The line scan frequency was close to 1 kHz and no algorithms where used during the image restoration process. For readout time of 1 ms the frame time, for a 50 lines image, is around 50 ms. Results show that under red irradiation or in dark (without optical bias) the photocurrent generated by a red scanner is independent on the applied voltage. Under blue/green irradiation it decreases as the applied voltage changes from reverse to forward bias being higher under blue than under green irradiation. As expected from Figure 5 the main difference occurs in the green spectral range. It is interesting to notice that around -2 V the collection with or without green optical image are the same, leading to the rejection of the green image signal. Taking the signal without bias as a reference, and tuning the voltages to -2 V, the red and blue signal are high and opposite and the green signal suppressed allowing blue and red color recognitions. The green information is obtained under slight forward bias (+1 V), where the blue image signal goes down to zero and the red remains constant. Readout of 1000 lines per second was achieved allowing continuous and fast image sensing, and color recognition without the need of the usual color filters or pixel architecture. By sampling the absorption region at different applied bias voltages it was possible to extract separately the RGB integrated information with a good rejection ratio.

Figure 9 shows the digital image using as optical image a graded wavelength mask (rainbow) to simulate the visible spectrum in the range between 400 and 700 nm. For image acquisition two applied voltages were used to sample the image signal: + 1V and -6 V. The line scan frequency was close to 500 Hz. For a readout time of 2 ms the frame time for a 40 lines image takes around 80 ms. The algorithm used for image color reconstruction took into account that at -6 V the positive signals correspond to the blue/green contribution and the negative ones to the red inputs. The green information was extracted from the image signal sampled at +2 V, where the blue signal is almost suppressed (Figure 8) and the green and red signals are negative. The combined integration of this information allows recording full range of colors at each location instead of just one color at each point of the captured image as occurs with the CCD image sensors.

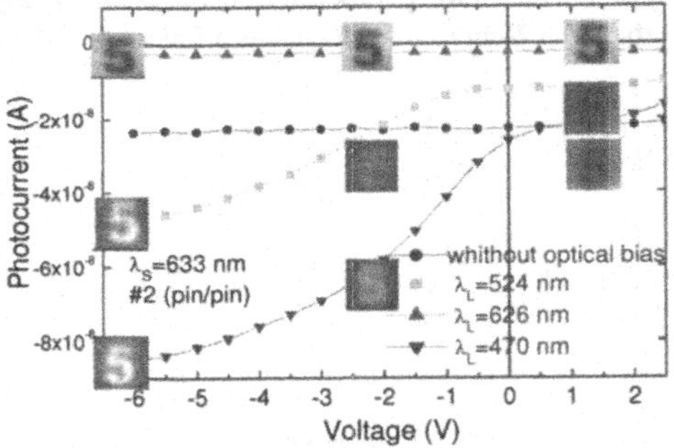

Figure 8. Photocurrent as a function of the applied voltage without and under blue, green and red irradiation. The same (5) RGB picture @ -6V, -2V and +1V are shown as inserts.

Figure 9. Digital image representation of the rainbow picture acquired with device #2a.

Wavelength-Division (de) Multiplexing transducer

Figure 10 displays the input and the multiplexed signals under negative (-8V) and positive (+1V) electrical bias. As expected from Figures 2, 5 and 8, the input red signal remains constant while the blue and the green ones decrease as the voltage changes from negative to positive. The output multiplexed signal, obtained with the combination of the three optical sources, depends on both the applied voltage and on the ON-OFF state of each input optical channel. Under negative bias, the multiplexed signal presents eight separate levels. The highest level appears when all the channels are ON and the lowest if they are all OFF. Furthermore, the levels ascribed to the mixture of three or two input channels are higher than the ones due to the presence of only one (R, G, B). Optical nonlinearity was detected as the sum of the input channels (R+B+G) is lower than the correspondent multiplexed signals (R&G&B). This optical amplification occurs when the red, the green or both channels are ON and suggests capacitive charging currents due to the time-varying nature of the incident lights. Under positive bias the levels were reduced to one half since the blue component of the combined spectra falls into the dark level, the red remains constant and the green component decreases (Figures 3, 5 and 8).

To recover the transmitted information at 2000bps per channel, the multiplexed signal was divided into time slots. A demux algorithm was implemented in Matlab that receives as input the measured photocurrent and derives the sequence of bits that originated it. The algorithm makes use of the variation of the photocurrent instead of its absolute intensity to minimise errors caused by signal attenuation. A single linkage clustering method is applied to find automatically eight different clusters based on the measured current levels in both forward and reverse bias. This calibration procedure is performed for a short calibration sequence. Each cluster is naturally bound to correspond to one of the known eight possible combinations of red, green and blue bits. Following this procedure the sequence of transmitted bits can be recovered in real time by sampling the photocurrent at the selected bit rate and finding for each sample the cluster with closest current levels. Using this simple key algorithm the independent red, green and blue bit sequences were decoded as: R[01111000], G[10011001] and B[10101010], as shown on the top of Figure 10, which are in exact agreement with the signals acquired for the independent channels.

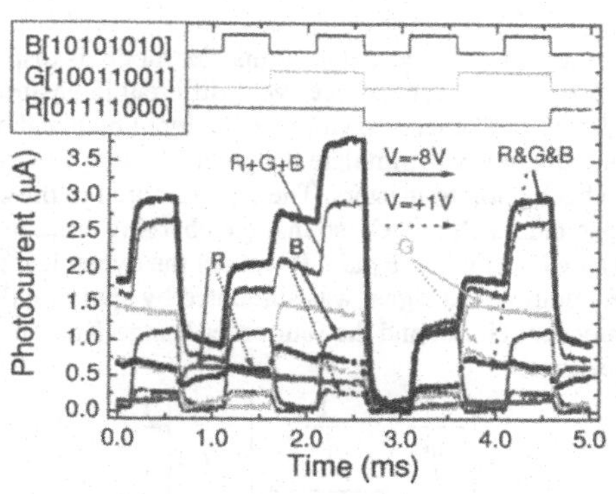

Figure 10. Single (R, G and B) and combined (R&G&B) signals under -8V (solid arrows) and +1V (dotted arrows).

THEORETICAL MODEL AND VALIDATION

Based on the experimental results and device configuration an electrical model was developed [15]. The silicon-carbon pi'npin device was considered as a monolithic double p-i-n photodiode with two optical connections for light triggering, respectively the front and the back diodes. Operation is explained in terms of the compound connected phototransistor equivalent model displayed in Figure 11a. Light color penetration across the sensor is also shown.

The two-transistor model (Q_1-Q_2) is obtained by bisecting the two middle layers in two separate halves that can be considered to constitute *pinp* (Q_1) and *npin* (Q_2) phototransistors separately.

The first step was to determine the *dc* biasing condition for each stage, followed by *ac* analysis using the proper equivalent circuits. When the pi'npin device is reverse-biased, the base-emitter junction of both transistors are inversely polarized and conceived as phototransistors, taking, in this way, advantage of the amplifier action of adjacent collector junctions which are polarized directly. This results in a current gain proportional to the ratio between both collector currents. Under positive bias the internal junction becomes reverse-biased. If not triggered ON it is nonconducting, when turned ON by light it conducts like a photodiode, for one polarity of current. The circuit

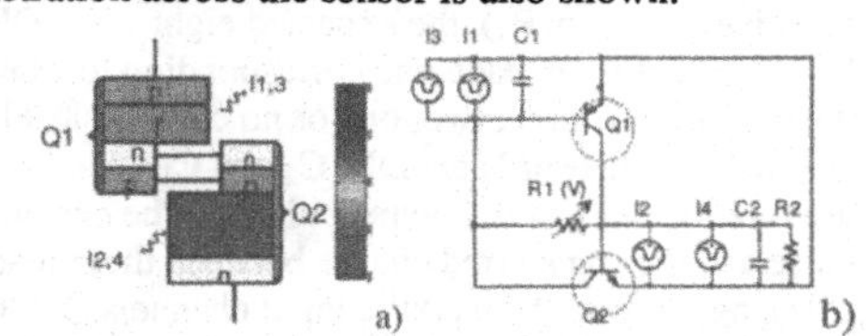

Figure 11. a) Two-transistor model; b) *ac* circuit representation.

representation of the *ac* two-transistor model is displayed in Figure 11b. To trigger the device four square-wave current sources with different intensities are used; two of which, I_1 and I_2, with different frequencies to simulate the input blue and red channels and other two, I_3 and I_4, with the same frequency to simulate the green channel due to its asymmetrical absorption across both front and back phototransistors. The charge stored in the space-charge layers is modeled by the capacitances C_1 and C_2, so that the charging currents across the reversed junctions ($i_{1,2} = C_{1,2}\,dv/dt$) are considered. R_1 and R_2 model the dynamical resistances under different *dc* bias conditions. The multiplexed signal was simulated by applying the Kirchhoff's laws for the simplified *ac* equivalent circuit and the four order Runge-Kutta method to solve the corresponding state equations.

$$\frac{dv_{1,2}}{dt} = \begin{bmatrix} -\frac{1}{r_1C_1} & \frac{1}{r_1C_1} \\ \frac{1}{r_1C_2} & -\frac{1}{r_1C_2}-\frac{1}{r_2C_2} \end{bmatrix} v_{1,2}(t) + \begin{bmatrix} \frac{1}{C_1} \\ \frac{1}{C_2} \end{bmatrix} i_{1,2}(t) \qquad i(t) = \begin{bmatrix} 0 & \frac{1}{r_2} \end{bmatrix} v_{1,2}(t) \qquad (3)$$

MATLAB was used as a programming environment and the input parameters chosen in compliance with the experimental results (Figure 10). The simulated transient currents (symbols) under negative and positive *dc* bias are displayed in Figure 12.

Good agreement between experimental and simulated data was observed. The eight expected levels, under reversed bias, and their reduction under forward bias (Figure 10) are clearly seen. The device behaves like a transmission system able to store and transport the minority carriers generated by the current pulses, through the capacitors C_1 and C_2. If not triggered ON (all the input channels OFF), both under positive and negative bias, the device is nonconducting (lowest level). Under negative bias (low R_1), the expected eight levels are detected, each one corresponding to the presence of three, two, one or no color channel ON. If I_1 or I_2 are ON, C_2 and C_1 are rapidly charged in inverse polarity. The current source keeps filling the capacitors during the pulse and the transferred charge between them reaches the output terminal as a capacitive charging current. With all the input channels ON, the packets of charge stored at C_1 (I_1, I_3) are sequentially transferred to C_2 and together with the minority carriers generated at the base of Q_2 (I_2, I_4) flow across the circuit. Under forward bias (high R_1) the device remains in its non conducting state unless a light pulse (I_2 or I_2+I_4) is applied to the base of Q_2. This pulse causes Q_2 to conduct because the reversed biased n-p internal junction behaves like a capacitor inducing a charging current (I_2+I_4) across both collector junctions. The collector of the conducting transistor pulls low, moving the Q_1 base toward its collector voltage, which causes Q_1 to conduct. The collector of the conducting Q_1 pulls high, moving the Q_2 base in the direction of its collector. This positive feedback (regeneration) reinforces the Q_2 already conducting state and a current I_2+I_4 will flow on the external circuit. The two transistor model also explains the use of the same device

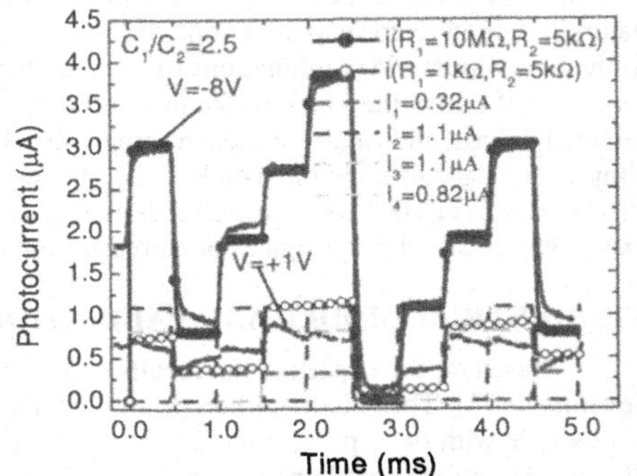

Figure 12. Simulated multiplexed (symbols), current sources (dash lines) and experimental (solid lines) signals under: a) positive (R_1=10MΩ; +1V) and negative (R_1=1KΩ; -8V) *dc* bias.

configuration in the Laser Scanned Photodiode (LSP) image and color sensor. For the color image sensor electrical model only the red channel is used (Figure 11a, $I_2 \neq 0$, $I_1=I_3=I_4=0$). To simulate a color image at the XY position, using the multiplexing technique, a low intensity red pulse (scanner, Φ_S, λ_S), impinges in the device in dark or under different red, green and blue optical bias (color pattern, Φ_L, $\lambda_{RGB,L}$, $\Phi_L > \Phi_S$. Figure 13 displays the experimental acquired electrical signals.

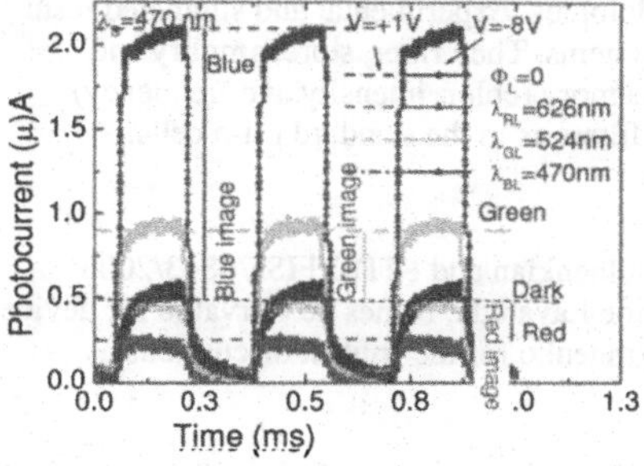

Figure 13. Experimental color recognition using the WDM technique.

Taking into account the results of Figures 8 and 10 a good agreement is observed. Without optical bias ($\Phi_L=0$) and during the red pulse, only the minority carriers generated at the base of Q_2 by the scanner, flow across the circuit (I_2) either in reverse or forward bias. Under red irradiation (red pattern, $\Phi \neq 0$, λ_{RL}) the base-emitter junction of Q_2 is forward bias, the recombination increases reducing I_2 thus, a negative image is observed whatever the applied voltage. Under blue ($\Phi \neq 0$, λ_{BL}) or green ($\Phi \neq 0$, λ_{BL}) patterns irradiations the signal depends on the applied voltage and consequently, on R_1. Under negative bias, the charge transferred from C_1 to C_2, reaches the output terminal as capacitive charging current. An optical enhancement is observed due to the amplifier action of adjacent collector junctions which are always polarized directly. Under forward bias the device remains in its non conducting state, unless the red pulse (I_2, dark level in Figure 12) is applied to the base of Q_2. Here, Q_2 acts as a photodiode for one polarity of the current. No amplification occurs and the red channel is strongly reduced when compared with its value under negative voltage. Under blue irradiation, the internal junction becomes reverse biased at +1 V (blue threshold) allowing the blue recognition. The behavior under a green pattern depends on the balance between the green absorption into the front and back diodes that determines the amount of charges stored in both capacitors. Under negative bias both the green component absorbed either in the front (blue-like) or at the back (red-like) diodes reaches the output terminal while for voltages at which the internal junction n-p becomes reversed (green threshold), the blue-like component is blocked and the red-like reduced. So, by using a thin a-SiC:H front absorber optimized for blue collection and red transmittance and a back a-Si:H absorber to spatially decouple the green/red absorption, the model explains why a moving red scanner (probe beam) can be used to readout RGB the full range of colors at each location without the use of a pixel architecture.

CCD digital image sensors are only capable of recording one color at each point. Since, under steady state illumination (optical image) each phototransistor acts as a filter this multiplexing readout technique can also be used if the stack device is embedded in silicon forming a two layer image sensor that captures full color at every point. Here, a demosaicing algorithm is needed for color reconstruction.

CONCLUSIONS

In this review paper single and stack pin heterojunctions based on a-SiC:H alloys were compared under different optical and electrical bias conditions using different readout techniques. Several applications are presented. A theoretical model gives insight into the physics of the device. Results show that we can think on three integrated transducers in a single photodetector. If a light scan with a fixed wavelength is used to readout the generated carriers it can recognize a color pattern projected on it, acting as a color and image sensor. If the photocurrent generated by different monochromatic pulsed channels or their combination is readout directly, the information is multiplexed or demultiplexed. Finally, when triggered by light with appropriated wavelengths, it can amplify or suppress the generated photocurrent working as an optical amplifier. Electrical models were presented to support the sensing methodologies. Experimental and simulated results show that the tandem devices act as charge transfer systems. They filter, store, amplify and transport the photogenerated carriers, keeping their memory (color, intensity and frequency) without the need for optical pre-amplifiers or optical filters as in the standard p-i-n cells.

ACKNOWLEDGEMENTS

This work supported by Fundação Calouste Gulbenkian and PTDC/FIS/70843/2006 project. The authors thank Markus Schubert, Guilherme Lavareda, Nunes de Carvalho for device depositions, and to Reinhard Schwarz and Yuriy Vygranenko for the fruitful discussions.

REFERENCES

1. S. Randel, A.M.J. Koonen, S.C.J. Lee, F. Breyer, M. Garcia Larrode, J. Yang, A. Ng'Oma, G.J Rijckenberg, and H.P.A. Boom, ECOC 07 (Th 4.1.4). (pp. 1-4). Berlin, Germany, 2007.
2. H.K. Tsai and S.C. Lee, IEEE electron device letters, EDL-8, (1987) pp.365-367.
3. G. de Cesare, F. Irrera, F. Lemmi, and F. Palma, IEEE Trans. on Electron Devices, Vol. 42, No. 5, May 1995, pp. 835-840.
4. A. Zhu, S. Coors, B. Schneider, P. Rieve, and M. Bohm, IEEE Trans. on Electron Devices, Vol. 45, No. 7, July 1998, pp. 1393-1398.
5. M. Topic, H. Stiebig, D. Knipp, F. Smole, J. Furlan, and H. Wagner, J. Non Cryst. Solids 266-269 (2000) 1178-1182.
6. M. Mulato, F. Lemmi, J. Ho, R. Lau, J. P. Lu, and R. A. Street, J. of Appl. Phys., Vol. 90, No. 3 (2001), pp. 1589-1599.
7. M. Vieira, M. Fernandes, J. Martins, P. Louro, R. Schwarz, and M. Schubert, IEEE Sensor Journal, 1, no.2 (August, 2001) pp. 158-167.
8. M. Vieira, M. Fernandes, P. Louro, R. Schwarz, and M. Schubert, J. Non Cryst. Solids 299-302 (2002) pp.1245-1249.
9. M. Vieira, A. Fantoni, M. Fernandes, P. Louro, and I. Rodrigues. Mat. Res. Soc. Symp. Proc 762@2003 A.18.13.
10. P. Louro, M. Vieira, Yu. Vygranenko, A. Fantoni, M. Fernandes, G. Lavareda, and N. Carvalho, Mat. Res. Soc. Symp. Proc., 989 (2007) A12.04.
11. P. Louro, M. Vieira, Yu. Vygranenko, M. Fernandes, R. Schwarz, and M. Schubert, Applied Surface Science 184, 144-149 (2001).
12. M. Vieira, M. Fernandes, P. Louro, A. Fantoni, Y. Vygranenko, G. Lavareda, and C. Nunes de Carvalho, Mat. Res. Soc. Symp. Proc., Vol. 862 (2005) A13.4.
13. M. Vieira, A. Fantoni, P. Louro, M. Fernandes, R. Schwarz, G. Lavareda, and C. N. Carvalho, Vacuum, Vol. 82, Issue 12, 8 August 2008, pp: 1512-1516.
14. M. Vieira, P. Louro, M. Fernandes, and A. Fantoni, Sensor and Actuators A 114/2-3 (2004), pp. 219-223.
15. M. A. Vieira, M. Vieira, M. Fernandes, A. Fantoni, P. Louro, and M. Barata, MRS Proceedings (2009) Vol. 1153, A08-03.

Mater. Res. Soc. Symp. Proc. Vol. 1245 © 2010 Materials Research Society 1245-A08-04

Micro-Channel Plate Detectors Based on Hydrogenated Amorphous Silicon

Nicolas Wyrsch[1], François Powolny[2], Matthieu Despeisse[2], Sylvain Dunand[1], Pierre Jarron[2], Christophe Ballif[1]
[1] Ecole Polytechnique Fédérale de Lausanne (EPFL), Institute of Microengineering (IMT), Photovoltaics and thin film electronics laboratory, Breguet 2, 2000 Neuchâtel, Switzerland,
[2] CERN, CERN Meyrin, 1211 Genève 23, Switzerland.

ABSTRACT

A new type of micro-channel plate detector based on hydrogenated amorphous silicon is proposed which overcomes the fabrication and performance issues of glass or bulk silicon ones. This new type of detectors consists in 80-100 μm thick layers of amorphous silicon which are micro-machined by deep reactive ion etching to form the channels. This paper focuses on the structure and fabrication process and presents first results obtained with test devices on electron detection which demonstrate amplification effects. Fabrication and performance issues are also discussed.

INTRODUCTION

As an alternative to particle detection using diodes, micro-channel plates (MCP) detectors are used for fast-time resolution. Such detectors consist in thick plates having very narrow micro-channel drilled throughout the plate. A high electrical field is applied between the two faces of the plate to create avalanche mechanisms in the micro-channels upon the hit of the channel wall with a primary electron. These avalanches lead to an amplification of the signal and are commonly used as image intensifying devices [1]. Such detector can be considered as an array of microscopic electron multiplier tubes. MCPs are conventionally fabricated from lead glass plates and treated to enhance secondary electron emission and to render the wall of the channels semiconducting (for charge replenishment). To overcome some of the performance limitations and manufacturing issues associated to glass MCPs, MCPs based on micro-machined Si wafers have been recently introduced [2,3]. However, the low bulk resistivity of c-Si requires an additional and critical deposition of a layer stack on the channel walls to achieve optimum MCP performances.

To overcome some critical limitations of current MCPs, we proposed a very innovative type using a-Si:H thick layers [4]. In this context, a-Si:H offers three important advantages for MCPs: A high bulk resistivity that can be tuned, the fact that a-Si:H can be micro-machined as c-Si material and the possibility to deposit the material (and the devices) on various type of substrates including CMOS readout chips [5]. The latter point implies that a-Si:H based MCP (AMCP) detectors could also be vertically integrated, greatly simplifying the cumbersome construction of usual MCPs, broadening the range of application including particle detection and imaging. The fact that AMCPs compared to glass MCPs do not require isolation/resistive layer stack greatly simplify the fabrication and should permit, thanks to an ideal bulk resistivity of a-Si:H, a fast charge recovery of the depleted micro-channels (after an avalanche event).

The typical structure of an AMCP and its functioning principle is given in Fig. 1. A primary electrons (coming directly on the device or generated by a photo-cathode or ionization converter) is multiplied by successive hits on the walls of the micro-channels and collected at the back of the MCP by a pixel electrode. Note here that we have in general many micro-channels on top of each individual pixel electrode. The latter can be part of an underlying readout chip or connected to adjacent electronic circuits. Such AMCP consists in a very thick (typically between 80 and 100 μm) a-Si:H deposited on the pixel electrode back plane (or readout chip) with at top common electrode. Micro-channels of a few microns in diameter are drilled by deep reactive ion etching (DRIE).

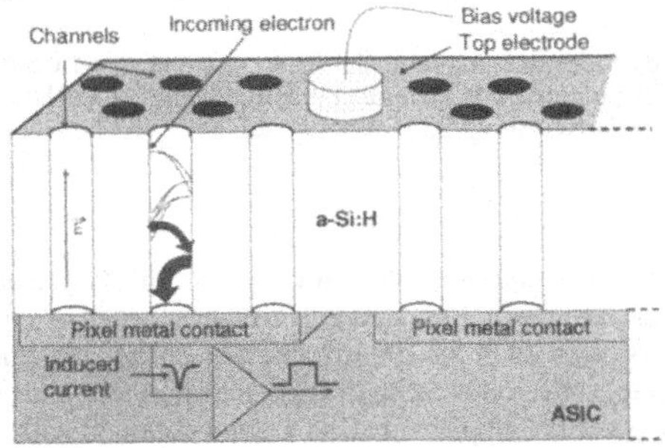

Figure 1. a-Si:H based MCP (AMCP) structure.

In this paper, we will describe and discuss the processes and issues related the fabrication of such AMCPs. First results on electron detection obtained using an electron beam induced current apparatus will be presented, demonstrating avalanche mechanisms in the micro-channels. Present performances, limitations and possible improvements will be discussed.

EXPERIMENT

AMCP structure and layout

First AMCP prototypes were fabricated on oxidized Si wafer covered with an Al back pixel electrodes. These devices comprise channels with a diameter of 3-5 μm drilled by DRIE into 80-100 μm thick a-Si:H layers. For testing purposes, 15x15 mm^2 reticles have been defined with 24 pixels with sizes of 0.5x0.5, 1x1 and 2x2 mm^2 (see Fig. 2). Various reticles were

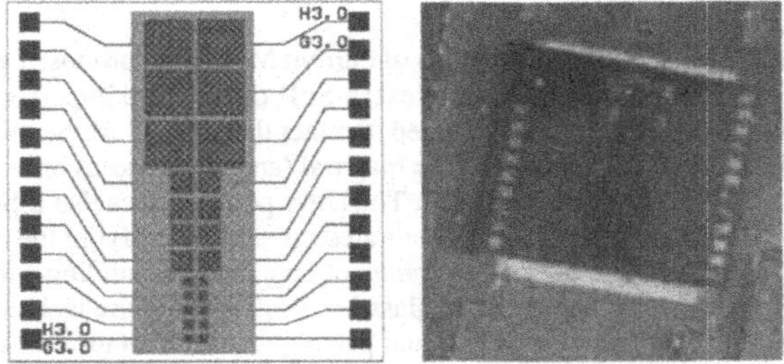

Figure 2. AMCP test reticle design (left) with the back electrodes and bonding pads in black; the colored area is covered by the thick a-Si:H layer while the patterned squares correspond to the area where the micro-channels are drilled. Picture of such a reticle (right).

designed with micro-channels (over each pixel) with a (nominal) diameter of 1.5 to 5 μm separated by a gap of (nominally) 2.5 to 3.5 μm.

AMCP fabrication

Deposition of these thick a-Si:H layers requires a careful optimization in order to achieve high deposition rate with reasonable material quality while maintaining a low internal mechanical stress. All a-Si:H layers were deposited by VHF PECVD (very high frequency plasma enhanced chemical vapor deposition) at a plasma frequency of 70 MHz on oxidized 4" Si wafers with an oxide layer thickness of 1-2 μm covered with the Al back contact plane. Even though almost stress free layers a-Si:H layer can be deposited by VHF PECVD at 170°C on c-Si [5], such temperature resulted in too high stress and delamination during the deposition of the a-Si:H layer. Successful growth of 80 to 100 μm with reasonable compressive stress (without delamination and allowing consecutive micro-machining process steps) was achieved at a deposition temperature of 235°C. With a deposition rate of approx. 2 nm/s, the 14 hours deposition time required for 100. As a front common electrode, a 100 nm thick a-Si:H n-type layer was also deposited by VHF PECVD.

Prior to the micro-fabrication of the channels, the a-Si:H was patterned over the pixel regions in order to release stress on the wafer and to get access to the bonding pads. This step was carried out using a hard mask and dry etching in O_2/SF_6 plasma. Drilling of the channels was performed by DRIE using a Bosh Process (DRIE machine from Surface Technology Systems). Micro-channels with aspect ratios over 25 could be obtained after optimization of the process. Part of this optimization process was done on c-Si wafer as the etching behavior of a-Si:H is very similar to the one of c-Si.Top and side views (Scanning Electron Microscope - SEM pictures) of holes drilled in thick a-Si:H layers are show in Fig. 3. SEM pictures of the details of a complete AMCP are given in Fig. 4.

AMCP characterization

Test devices were characterized using either a 10 kV electron gun or by electron beam

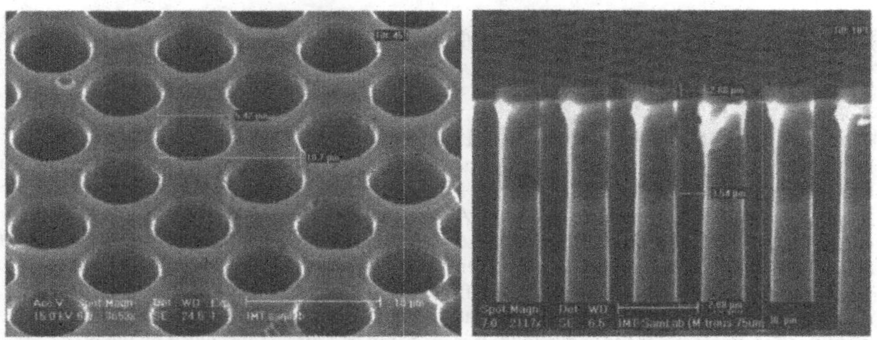

Figure 3. SEM pictures of the surface and cross section of thick a-Si:H layers with holes drilled by DRIE.

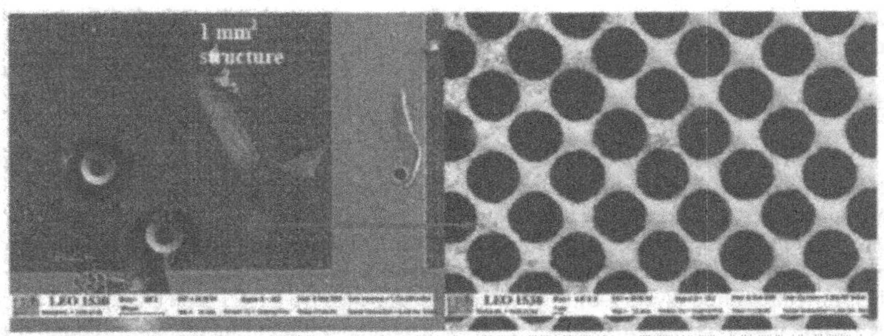

Figure 4. SEM pictures showing the corner of one AMCP pixel as well as a zoom (right) over that pixel. One may also observe large defects (left picture) created during the DRIE process.

induced current (EBIC) mapping. The latter was performed at the European Synchrotron Radiation Facility (ESRF) in Grenoble, on a Scanning Electron Microscope (SEM) from Carl ZEISS (named "Leo 1530 Gemini"). A schematic view of the electrical set-up is given in Fig. 5. A Keithley 2410 was used for the application of the bias voltage and for the measurement of the current. Samples have been measured as a function of the bias voltage and e-beam intensity. For EBIC imaging the AMCP pixel were connected to a custom fast current/voltage converter and amplifier for leakage compensation and signal conditioning. Samples were characterized as a function of the bias voltage, beam incidence angle and beam energy (between 4 and 30 keV).

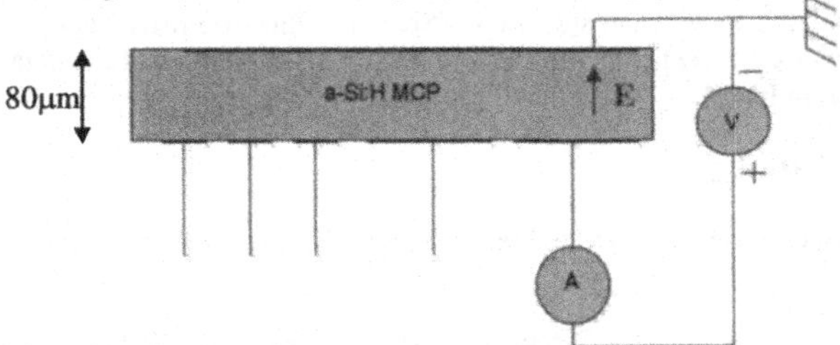

Figure 5. Electrical set-up for AMCP signal/leakage measurements.

RESULTS AND DISCUSSION

Several reticles were mounted on test circuits for electrical characterization (cf. Fig. 2) and exposed to an electron beam of 10 keV of variable intensity. Preliminary results obtained on one of the samples are shown. Figure 6 shows the current measured on a 1x1 mm^2 pixel as a function of bias voltage and beam intensity. As one can observe, above 400 V bias voltage, the current induced by the electron beam increase proportionally with the bias voltage (amplification effect). A clear dependency on the beam intensity is also observed. Rather large discrepancies were observed between structures and the difficulties to evaluate a gain values called for more detailed analysis of the detector response. The presence of a threshold voltage for the amplification effect is not yet understood. It is probably an artifact introduced by the poor uniformity of the response of the present device as discussed below.

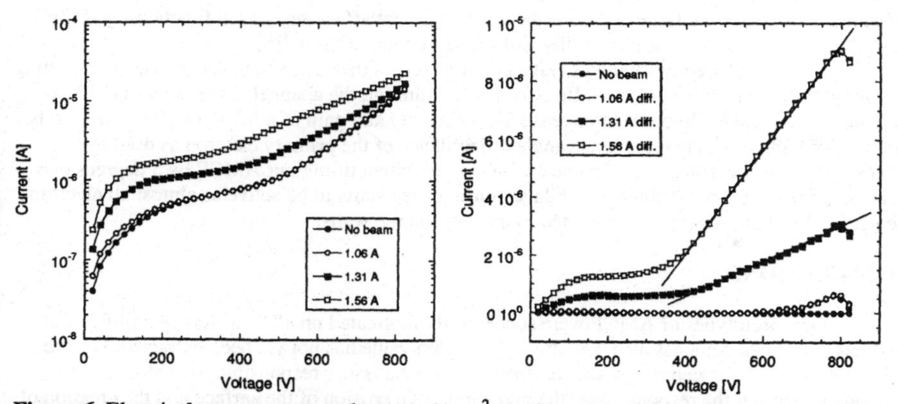

Figure 6. Electrical current measured on a 1x1 mm^2 pixel as a function of bias voltage and beam intensity (indicated here as a function of the e-gun filament current): (left) Raw current value, (right) current after subtraction of the leakage current (current without beam).

EBIC images were taken under various electron beam energy to evaluate the local response of the detector. The DC signal was here compensated to reveal the "active" area of the AMCP area where signal amplification takes place. Three EBIC images of a 0.5x0.5 mm^2 pixel are shown in Fig. 7. One can first observe (left) that the response is not homogeneous over the pixel area. Only the channels at the periphery are responding while the central area of the pixel is inactive. Detailed analysis revealed that an important erosion of the surface of the n-doped top layer, over the entire area of the pixel back electrode, took place during the DRIE process. We believe that this erosion process is also related to the formation of the large defects (seen in Fig. 2) and takes place at the end of the DRIE process when the channels start to open down to the back electrodes. The effect of this erosion is to strongly increase the surface resistivity which leads in a drop of the electrical field for channels far from the pixel edge. One can also observes

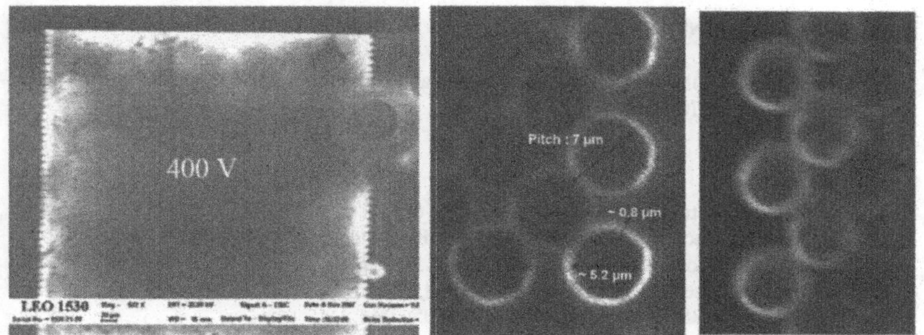

Figure 7. EBIC pictures of a 0.5x0.5 mm^2 pixel biased with 400 V: (left) overall pixel image, (centre) zoom on the corner, (right) zoom on the periphery with a sample tilting of 5°.

that the presence of the large defects seen in Fig. 2 do not affect the overall functioning of the AMCP, those defects being electrically isolated from their surrounding.

A zoom over a few channels (Fig. 7 centre) shows that, as expected, only electrons hitting walls of the channels are detected. Electrons going through the channel in the center do not create any avalanche. In order to increase the sensitivity and gain of a MCP, channels have to be at a certain angle with respect to the angle of incidence of the primary electron as used for example in the chevron deign of standard MCPs [1]. When tilting our AMCP by 5 degrees one can observe (Fig. 7 right) that a much lager surface area starts to be active as almost all electrons entering the channels are now expected to impinge on the walls.

CONCLUSIONS

First prototypes of AMCP were successfully fabricated on 4" Si wafers. Amplification effects have been demonstrated even though the gain value has not yet been extracted. For the time being the fabricated devices suffer from fabrication issues responsible for a strong inhomoheneity of the response over the pixel areas. An erosion of the surface and the creation of local defects are observed which will require an optimization of the DRIE process as well as an adaptation of the device structure. Replacement of the n-doped a-Si:H top contact layer by a metallic contact (not implemented so far to avoid any contamination of the DRIE system with metallic impurities) could also be applied.

Beside the improvement of the fabrication process, detailed investigation are needed to first understand and, secondly, to improve the avalanches process in the micro-channels. This should help define the limitations and potential of this new technology. Nevertheless, the high bulk resistivity of a-Si:H as used in AMCP should eliminate the issue of charge replenishment in conventional glass MCP. The fact that AMCP performance can be easily imaged by EBIC is here a first evidence of this advantage. Compared to c-Si MCP, this high bulk resistivity considerably simplifies the device structure and eases the fabrication.

These first encouraging results support the expectations for AMCP. The devices can be fabricated using known and mastered technology over 4" wafers and the processing steps are compatible with CMOS technology. Vertical integration of MCP over readout chip should be possible allowing higher performance and reliability.

ACKNOWLEDGMENTS

The authors would like to thank John Morse and Irina Snigireva at the ESRF in Grenoble for helping with the EBIC imaging.

REFERENCES

1. J. L. Wiza, Nucl. Instr. and Meth. 162, 1979, 587-601.
2. C. P. Beetz et al., Nucl. Instr. and Meth. In Phys. Res. A 442, 2000, 443-451.
3. Q. Duanmu et al., Proc. of SPIE Vol. 4601, 2001, 284-287.
4. F. Powolny, Ph.D. thesis, University of Neuchâtel, 2009.
5. N. Wyrsch et al., MRS Proc. Symp. Vol. 869, 2005, 3-14.

Nanostructured Silicon

Mater. Res. Soc. Symp. Proc. Vol. 1245 © 2010 Materials Research Society 1245-A09-03

Optical absorption in co-deposited mixed-phase hydrogenated amorphous/nanocrystalline silicon thin films

L. R. Wienkes,[1] A. Besaws,[1,4] C. Anderson,[2] D. C. Bobela[3], P. Stradins,[3] U. Kortshagen,[2] and J. Kakalios[1]

1 School of Physics and Astronomy, University of Minnesota, Minneapolis, MN 55455
2 Department of Mechanical Engineering, University of Minnesota, Minneapolis, MN 55455
3 National Renewable Energy Laboratory, Golden, CO 80401
4 College of Menominee Nation, Keshena, WI

ABSTRACT

The conductivity of amorphous/nanocrystalline hydrogenated silicon thin films (a/nc-Si:H) deposited in a dual chamber co-deposition system exhibits a non-monotonic dependence on the nanocrystal concentration. Optical absorption measurements derived from the constant photocurrent method (CPM) for similarly prepared materials are reported. The optical absorption spectra, in particular the subgap absorption, are found to be independent of nanocrystalline density for relatively small crystal fractions (< 4%). For films with a higher crystalline content, the absorption spectra indicate broader Urbach slopes and higher midgap absorption. These data are interpreted in terms of a model involving electron donation from the nanocrystals into the amorphous material.

INTRODUCTION

The unique properties of mixed-phase materials, in which nanoscale particles are embedded within a host semiconductor or insulator matrix, have attracted interest for such applications as high efficiency solar cells [1,2], non-volatile memory and electron emitters [3,4], bandgap engineering [5], and electroluminescent devices [5]. In recent years, interest has focused on the system of silicon nanocrystallites within hydrogenated amorphous silicon thin films (a/nc-Si:H) for photovoltaic applications. Hydrogenated amorphous silicon (a-Si:H) has long been known to exhibit light-induced defects, which degrade the electronic properties of the film and reduce the efficiency of a-Si:H-based solar cells (traditionally referred to as the Staebler-Wronski effect (SWE)) [6,7]. The interest in nanostructured a/nc-Si:H films is due in part to reports of an enhanced resistance to light-induce defect formation in a/nc-Si:H [8]. These mixed-phase materials are typically synthesized in a single chamber plasma deposition system. In this paper, we describe measurements on materials deposited in a dual chamber co-deposition system.

MATERIALS PREPARATION

The synthesis of these mixed-phase a/nc-Si:H films in the dual chamber co-deposition system has been described previously [9]. Briefly, the films are deposited in a system where silicon nanocrystals are grown in a separate plasma reactor, and then are entrained by an inert argon carrier gas and injected into a second plasma enhanced chemical vapor deposition (PECVD) chamber where the amorphous silicon film is grown. The silicon nanocrystallites are then incorporated into the growing a-Si:H film. Convection of the carrier gas as it enters the second PECVD chamber can be used to deliberately vary the nanocrystallite concentration in the mixed phase films, by varying the distance of the substrates from the injection tube, as shown in figure 1. The benefits of this process are twofold; the growth conditions can be optimized for each phase separately and films with varying nanocrystalline concentrations can be grown simultaneously, allowing for a more direct comparison of the effects of nanocrystalline inclusions. Typically three films are grown at once; film C which is directly under the injection tube, film B which is 2.5 cm from film C, and film A, which is 5 cm from film C. Cross-sectional TEM images indicate that the nanocrystalline inclusions are approximately 5 nm in diameter and are present uniformly throughout the amorphous silicon film thickness. The films were grown on Corning 7059 substrates and are approximately 200 - 700 nm thick. Raman spectroscopy was used to characterize the crystal fraction as described in Adjallah et al [9].

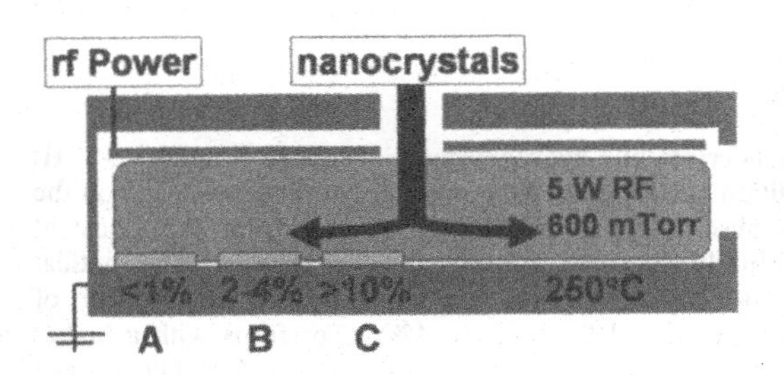

Figure 1: Sketch of the dual plasma deposition system; the nanocrystals are grown in a separate plasma chamber above the main PECVD system and then injected using argon as a carrier gas.

EXPERIMENTAL RESULTS

Optical absorption measurements were carried out on two series of films from reference [9]; for the duration of this paper, we shall denote them as run 1 and run 2 (these correspond to run 2 and 3 in reference [9] respectively). The films displayed a non-monotonic dependence of the dark conductivity on crystal fraction; this can be seen for run 1 in figure 2 (similar results are found for run 2). Film B consistently showed the highest conductivity and lowest activation energy, while film A had the lowest conductivity and film C was somewhere in between the other two films. This trend was quite robust as it was observed across six separate film depositions [9]. Run 1 had activation energies of 0.90, 0.64 and 0.91 eV for films A, B, and C respectively. A proposed explanation for this non-monotonic variation of the conductivity with crystal fraction involves charge donation from the nanocrysalline inclusions, coupled with variations in the dangling bond density in these films as a function of crystalline content. The optical absorption spectra for these films were measured using the constant photocurrent method (CPM) [10] in order to ascertain any variation of the defect density in these mixed-phase films.

Briefly, in CPM as the wavelength of light incident on the sample is varied, one measures the light intensity required to maintain the photocurrent at a constant value; the inverse of the light intensity provides the absorption coefficient in relative terms. By maintaining a uniform photocurrent throughout the spectral region of interest, the quasi-Fermi level, and thus the recombination center occupation, is unchanged for differing light absorption. This helps to keep the mobility-lifetime product ($\mu\eta\tau$) constant, which is true over the range we are measuring, as shown by Jackson et al [11]. A direct measurement of the absorption, typically using transmission/reflection spectroscopy, sets the absorption spectra to an absolute scale. In this report, all CPM spectra reported are the result of dc measurements.

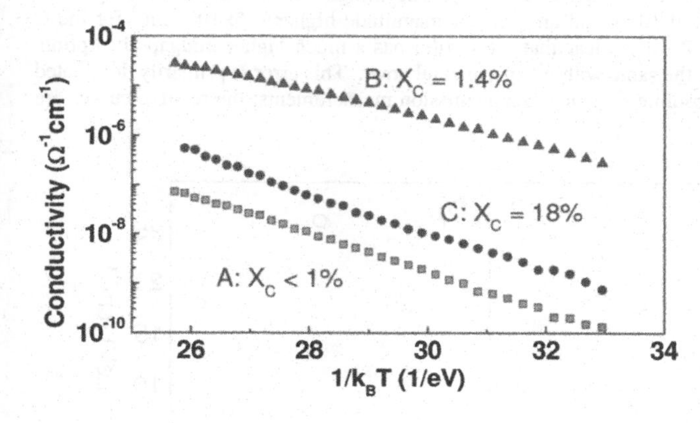

Figure 2: Arrhenius plot of the conductivity of the films from run 1. The dark conductivity exhibits a non-monotonic dependence on crystal fraction. The activation energies were 0.90, 0.64, and 0.91 eV for films A, B, and C respectively.

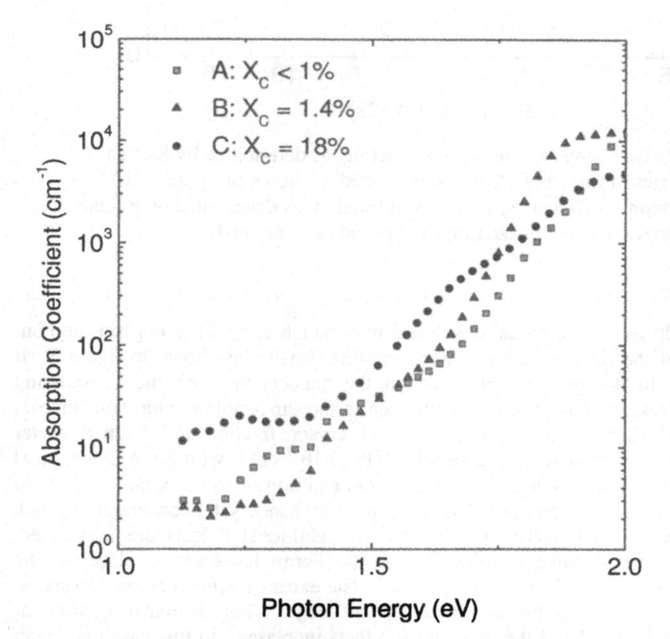

Figure 3: Plot of the spectral dependence of the optical absorption coefficient as determined by CPM for the three mixed-phase films for run 1. The C film, deposited directly underneath the particle injection tube (as in figure 1) clearly has a higher subgap absorption, while the optical absorption spectra for the A and B films (with lower nanocrystalline contents) are very similar.

The CPM absorption curves are plotted in figure 3. The Urbach slopes were calculated and are shown versus crystal fraction in figure 4a, along with the CPM absorption values at 1.2 eV in figure 4b. Both figures 4a and 4b include data from both run 1 and 2. Using the absorption values and the conversion developed by Wyrsch et al [12], we have also calculated the expected dangling bond densities based on their calibration. The midgap defect density is roughly 6-7×10^{16} cm^{-3} for the A and B films and an order of magnitude higher ~ 5×10^{17} cm^{-3} for the C films for both runs 1 and 2. It is clear that the C film has a much higher midgap absorption, while the A and B film are the same within experimental error. This error is primarily dominated by the conversion to an absolute scale using transmission measurements; the relative curves are quite reproducible

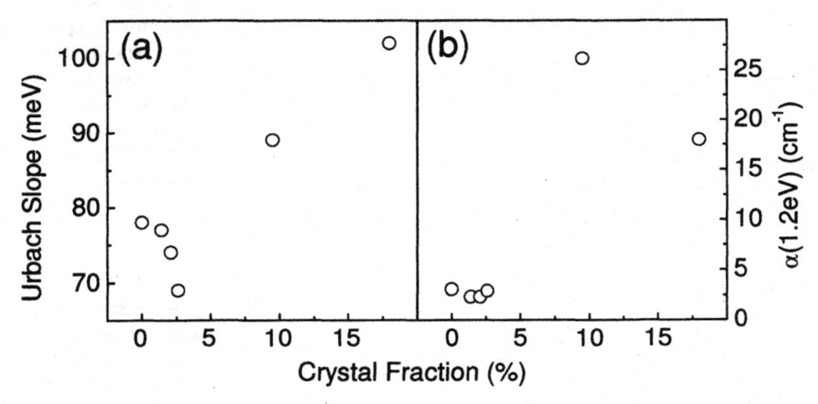

Figure 4: (a) Plot of the Urbach slope against crystal fraction, as determined by Raman measurements, for two series of mixed phase films, deposited as shown in figure 1. (b) Plot of the midgap optical absorption coefficient against crystal fraction, as determined by Raman measurements, for two series of mixed phase films, deposited as in figure 1.

DISCUSSION

The CPM results help refine the model developed in Adjallah et al [9] to explain the non-monotonic dependence of the conductivity on nanocrystallite density, as shown in figure 5. It was postulated that due to the band offsets between the nanocrystals and the surrounding amorphous material, each nanocrystal donates an electron to the surrounding amorphous matrix, effectively "doping" the material. If one assumes a 2% crystal fraction and 5 nm diameter particles, the density of nanocrystals/cm^3 is approximately 3×10^{17} cm^{-3}, with a presumed equal density of donated charges. At the same time, however, the addition of nanocrystals is likely to be accompanied by additional dangling bond defects. For a low nanocrystal concentration, such as the B films in figure 1, a relatively small number of additional defects are introduced, insufficient to trap all of the donated charges, and so the Fermi level moves closer to the conduction band, increasing the conductivity (figure 5b). The excess electrons fill in gap states, pushing the Fermi level closer to the conduction band edge. For a higher density of nanocrystalline inclusions (the C film) the number of defects increases. In this case the defect density can accommodate all of the excess donated charges, and the Fermi level moves back towards midgap (figure 5c). The increased rate of defect production in the higher crystal fraction material is possibly due to crystal agglomeration. These agglomerates would create voids in the amorphous material as a result of shading, resulting in more defects from the void surfaces, though more work is needed to verify this hypothesis.

The similarity of the absorption at 1.2 eV and the Urbach slopes for films A and B support the picture that has been developed. The additional dangling bond states in film B should be detected by CPM, however the change may be relatively small and obscured by the experimental error. The broadening of the Urbach slope, indicative of greater disorder and strained Si-Si bonds in the film, and the enhanced low level absorption in the C film is also consistent with this picture.

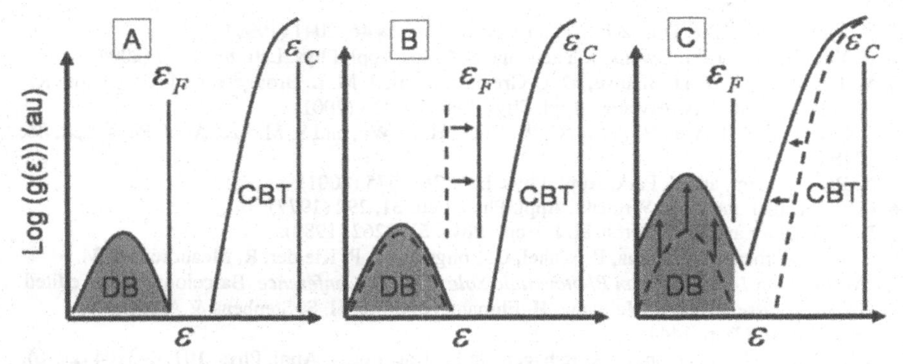

Figure 5: A sketch of the density of states, denoted g(ε) in the figure, of the amorphous portion of the mixed phase material. DB represents a band of dangling bond states, while CBT stands for conduction band tail. Filled states are in gray. The dashed lines in (b) and (c) represent the density of states and Fermi level of the A film, showing how the addition of nanocrystals changes its features. The nanocrystals in the B film create only a small number of extra defects so that the excess charges from the nanocrystals fill in the gap states, pushing the Fermi level towards the conduction band tail. In the C film, sufficient additional defect states are created to accommodate the electrons donated by the nanocrystals and the Fermi level returns to its original position.

CONCLUSIONS

Optical absorption measurements have been performed on mixed-phase hydrogenated amorphous silicon containing silicon nanocrystallite inclusions, grown in a dual chamber co-deposition process. The optoelectronic properties of these films are studied in order to understand the charge transport processes and density of states in these mixed-phase films. The optical absorption and conductivity results are understood using a simple charge donation and defect creation picture. Currently, work is underway to further study the defects states using electron spin resonance (ESR) and a separate optical absorption technique, photothermal deflection spectroscopy (PDS) [13], both of which provide complementary information to CPM.

ACKNOWLEDGEMENTS

This work was partially supported by the NSF under Grant Nos. NER-DMI-0403887 and DMR-0705675, the NINN Characterization Facility, the Xcel Energy grant under RDF Contract No. RD3-25, the National Renewable Energy Laboratory, the University of Minnesota Center for Nanostructure Applications, and the NSF MRSEC REU program. The authors thank Yves Adjallah for the conductivity and Raman data.

REFERENCES

1. K. Yamamoto, IEEE Transactions on Electron Devices **46**, 2041 (1999).
2. B. Yan, G. Yue, J. M. Owens, J. Yang, and S. Guha, Appl. Phys. Lett. **85**, 1925 (2004).
3. M. L. Ostraat, J. W. D. Blauwe, M. L. Green, L. D. Bell, M. L. Brongersma, J. Casperson, R. C. Flagan, and H. A. Atwater, Appl. Phys. Lett. **79**, 433 (2001).
4. T. C. Chang, S. T. Yan, P. T. Liu, C. W. Chen, H. H. Wu, and S. M. Sze, Appl. Phys. Lett. **85**, 248 (2004).
5. N. Park, T. Kim, and S. Park, Appl. Phys. Lett. **78**, 2575 (2001).
6. D. L. Staebler and C. R. Wronski, Appl. Phys. Lett. **31**, 292 (1977).
7. D. L. Staebler and C. R. Wronski, J. Appl. Phys. **51**, 3262 (1980).
8. P. R. Cabarrocas, S. Hamma, P. St'ahel, C. Longeaud, J. P. Kleider, R. Meaudre, and M. Meaudre, in *14th European Photovoltaic Solar Energy Conference*, Barcelona, Spain, edited by H. A. Ossenbrink, P. Helm and H. Ehmann (Bedford: H. S. Stephens & Associates, Bedford, 1997), p. 1444.
9. Y. Adjallah, C. Anderson, U. Kortshagen, and J. Kakalios, J. Appl. Phys. **107**, 043704 (2010).
10. M. Vaněček, J. Kocka, J. Stuchlík, and A. Tríska, Solid State Commun. **39**, 1199 (1981).
11. W. B. Jackson, R. J. Nemanich, and N. M. Amer, Phys.Rev.B **27**, 4861 (1983).
12. N. Wyrsch, F. Finger, T. J. McMahon, and M. Vaněček, J. Non Cryst. Solids **137-138**, 347 (1991).
13. W. B. Jackson, N. M. Amer, A. C. Boccara, and D. Fournier, Appl. Opt. **20**, 1333 (1981).

Mater. Res. Soc. Symp. Proc. Vol. 1245 © 2010 Materials Research Society 1245-A09-06

Synthesis of Silicon Nano-particles for Thin Film Electrodes Preparation

David Munao[1], Jan van Erven[1], Mario Valvo[1], Vincent Vons[1], Alper Evirgen[1], Erik Kelder[1]
[1]NanoStructured Materials / Delft ChemE, TUDelft, Julianalaan 136 BL, Delft, Netherlands

ABSTRACT

The present work concerns novel approaches to fabricate silicon-based electrodes. In this study silicon nano-particles are synthesized via two aerosol routes: Laser assisted Chemical Vapour Pyrolysis (LaCVP) and Spark Discharge Generation (SDG). These two techniques allow the generation of uniformly sized particles, with a good control over the composition and the range of sizes. Herein, particles with size ranging from 2 to 70 nm were obtained. Starting from these nano-particles, nano-structured porous thin films are produced either by electrospraying a polymeric solution in which LaCVP-produced particles were previously dispersed, or by inertial impacting the SDG-produced particles directly from gas phase onto a substrate. The Electro-Spray (ES) is used to produce composite films for Li-ion battery applications, whereas the Inertial Impaction (II) is used to produce pure silicon films for other purposes (i.e.: sensors).

INTRODUCTION

During the last decades the interest in Si-based devices has risen significantly due to the large number of applications for this material, especially in its nano-structured form. As new mechanical, optical and electrical properties arise from quantum confinement effects, a whole range of new applications is emerging. These span from optoelectronics to sensors, as well as energy storage and biomedicals [1-3]. The synthesis and the assembly of Si nano-particles are herein carried out via aerosol-based approaches for the production of thin film structures. One of the most promising applications of thin films electrodes is related to energy storage via Li-ion batteries. Silicon-based negative electrodes are considered the best alternative to the commercially used graphite or carbon anodes (i.e. 372 $mAhg^{-1}$) due to the fact that their theoretical capacity (i.e. 4200 $mAhg^{-1}$) is roughly ten times higher . However, severe capacity fading still represents a limiting factor for the commercialization of Si-based anodes. The capacity fading is caused by the volume change of the host Si structure upon alloying/de-alloying with lithium. This results in fractures and loss of electrical contact between the various parts of an electrode. In particular, significant fractions of the active material are no longer available for the electrochemical process and the battery rapidly loses its performance. In this work three approaches to increase the mechanical stability of the films are considered: reducing the size of the host particles, nano-structuring the electrode film and using different binders to form a nano-composite structure. In this way practical shortcomings of silicon are considerably reduced.

EXPERIMENTAL DETAILS

Silicon nano-particle syntheses

Silicon nano-particles were produced via two different aerosol routes: Laser assisted Chemical Vapor Pyrolysis (LaCVP) and Spark Discharge Generation (SDG).

Haggerty and Cannon developed the synthesis of nano-particles by laser pyrolysis at the Massachusetts Institute of Technology (MIT) at the beginning of the 1980s [4-6] and the method was later on adopted in the Technische Universiteit of Delft (TUDelft) for producing technical ceramics [7]. In this work SiH_4 is used as source of Si and N_2 as sheath gas. The reactant is thermally decomposed to form Si and H_2 in an oxygen-free environment. Aerosol particle generation from spark discharge has been developed by Schwyn et al. at the end of the 1980's [8]. In our case, two p-type boron-doped Si rods (SiMat) are used as electrodes. The cylindrical electrodes have a diameter of 4 mm, a length of 40 mm and a resistivity of less than 10^{-2} Ωcm. Detailed descriptions of the setup geometries are given in references [9] and [10] for the LaCVP and the SDG, respectively. Experimental conditions used in the particle synthesis are listed in the relevant figures.

Transmission electron microscopy (TEM, coupled with Energy Dispersive Spectroscopy, EDS) was carried out to study the morphology, size and cristallinity of the as-produced nano-particles. TEM was performed using a FEI Tecnai TF20 electron microscope with a field emission gun as the source of electrons operated at 200 kV. TEM samples were prepared by placing a few droplets of a suspension of ethanol and ground silicon on a Quantifoil® carbon polymer supported by a copper grid, followed by drying at ambient conditions in order to evaporate the ethanol. EDS elemental analysis was performed using an Oxford Instruments EDX system. The phase analyses of synthesized powders were performed by BRUKER-AXS D8-Advance X-Ray Diffractometer that was operated at 40 kV and 40 mA equipped with a Cu K_α radiation source. The scan range was chosen as $25° < 2\theta < 60°$ with a scan speed of 10 and 60 sec/step, 0.05 and 0.015 step size for LaCVP and SDG particles, respectively. The surface properties of the LaCVP produced particles were investigated via infrared spectroscopy (Perkin Elmer Spectrum one FTIR), whereas thermal gravimetric analysis (PerkinElmer TGA7) was carried out to study the oxidation rate of the LaCVP-produced nano-particles in air. TGA was carried out in air at a heating rate of 10°C/min.

Thin film preparation

Thin films were prepared following two different aerosol routes. The first one combines LaCVP-produced particles with the ES technique while the second route combines the SDG-generated particles with II technique.

Particles produced via LaCVP were suspended into a polymer solution in order to form so-called precursor ink. The ink was loaded into a glass syringe and electrosprayed onto a stainless steel substrate (see figure 1, right). Two composite electrodes were prepared via ES, namely Si-PolyVinylidene Fluoride (PVdF-Solef 1015) and Si-CarboxyMethylCellulose (CMC-Alpha Aesar). The weight ratio between the Si nano-particles and the polymer was set to 3:7 and 1:3 for the Si-PVdF and the Si-CMC samples, respectively. For each sample the distance between the nozzle (EFD Ultra, 0.8 mm in diameter) and the substrate was kept constant at 20 mm, as well as the flow rate, which was 0.1 ml/h. Si-PVdF and Si-CMC samples differ respectively for: the voltage applied (8.3 and 10.5 kV), the volume sprayed (0.03 and 0.06 ml) and the temperature of the substrate (280 and 75 °C).

The electrochemical behavior of the ES samples was also tested via a galvanostatic test. In particular, coin cells (CR2320 Hohsen) were assembled inside a He filled glove box using metallic Li as reference and counter electrode, $LiPF_6$ in EC:DMC 1:1 mixture (Mitsubishi Chemical) as electrolyte and the Si-composite thin film as the working electrode. The galvanostatic procedure was performed on a Maccor (S4000) battery tester in order to assess the electrode performance upon cycling. The morphology of the electrodes before and after cycling was further studied via Scanning Electron Microscopy (SEM-Philips XL 20) and Atomic Force Microscopy (AFM-NT-MDT NTEGRA).

The particle produced via SDG were directly conveyed from the gas phase to form Porous Silicon (PSi) thin films via a one-step process. The particles were impacted onto a stainless steel substrate. The II setup is depicted in figure 1 (left). These particles were accelerated through a critical orifice with a diameter of 200 μm towards the substrate by a pressure gradient. The pressure inside the system was kept at 0.5 mbar, by a vacuum pump (Alcatel 2063 SD). The impaction nozzle diameter was 2.5 cm. In the current setup only the substrate holder can be moved along the impaction direction. However, it has been proven that moving the nozzle in one or two directions parallel to the substrate surface can produce lines and layers of impacted particles [11]. The distance between the nozzle and the substrate was kept constant during the impaction process. Impaction took place for 45 minutes. It is important to remark that this technique is very flexible and can be applied, in principle, to every aerosol source (i.e.: LaCVP) in order to grow thin layers. The morphology of the PSi layers was also investigated by SEM and AFM.

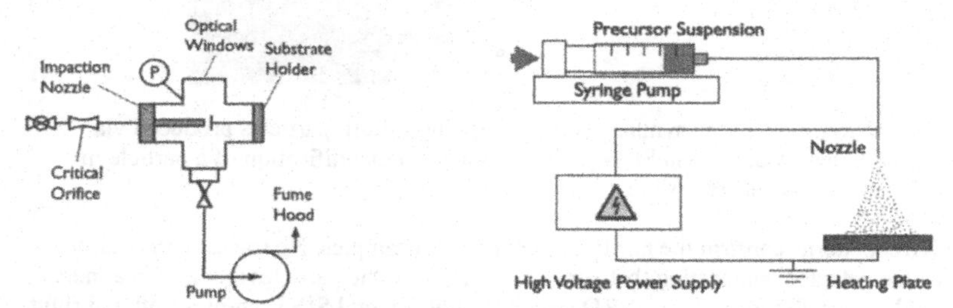

Figure 1. Schematic representation of the Inertial Impaction (II) setup (left) and of the Electro-Spray (ES) setup (right).

DISCUSSION

Si nanoparticles characterization

Two different TEM images of the as-produced Si nano-particles are shown in figure 2. The particles resulting from the LaCVP process (figure 2, left) show a spherical shape, are fairly monodisperse and have a typical size of approximately 20 nm. They show a crystalline core surrounded by an amorphous layer. Furthermore, EDS analysis shows the presence of oxygen. In particular, the oxygen is expected at the surface of the nano-particles. The oxide surface termination has been confirmed by IR spectroscopy. Indeed, the spectra show an intense broad band at about 1070 cm^{-1}, due to the presence of mainly substoichiometric SiO_x ($x < 2$). The observed amorphous layer likely constitutes a native oxide that immediately passivates the particles and accounts for their chemical stability in air. Their stability against oxidation has been confirmed by thermal analysis. TGA shows that no mass gain due to the oxidation process occurs up to 200 °C. In addition, no visible color change is observed on a time-scale of 6 months after synthesis.

Particles from SDG are smaller that the ones produced via LaCVP. Their primary particle size varies between 2 and 5 nm, however, some bigger (50-200 nm) particles are also observed. Their presence is attributed to the solidification of liquid droplets of silicon that are sometimes expelled from the high temperature region were the spark interacts with the electrodes. The silicon nano-particles show a fractal-like structure that is typical for SDG [12]. The clearly visible lattice lines confirm their crystalline nature; the spacing distance between the (111) Si crystalline planes is observed in figure 2 (see inset of the picture on the right).

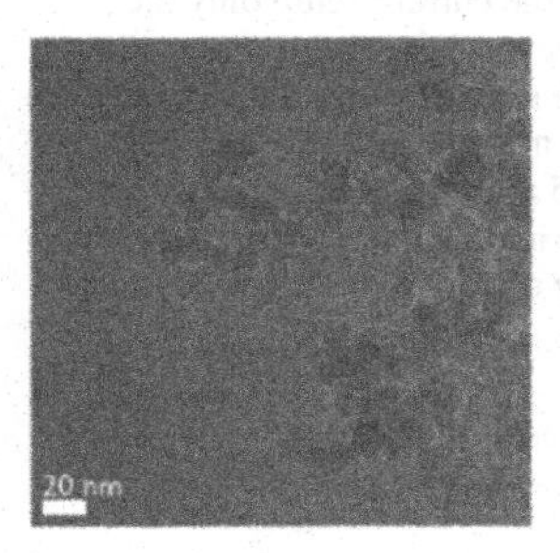

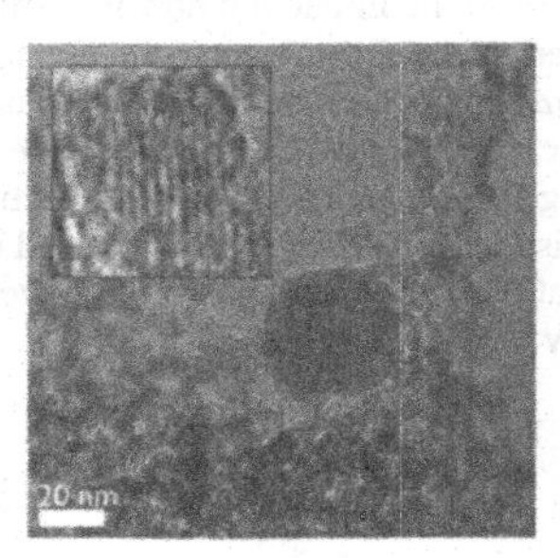

Figure 2. TEM micrographs of the synthesized nano-particles. Left: particles produced via LaCVP. Right: particles produced via SDG. The inset shows a magnification of a particle in which the (111) crystal planes are visible.

The XRD patterns confirm the purity of the produced samples. No traces of crystalline silicon oxides are detected; suggesting that surface oxidation of the particles resulted in a mere amorphous oxide layer. Figure 3 shows XRD spectra of LaCVP and SDG samples (left and right respectively). For the LaCVP particles it is possible to apply the Scherrer formula to calculate the crystallite size. An average crystallite size of 60 Å is estimated for LaCVP particles.

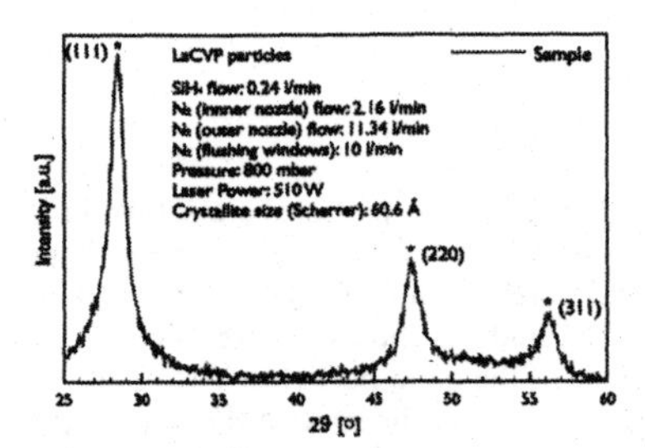

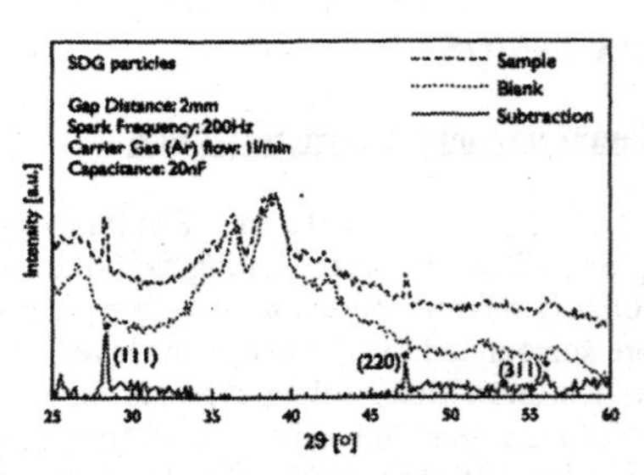

Figure 3. XRD spectra of the synthesized nano-particles. Left: particles produced via LaCVP. Right: particles produced via SDG. Synthesis parameters are specified in the legend. It is important to remark that the limited intensity in the second case is due to the small amount of powder available for the measurement.

For the SDG particles, the analysis was performed on the powder deposited on the filter membrane, due to the fact that the amount of the collected material was rather limited (i.e. ~1mg). Subtraction of the spectrum of the "blank" filter yields a clearer visualization of the Si peaks corresponding to (111), (220) and (311) crystallographic planes in the graph (see figure 3, right).

Thin film characterization

Two lateral views of the cross sections of the deposited Si layers are shown in figure 4. The ES layer (figure 4, left) is about 17 μm thick and shows a compact porous structure. The layer is homogeneous in thickness over the entire sprayed surface. Inertial impacted layers show a variable thickness that ranges from 6 to 7 μm depending on the position with respect to the nozzle center. The appearance of the impacted film is porous, with a loose-like structure, partly fibrous. The fibers are the result of the impaction of particles with the same kinetic energy on the same spot. Therefore these are orientated in the impaction direction (see figure 4, right).

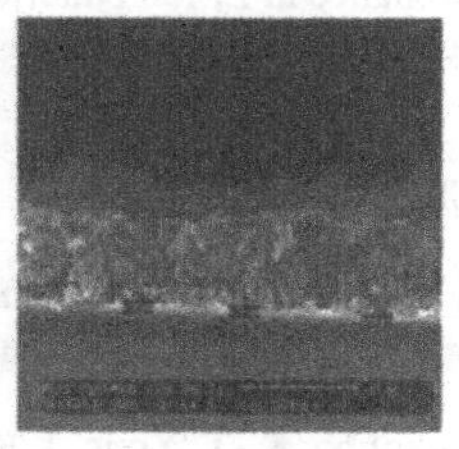

Figure 4. SEM lateral view of the cross-section for the deposited thin layers. ES Si-CMC layer (left) and II layer (right). Note that the layers are deposited on aluminum and silicon wafer respectively.

The morphology of the thin films, investigated by AFM, are shown in figure 5. Both layers are nano-structured, highly porous, with quasi-spherical agglomerates whose size ranges from 100 to 200 nm for the ES Si-CMC sample and from 40 to 70 nm for the II sample.

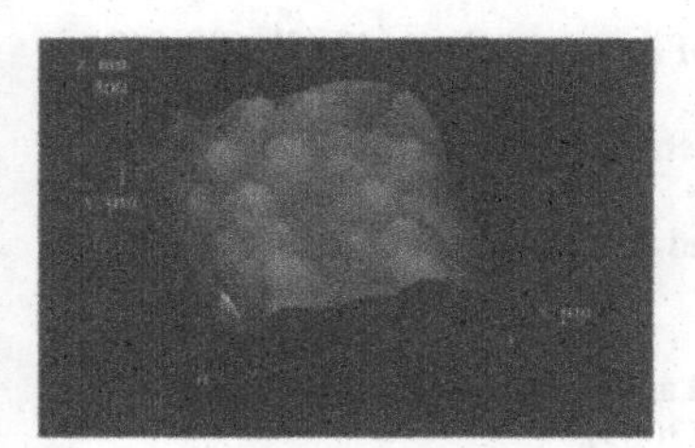

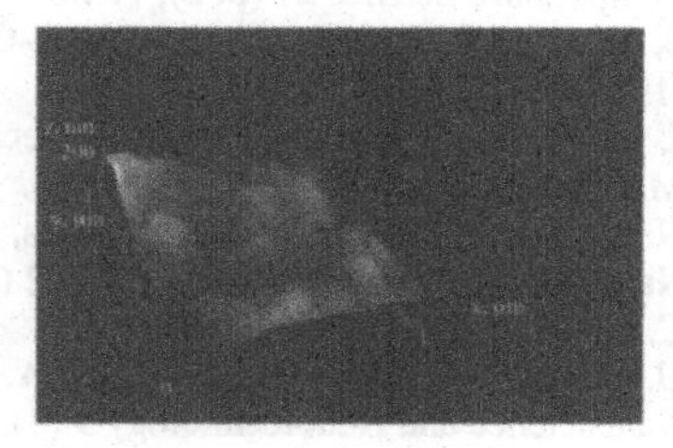

Figure 5. AFM images of the deposited thin layers. ES Si-CMC layer (left) and II layer (right).

It is worth noting that it is also possible to observe the influence of the Li alloying/de-alloying process on the electrosprayed composite electrodes before and after cycling. The SEM analyses show that in the case of the Si-PVdF sample there is no evidence of macro-fracture on the electrode surface, showing the layer is still intact even after 50 cycles. Both samples, Si-PVdF and Si-CMC, are not locally damaged in their sub-micrometric structure upon cycling. SEM and AFM show that only the agglomerates tend to grow in size. As a result the capacity retention of the system is raised and the mechanical resistance towards strain is significantly enhanced when compared to Si-based anodes based on bigger particles (i.e. sub-micron sized particles).

CONCLUSIONS

Two novel techniques to generate thin films with a porous structure made of silicon nano-particles are presented. The resulting films are nano-structured and a good control over the morphology can be achieved. These structures are attractive candidates for applications to Li-ion battery electrodes or gas sensor due to their high surface to volume ratio, which enhances the kinetics of the specific reactions. In particular, this type of nano-composite electrodes shows a superior performance in Li-ion batteries when compared to micron-sized, silicon-based materials. Their cycle life is enhanced due to a high mechanical stability of the thin film.

REFERENCES

1. X. Li, Y. He, S. S. Talukdar and M. T. Swihart, Langmuir **19** (20), 8490-8496 (2003).
2. J. Graetz, C. Ahn, R. Yazami and B. Fultz, Electrochemical And Solid State Letters **6** (9), A194-A197 (2003).
3. F. Erogbogbo, K.-T. Yong, I. Roy, G. X. Xu, P. N. Prasad and M. T. Swihart, ACS Nano **2** (5), 873-878 (2008).
4. U. Kasavajjula, C. Wang and A. J. Appleby, J Power Sources **163** (2), 1003-1039 (2007).
5. W. Cannon, S. Danforth, J. Flint, J. Haggerty and R. Marra, Journal Of The American Ceramic Society **65** (7), 324-330 (1982).
6. W. Cannon, S. Danforth, J. Haggerty and R. Marra, Journal Of The American Ceramic Society **65** (7), 330-335 (1982).
7. R. A. Bauer, F. E. Kruis, J. G. M. Becht, B. Scarlett and J. Schoonman, High Temperature Science **27** (pt 2), 77-88 (1988).
8. S. Schwyn, E. Garwin and A. Schmidt-Ott, Journal of Aerosol Science **19** (5), 639-642 (1988).
9. J. van Erven, D. Munao, Z. Fu, T. Trzeciak, R. Janssen, E. Kelder and J. C. M. Marijnissen, Kona **27**, 157-173 (2009).
10. N. S. Tabrizi, M. Ullmann, V. A. Vons, U. Lafont and A. Schmidt-Ott, Journal of Nanoparticle Research **11** (2), 315-332 (2009).
11. C. Peineke, PhD, TUDelft, 2008.
12. U. Lafont, L. Simonin, N. S. Tabrizi, A. Schmidt-Ott and E. M. Kelder, Journal of Nanoscience and Nanotechnology **9** (4), 2546-2552 (2009).

Films and Growth

Mater. Res. Soc. Symp. Proc. Vol. 1245 © 2010 Materials Research Society 1245-A10-01

Chemical Transport Deposition of Purified Poly-Si Films from Metallurgical-Grade Si Using Subatmospheric-Pressure H_2 Plasma

Kiyoshi Yasutake[1,2,3], Hiromasa Ohmi[1,2,3] and Hiroaki Kakiuchi[1,3]

[1] Department of Precision Science and Technology, Graduate School of Engineering, Osaka University, 2-1 Yamadaoka, Suita, Osaka 565-0871, Japan

[2] Research Center for Ultra-Precision Science and Technology, Graduate School of Engineering, Osaka University, 2-1 Yamadaoka, Suita, Osaka 565-0871, Japan

[3] Japan Science and Technology Agency, CREST, 5 Sanbancho, Chiyoda-ku, Tokyo, 102-0075, Japan

ABSTRACT

Purified Si film is prepared directly from metallurgical-grade Si (MG-Si) by chemical transport using subatmospheric-pressure H_2 plasma. The purification mechanism is based on the selective etching of Si by atomic H. Since most metals are not etched by H, this process is efficient to reduce metal impurities in Si films. It is demonstrated that the concentrations of most metal impurities (Fe, Mn, Ti, Co, Cr, Ni, etc.) in the prepared Si film are in the acceptable range for applying it to solar-grade Si (SOG-Si) material, or below the determination limit of the present measurements. On the other hand B and P atoms, which make volatile hydrogen compounds such as B_2H_6 and PH_3, are difficult to eliminate by the present principle. From the infrared absorption measurements of the etching product produced by the reaction between H_2 plasma and MG-Si, it is found that the main etching product is SiH_4. Therefore, a remote-type chemical transport process is possible to produce SiH_4 gas directly from MG-Si. Combining other purifying principle (such as a pyrolysis filter), this process may have an advantage to eliminate B_2H_6 and PH_3 from the produced SiH_4 gas.

INTRODUCTION

The recent explosive increase in photovoltaic (PV) market causes the shortage of high-purity Si material for solar cell production [1,2]. To solve this problem and make Si solar cells less expensive, reducing the cost of polycrystalline Si for solar cells continues to be a high priority issue in PV industries. Hence, there have been many efforts to produce SOG-Si feedstock from MG-Si at a high efficiency by substituting metallurgical processes such as directional solidification for conventional chemical purification processes such as Siemens process [3–5]. In the metallurgical process [6,7], not only long solidification time but also high temperature to melt Si is necessary, and the elimination rates for metal impurities typically range between 10^{-2} and 10^{-3} in a single solidification process.

On the other hand, in terms of reducing the consumption of high-purity Si material, thin-film Si solar cells are considered to be preferable [8]. Generally, Si thin films are fabricated by various chemical vapor deposition (CVD) techniques [9,10]. A high purity feedstock gas, such as SiH_4, is needed in the CVD process. SiH_4 gas is also produced from MG-Si through the classical chemical purification process. SiH_4 gas is not only toxic, but also a precious compound. Therefore, a simple, high-efficiency method for preparing high-purity Si film directly from cheap MG-Si would be attractive for overcoming the Si feedstock problem.

For efficient fabrication of Si films, Ohmi et al. [11–15] have proposed atmospheric-pressure plasma enhanced chemical transport (APECT) method in which no source gas is needed. In this method, high-pressure (100 – 760 Torr) stable glow plasma of pure H_2 is generated between two parallel electrodes (less than 2 mm apart) by supplying a 150 MHz very high frequency (VHF) power. One of the electrodes is composed of cooled Si solid source and the other the heated substrate (200 – 400°C). The temperature difference between the Si solid source and the substrate is important for achieving unidirectional transport from the source to the substrate. Since the etching rate increases with decreasing temperature [16–18], the etching process dominates at cooled solid source side while deposition dominates at heated substrate side.

The hydrogenation reactions and properties of hydrogenated species strongly depend on the kinds of elements that are present in the MG-Si. For instance, it is known that the hydrides of typical metal atoms (Fe, Ti, Al) are unstable or difficult to volatilize by hydrogenation near room temperature [19–21]. Consequently, the Si matrix is preferentially etched by atomic H compared to impurity metal atoms. So, this process can be applied to reduce the concentration of impurities in a film deposited on a substrate.

In this article, we describe two new processes to produce purified Si material directly from MG-Si based on the APECT principle. One is the formation of purified Si film using 150 MHz VHF H_2 plasma which is generated between a narrow plasma gap at near atmospheric pressure (proximity-type APECT). The other is a new SiH_4 formation process directly from MG-Si by using 2.45 GHz microwave plasma of H_2 at near atmospheric pressure (remote-type APECT), which is briefly mentioned in the later part.

EXPERIMENTAL

Figure 1 schematically shows the experimental setup for proximity-type APECT deposition used in this study. The pure hydrogen plasma at near atmospheric-pressure was generated in the gap (1 mm) between the solid Si source and the substrate by the 150 MHz VHF power supply. The Si solid source was attached to the electrode made of copper, which was cooled by 20°C cooling water at a circulation rate of 2 L/min. A 2-mm-thick MG-Si plate sawn from an ingot (98% purity) was used as the solid source. P-type CZ-Si (1–10 Ωcm) and glass (Corning 1737) were used as substrates. In the present experiments, the H_2 pressure was varied in the range $P_{H2} = 100 – 760$ Torr, and the substrate temperature $T_{sub} = 200 – 600$°C and the VHF power $W_{VHF} = 16 – 160$ W/cm^2.

The volatile hydrogenated species generated from MG-Si were analyzed by gas phase Fourier-transform infrared (FTIR) absorption spectroscopy. The impurity concentrations in the prepared Si films were determined by inductively coupled plasma mass spectrometry (ICPMS: Agilent, 7500 ICPMS) and glow discharge mass spectrometry (GDMS: VG Elemental, VG9000). ICPMS measurements were conducted by dissolving the sample in HNO_3 :HF solution. In the

GDMS measurements, an area of 10 mm in diameter on the sample surface was sputtered by dc Ar glow discharge. The surface and cross-sectional morphologies of the Si films were observed by scanning electron microscopy (SEM; Hitachi S-800). The crystallinity of the film was characterized by cross sectional transmission electron microscopy (XTEM; JEOL JEM-2000FX).

Figure 2 shows the GDMS depth profiles of several metal impurities in one of the MG-Si plates used in this study. Since the profiles indicate rather smooth impurity distributions and the concentration difference in the two measured points (1 and 2) is below the measurement accuracy, significant impurity segregation may not be existent in the present MG-Si material.

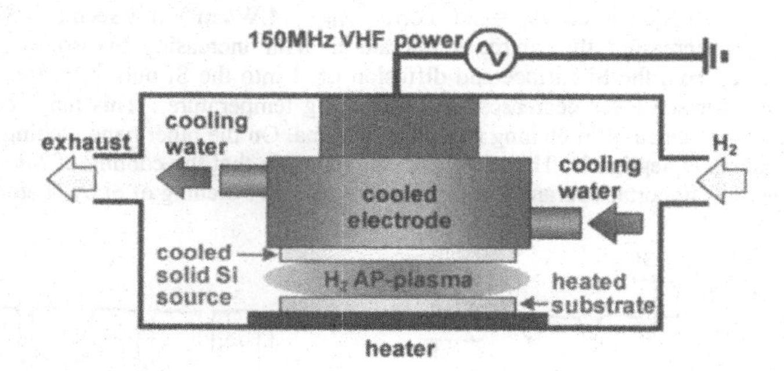

Figure 1. Experimental setup for proximity-type APECT deposition. Solid Si source is attached on water-cooled electrode by ceramic screws. The chamber has a load-lock system (not shown). The substrate heater is set on X-Z stage. The plasma gap is adjusted by Z stage.

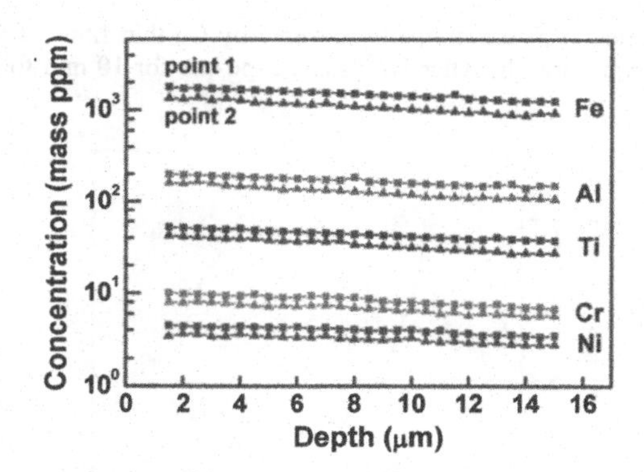

Figure 2. Depth profiles of metal impurity concentrations in MG-Si used in this study. GDMS measurements were performed at the two positions on MG-Si plate (points 1 and 2).

RESULTS AND DISCUSSION

Selective etching of Si by H_2 plasma

To confirm the selectivity of H etching, we prepared a Si specimen partly covered by Cu thin films as a solid source, which was attached to the water-cooled electrode. Figure 3 shows surface profiles of the specimen before and after the H_2 plasma exposure at P_{H2} = 200 Torr for 20 min. It is clear that etched parts correspond to the bare Si and the portions covered with Cu films are not etched by H_2 plasma. Figure 4 shows the temperature dependence of the etching rate for Si(001) wafer as a solid source (P_{H2} = 200 Torr, W_{VHF} = 44 W/cm^2). It is seen that Si etching rate increases with decreasing the source temperature. With increasing the source temperature, desorption of H from the Si surface and diffusion of H into the Si bulk increase. Then the H content in the surface layer decreases with increasing temperature. This may relate with the temperature dependence of Si etching rate by H_2 plasma. On the other hand etching rate of Fe at low temperatures is negligible. Therefore, we can conclude that the cooling of MG-Si feedstock is one of the most important parameters for efficient selective etching of Si by H atoms.

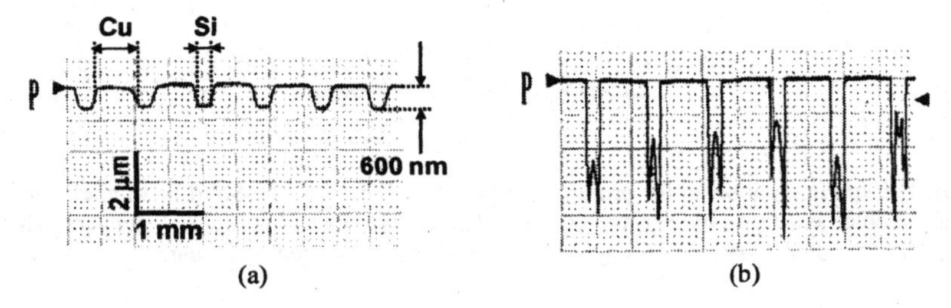

Figure 3. Surface profiles of Si sample partly covered by Cu thin films of 600 nm thickness. (a) Before H_2 plasma exposure. (b) After H_2 plasma exposure for 10 min followed by Cu film removal.

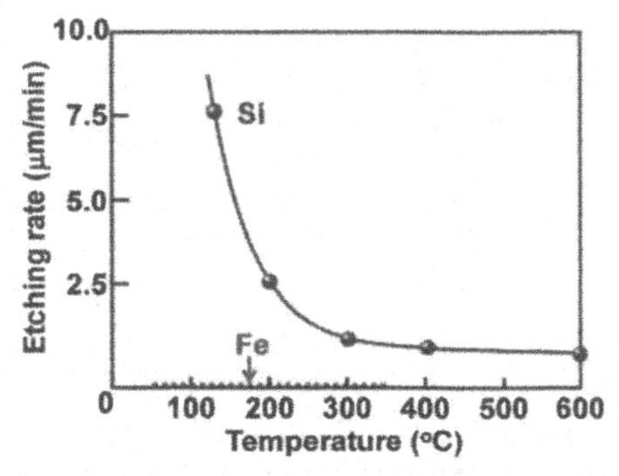

Figure 4. Temperature dependence of Si etching rate (P_{H2} = 200 Torr, W_{VHF} = 44 W/cm^2). Etching rate for Fe is negligible.

Identification of the etching product

The FTIR analysis of the generated stable volatile gas during APECT process was performed by replacing the substrate with a measurement jig on the heater stage. The jig has a 1 mm pinhole through which the gas in the plasma was suctioned and introduced into the FTIR gas cell located about 1 m away from the plasma. Figure 5 shows FTIR absorption spectra for various heater temperatures. The absorption bands around 976 and 2190 cm^{-1} are clearly visible, and these peaks are coincident with those of pure SiH_4 gas [22]. Absorption peaks belonging to higher-order silane (such as Si_2H_6 at 843 cm^{-1} [23]) cannot be seen. So, it is confirmed that the Si etching proceeds mainly by generating SiH_4 gas even at high pressure conditions at around 200 Torr. The reduction in the SiH_4 peak height with temperature is due to the reduction in the etching rate of Si. These results indicate that the SiH_4 generation is governed mainly by a chemical reaction between Si surface and H atoms.

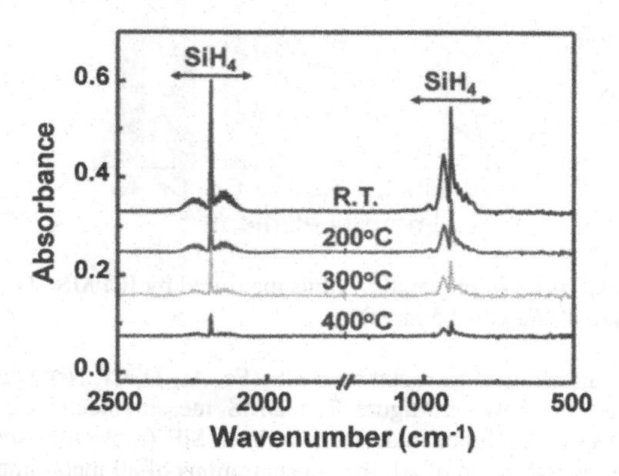

Figure 5. FTIR absorption spectra of generated gaseous species from MG-Si source. Temperature was varied by substrate heater. (P_{H2} = 200 Torr, W_{VHF} = 20 W/cm^2).

Deposition of purified Si films from MG-Si

Using MG-Si as a solid source for proximity-type APECT process, Si films with the thickness of 20 – 30 μm were deposited on Si substrate with the condition T_{sub} = 400 °C, P_{H2} = 200 Torr, W_{VHF} = 55 W/cm^2 and the deposition rate of R_{depo} = 3.5 nm/s. Figure 6 show the metal impurity concentrations in the deposited Si films measured by ICPMS. The solid and hatched bars indicate the impurity concentrations in MG-Si source and deposited Si film, respectively. The solid diamonds indicate the determination limits for each element, which depend on the background signal that might be caused by contamination introduced during sample preparation. The impurity concentrations of all the elements are reduced in the prepared Si film compared to the MG-Si source. In particular, the reduction rate of Fe concentration is about five orders of magnitude, and decreases below the determination limit. The Mn concentration also decreases below the determination limit. The Co, Cr, and Ni concentrations in the Si film decrease to either nearly equal to or less than the determination limits, but their initial concentrations in MG-Si are low. The Al concentration also decreases, but the obtained concentration is not below the determination limit. In the deposited film, Cu is detected at a relatively high concentration, which may not indicate the net concentration in the Si film. The obtained Cu concentration seems to be affected by chemical complexes (HNO_3, etc.) with the same molecular weight as Cu (mass number: 63).

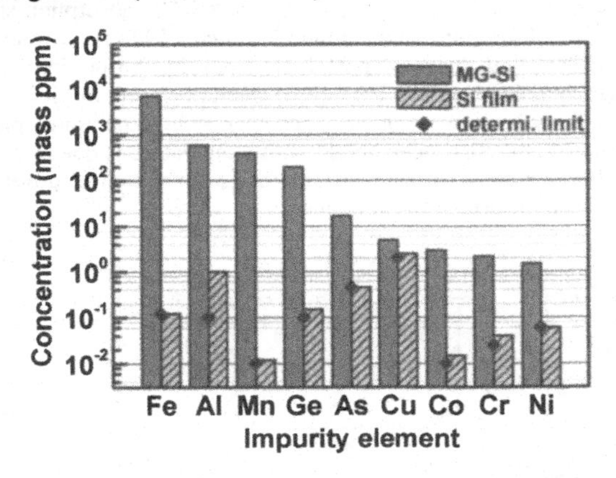

Figure 6. Impurity concentrations in prepared Si films measured by ICPMS. (T_{sub} = 400 °C, P_{H2} = 200 Torr, W_{VHF} = 55 W/cm^2, R_{depo} = 3.5 nm/s)

The impurity concentrations of some metal elements (Fe, Al, Ti, Cr, Ni) were also evaluated by GDMS and the result is shown in figure 7. GDMS measurements are less subject to interfering ions arising from dissolution reactions, such as SiF (molecule weight: 47) for Ti (mass number: 47). With the exception of Al, the concentrations of all metal impurities decrease below the determination limits by chemical transport process. The determination limits of Fe, Ti, and Ni concentrations are 0.02, 0.005, and 0.03 mass ppm, respectively. On the other hand, although Al concentration is reduced by chemical transport, its removal rate is lower than that of Fe. The Al concentration in the Si film measured by GDMS is about 0.5 mass ppm, whereas the determination limit is about 0.05 mass ppm. This implies that some of the Al volatilized to the plasma by forming its hydride (AlH_3).

It is supposed that most metal impurities are not etched by H_2 plasma and they remain in MG-Si source. To confirm this hypothesis, we measured the surface concentration of metal impurities in the MG-Si source by total reflection X-ray fluorescence (TREX) method. Figure 8(a) shows the photograph of MG-Si surface after H_2 plasma exposure for 30 min (P_{H2} = 200 Torr, W_{VHF} = 55 W/cm^2). A shiny part (6 × 3 cm^2) is produced by the H_2 plasma exposure, where the TREX measurements were performed. Figure 8(b) shows the surface impurity concentrations measured before and after the plasma exposure. Each datum indicates the surface concentration value averaged over 24 sample points. Surface concentrations of all the measured metal impurities in the MG-Si source increase after the H_2 plasma exposure. This may suggest that the metal impurities remain in MG-Si by the selective etching of Si. On the other hand, no metallic particle was detected on the deposited Si surface with the film thickness of up to 50 μm.

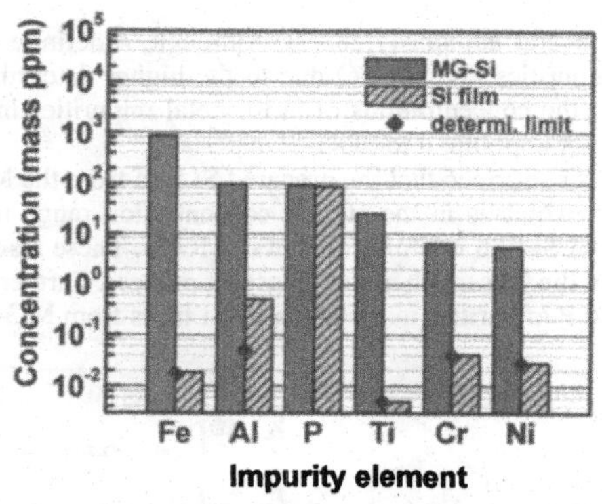

Figure 7. Impurity concentrations in prepared Si films measured by GDMS. (T_{sub} = 400 °C, P_{H2} = 200 Torr, W_{VHF} = 55 W/cm^2, R_{depo} = 3.5 nm/s)

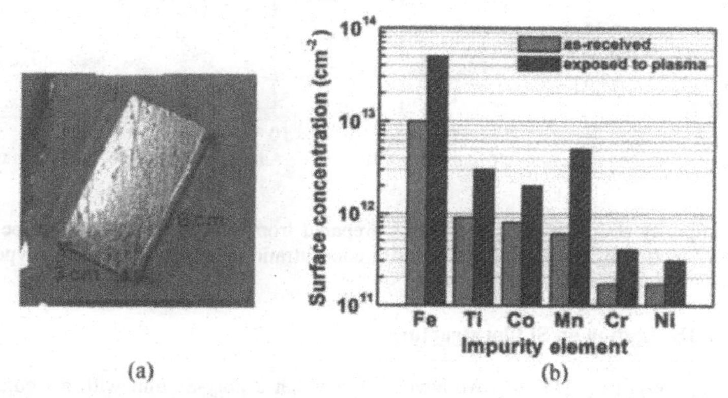

Figure 8. Photograph of MG-Si source (a), and surface impurity concentrations in MG-Si measured by TREX before and after the H_2 plasma exposure for 30 min (b). (P_{H2} = 200 Torr, W_{VHF} = 55 W/cm^2)

However, to eliminate the possibility of falling down of the metal particles to the film surface, it may be better to set the MG-Si source at the bottom and the substrate at the top position.

In SOG-Si, the concentrations of similar impurities to those in SEG-Si should be limited, but the acceptable levels are substantially higher. The admissible impurity levels in SOG-Si were defined by studying the conversion efficiency of solar cells prepared from Czochralski grown Si single crystals as a function of impurity concentration [24,25]. The results can be used to determine the degradation threshold value for solar cell efficiency. Figure 9 shows our measured impurity concentrations in Si films prepared from MG-Si by APECT process (figures 6 and 7) as a function of the admissible impurity levels in SOG-Si based on the data reported by Davis et al.

[24,25] and reviewed by several authors [1,26,27]. Although, a definite conclusion cannot be drawn on several metal impurities (Ti, Cr, Fe) due to the higher determination limits than the admissible impurity levels, the concentrations of most metal impurities in the prepared Si film are in the acceptable range for applying it to SOG-Si material.

At the present stage, it is confirmed that the prepared Si film from the MG-Si source contains about 100 mass ppm of P, which is in the similar concentration range in MG-Si. Also the B concentration in the prepared Si film is similar to that in MG-Si. These results mean that B and P atoms are not eliminated by the present chemical transport process. Further study is necessary on the elimination of Al, B and P impurities in the prepared Si films from MG-Si.

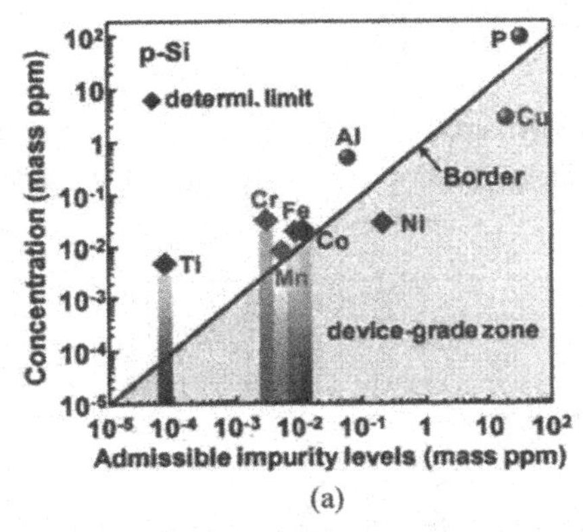

(a)

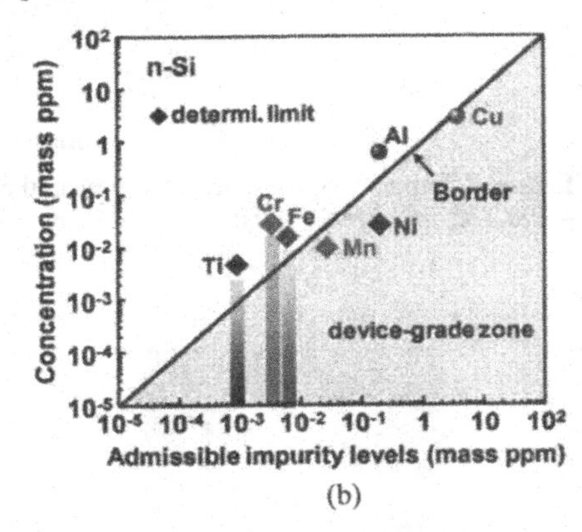

(b)

Figure 9. Impurity concentrations in Si films prepared from MG-Si by proximity-type APECT process as a function of the admissible impurity concentrations in SOG-Si for (a) p-type and (b) n-type Si.

Influence of He dilution on Si film structure

In the previous study [12–14], we have reported that a poly-Si film with a wedge-shaped columnar structure was formed at a deposition rate of 15 nm/s. The average grain size in the prepared Si film with the thickness of 4.5 μm was about 2 μm. The Si film was defective, and was not readily applicable to the base layer of the Si solar cell. However, it has been reported that the noble gas dilution of H_2 plasma dramatically improves the surface morphology and the film structure, while the process pressure slightly affects the film structure [14]. Figure 10 shows surface SEM images of Si films prepared by a 100 % H_2 plasma (a) and by a He–25% H_2 plasma (b). RHEED patterns are shown as insets in the corresponding SEM images. These figures show that the grain morphology changed dramatically upon He dilution. Without He dilution, the grains reveal anisotropic rhombus morphology, which preferentially grows along the <110> directions. With He dilution, the grains become square-shaped.

Figure 11 shows the XTEM bright field images of the corresponding Si films to figure 10. Each insertion shows the transmission electron diffraction (TED) pattern of the film. In the case of pure H_2 plasma (figure 11(a)), fine grains are observed near the interface between the film and the substrate. When the film grows thicker than about 250 nm, each grain becomes clear and enlarges its lateral size with increasing film thickness. Each grain contains many defects with random orientation. The diffraction pattern shown in the inset exhibits a ring pattern with an overlapped spot pattern. Therefore, it is concluded that the Si film prepared without He dilution show a fine-grained poly-Si structure from the beginning of the film growth. On the other hand, in the case of Si film prepared with He dilution (figure 11(b)), the grain size increases and the density of the dark area decreases considerably compared to the Si film prepared with 100% H_2. The sides of the grain are nearly along <110>. Although each Si grain contains defects parallel to {111} planes (stacking faults or hydrogen-induced platelet defects), the defect density in the Si film decreases upon He dilution. The TED pattern shown in the inset in figure 11(b) exhibits the same diffraction pattern as the substrate, which means that the Si film grew epitaxially on the substrate. It has been also revealed by FT-IR measurements (data not shown) that He diluted APECT method can suppress the post oxidation of the Si films. Plasma diagnostic results showed that the gas temperature is lower and electron temperature is higher in He diluted plasma than those in pure H_2 case. So, the kind and/or energy of precursors in these plasmas are different, which may relate with the difference in surface migration of precursors between these plasmas.

Figure 10. Surface SEM images of Si films prepared with 100% H_2 plasma (a) and He–25% H_2 plasma (b). RHEED patterns are shown as insets in the corresponding SEM images.

Figure 11. XTEM images of Si films prepared with 100% H_2 plasma (a), and He–25% H_2 plasma (b). TED patterns are shown as insets in the corresponding XTEM images.

By diluting H_2 with He, deposition rate of Si films (R_{depo}) reduced gradually. The reduction ratio of R_{depo} by 50 and 75% He dilution is about 0.78 and 0.52 from that of 100% H_2 plasma, respectively. So, it can be said that without much reduction of R_{depo}, crystallinity of the Si films can be dramatically improved by He dilution. A detailed characterization of defects in the prepared Si film is needed to evaluate its applicability to the base layer of the Si solar cell.

High-rate SiH_4 formation using 2.45 GHz microwave plasma

By the proximity-type APECT method (figure 1), B and P impurities are difficult to be eliminated from the deposited Si films. To reduce B and P atoms by combining the other purification principle than the Si selective etching, we are studying a remote-type APECT method. In this method, SiH_4 gas is generated with high-rates by selective etching of MG-Si feedstock by H atoms generated by 2.45 GHz microwave plasma. Preliminary experiments were performed with a simple electrode configuration made of a stainless steel pipe through which H_2 gas was supplied to the H_2 plasma. The plasma gap between the electrode and Si feedstock was 0.6 mm.

To increase Si etching rate, we performed the etching experiments under various plasma conditions. The H_2 pressure dependence of Si etching rate revealed that it increases with H_2 pressure until about 400 Torr. However, by further increasing H_2 pressure, the decrease in Si etching rate due to the 3 body recombination of H atoms in the gas phase may become apparent. The recombination loss rate increases to the 3rd power of the pressure. So, it is necessary to further study on the optimum operating pressure for high-efficiency H atom generation. With an optimized condition at present, we have obtained the highest Si etching rate of 38 μm/min [28]. This very high etching rate is more than 2 orders of magnitude higher than the values obtained by hot wire or low-pressure plasma methods [29,30]. This may relate with the high H atom generation by microwave plasma.

To reduce the production cost of SiH_4, the Si recovery as SiH_4 is an important parameter as well as the high Si etching rate. To estimate the Si recovery, we measured the SiH_4 concentration in the etching product by FT-IR and divided it with the etched Si amount. From the estimated Si recovery as a function of the gas residence time in the plasma, it was found that when the gas velocity is small, the generated SiH_4 gas is re-decomposed in the plasma to make particles. So, the gas residence time should be short to get high Si recovery. From above results, we can get some guiding principles to develop high-rate SiH_4 generation process. Namely, i) high-power microwave plasma is efficient for H atom generation but the temperature of MG-Si should be kept low, ii) the higher H_2 pressure is better to generate high-density H atoms as long as 3-body recombination loss is not apparent, and iii) high gas velocity is needed to obtain high Si recovery.

Elimination method of B and P impurities

Concerning the impurities that make volatile hydrogen compounds, such as B_2H_6 or PH_3, they are difficult to be eliminated by the present refining principle. Especially, since B and P are electrically active in Si, they should be eliminated from SOG-Si to the level below 0.1 ppm. Therefore, for producing high-purity Si material, a combination with another purifying principle is indispensable. So, firstly, we are developing a pyrolysis filter to eliminate B [31,32]. Since B_2H_6 decomposes at lower temperatures than SiH_4, we can use the appropriate filter temperature to eliminate B_2H_6 from SiH_4. The pyrolysis filter is made of heated porous carbon material. From the temperature dependence of gas transmission rate measured by the FT-IR absorption spectroscopy, decomposition temperature for SiH_4 is about 450°C, while that for B_2H_6 is about 200°C. Therefore, we can purify SiH_4 gas by setting the temperature of pyrolysis filter about 400°C. A similar study to eliminate P is under way.

CONCLUSIONS

In summary, the ICPMS and GDMS results demonstrate that a purified Si film can be prepared directly from MG-Si by chemical transport in 150 MHz VHF H_2 plasma (proximity-type APECT). The elimination rates for metal impurities depend on the kind of elements, and they either equal to or surpass those of the conventional metallurgical solidification method. The prepared Si film from MG-Si is polycrystalline. The surface morphology and the crystallinity of the Si film can be drastically improved by He dilution of the H_2 plasma. Preliminary results on the direct SiH_4 gas formation from MG-Si using 2.45 GHz microwave H_2 plasma (remote-type APECT) show that this method is capable of reducing B atoms efficiently, which is promising as a new process for SOG-Si production. To lower the manufacturing cost of Si by the present process to compete with the classical trichlorosilane process, further study is needed on the utilization efficiencies of applied plasma power, H_2 gas and MG-Si material based on atmospheric-pressure plasma science.

ACKNOWLEDGMENTS

The authors are grateful to Professor H. Mori of Research Center for Ultra-High Voltage Electron Microscopy, Osaka University, for the support in the TEM observations. The technical assistance of Messrs A. Takeuchi and N. Nishimoto of Osaka University is also greatly appreciated. This work was partially supported by the Industrial Technology Research Grant Program in 2006 from the New Energy and Industrial Technology Development Organization (NEDO) of Japan (Grant No. 06A31013d).

REFERENCES

1. S. Pizzini, *Appl. Phys.* **A96**, 171 (2009).
2. K. Hesse, E. Schindlbeck, E. Dornberger and M. Fischer, in *Proc. 24th Eur. Photovltaic Solar Energy Conf., Hamburg*, 2009, p.883.
3. G. Flamant, V. Kurtcuoglu, J. Murray and A. Steinfeld, *Sol. Energy Mater. Sol. Cells* **90**, 2099 (2006).
4. J. Degoulange, I. Perichaud, C. Trassy and S. Martinuzzi, *Sol. Energy Mater. Sol. Cells* **92**, 1269 (2008).
5. Y. Wan, P. S. Raghavan, C. Chartier, J. Talbott and C. Khattak, *Proc. 2006 IEEE 4th World Conf. on Photovoltaic Energy Conversion, Waikoloa, HI* (IEEE, Piscataway, 2006), p. 1342.
6. J. C. S. Pires, J. Otubo, A. F. B. Baga and P. R. Mei, *J. Mater. Process. Technol.* **169**, 16 (2005).
7. N. Yuge, H. Baba, Y. Sakaguchi, K. Nishikawa, H. Terashita and F. Aratani, *Sol. Energy Mater. Sol. Cells* **34**, 243 (1994).

8. S. Hegedus, *Prog. Photovolt. Res. Appl.* **14**, 393 (2006).
9. A. Illiberi, K. Sharma, M. Creatore and M. C. M. van de Sanden, *Mater. Lett.* **63**, 1817 (2009).
10. I. T. Martin, H. M. Branz, P. Stradins, D. L. Young, R. C. Reedy and C. W. Teplin, *Thin Solid Films* **517**, 3496 (2009).
11. H. Ohmi, K. Yasutake and H. Kakiuchi, WO/2007/049402, 3 May 2007.
12. H. Ohmi, H. Kakiuchi, Y. Hamaoka and K. Yasutake, *J. Appl. Phys.* **102**, 023302 (2007).
13. H. Ohmi, K. Kishimoto, H. Kakiuchi and K. Yasutake, *J. Phys.* **D41**, 195208 (2008) .
14. D. Kamada, K. Kishimoto, H. Kakiuchi, K. Yasutake and H. Ohmi, *Surf. Interface Anal.* **40**, 979 (2008).
15. H. Ohmi, A. Goto, D. Kamada, Y. Hamaoka, H. Kakiuchi and K. Yasutake, *Appl. Phys. Lett.* **95**, 181506 (2009).
16. A. P. Webb and S. Veprek, *Chem. Phys. Lett.* **62**, 173 (1979).
17. H. Matsumura, K. Kamesaki, A. Masuda and A. Izumi, *Jpn. J. Appl. Phys.* **40**, L289 (2001).
18. H. Ohmi, H. Kakiuchi, K. Nishijima, H. Watanabe and K. Yasutake, *Jpn. J. Appl. Phys.* **45**, 8488 (2006).
19. J. V. Badding, R. J. Hemley and H. K. Mao, *Science* **253**, 421 (1991).
20. C. Borchers, T. I. Khomenko, A. V. Lenov and O. S. Morozova, *Thermochim. Acta* **493**, 80 (2009).
21. G. Sandrock, J. Reilly, J. Graetz, W.-M. Zhou, J. Johnson and J. Wegrzyn, *Appl. Phys.* **A80**, 687 (2005).
22. B. Bartlome, A. Feltrin and C. Balif, *Appl. Phys. Lett.* **94**, 201501 (2009).
23. P. W. Morrison, Jr. and J. R. Haigis, *J. Vac. Sci. Technol.* **A11**, 490 (1993).
24. J. R. Davis, A. Rohatgi, R. H. Hopkins, P. D. Blais, P. Rai-Choudhury, J. R. McCormick, H. C. Mollenkopf, *IEEE Trans. on Electron Dev.* **ED-27**, 677 (1980).
25. R. H. Hopkins, J. R. Davis, A. Rohatgi, M. H. Hanes, P. Rai-Chaudhury and H. C. Mollenkopf , *Final Report, DOE/JPL-* 954331-82/13 (1982).
26. J. Dietl, *Sol. Cells* **10**, 145 (1983).
27. B. G. Gribov and K. V. Zinov'ev, *Inorg. Mater.* **39**, 653 (2003).
28. T. Funaki, Y. Nakahama, M. Kadono, K. Yasutake and H. Ohmi, *Ext. Abst. 70th Autumn Meeting, 2009, Jpn. Soc. Appl. Phys.*, 10p-N-3.
29. Izumi, H. Sato, S. Hashioka, M. Kubo and H. Matsumura, *Microelectron. Eng.* **51-52**, 495 (2000).
30. K. Sasaki and T. Takada, *Jpn. J. Appl. Phys.* **37**, 402 (1998).
31. H. Ohmi, K. Yasutake, Y. Nakahama and T. Funaki, WO/2009/154232, 23 Dec. 2009.
32. H. Ohmi, K. Yasutake, Y. Nakahama, T. Funaki and M. Kadono, *Ext. Abst. 57th Spring Meeting, 2010, Jpn. Soc. Appl. Phys.*, 19a-ZB-11.

Mater. Res. Soc. Symp. Proc. Vol. 1245 © 2010 Materials Research Society 1245-A10-04

Ultrafast Deposition of Crystalline Si Films Using a High Density Microwave Plasma

Haijun Jia[1,2] and Michio Kondo[1]
[1]Research Center for Photovoltaics, National Institute of Advanced Industrial Science and Technology (AIST), Tsukuba, Ibaraki 305-8568, Japan
[2]Baoding TianWei SolarFilms Co., Ltd. Baoding, Hebei 071051, People's Republic of China

ABSTRACT

A multi-pressure microwave plasma source is developed and is applied for the fast deposition of crystalline silicon films. In this paper, the plasma source is diagnosed firstly. Electron density, electron temperature and discharge gas temperature of the plasmas generated in ambient air are studied using optical emission spectroscopy (OES) method. By using the high density microwave plasma source, depositions of crystalline silicon films from SiH_4+He mixture at reduced pressure conditions are investigated systematically. After optimizing the film deposition conditions, highly crystallized Si films are deposited at a rate higher than 700 nm/s. We also find that the deposited films are fully crystallized and crystalline structure of the deposited film evolves along the film growth direction, i.e. large grains in surface region while small grains in the bottom region of the film. Based on the observed results, a possible mechanism, the annealing-assisted plasma-enhanced chemical vapor deposition, is proposed to describe the film growth process.

INTRODUCTION

Due to the high carrier mobility, strong stability against light exposure and wide-range spectral sensitivity, high quality crystalline silicon thin films have attracted much attention as a promising material for electronic device applications such as thin film transistors and solar cells [1,2]. Among the various approaches for the preparation of crystalline Si films, the most successful technique thus far is the plasma based chemical vapor deposition (PECVD) process either ignited by radio frequency (rf) or by very high frequency (VHF) using SiH_4 highly diluted by H_2 as reactive source gases [3,4]. For the cost-effective applications of the material, special attention has been addressed in the recent year to increase its deposition rate while maintaining (reasonable) film quality [5].Recently, there is a growing interest in transferring the low-pressure plasma technologies to high pressure or even atmospheric pressure conditions [6]. The enhanced plasma chemistry in the latter case, especially the higher electron density and the lower electron temperature, offers a high potential for material preparation and modification.

Taking into account the above points, in this study, a microwave plasma source, which allows the plasmas to be generated in a wide pressure range from ~1 Torr to atmospheric pressure, is used for the fast deposition of crystalline Si films. In this paper, firstly, we diagnose the properties of the plasma source at atmospheric pressure condition. Moreover, we perform the deposition of crystalline Si films from SiH_4+He mixture by using this microwave plasma source.

Through optimizing the deposition parameters, a deposition rate over 700 nm/s is achieved for the highly crystallized Si films. Mechanisms behind the results are discussed in detail. We find that the film crystalline structure evolves during the deposition process. Annealing-assisted plasma enhanced chemical vapor deposition is proposed to describe the evolution.

EXPERIMENTS

The microwave-induced plasma source is shown schematically in Fig. 1 [7]. It is a plasma source (AET Japan, Inc.) featuring a hybrid-mode resonator. The cylindrical brass cavity possesses dimensions designed to permit internal resonant oscillation of an electromagnetic wave (2.45 GHz). Microwave power is supplied through a coaxial cable and is coupled into the cavity through a rod antenna. A quartz glass tube with a diameter of 10 mm is positioned in the center of the cavity. The plasma is generated inside the hollow quartz tube around a spiral tungsten antenna. Source gases are introduced from one end of the tube and flow towards the other open end. By using the plasma source, helium and argon plasmas can be generated either in an open atmosphere environment or at a reduced-pressure condition. The entire plasma generation unit is installed inside a stainless steel chamber connected to pumps, a gas supply system and an exhaust gas absorption system.

Characteristics of the microwave-induced plasma source were diagnosed at atmospheric pressure. Optical emission spectroscopy (OES) is used for the diagnoses instead of the electrostatic Langmuir probe that is usually used for the low-pressure plasmas.

Si film depositions using the microwave plasma source were performed with SiH_4 and He as source gases. Flow rate of SiH_4 and microwave power were used as the variable deposition parameters. The SiH_4 flow rate varied from 5 sccm to 50 sccm and the microwave power changed from ~45 W to ~130 W. Flow rate of the He gas was fixed at 400 sccm and a working pressure of 2 Torr was employed during the depositions. The depositions were continued for 30 seconds. A quartz fiber (0.3 mm in diameter) used as substrate was placed in the center of the quartz tube. A thermal stable and flexible fiber can be applied to fabricate a fiber-type thin film solar cell with advantages of light weight, high flexibility and possibility of roll-to-roll fabrication process. No intentional heating of the substrate was performed during the depositions. The deposited Si films were characterized using scanning electron microscope (SEM) and transition electron microscope (TEM), Raman spectroscopy and electron spin resonance (ESR) to evaluate the film deposition rate R_d, crystallinity I_c/I_a and defect density N_s. Here, the film Raman crystallinity I_c/I_a was defined as a ratio of the integrated intensity of the Raman peak centered at 520 cm^{-1} from crystalline phase, I_c, to the intensity of the peak centered at 480 cm^{-1} from amorphous phase, I_a.

RESULTS

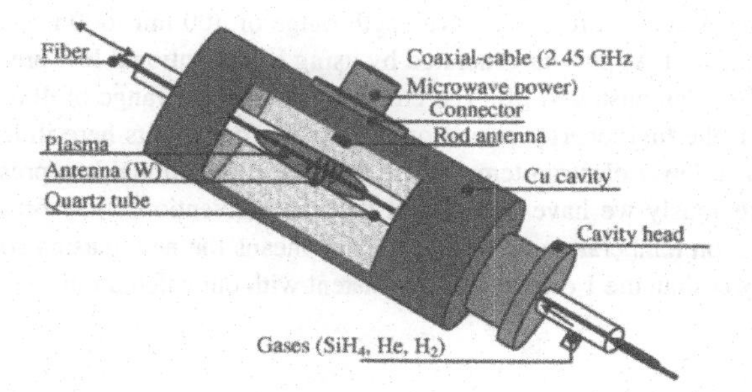

Fig. 1 (a) Schematic diagram of the microwave induced plasma source used in this study.

Characteristics of the microwave-induced plasma source

As the most important plasma parameters, the electron density (n_e), electron temperature (T_e) and discharge gas temperature (T_{gas}) of the microwave-induced plasma source were investigated. [7] The measurements were performed in He (for n_e and T_{gas} measurements) plasmas and Ar (for T_e measurement) plasmas generated in ambient air as function of the microwave power and the source gas flow rate.

Electron density

The electron n_e density was estimated by using a method based on the Stark broadening of the H_β spectral line (486.13 nm, 4d-2p) originating from the decomposition of H_2O.

As has been reported in our previous work, a n_e larger than 10^{15} cm^{-3} is obtained even at a microwave power of 30 W. With increasing the microwave power, the n_e also increases. A n_e as high as $>2.5\times10^{15}$ cm^{-3} can be achieved at the microwave power of 50 W. For the gas flow rate dependence of n_e, on the other hand, a slight decrease in n_e upon increasing the He flow rate is observed, which may be originated from the accelerated diffusion loss of the electrons to the air environment.

Electron temperature

The electron temperature T_e was also estimated by using the OES method by employing the Ar atmospheric pressure plasmas by using the Boltzmann plot method. Similar with the n_e, the T_e increases upon increasing the microwave power. However, even under a high power of 70 W, the T_e is still lower than 1 eV. Also similar with the changing tendency of the n_e, a decrease in the T_e

happens with increasing the Ar gas flow, which may due to the faster transition of kinetic energy of the electrons to the neutral species.

Figure 2 shows a typical optical emission spectrum of the Ar atmospheric pressure plasma generated using the microwave plasma source in a wavelength range of 400 nm -600 nm. As a comparison, an emission spectrum of Ar plasma generated by using a conventional low pressure microwave plasma source is also demonstrated [8]. The emission lines in the range of 400~600 nm are usually originated from the high energy excitation. Therefore, the results here illustrate that the new plasma source has a lower electron temperature than the conventional low pressure microwave plasma source. Previously we have determined that the conventional low pressure microwave plasma had an electron temperature of 1-1.2 eV. Than means the new plasma source has an electron temperature lower than the 1 ev, which is consistent with our calculation.

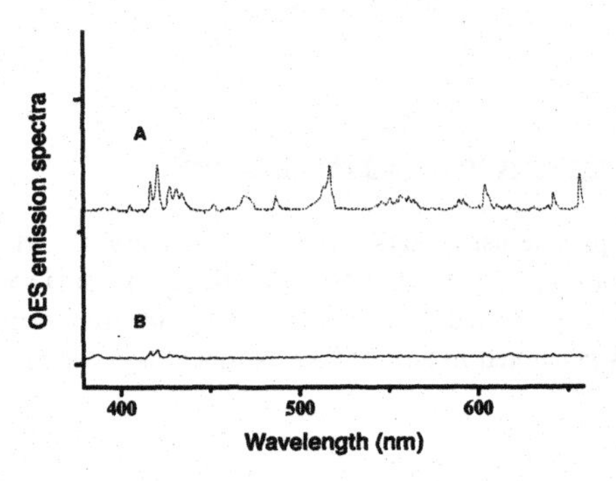

Fig. 2 Emission spectra of Ar plasmas of (A) generated by a conventional low pressure microwave plasma source; and (B) generated by the new microwave plasma source developed in this work at atmospheric pressure condition.

Discharge gas temperature

The gas temperature was measured by using a calibrated thermocouple. The sensor was inserted directly into the plasma without any samples in front of it. At atmospheric pressure conditions, the gas temperature in the plasma regions increases with increasing the microwave power. For example, at a microwave power of 50 W, a gas temperature higher than 400 ^{o}C can be detected. This value is a little bit higher than the conventional low pressure plasma source. While the new plasma source is still not a thermal plasma source since the gas temperature is much lower than the electron temperature. The gas temperatures under the plasma conditions used for the Si film depositions (the process temperature) were also measured and will be shown later.

Depositions of crystalline Si films

By using the SiH_4+He microwave plasmas, Si film depositions were performed as functions of input microwave power and SiH_4 flow rate [9]. The depositions were performed by using a working pressure of 2 Torr. Properties of the prepared films were investigated systematically. According to the observed results, Si film depositions were then performed under optimized plasma conditions.

Si film depositions as a function of the microwave power

Figure 3 shows the film deposition rate, Raman crystallinity I_c/I_a of the Si films fabricated under different microwave power conditions. SiH_4+He mixture with respective flow rate of 5 sccm and 400 sccm was used for the depositions. It is shown that the film deposition rate increases with increasing the microwave power and a rate higher than 250 nm/s is achieved when using a microwave power of ~130 W. Raman spectra of the corresponding Si films (not shown here) illustrated that the prepared Si film changed from amorphous to crystalline at a certain microwave power range (~90 W) and the crystallinity improved with increasing the power further. As shown in this figure, the film Raman crystallinity increases from I_c/I_a <3 to I_c/I_a >12 when the microwave power increases from ~90 W to ~120 W.

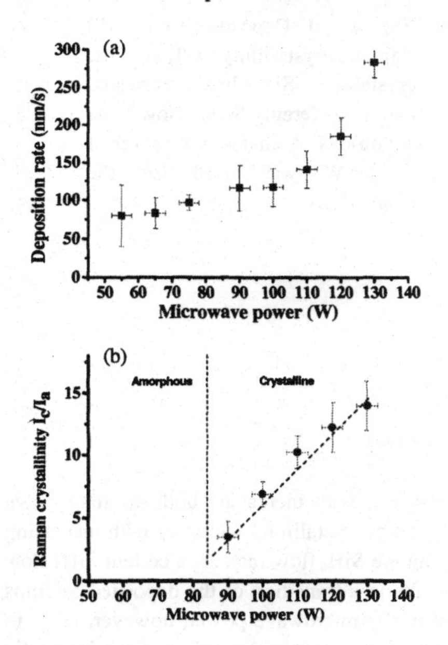

Fig. 3 (a) Deposition rate, (b) Raman crystallinity I_c/I_a of the crystalline Si films fabricated under different microwave power conditions. A SiH_4 flow rate of 5 sccm was used for the depositions.

Si film depositions as a function of the SiH_4 flow rate

Results of the film deposition rate, Raman crystallinity of the Si films deposited at different SiH_4 flow rate are shown in Fig. 4. For these depositions, a microwave power of ~110 W was used. Other deposition parameters were the same with those used in Fig. 3. As demonstrated in this figure, the film deposition rate increases significantly with increasing the SiH_4 flow rate. A very fast rate of larger than 500 nm/s is realized by using a SiH_4 flow rate of 30 sccm. However, corresponding Raman spectra of these films show that the contribution of amorphous component in these films also increases with the SiH_4 flow rate. As shown here, the film Raman crystallinity decreases from I_c/I_a=~10 to $I_c/I_a < 3$, and a SiH_4 flow rate larger than 30 sccm results the fabricated Si film to be amorphous.

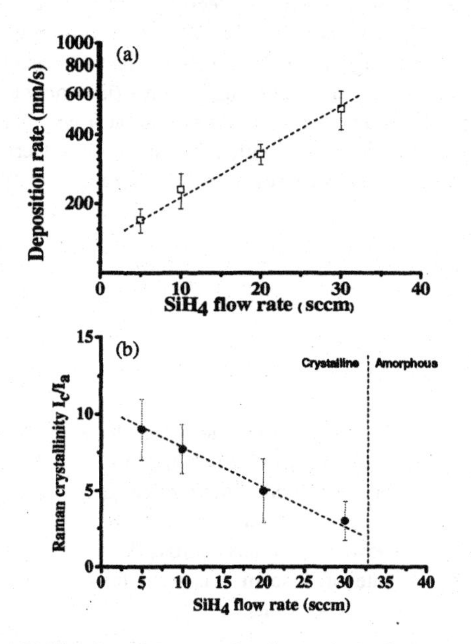

Fig. 4 (a) Deposition rate, (b) Raman crystallinity I_c/I_a of the crystalline Si films fabricated under different SiH_4 flow rate conditions. A microwave power of ~110 W was used for the depositions.

Si film depositions under the optimized plasma conditions

The above results show the film growth rate increases with increasing both the microwave power and the SiH_4 flow rate. On the other hand, the film crystallinity increases with increasing the microwave power while decreases with increasing the SiH_4 flow rate. At a certain SiH_4 flow rate condition, proper microwave power supply results in a transition of the deposited Si films from amorphous to crystalline. An excessive input of the microwave power, however, leads to the deposition of powder-like films. To realize the high film deposition rate and the high film crystallinity spontaneously, proper balance between the microwave power and the SiH_4 flow rate is needed. A high SiH_4 flow rate along with an appropriate input microwave power are expected to be the suitable conditions.Consequently, we performed Si film depositions by using a SiH_4 flow rate of 30 sccm and an input power of ~120 W. Other deposition parameters were the same with previous cases. A film deposition rate over 700 nm/s was achieved for the compact Si films with a Raman crystallinity of I_c/I_a>15.

Investigation on film growth process

To investigate the film growth process under high deposition rate conditions, the Si film depositions were performed as a function of the deposition time. A SiH_4 flow rate of 15 sccm was used in this series experiments. Also, different microwave powers ranging from 70 W to 125 W were applied for the investigation. Other deposition parameters were the same with previous cases. Figure 5 shows the Raman crystallinity of the Si films deposited for different time at different microwave power conditions. Here, when the film was deposited by using the power of 70W, no crystalline phase appears even for long deposition time. On the other hand, in the case of power of 110W, the films deposited for time shorter than 15 seconds were amorphous. For these amorphous films, the value of I_{LO}/I_{TO} calculated from the Raman spectra of amorphous Si film was used to demonstrate the degree of the structural disorder of the Si net work. For the film deposited using the power of 70W, similar structural order was observed for different deposition time. While for the film deposited using the power of 110W, with increasing the deposition time and before the film becoming crystalline, the I_{LO}/I_{TO} decreases, indicating a reduced structural disorder degree. This could be a sign for the beginning of the crystalline growth. After the film becomes crystalline, the I_c/I_a further increases with increasing the deposition time. For the film deposited by using the power of 125W, the crystallization appears even when the deposition time is short, i.e. the initial growth stage. These results demonstrated that when the deposition power is higher than a certain value, the film structural order, and/or the film crystallinity, improves with increasing the deposition time.

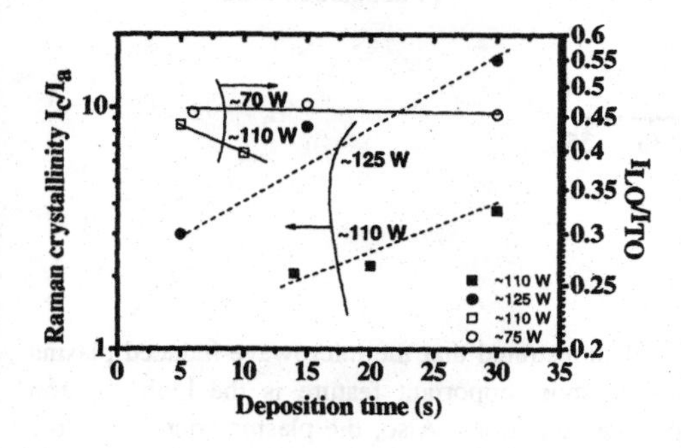

Fig. 5 Degree of structural order, and/or film crystallnity, of the films deposited under different microwave power for different time.

Depth distribution of the film crystalline structure for the film deposited for long time using high microwave power was also checked by using the reactive ion etching for film etching together with Raman measurement after each etching turn [10]. The results showed that after long time deposition, the film near substrate (the initially deposited film) also presents quite high film crystallinity. The results are different with that shown in Fig. 5, which is the initially deposited film show low film crystallinity. These suggest that there is a crystalline evolution process during the deposition.

Figure 6 shows the H concentration in the films corresponding to those shown in Fig. 5 as the function of deposition time. For the amorphous films (deposited using power of 70W), the H concentration shows almost no changes or even a slight increase. While for the crystalline Si films, the H concentration decreases significantly with increasing the deposition time. The decrease in the H concentration and the increase in the film crystallinity imply that there is an annealing process that could promote the film crystallinty throughout the entire film and leads to the evolution of film crystalline structure during the deposition. In addition, the rather low H concentration suggests a high process temperature during the deposition. To approve this, the process temperature (at reduced pressure conditions) was estimated by setting a thermal couple in the position of the substrate. The measured temperature increases quickly with increasing the plasma generation time. For example, when a microwave power of 120W was used, the process temperature increased up to 650°C within 60 seconds [10], confirming the high process temperature, which may be the main factor that is responsible for the annealing process.

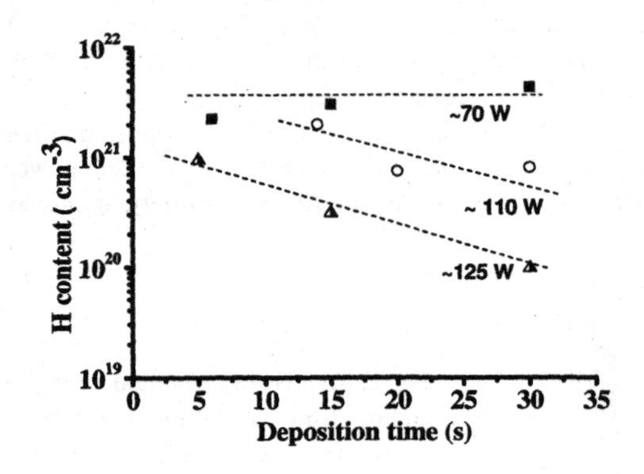

Fig. 6 H concentration in Si films deposited using different microwave power as a function of deposition time.

DISCUSSION

Through the plasma diagnoses, it has been demonstrated that the microwave-induced plasma source has several specific characteristics. The most important feature is the high electron density of >10^{15} cm^{-3} in a wide range of plasma conditions. Also, the plasma source has low electron temperature of <1 eV. These features are suggested to be remained at the reduced pressure conditions used for the Si film depositions. On the one hand, OES observations of the plasma used for the film depositions shows a quite strong emission signal from excited H atoms even without using the H_2 dilution gas, demonstrating a rather high dissociation efficiency of the source gases [6]. This should be mainly resulted from the high electron density achieved in the plasma. On the other hand, comparison of emission spectra of Ar plasmas generated at reduced pressure conditions by using this new plasma source and the conventional microwave plasma source (with electron temperature of ~1 eV) illustrated that the new plasma source shows weak emission from the high energy excitation, which is an indication of low electron temperature (lower than 1 eV).

In the SiH_4+He microwave plasma used in this study, high density film deposition precursors as well as H atoms are generated due to the high electron density, which contribute to the achievements of high deposition rate and high film crystallinity. The electron density increases with increasing the microwave power, which can be used to partially explain the increases in the film deposition rate and the crystallinity with the input power as shown in Fig. 3.

On the film growing surface, the diffusion of the film precursors on the growing surface is suggested to be markedly enhanced due to several effects in this work. Firstly, the film precursors themselves possess high energy to perform the surface diffusion. Through frequent collisions with the energized radicals in the gas phase, the film precursors gain sufficient energy that finally can promote their surface diffusions. Next, the high temperature discharge gas also heats up the growing surface. The growing surface with high temperature could promote the surface diffusion of the precursors as well. As a result, formation of the highly crystallized films is facilitated. The above effects are more pronounced at high microwave power conditions and can be used as another explanation for the higher film crystallinity achieved at high powers.

As shown in Figs. 5 and 6, the film crystallinity increases while the H concentration decreases with increasing the deposition time. These behaviors are also generally observed in the crystallization process of a-Si:H films by solid phase crystallization. Also the previous results demonstrated that after long time deposition, complete crystallization is realized even for the initially deposited low-crystallized Si film [10]. According to the analysis of the film growth process and the TEM observation of the highly crystallized Si film [10], a possible mechanism for structural evolution, the annealing-assisted plasma-enhanced chemical vapor deposition, can be proposed, as illustrated in Fig. 7. At the initial growth stage, Si film with low crystallinity is deposited. Along with the deposition continuing, annealing effect promotes the film crystallinity throughout the entire film. Finally, highly crystallized Si film and high film crystallinity through the entire film were achieved with large grains in the upper part of the film. While, since the deposition involves the plasma process, other factors such as ion bombardment may also contribute to the crystallization process. Further investigation, however, is needed to understand the phenomenon more clearly.

CONCLUSIONS

A microwave induced plasma source has been installed and applied to crystalline Si film deposition. By using this plasma source, plasmas can be generated both at atmospheric pressure and at reduced pressure conditions. Important parameters of the plasma source, i.e. the electron density n_e, electron temperature T_e and discharge gas temperature T_{gas}, were studied at atmospheric pressure conditions by using the OES measurement. It was found that high microwave power increased the n_e, T_e and T_{gas}, while high gas flow rate decreased these parameters. A high $n_e > 10^{15}$ cm^{-3} and a low $T_e < 1$ eV were obtained by this microwave plasma source with a relatively high discharge gas temperature of 400 ~ 800 °C.

Systematic study of crystalline Si film deposition from SiH_4+He was performed by using the high density microwave plasma. Results demonstrated that microwave power and SiH_4 gas flow rate were important parameters which could affect the deposition process. By using the optimized plasma conditions, highly crystallized Si films could be deposited at a rate higher than 700 nm/s. According to the investigations about the film growth process, a possible mechanism, the annealing-assisted plasma enhanced chemical vapor deposition, was proposed to describe the deposition process.

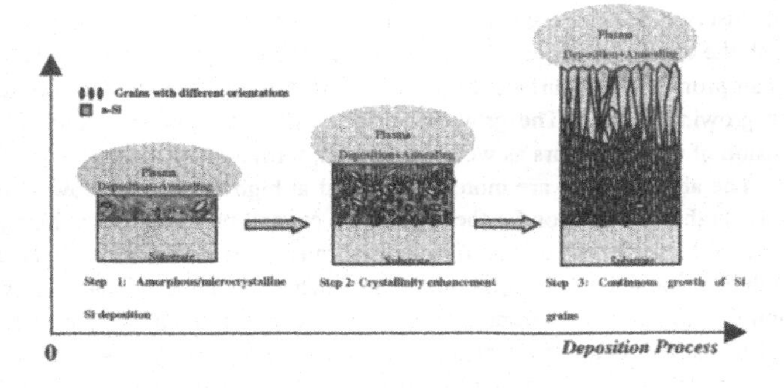

Fig. 7 Possible growth mechanism, the annealing assisted plasma enhanced chemical vapor deposition, which can be used to present the growth process of crystalline Si film by the high-density microwave plasma.

ACKNOWLEDGEMENT

This work was supported by the New Energy and Industrial Technology Development Organization (NEDO) of Japan.

REFERENCES

1. J. Meier, R. Fluckiger, H. Keppner and A. Shah, Appl. Phys. Lett. **65**, 860 (1994).
2. R. E. I. Schropp, R. Carius, G. Beaucarne, MRS Bulletin **32**, 219 (2007).
3. A. Matsuda, T. Yoshida, S. Yamasaki and K. Tanaka, Jpn. J. Appl. Phys. **6**, L439 (1981).
4. B. Rech, T. Roschek, J .Muller, S. Wieder and H. Wagner, Sol. Energy Mater. Sol. Cells **66**,

267 (2001).
5. T. Matsui, M. Kondo, A. Matsuda, Jpn. J. Appl. Phys. **42**, L901 (2003).
6. K. Yamakata, M. Hori, T. Goto, S. Den, T. Katagiri and H. kano, J. Appl. Phys. **98**, 043311 (2005).
7. H. Jia, H. Kuraseko, M. Kondo, J. Appl. Phys. **103**, 024904 (2008).
8. H. Jia, H. Shirai, M. Kondo, J. Appl. Phys. **101**, 114912 (2007).
9. H. Jia, M. Kondo, J. Appl. Phys. **105**, 104903 (2009).
10. H. Jia, H. Kuraseko, H. Fujiwara and M. Kondo, Mater. Res. Soc, Symp. Proc., 1066-A01-04 (2008).

Flexible Electronics

Mater. Res. Soc. Symp. Proc. Vol. 1245 © 2010 Materials Research Society 1245-A11-03

Float Foil Growth of Si-Foils for Solar Cell Applications

Uri Cohen[1] and Michael Roitberg[1]
[1]Ribbon Technology LLC, 4147 Dake Avenue, Palo Alto, CA 94306, U.S.A.

ABSTRACT

A new Float Foil Growth (FFG) technique has been demonstrated for growing thin Si-foils from molten metal solvent, such as molten indium (In) or tin (Sn), at temperatures below 1,000°C. Si-source is first dissolved to saturation (or close to saturation) in a molten metallic bath (or solvent) at a temperature T_2 ($T_2 \leq 1,000°C$), and the molten bath is then cooled to T_1, where $T_2 \gg T_1$. Due to lower solubility of Si at T_1 than at T_2, Si separates (or is driven) out of solution and, due to its much lower density than that of the molten metallic bath, it floats to the top of the melt to form a floating thin Si-foil. The thickness of the Si-foil is determined primarily by T_2, the dissolution temperature (i.e., Si solubility at T_2), and the depth of the molten bath. This paper reports preliminary results demonstrating the utility of the FFG technique for growing Si-foils. Si-foils with thickness range of 50-200μm were obtained from molten In baths. The Si-foils were multicrystalline with crystalline (or grain) size of several millimeters, having a strong <111> preferred orientation. The Si-foils were very pure; with In (solvent) content as low as 14ppb. Other metallic impurities were below 0.1ppm, oxygen content was as low as 1.8ppm, and carbon content was below the detection level (50ppb). It is expected that large FFG thin Si-foils, when produced on large scale, will offer significant Si material cost and energy savings (> 80%), compared with conventional sliced Si wafers, with similar photovoltaic conversion efficiency.

INTRODUCTION

Silicon solar cell panels are usually fabricated from polished round wafers, sliced from single-crystalline or polycrystalline Si ingots. The sawing, polishing, and etching of the wafers results in a costly loss, or "kerf", of Si material. Also, the process of growing single-crystalline or polycrystalline ingots is energy-intensive and costly. In addition, the size of the round wafers is limited, having a typical size of less than 200mm or 300mm in diameter. Therefore, many wafers are required to assemble one panel. To minimize efficiency losses due to unused area between round wafers on a panel, the round ingots are first machined into elongated semi-square rods prior to slicing them into semi-square wafers. This machining results in a further Si material loss. Growing large (single-crystalline, or large-grain polycrystalline) thin Si-foils, at lower temperatures (≤ 1,000°C) than conventional Si growth from its melt (.1,414°C), will offer significant (> 80%) Si material cost reduction and energy savings (compared with conventional Si wafers), along with high solar cell conversion efficiency.

This paper reports preliminary results obtained in a feasibility study of a new Floating Foil Growth (FFG) technique [1-2] for growing thin Si-foils at temperatures below 1,000°C. Si-source is first dissolved to saturation (or close to saturation) in a molten metallic bath at a temperature T_2 ($T_2 \leq 1,000°C$), and the molten bath is then cooled to T_1, where $T_2 \gg T_1$. Due to the lower solubility of Si at T_1 than at T_2, Si separates out of solution and, due to its lower density (than that of the metallic bath), floats to the top of the molten bath to form a thin Si-foil. The Si-foil is then pulled-off from the top surface of the molten metallic bath. According to the

In-Si phase diagram [3], the solubility of Si in molten indium (In) is about 0.5 wt% at 1,000°C, about 0.2 wt% at 900°C, and essentially 0.0 wt% at 600°C. Therefore, dissolving Si in molten In at an elevated temperature T_2 (such as between 900-1,000°C), and then cooling the melt to a much lower temperature T_1 ($\leq$ 600°C), drives the dissolved Si out of solution. Due to its lower density the driven out Si floats and grows at the top surface of the molten bath, as a floating Si-foil. For comparison, Si has a density (or specific gravity) of 2.33 g/cm^3, and In has a density of 7.31 g/cm^3. During cooling, Si grows on the bottom surface of the floating Si-foil to its final thickness (until dissolved Si is substantially depleted out of solution).

Since the growth of FFG Si-foil takes place across the entire foil/bath interface, in the direction of the foil's thickness, the overall foil production throughput can be exceptionally high (since the foil needs only grow to its final thickness). The authors are not aware of any theoretical limits on the width and length of FFG ribbons or foils. FFG Si ribbons or foils having width of about 1.0 meter, and length exceeding several meters, should be feasible in large scale production. Pulling rate of FFG Si ribbon is anticipated to be about 0.2-1.0 meter/min. Thus, it is expected that Si-foil or ribbon production rate (in a single line) will be about 10-50 square meters per hour. Standard width (5" or 6") Si ribbons or foils should be even simpler to grow by the FFG technique. Standard wafers sizes will then be cut from the Si ribbons. Using FFG Si ribbon or foil thickness in a range of 50-100µm, Si material cost saving is expected to be more than 80%. Furthermore, due to the extremely small segregation coefficients of most impurities, the FFG process is inherently self-purifying. As such, it is expected that lower grade Si-source, such as metallurgical Si, could be used in FFG Si-foil growth, thereby further reducing the Si material cost for solar cell applications. The present paper describes preliminary, small scale, proof-of-concept experimental results of growing Si foils by the FFG technique.

EXPERIMENTAL

A Quartz crucible was loaded with about 400g of high purity (99.9999%) In ingots, and placed on a quartz pedestal inside a 3" quartz tube, in a vertical tube furnace, as seen in Figure 1. The 3" vertical quartz tube had water-cooled top and bottom flanges. The top and bottom flanges had vacuum and gas inlets and outlets, along with fittings for lowering and raising a crucible pedestal stand, a thermocouple, a Si-source holder, a quartz plate strainer, and a gas bubbler/cooling tube. A Si-source strip was cut from a Si wafer and attached to a quartz holder by tungsten wire. Figure 2 shows a Si-source attached to its holder (after a completed run), and Figure 3 shows the quartz strainer.

Figure 1. Experimental System

Figure 2. Si-source

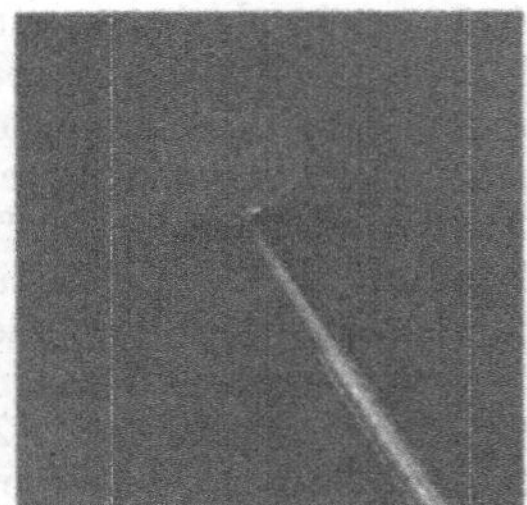

Figure 3. Quartz strainer

Following the loading of a quartz crucible with In ingots inside the vertical 3" quartz tube, the top and bottom flanges are sealed, and the system is evacuated. Inert gas ($Ar-4\%H_2$) is introduced, and the furnace is turned on. Several cycles of vacuum/flushing are used during the heating to expel moisture and residual air. Following the melting of the In ingots, the temperature of the In melt is raised to T_2 (~900-1,000°C), the quartz strainer is pushed downward inside the melt, toward the bottom of the crucible, and the Si-source is lowered into the In melt for dissolution. The Si-source is left inside the In bath for about 45-60 minutes to ensure dissolution to saturation (or close to saturation). The remainder of the Si-source is then pulled up above the melt (as seen in Figure 2), and programmed cooling is commenced. Once the molten bath reaches T_1(~500-600°C), the quartz strainer with the floating Si-foil is pulled up above the level of the molten bath. The system is then cooled to room temperature, the flanges opened, and the crucible with the frozen bath are removed.

RESULTS

Figure 4 shows a quartz crucible, with frozen In bath following the completion of a run, placed on a pedestal stand inside the 3" quartz tube. The 3" quartz tube is located inside the vertical furnace. Figure 5 is a top view of the frozen bath and floating crystallized Si in an early experiment, when no strainer was used to separate the Si foil.

Figure 4. Crucible and frozen bath

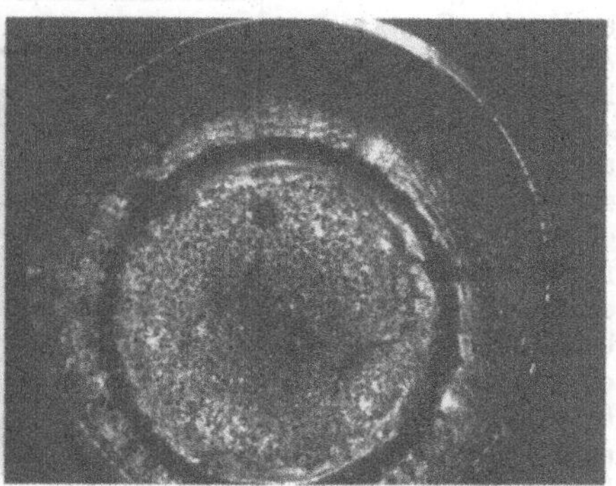

Figure 5. Top view of frozen bath

Free-standing flat thin Si-foils with thickness range of 50-200μm, were obtained from molten indium (In) bath. The obtained Si-foils were multicrystalline with grain size (single-crystals) of several millimeters. The flat Si-foils had a strong <111> preferred orientation. The Si-foils were extremely pure; with In (solvent) content as low as 14ppb. All other metallic impurities were below 0.1ppm. Oxygen content was about 1.8ppm, and carbon content was below detection level (50ppb).

Figure 6 shows a flat single-crystal Si-foil, about 7.4mm long, and about 75μm thick. The Si-foil was obtained from a molten indium (In) bath, saturated with Si at $T_2 = 999°C$, slowly cooled to $T_1 = 600°C$, and separated from the molten bath at T_1. The grid lines are spaced 0.25" apart from each other. Figure 7 shows a flat Si bicrystal about 7.0mm long.

Figure 6. Flat Si crystal; ~7.4mm long; area ~ 0.18cm^2; ~75μm thick.

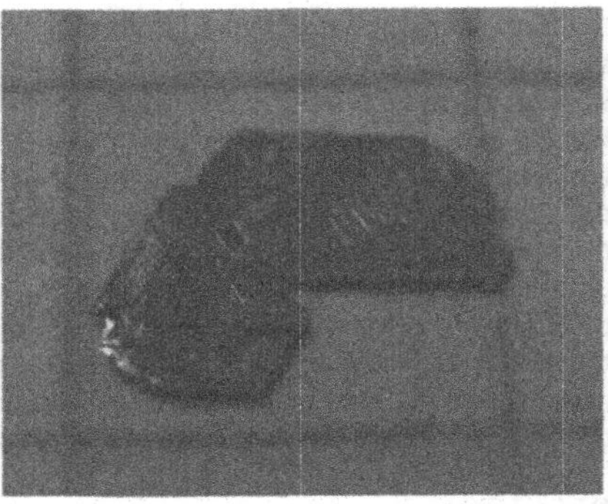

Figure 7. Flat bicrystal; ~7.0mm long; area ~ 0.26cm^2; ~96μm thick.

The flat Si crystal shown in Figure 6 was analyzed by Evans Analytical Group for impurities by Laser Ablation ICP-MS technique. Table 1, shows the trace elemental impurities found. It is noted that, except to Fe (which was below 1.0 ppm), all other metals were below 0.1 ppm (7N purity). Such purity level already exceeds the required level for photovoltaic applications. Iron (Fe) contamination in the foil may have originated from the experimental setup or from the metallic (In) bath. Iron impurity level should also be reduced to below 0.1 ppm.

Table 1. Trace elemental concentrations found in silicon sample #123 RTL-123.

Element	Concentration (ppm wt)	Element	Concentration (ppm wt)
Li	< 0.1	In	* Interference
Be	< 0.1	Sn	< 0.1
B	< 0.1	Sb	< 0.1
Na	< 1	Te	< 0.1
Mg	< 0.1	Cs	< 0.1
Al	< 0.1	Ba	< 0.1
Si	Matrix	La	< 0.1
P	< 10	Ce	< 0.1
K	< 1	Pr	< 0.1
Ca	< 1	Nd	< 0.1
Sc	< 0.1	Sm	< 0.1
Ti	< 0.1	Eu	< 0.1
V	< 0.1	Gd	< 0.1
Cr	< 0.1	Tb	< 0.1
Mn	< 0.1	Dy	< 0.1
Fe	< 1	Ho	< 0.1
Co	< 0.1	Er	< 0.1
Ni	< 0.1	Tm	< 0.1
Cu	< 0.1	Yb	< 0.1
Zn	< 0.1	Lu	< 0.1
Ga	< 0.1	Hf	< 0.1
Ge	< 0.1	Ta	< 0.1
As	< 0.1	W	< 0.1
Se	< 0.1	Re	< 0.1
Rb	< 0.1	Os	< 0.1
Sr	< 0.1	Ir	< 0.1
Y	< 0.1	Pt	< 0.1
Zr	< 0.1	Au	< 0.1
Nb	< 0.1	Hg	< 0.1
Mo	< 0.1	Tl	< 0.1
Ru	< 0.1	Pb	< 0.1
Rh	< 0.1	Bi	< 0.1
Pd	< 0.1	Th	< 0.1
Ag	< 0.1	U	< 0.1
Cd	< 0.1		

* Unable to determine value due to contamination from previous testing.

Other Si crystals were analyzed by SIMS for other impurities. In (solvent) content was as low as 14ppb, oxygen content was about 1.8ppm, and carbon content was below detection level (50ppb).

Figures 8(a)-8(d) show SEM photos of the flat Si bicrystal of Figure 7.

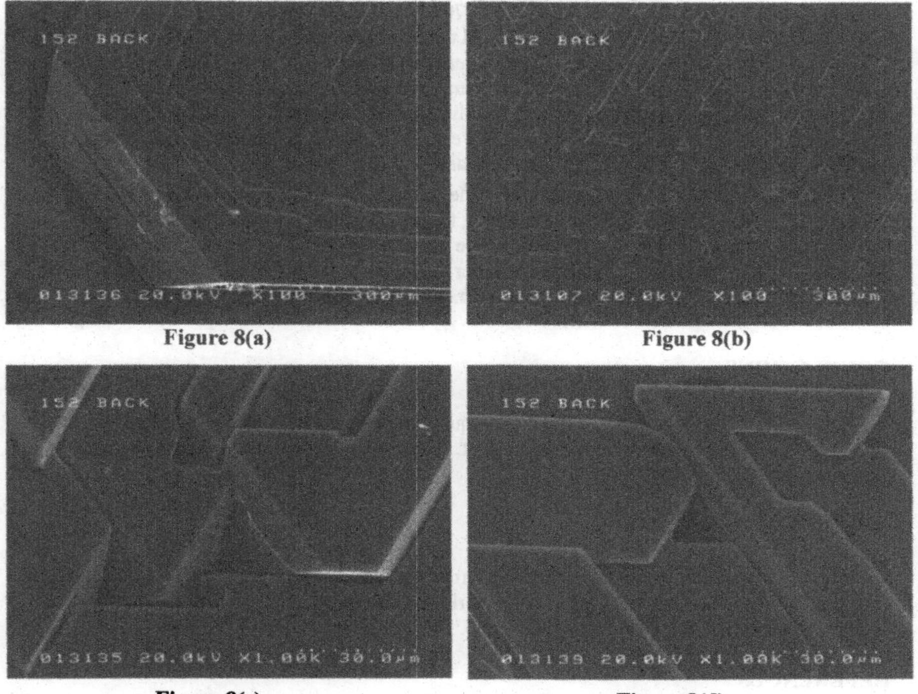

Figure 8(a) **Figure 8(b)**

Figure 8(c) **Figure 8(d)**

DISCUSSION

The dissolution temperature T_2 may be optimized when other parameters are known (or estimated). For example, using the solubility data from the In-Si phase diagram [3], and assuming a Si-foil thickness of 100μm (0.01cm) is required, T_2 can be estimated. Using a 5cm deep molten indium bath, T_2 of about 770°C is calculated. The estimated T_2 is calculated as follows:

Weight of a molten indium column of 5cm deep and 1 cm^2 area: $h_{In}*\rho_{In} = 5*7.31 = 36.55g$
Weight of required of 0.01cm thick Si foil and 1 cm^2 area: $h_{Si}*\rho_{Si} = 0.01*2.33 = 0.0233g$
Weight percentage of dissolved Si in molten In: $100*0.0233/36.55 = 0.06$ wt%
T_2 at which Si solubility in molten In is 0.06 wt% (from phase-diagram): $\mathbf{T_2 = 770°C}$.

Here h_{In} is the depth of the molten indium bath, and ρ_{In} is indium density. If a Si-foil thickness of about 200μm is required, either the depth of the In molten bath is doubled to 10cm (using the same T_2 of 770°C), or T_2 can be raised to about 850°C (using the same h_{In} of 5cm). Similarly, if a Si-foil thickness of about 50μm is required, either the depth of the In molten bath can be halved to 2.5cm (using the same T_2 of 770°C), or T_2 can be lowered to about 730°C (using the same h_{In} of 5cm). It should be noted that the above calculations assume full saturation at T_2, and complete depletion of the entire content of the dissolved Si from the molten bath. In reality, however, due to non-equilibrium dissolution and/or foil growth, dissolution at T_2 may not reach saturation, and some Si may be left dissolved in the molten bath, so that the actual Si-foil thickness is less than the calculated value. Nevertheless, such calculations are useful and convenient guidelines for the design of the apparatus and its operation.

Of particular importance is the fact that, due to lack of interfacial stress during growth between the floating Si-foil and the liquid bath, the Si-foil can grow with essentially no dislocations. This is similar to weightless space growth. The FFG growth mechanism should be advantageous for obtaining high quality monocrystalline (or quasi-monocrystalline) flat Si-foils having long minority lifetimes, and high conversion efficiency.

CONCLUSIONS

The feasibility of using a new FFG technique for obtaining floating Si-foils has been demonstrated on a small scale proof-of concept. The quality and purity of the flat Si-foils are adequate for solar cell applications.

ACKNOWLEDGMENTS

The authors would like to thank Dr. Richard Swanson, Founder and President of SunPower Corporation, for his interest and help. This work was supported in part by the US Department of Energy, Subcontract No. DE-FC36-7GO17043.

REFERENCES

1. U. Cohen, Patent Pending.
2. U. Cohen and M. Roitberg, Patent Pending.
3. ASM Alloy Phase Diagrams, Diagram 901409 (Olesinski 1990), ASM International, 2006.

Thin Film Transistors and Materials

Mater. Res. Soc. Symp. Proc. Vol. 1245 © 2010 Materials Research Society 1245-A12-01

Influence of Embedded a-Si:H Layer Location on Floating-gate a-Si:H TFT Memory Functions

Yue Kuo and Mary Coan
Thin Film Nano &Microelectronics Research Laboratory, Texas A&M University, College Station, TX, 77843-3122

ABSTRACT

The influence of the location of the embedded a-Si:H layer in the gate dielectric film of the floating-gate a-Si:H TFT on the charge trapping and detrapping mechanisms has been investigated. The thin channel-contact SiN_x gate dielectric layer favors both hole and electron trappings under the proper gate voltage condition. The sweep gate voltage affect the locations and shapes of forward and backward transfer characteristics curves, which determines the memory function. In order to achieve a large memory window, both the location of the embedded a-Si:H layer and the gate voltage sweep range need to be optimized.

INTRODUCTION

Nonvolatile memory devices based on the floating-gate a-Si:H thin film transistor (TFT) or MOS capacitor have been demonstrated [1-4]. A thin a-Si:H film is embedded in the gate dielectric film, e.g., sandwiched between a "channel-contact" layer adjacent to the a-Si:H channel film and a "control" layer adjacent to the gate electrode, to serve as the charge trapping medium. The memory function can be estimated from its capacitance-voltage (CV) hysteresis curve in the capacitor or the transfer characteristics, i.e., drain current-gate voltage (I_d-V_g), hysteresis curve in the TFT. The floating-gate TFT's hysteresis curve is composed of the forward and backward transfer characteristics curves. The former determines the charge trapping mechanism and capacity. The latter shows the charge detrapping or erasing characteristics. The "memory window" is determined by the threshold voltage difference (ΔV_t) between the two curves [5]. Since the floating-gate a-Si:H TFT is made of the same a-Si:H and SiN_x thin film materials as those of the conventional a-Si:H TFT, the whole device can be fabricated at a low temperature on various types of substrates for a wide range of applications including rigid and flexible displays, sensors, imagers, etc.

There were studies on various factors affecting the charge trapping and detrapping mechanisms. For example, the polarity and magnitude of the applied V_g directly influence the memory window. The drain voltage (V_d) also plays an important role in the charge storage process [3,4]. The stored charges can be released with non-electrical methods, such as thermal annealing or light exposure [4]. The charge storage and retention capabilities are affected by bending of the substrate, which is critical to the flexible electronics application [6]. Currently, this is no report on the influence of the gate dielectric structure on the charge and discharge mechanisms as well as the memory window. In this paper, authors carried out experiments to investigate how the location of the embedded a-Si:H layer controls the memory function of the floating-gate a-Si:H TFT.

EXPERIMENTAL

Floating-gate a-Si:H TFTs were fabricated on a Corning 1737 glass substrate using a 2-photomask process [1,2]. The first mask was used to define the gate line. The source/drain contact areas were defined using a backlight exposure step [7]. The second mask was used to define the source/drain electrode including the n^+ ohmic contact region. The final n^+ layer etch was done with a highly selective n^+-to-SiN_x RIE process [8]. The complete TFT was annealed at 250°C to remove the RIE-induced plasma damage [8]. The 5-layer control SiN_x/embedded a-Si:H/channel-contact SiN_x/channel a-Si:H/ passivation SiN_x structure was deposited by PECVD using a one-pump-down process without breaking the vacuum. The thickness of the combined channel-contact and control SiN_x gate dielectric layers was fixed at 300 nm. The thickness of the embedded a-Si:H layer was 15 nm. Floating-gate TFTs with three different channel-contact SiN_x thicknesses, i.e., 100 nm, 150 nm, and 200 nm, were prepared. A control TFT, i.e., without the embedded a-Si:H layer, was also fabricated. The hysteresis of the transfer characteristics of the TFT was measured (at V_d = 10V) from a negative V_g value to a positive V_g value, i.e., the forward characteristics curve, and then back to the negative V_g value, i.e., the backward characteristics curve. All electrical measurements were performed in a black-box using an Agilent 4155C Semiconductor Parameter Analyzer.

RESULTS AND DISCUSSION

Charge Trapping-free Control TFT

Figure 1 shows transfer characteristics hysteresis curves of the control TFT with various V_g sweeping ranges, i.e., -10V to 10V to -10V, -20V to 20V to -20V, and -30V to 30V to -30V, separately. In spite of the large variation of V_g sweep ranges, there is almost no difference between the forward and the backward characteristics curves. The lack of charge trapping capacity of the control TFT is contributed by the slightly nitrogen-rich SiN_x gate dielectric, which contains a proper SiH/NH ratio [9].

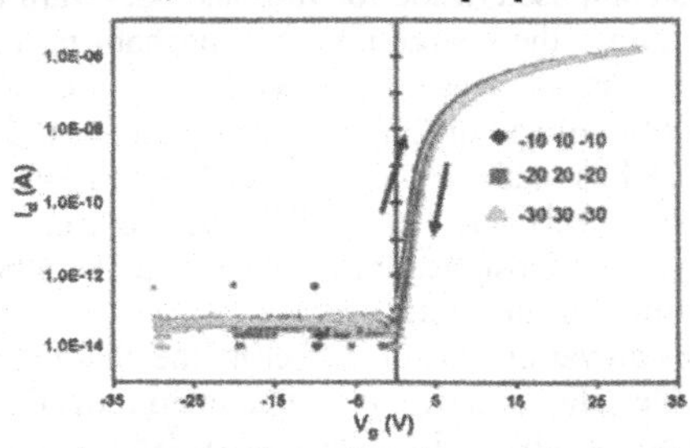

Fig. 1. Hysteresis curves of transfer characteristics of control TFT.

Influence of Location of Embedded a-SiH Layer on Forward Characteristics Curves

Figure 2 shows the forward characteristics curves of three floating-gate a-Si:H TFTs, i.e., with the channel-contact SiN_x thicknesses of 100 nm, 150 nm, and 200 nm, separately. The V_t and the shape of each curve are dependent on the magnitude of the starting V_g. For example, when the starting V_g value is small, i.e., -5V or -10V as shown in Fig. 2(a), the curve shifts slightly toward the negative V_g direction with the thinning of the channel-contact SiN_x layer. However, when the starting V_g value is -30V, as shown in Fig. 2(b), the shift of the curve is more pronounced. In addition, the influence of the channel-contact SiN_x thickness to the separation of adjacent curves becomes non-negligible. These results are related to the charge trapping mechanism of the a-Si:H embedded gate dielectric layer. When the TFT is biased with a negative V_g, a hole-rich inversion layer is induced in the a-Si:H channel layer near the channel-contact SiN_x interface. Holes are injected into the gate dielectric layer when the $-V_g$ is large enough. Since the SiN_x film does not trap charges, as shown in Fig. 1, these injected holes are retained at the embedded a-Si:H site. For the same total gate dielectric thickness, under the same $-V_g$, the thinner the channel-contact SiN_x layer is, the easier holes are injected from the channel region to the embedded a-Si:H layer due to the larger electric field. Therefore, the three curves are aligned in the order of the channel-contact SiN_x layer thickness of 100 nm < 150 nm < 200 nm.

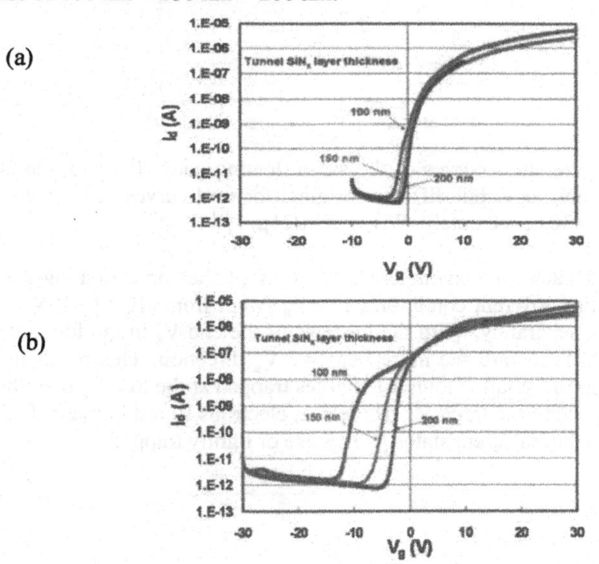

Fig. 2. Floating-gate TFTs with channel-contact gate SiN_x thickness of 100 nm, 150 nm, and 200 nm, separately. Starting V_g are (a) -10V and (b) -30V. W/L = 59μ/21μ.

Previously, it was observed that some of the charges stored in the embedded dielectric structure were loosely trapped, i.e., they were quickly released upon the removal of the stress V_g. Other charges were more tightly trapped, i.e., they were slowly released afterwards (1-4). The large bump in the Fig. 2 (b) curve with the 100 nm thick channel-contact SiN_x layer can be explained by the above mechanism, i.e., the release of a large number of loosely trapped holes with the reduction of the $-V_g$.

Influence of Location of Embedded a-SiH Layer on Backward Characteristics Curves

The charge detrapping characteristics of the floating-gate TFT is strongly influenced by its original charging state. The final V_g in the forward characteristics curve is important. Figure 3 shows backward characteristics curves of the same floating-gate TFT previously swept at V_g -30V to 30V, -20V to 30V, and -10V to 30V, separately. The same amount of charges were stored in the TFT once the end V_g's in the forward curves are the same. Since all curves in Fig. 3 coincide, their discharge characteristics are the same.

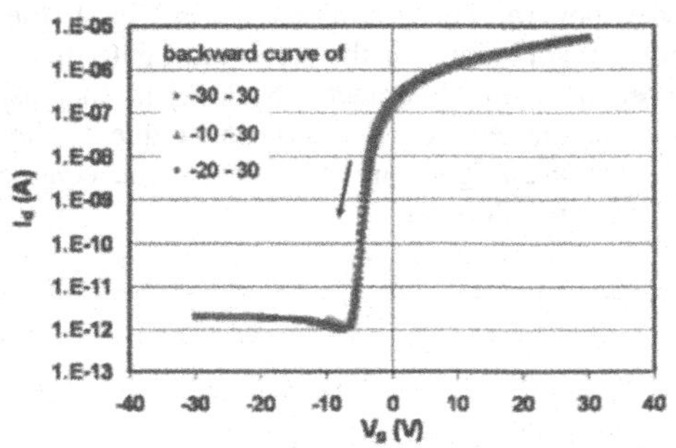

Fig. 3. Backward characteristics curves of the same floating-gate TFT with the 200 nm thick channel-contact SiN_x layer but different original forward curves of -30V to 30V, -20V to 30V, and 10V to 30V, separately. W/L = 59μ/21μ.

Figure 4 shows backward characteristics curves of the same floating-gate TFT previously charged under different conditions, i.e., V_g swept from -10V to 10V, -20V to 20V, and -30V to 30V, separately. With the increase of the end V_g in the forward curve, the backward curve shifts toward the more negative V_g direction. This result indicates that it is easier to detrap the small amount of charges trapped at the low V_g than the large amount of charges trapped at the large V_g. Therefore, electrons stored in the embedded a-Si:H site may be retained in different states, e.g., loose or tightly trapped.

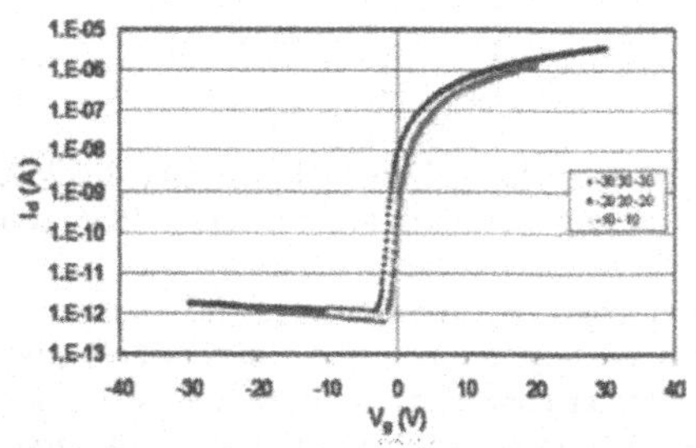

Fig. 4. Backward characteristics curves of the same floating-gate TFT with the 100 nm thick channel-contact SiN_x layer but different original forward curves of V_g sweep from -30V to 30V, -20V to 20V, and -10V to 10V. W/L = 59μ/21μ.

Influence of Location of Embedded a-SiH on Hysteresis of Transfer Characteristics

As discussed in previous sections, the charge storage and release mechanisms in the floating-gate TFT are strongly affected by the location of the embedded a-Si:H film in the gate dielectric layer. The operation conditions, e.g., the V_g sweep range and direction, also play important role in the process. For example, in the forward transfer characteristics curve, the starting V_g dominates the hole trapping-detrapping process and therefore, the V_t and shape. The end V_g controls the total amount of trapped electrons. The starting V_g of the backward transfer characteristics curve controls the release of the trapped electrons, e.g., its V_t. The memory window of the floating-gate TFT is determined by the relative position of these two curves. Two types of transfer characteristics hysteresis curves have been observed, as shown in Figure 5 (a) and (b). When the amount of stored charges is small, the backward curve is located on the positive V_g direction of the forward curve, as shown in Fig. 5(a). This is due to the quick drainage of the stored electrons. It occurs in cases, such as the TFT with a thick, e.g., 200 nm thick, channel-contact SiN_x layer or the forward curve with a small end V_g. However, when the amount of stored charges is large, the backward curve is located on the negative V_g direction of the forward curve, as shown in Fig. 5(b). This is due to the gradual release of the trapped electrons. It occurs in cases, such as the TFT with a thin, e.g., 100 nm thick, channel-contact SiN_x layer or the forward curve with a large end V_g.

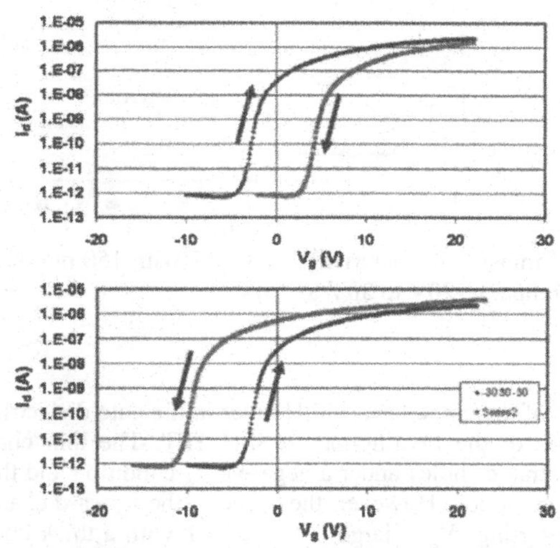

Fig. 5. (a) Clockwise and (b) counterclockwise hysteresis curves of floating-gate TFTs.

Since the memory function is determined by the relative location of the forward and the backward transfer characteristics curves, the direction of the hysteresis transfer characteristics curve is not a major concern as long as a large ΔV_t is achieved. However, the shape of the transfer characteristics curve can affect the determination of the ΔV_t. For example, Figure 6 shows a floating-gate TFT with the forward and backward transfer curves crossed due to the large distortion of the forward transfer curve. In another case,

Figure 7 shows a TFT with the forward and backward curves overlapped due to a certain combination of the channel-contact SiN_x layer thickness and the V_g sweep range. In both cases, electrons were stored at the end of the forward transfer curve and detrapped at the end of the backward transfer curve. However, it is difficult to determine their memory windows from these hysteresis curves. Therefore, in order to achieve a large memory window, these types of hysteresis curves need to be avoided, e.g., by changing the location of the embedded a-Si:H layer or the V_g sweep range.

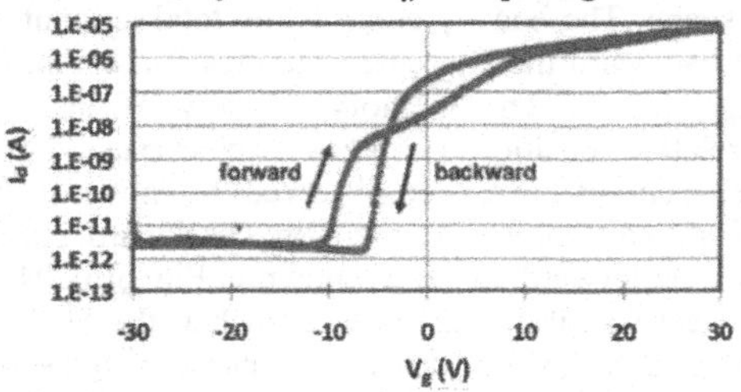

Fig. 6. Hysteresis of transfer characteristics of a TFT with 100 nm thick channel-contact SiN_x and V_g sweep range of -30V to 30V to -30V.

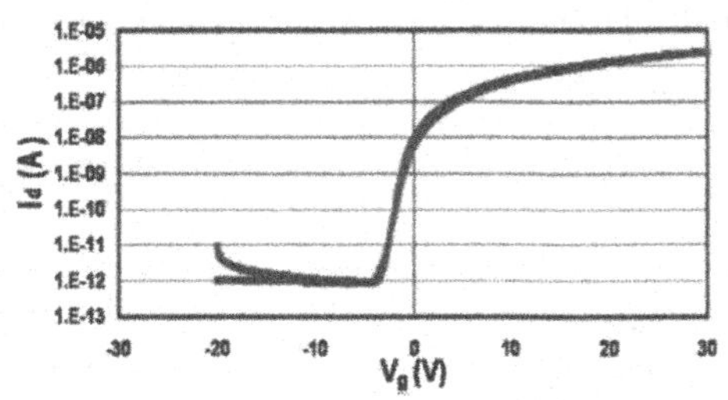

Fig. 7. Hysteresis of transfer characteristics of a TFT with 150 nm thick channel-contact SiN_x and V_g sweep range of -20V to 30V to -20V.

CONCLUSIONS

The location of the embedded a-Si:H layer in the gate dielectric film is critical to the memory function of the floating-gate a-Si:H TFT. The thin channel-contact SiN_x dielectric favors the trap of holes under a negative V_g condition and the trap of electrons under a positive V_g condition. However, the shape of the forward characteristics curve is distorted when the starting $-V_g$ is large. For the TFT with a thick channel-contact SiN_x dielectric layer, it is more difficult to trap charges to the embedded a-Si:H site than that with a thin channel-contact SiN_x dielectric layer. Electrons trapped by the small positive V_g are easier to detrap than those trapped by the large positive V_g are. The hysteresis curve can be clockwise or counterclockwise depending on the location of the embedded a-Si:H layer and the V_g sweep range. In order to achieve a large memory window, it is necessary to optimize both of them. In addition to V_g, time is another important operation parameter that can influence the discharge process and therefore, the memory window. More studies are required on that subject.

ACKNOWLEDGMENTS

This work was partially supported by DHS Grant No. 2009-DN-077-ARI018-03. Authors acknowledge Alou C.-H. Lin for sputter deposition of metal electrodes.

REFERENCES

1. Y. Kuo and H. Nominanda, *Appl. Phys. Lett.* **89**, 1 (2006).
2. Y. Kuo and H. Nominanda, Mat. Res. Soc. Symp. Proc. Vol. 989, Warrendale, PA, A10-03 (2007).
3. Y. Kuo and H. Nominanda, *J. Korean Phys. Soc.* **54**(1), 409 (2009).
4. Y. Kuo and H. Nominanda, Mat. Res. Soc. Symp. Proc. Vol. 1066, Warrendale, PA, 1066-A08-02 (2008).
5. D. Kahng and S. M. Sze, *Bell Syst. Tech. J.* **46** 1283 (1967).
6. Y. Kuo and M. Coan, Proc. 16th Intl. Workshop on Active-Matrix Flatpanel Displays and Devices, 259 (2009).
7. Y. Kuo, SPIE Proc. **1456**, 288 (1991).
8. Y. Kuo and M. Crowder, *J. Electrochem. Soc.* **139**(2), 548 (1992).
9. Y. Kuo, *J. Electrochem. Soc.* **142**(1), 186 (1995).

Mater. Res. Soc. Symp. Proc. Vol. 1245 © 2010 Materials Research Society 1245-A12-03

On the Mechanism of Nucleation in Pulsed-Laser Quenched Si Films on SiO_2

Y. Deng, Q. Hu, U. J. Chung, A. M. Chitu, A. B. Limanov, and James S. Im
Program in Materials Science and Engineering, Department of Applied Physics and Applied Mathematics, Columbia University, New York, NY 10027, U.S.A.

ABSTRACT

We have investigated the solid nucleation mechanism in laser-quenched Si films on SiO_2. Previously neglected experimental steps, consisting of BHF-etching and irradiation in vacuum, were implemented to reduce potential extrinsic influences. The resulting experimental findings and computational analysis lead us to conclude that solid nucleation consistently takes place heterogeneously at, and only at, the bottom liquid Si-SiO_2 interface.

INTRODUCTION

Irradiating thin Si films on SiO_2 using a short-duration laser pulse can lead to various melting and solidification scenarios [1]. For those cases in which the incident laser energy density exceeds the critical value needed to induce full melting of the films (i.e., above the complete melting threshold (CMT)), the films are found to eventually transform, after being rapidly quenched to temperatures well below the equilibrium point, via nucleation and growth of solids [2]. The precise mechanism through which solid nucleation takes place within the deeply supercooled elemental liquid phase constitutes one of several fundamentally meaningful topics encountered within the field. Additionally, the topic can also be identified as being technologically relevant in that the presence of nucleated grains are found to substantially degrade the microstructural quality of the excimer-laser-crystallized polycrystalline Si films that are presently utilized for fabricating low-temperature polycrystalline Si films-based thin-films transistors [3].

Up until now, it has been predominantly viewed as being the case within the community [2, 4, 5] that homogeneous nucleation corresponds to the "intrinsic" mechanism of nucleation involving "pure" Si films on "clean" SiO_2 (i.e., the nucleation mechanism manifested in those cases for which various extrinsic factors and influences, such as chemical contaminants and/or surface layers and reactions, have been eliminated). This paper focuses on reexamining this particular topic regarding the mechanism of nucleation in pulsed-laser-quenched Si films on SiO_2. Our present investigation reveals preliminary findings that, when steps are taken to ensure that clean Si films are irradiated in vacuum, heterogeneous nucleation of solids proceeding at the bottom interface between liquid Si and SiO_2 corresponds to the singularly prevalent mechanism of nucleation.

EXPERIMENT

The samples used in the current study consisted of dehydrogenated amorphous or polycrystalline Si films (prepared by PECVD as well as LPCVD) on SiO_2-layer covered glass, quartz, or Si-wafer substrates. The thickness of the Si films ranged from 50 nm to 300 nm. An excimer-laser-based system (308 nm wavelength, ~30 ns FWHM) was employed to irradiate the films at various energy densities within the complete melting regime. The laser beam was homogenized and projected onto the sample surface.

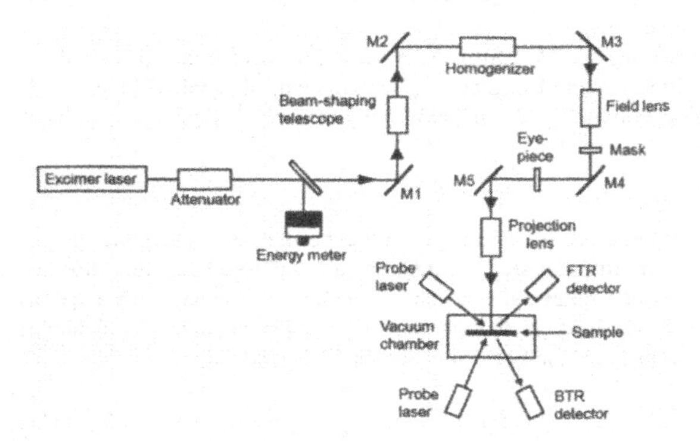

Fig 1. Schematic diagram illustrating the front-side and backside transient reflectance set up.

A significant component of the present experimental investigation comprises a couple of experimental steps that were taken to reduce the possibility of the transitions from being affected by potential extrinsic elements and factors. Specifically: (1) the native oxide layer on the sample surface was removed by BHF solution just prior to being irradiated, and (2) the samples were transferred into and irradiated within a vacuum chamber. Single-pulse laser irradiation as well as *in situ* front-side and backside transient reflectance (FTR and BTR) analysis were performed through quartz windows in the vacuum chamber. Laser diodes (670 nm) were used for FTR and BTR analysis. Microstructure characterization of the irradiated samples was conducted primarily using both planar and cross-sectional transmission electron microscopy. Numerical analysis was also performed using a nonequilibrium 3-D model capable of accounting for the stochastic nature of nucleation and interfacial undercooling [6].

RESULTS

Figure 2 shows the FTR and BTR signals that are obtained from irradiating a 150-nm-thick a-Si film on a glass substrate at 1.45 times the CMT energy density. Because liquid Si is metallic and has a much higher reflectivity than solid Si, the FTR signal quickly rises to form a plateau when melting is initiated at the surface [7]. This plateau remains essentially flat as long as the near-surface region is fully liquid. The BTR signal, on the other hand, shows rapid oscillations at the beginning of melting due to the interference effect resulting from the fast-moving melting front. When the primary melting front reaches the bottom interface (i.e., when the Si film is completely melted), the BTR signal also exhibits, as expected, a plateau. As is the case with the FTR signal, the BTR signal remains essentially featureless until the near-interface region is no longer fully liquid. In other words, the drops in the FTR and BTR signals indicate the moments at which the surface and interface regions, respectively, can no longer be identified as being fully liquid (i.e., containing at least some solids).

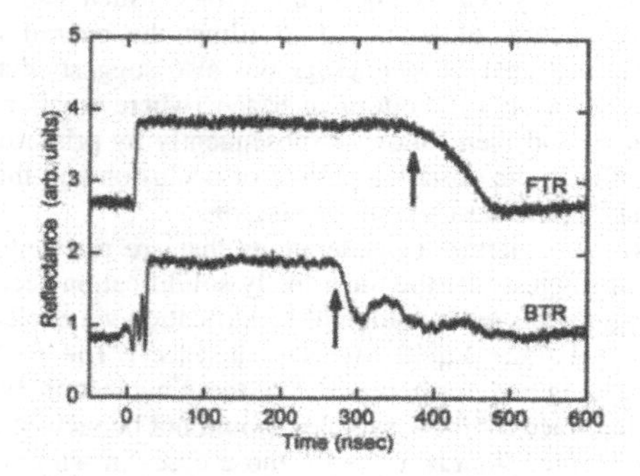

Figure 2. Front-side and backside transient reflectance signals of a 150 nm a-Si film on a SiO_2-coated glass substrate irradiated at 1.45 times the CMT energy density.

The moments at which the signals drop are marked by the arrows in figure 2; it can be readily discerned that the plateau in the BTR signal ends earlier than the plateau in the FTR signal. This particular trend, which reveals that solids appeared substantially earlier at the back interface than at the front surface, was consistently and reproducibly observed for all other energy densities and samples, *provided that they were BHF-etched and irradiated in vacuum.*

Figure 3 shows the cross-sectional TEM image of a 150 nm a-Si film on a glass substrate irradiated at 1.45 times the CMT energy density. In general, it was consistently observed to be the case that grains near the bottom interface are much smaller and / or more defective than grains near the surface region.

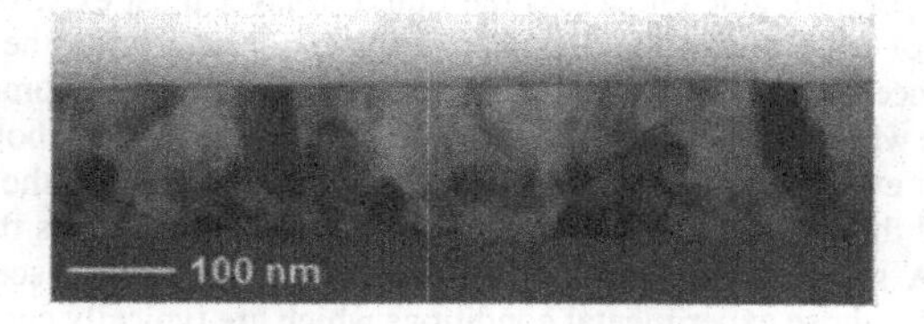

Fig 3. Cross-sectional TEM image of a 150 nm a-Si film on a SiO_2-coated glass substrate irradiated at 1.45 times the CMT energy density.

DISCUSSION

We would like to start our discussion by focusing on the FTR and BTR signals that specifically reveal how the plateau in the BTR signal ends considerably and consistently earlier than the plateau in the FTR signal. This substantial temporal difference indicates that solidification is always detected notably earlier at the bottom interface than at the surface. This, in turn, means that the initiation of solidification within the supercooled liquid Si must have likely taken place at, or at least very near, the bottom interface. Such a solidification scenario is also reflected in the microstructure of solidified Si films: the overall pattern of the grain boundaries revealed in cross-sectional TEM micrographs also suggests that solidification must have been first initiated at, or near, the bottom interface (where smaller and more defective "equiaxed" grains are located) and then followed subsequently by primarily upward growth of the grains, resulting eventually in the observed pattern of occlusion and the formation of larger and less-defective "columnar" grains near the surface region.

Based on the kinetic and thermal considerations that are relevant to our experimental conditions, we conclude and suggest that the most likely solidification scenario revealed by the present experimental findings entails the initiation of solidification via nucleation of solids taking place heterogeneously at the bottom liquid Si–SiO_2 interface. The fact that heterogeneous nucleation of solids is proceeding and, furthermore, that these nucleation events are taking place at, and only at, the bottom interface for these samples should not be viewed as being unexpected: (1) solid nucleation in supercooled liquids, even for those cases in which deep supercooling is achieved and in which substantial efforts are made to specifically avoid heterogeneous nucleation [8, 9], is found to be essentially always dominated by the heterogeneous mechanism, (2) the thin film configuration can be viewed as favoring heterogeneous nucleation as the samples are interface-rich and volume-deprived, (3) these electronic device quality films prepared through chemical vapor-deposition processes are considered to be clean and devoid of heterophase nucleants, (4) a pristine liquid surface is viewed thermodynamically as being ineffective in catalyzing heterogeneous nucleation of solids, and (5) our previous experimental work [10] on measuring the rates of heterogeneous nucleation of solid Si at elevated substrate temperatures suggests that the heterogeneous mechanism may also dominate for the conditions that are typically encountered in pulsed-laser quenching experiments.

An alternative scenario for which at least a qualitative argument can be constructed corresponds to the situation in which homogeneous nucleation takes place preferentially near the bottom interface [4, 5, 11]. This argument recognizes and builds on the fact that (1) a small but non-zero vertical temperature gradient across the liquid Si layer must exist within the film [11], and (2) that the rate of solid nucleation can depend hyper-sensitively on the liquid temperature [12]. As a consequence, it is conceivable to envision spatially localized homogenous nucleation of solids taking place within the colder part of the bulk of the film near the bottom interface. The most rigorous way to evaluate the validity of this argument is to analyze the situation using the previously developed three-dimensional numerical model that incorporates the stochastic nature of nucleation [6]. A set of preliminary "homogeneous-nucleation-only-scenario" simulations reveal that – at least for those experimental conditions which are typically encountered in pulsed-laser quenched thin Si films on SiO_2 – homogeneous nucleation would actually take place relatively volumetrically uniformly throughout the thickness of the film. This finding indirectly but strongly discounts the possibility of spatially localized nucleation of solids taking place homogeneously near the bottom interface.

Finally, we would like to comment on the relevance and meaning of the extra experimental steps that were taken in the present investigation (i.e., BHF etching of the samples and subsequent irradiation in vacuum). Following these procedures led to quantitatively reproducible and physically consistent experimental results that are reported in this paper for all the samples that were evaluated. In contrast, more varied signals and microstructures were observed when these steps were not taken. Since the removal of the surface oxide layer/contamination via BHF etching and the prevention of the surface reactions from taking place via irradiation in vacuum can essentially only lead to the transformations taking place in a more pristine chemical and physical environment, we suspect that (1) a number of previous investigations that did not employ these procedures may have suffered consequently from the participation of more complex chemical and physical complications (e.g., encountering additional heterogeneous nucleation from being catalyzed at the native/newly formed oxide interface), and (2) heterogeneous nucleation proceeding at the bottom interface may correspond to the "intrinsic" nucleation mechanism. This in turn, at least according to classical nucleation theory, means that homogeneous nucleation of solids, under similar or at lower cooling rates, cannot be the dominant mechanism of nucleation even for those cases where various extrinsic factors are encountered. Additional and more detailed experimental results and discussions addressing and substantiating these points will be provided in a forthcoming paper [13].

CONCLUSIONS

We have investigated and analyzed the mechanism of solid nucleation in laser-quenched liquid Si films on SiO_2. When the experimental measures were taken to reduce extrinsic experimental influences, the resulting FTR and BTR signals, as well as the cross-sectional TEM micrographs indicate that the solidification was initiated at or near the bottom liquid Si-SiO_2 interface. These observations, together with the result from three-dimensional numerical analysis that quantitatively discounts the possibility of spatially localized homogeneous nucleation taking place near the bottom interface enable us to conclude that heterogeneous nucleation of solids must have taken place at the bottom interface when clean Si films are irradiated in vacuum.

ACKNOWLEDGMENTS

We would like to thank Prof. Mike O. Thompson at Cornell University for valuable discussions. This work was supported, in part, by the U.S. Department of Energy, Basic Energy Sciences and Division of Materials Science under grant DE-FG02-94ER45520. This research was also supported, in part, by KAIST WCU (World Class University) program through the National Research Foundation of Korea funded by the Ministry of Education, Science and Technology under grant R32-10051.

REFERENCES

1. J. S. Im, H. J. Kim, and M. O. Thompson, Appl. Phys. Lett. **63**, 1969 (1993).
2. S. R. Stiffler, M. O. Thompson, and P. S. Peercy, Phys. Rev. Lett. **60**, 2519 (1988).
3. J. S. Im, M. A. Crowder, R. S. Sposili, J. P. Leonard, H. J. Kim, J. H. Yoon, V. V. Gupta, H. Jin Song, and H. S. Cho, Physica Status Solidi A **166**, 603 (1998).
4. F. C. Voogt, and R. Ishihara, Thin Solid Films **383**, 45 (2001).
5. J. Boneberg, J. Nedelcu, H. Bender, and P. Leiderer, Mater. Sci. Eng. A, Struct. Mater., Prop. Microstruct. Process. **A173**, 347 (1993).
6. J. P. Leonard, and J. S. Im, Appl. Phys. Lett. **78**, 3454 (2001).

7. D. H. Auston, C. M. Surko, T. N. C. Venkatesan, R. E. Slusher, and J. A. Golovchenko, Appl. Phys. Lett. **33**, 437 (1978).
8. G. Devaud, and D. Turnbull, Appl. Phys. Lett. **46**, 844 (1985).
9. Y. Shao, and F. Spaepen, J. Appl. Phys. **79**, 2981 (1996).
10. J. P. Leonard, Doctorate Thesis, Columbia University (2000).
11. S. R. Stiffler, M. O. Thompson, and P. S. Peercy, Appl. Phys. Lett. **56**, 1025 (1990).
12. J. W. Christian, *Theory of Transformations in Metals and Alloys: Part I Equilibrium and General Kinetics Theory* (Pergamon, New York, 1975), p. 418.
13. Y. Deng, Q. Hu, U. J. Chung, A. M. Chitu, A. B. Limanov, and J. S. Im (to be published).

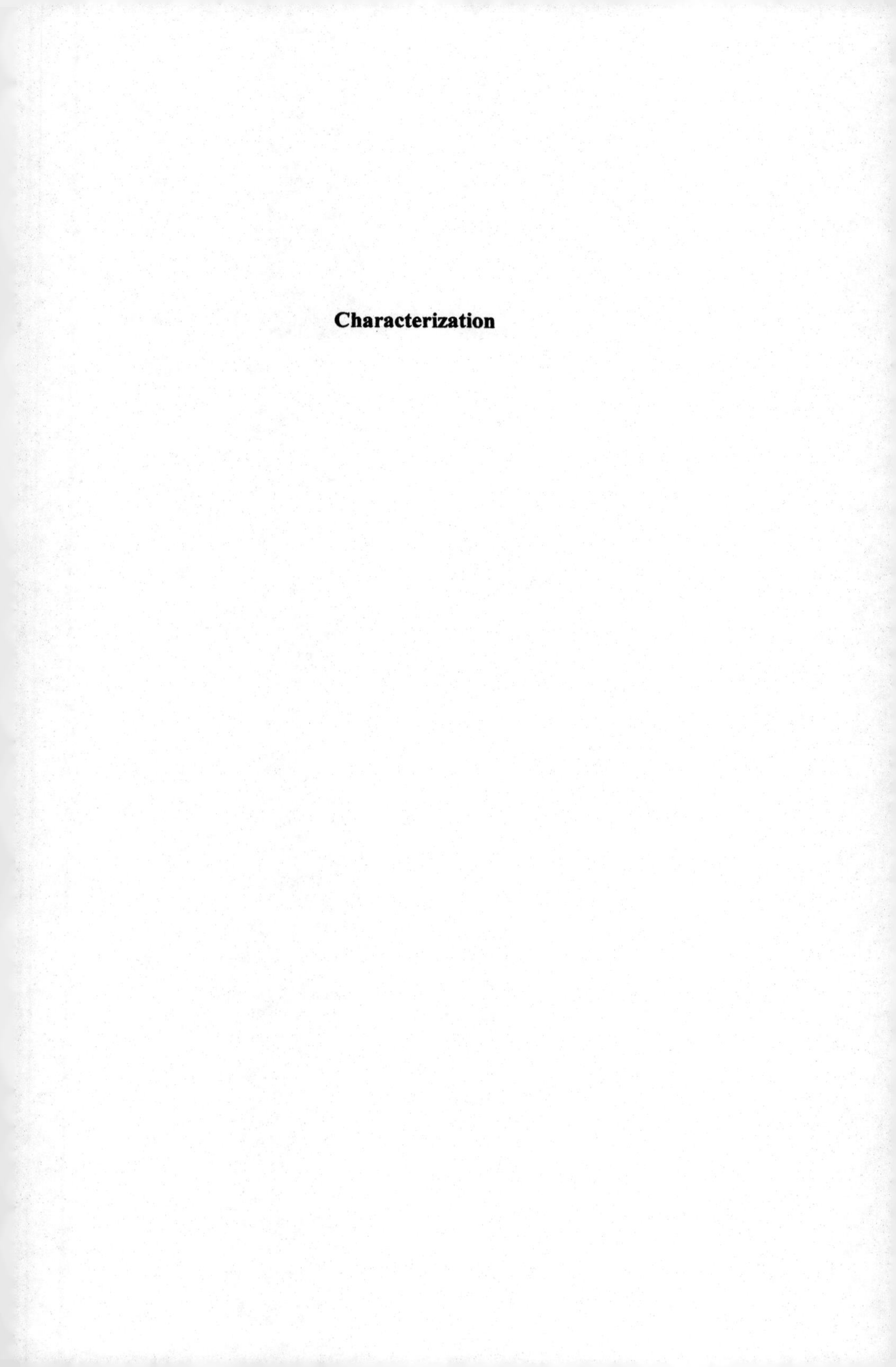

Characterization

Mater. Res. Soc. Symp. Proc. Vol. 1245 © 2010 Materials Research Society **1245-A13-01**

Investigation of Near-IR Emission from Hydrogenated Nanocrystalline Silicon – The Oxygen Defect Band

J.D. Fields [1], P.C. Taylor [1], D.C. Bobela [2], B. Yan [3], G. Yue [3]

[1] Colorado School of Mines – Golden, CO
[2] National Renewable Energy Laboratory – Golden, CO
[3] United Solar Ovonic LLC, 1100 West Maple Road, Troy, MI 48084

ABSTRACT

Hydrogenated nanocrystalline silicon (nc-Si:H), a mixture of nanometer sized crystallites and amorphous silicon tissue, demonstrates a photoluminescence band centered at ~ 0.7 eV, which emerges in response to annealing at an onset temperature of ~ 200–300 °C. This temperature range correlates well with hydrogen effusion spectroscopy studies, and evidence suggests thermal liberation of hydrogen from grain boundary regions allows oxidation of crystallite surfaces during annealing. We tentatively attribute the 0.7 eV PL in nc-Si:H to deep donor defect states related to oxygen precipitates, and argue for the possible involvement of dislocations inside of crystallites to accompany these precipitates.

INTRODUCTION

Oxygen impurities create electronic defects in silicon systems. For example, Czochralski (Cz-) and float zone (Fz-) crystalline silicon (c-Si) systems possess oxygen thermal double donor defect states (TDDs) [1], and hydrogenated amorphous silicon (a-Si:H) films contain O_3^+ defects [2]. These centers constitute shallow donor levels, and when present unintentional n-type doping results. Some argue that oxygen impurities impact the electronic structure of hydrogenated nanocrystalline silicon (nc-Si:H), a mixture of nanometer sized crystallites and amorphous silicon tissues, more strongly than any other defect commonly possessed by the system [3]. Dalal *et al.* report shallow donor defects attributed to oxygen in nc-Si:H films with density $\sim 8 \times 10^{15}$ cm^{-3} [4]. In addition, photoluminescence and electrical characterization suggest oxygen involvement in the formation of deep electronic levels in nc-Si:H [4]. In this study we focus our attention on these proposed oxygen related deep defects.

Despite extensive investigation of the behavior of oxygen in silicon, the microscopic origin of oxygen related electronic centers remains elusive. In c-Si systems, formation of shallow and deep states attributed to oxygen typically accompanies agglomeration of silicon–oxide complexes [5, 6]. However, controversy lingers regarding whether the electronic defects exist in the oxides themselves, at c-Si/oxide interfaces, or in extended defects surrounding the precipitates. To study these effects in Cz-Si and Fz-Si researchers commonly employ intensive annealing treatments involving temperatures near 450 °C for durations of ~ 64 hours [5,7, 8], and some argue oxide complexes form in nc-Si:H mixtures at substantially lower temperatures (~200°C) due to increased diffusion along grain boundaries [3].

For these reasons we look to oxygen to attribute a luminescence band observed in thermally treated nc-Si:H materials. This *defect band* emerges in response to annealing the films at ~ 200–300 °C, and shows a strong resemblance to that observed by Tajima in Cz-Si [5]. Tajima observed the emergence of a ~0.7 eV luminescence band after subjecting a Cz-Si sample to the aforementioned annealing treatment, and attributed the emission to deep levels associated with oxygen TDDs and oxygen agglomerates. The defect band we observe in nc-Si:H probably involves the same electronic states which caused the 0.7 eV PL in the Cz-Si experiment, however it is not clear whether the states are TDD type defects or something of a different nature. The coincidence between deep defect energy levels attributed to dislocations in oxide containing c-Si systems and those observed in nc-Si:H materials suggests that some of the crystallites may contain dislocations.

EXPERIMENT

United Solar Ovonics provided nc-Si:H films of varying crystalline volume fraction with crystallite dimensions averaging 6 nm x 20 nm [9]. All of the samples were VHF-PECVD grown, some on Ag/ZnO coated stainless steel substrates, others directly on stainless steel. Several samples were capped with a transparent conducting oxide layer. This difference did not appear to affect the 0.7 eV defect luminescence. Sample subsets where either annealed in air at atmospheric pressure, in an argon over-pressure, or under vacuum of ~ 10^{-6} Torr. Air and argon ambient annealing treatments exposed samples for 30 minutes to the specified temperatures, with 30 minutes of ramp-up and several hours of cool down time for each treatment. The vacuum annealed sample experienced a 120 minute exposure in an evacuated and preheated chamber. Secondary ion mass spectroscopy measurements performed previously on nc-Si:H mixtures like those used in the present study revealed ~10^{19} cm^{-3} oxygen impurities in the films initially, which increases and redistributes in response to annealing in air [10].

Crystalline volume fraction measurements employed a WiTech confocal Raman system using doubled YAG 532 nm excitation in the back-scattering configuration. Photoluminescence (PL) experiments utilized a Nicolet interferometer equipped with InGaAs and MCT detectors, and an Inova Ar^+ cw-laser set to the 514.5 nm line at power density less than 500 mW/cm^2, with one exception. Higher excitation power (~1.25 W/ cm^2) produced a greater signal to noise ratio during the defect band temperature dependence experiment. Interference fringes were subtracted from the spectra, as is customarily done in FTPL spectroscopy. Hydrogen effusion data were obtained at NREL using a turbo pumped vacuum system, a calibrated heating system, and a secondary ion mass spectrometer.

RESULTS

Initially we observe two bands centered at 1.3 eV and 0.95 eV (Figure 1 (a)), which we attribute to band-tail to band-tail transitions in the amorphous regions and nanocrystallite grain boundary (nc-Si g.b.) regions, respectively. For more information regarding the behavior of these bands, see reference [10].

In this experiment, a pronounced onset of the defect band occurred after annealing the nc-Si:H film in air at 300°C. Figure 1 (a) shows a shoulder on the low energy side of the annealed sample spectrum, cut-off by the edge of the InGaAs detector spectral operating range. Using an MCT detector (Figure 1 (b)) with higher sensitivity in the near-IR, we see the defect band fully resolved and centered at ~ 0.7 eV. The defect band also emerged for samples annealed in argon overpressure, and under vacuum. Thus, we conclude that the onset occurs regardless of annealing ambient.

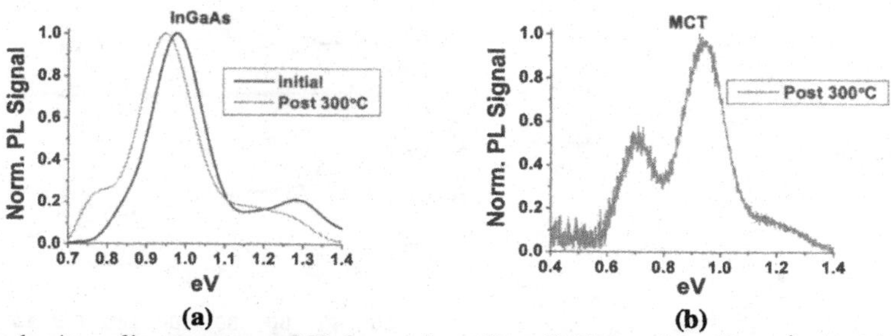

Figure 1: Annealing response of PL from a "good" nc-Si:H material collected with (a) InGaAs, and (b) MCT detectors. The defect band is centered at ~ 0.7 eV.

Figure 2 (a) shows Raman spectra for three nc-Si:H films of varying degrees of crystallinity. We assign comparative crystalline volume fractions by noting the strength of 480 cm^{-1} and 520 cm^{-1} Raman signals, attributable to amorphous and crystalline regions, respectively. We see that Sample A contains predominantly amorphous regions, Sample C contains mostly crystalline silicon, and Sample B possesses appreciable amounts of both phases. We omit grain boundary contributions for simplicity, since qualitative comparisons suffice to illustrate the correlation with PL annealing response. The sample with highest crystalline volume fraction demonstrates the most pronounced 0.7 eV PL emergence (Figure 2 (b)), the highly a-Si:H sample shows no detectable 0.7 eV PL, and Sample B shows a subtle defect band onset.

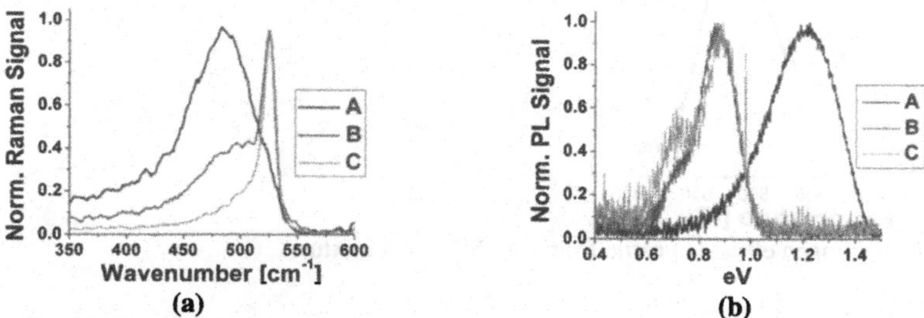

Figure 2: (a) Raman and (b) post 300 °C anneal PL observed for 3 nc-Si:H films of varying crystallinity.

Luminescence attributed to (nc-Si g.b.) regions, a-Si:H regions, and that believed to involve oxygen defects quenches with increasing temperature. In this experiment the defect PL was identifiable from 18 K up to 160 K. Over this temperature range, the defect PL band intensity decays less rapidly than that previously observed for nc-Si surfaces, and more rapidly than that attributed to a-Si:H regions [10]. Above 100 K the defect PL decay follows the a-Si:H band more closely than that attributed to nc-Si grain boundaries In figure 3 (b) we see that within the error bars, the defect band luminescence energy remains approximately constant, while the luminescence peaks attributed to transitions involving band tails clearly red-shift over this temperature range. We also provide the temperature dependence of the c-Si band gap over this range for comparison.

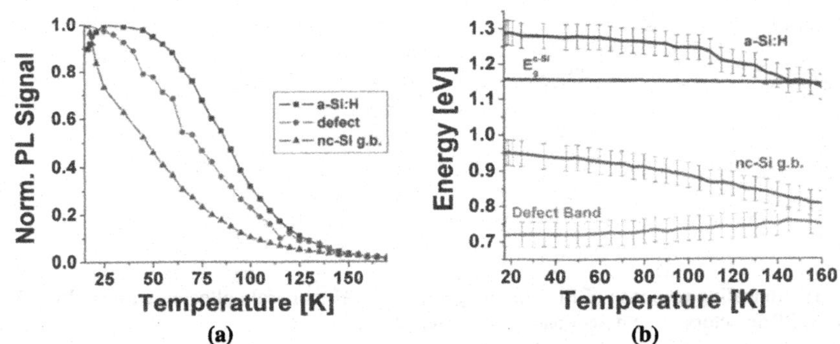

(a) (b)

Figure 3: Temperature dependent defect PL shown with previously published data for emission attributed to amorphous phase (a-Si:H) and grain boundary regions (nc-Si) for comparison [10].

Hydrogen effusion data shows hydrogen beginning to evolve at ~ 200 °C and readily leaving the film approaching 300 °C (Fig. 4), consistent with the observed defect band onset temperature. As the temperature increases, hydrogen effusion rates peak in two temperature regions, ~ 310 °C and ~ 550 °C. The former we attribute to hydrogen leaving grain boundary regions and the latter to hydrogen evolving from a-Si:H bulk.

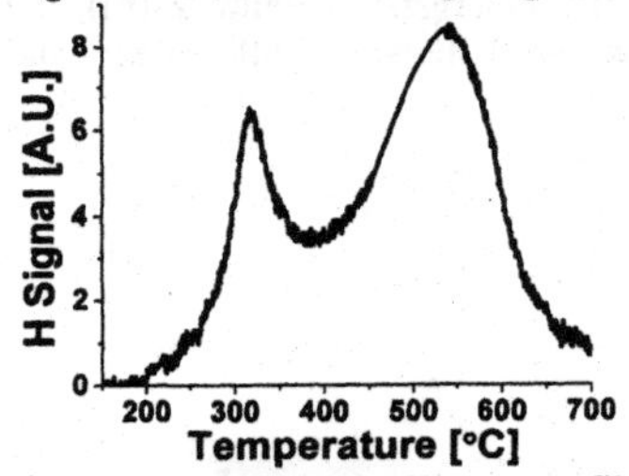

Figure 4: Hydrogen effusion profile from a nc-Si/a-Si:H mixture.

DISCUSSION

In one experiment, a 180 °C anneal caused a subtle onset of the defect band in a sample produced by an underperforming recipe, while films characteristic of "best" devices require higher temperatures (230–300 °C) to produce the emission. From this we expect that there may be a correlation between device performance and susceptibility to deep defect formation. Observing the coincidence between the defect band onset temperature range of ~ 200–300 °C, and the temperatures believed to deplete hydrogen from grain boundary regions, we expect that the deep defects form at interfaces. We consider that as Si–H bonds break on the crystallite surfaces and in amorphous regions close to grain boundaries, dimerization and other reconstruction events occur and dangling bonds also form. If oxygen resides in grain boundary regions as we expect because of segregation behavior observed for dopants and various impurities in polycrystalline Si systems, then it would follow that oxygen might react with the dangling bonds.

The varying crystalline volume fraction sample set clearly shows that 0.7 eV PL emerges more strongly the higher the crystallinity, which suggests that the defect luminescence originates in the crystalline phase. Hydrogenated microcrystalline silicon (μc-Si:H) films demonstrate the same trend [11], and 0.7 eV PL also occurs in laser annealed polycrystalline Si [12]. Based on these findings and the behavior we observed in nc-Si:H films, it seems very likely that the 0.7 eV defect luminescence from nc-Si:H involves the same oxygen related c-Si defect states which produced 0.7 eV PL in Tajima's Cz-Si experiment [8].

Contemplating the types of electronic states which could produce a 0.7 eV transition in these systems, one quickly recognizes a discrepancy between the energy levels we believe these materials to possess and popular 0.7 eV oxygen defect hypotheses. Several authors accept Tajima's conclusion that the 0.7 eV PL involves deep levels associated with oxygen TDDs and oxygen agglomerates [11, 12]. However, the deepest known oxygen TDD defect state resides only 156 meV below the c-Si conduction band edge [5]. Thus, radiative recombination with 0.7 eV emission requires other types of defect states. For reference in further discussion we provide density of states cartoons representative of the nano-crystallite phase (Fig. 5):

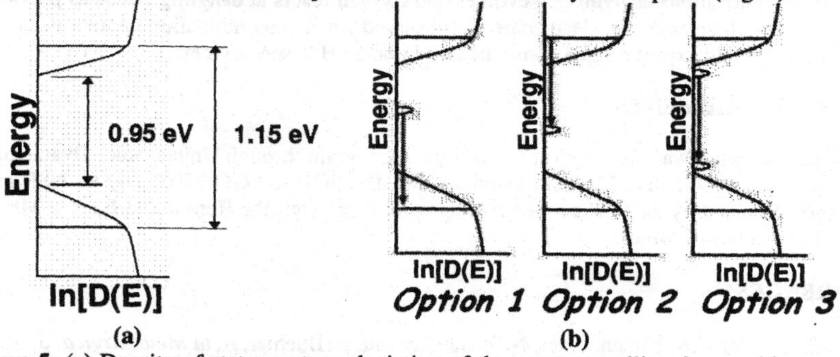

Figure 5: (a) Density of states cartoon depiction of the nano-crystallite phase, with exponential band tails extending into the band gap; (b) options we consider for types of defect states which could enable a 0.7 eV transition.

Since the defect band does not red-shift with increasing temperature like the PL attributed to band tail transitions in a-Si:H and nano-crystallite surface regions, we eliminate the possibility that the 0.7 eV PL involves band tail states. Most likely the defect PL originates from deep states, with transitions to or from extended states represented by flat lines in the cartoons. In the crystalline regions, the valence band extended state to conduction band extended state energy separation should be that of bulk crystalline silicon. We also include the possibility of transitions between two defect states in the gap (Option 3).

Defect states like those shown in Option 1 in Figure 5 (b) seem most likely to explain the 0.7 eV PL in nc-Si:H films based on electrical characterization results. DLTS experiments on nc-Si;H systems reveal a population of deep defect states at energies 0.35 to 0.50 eV below the conduction band edge [4]. Since the deep defects involved in 0.7 eV transitions probably reside in the crystalline phase, we also look to c-Si research results. When Claeys *et al.* investigated oxygen defects in Cz-Si and Fz-Si, they discovered a population of deep states with energy E_c–0.43 eV, attributed to dislocations surrounding oxygen defects, as opposed to the oxides themselves [7]. This energy value matches well with what we expect to exist in our nc-Si::H films from PL and DLTS measurements. Additionally, temperature dependent dark conductivity measurements on nc-Si:H films performed recently, which will be published later, show an activation energy of 0.4 eV for an annealed nc-Si:H film. All of these observations suggest the existence of a population of deep defects with energy ~ (E_c–0.4 eV) in annealed nc-Si:H materials, consistent with Option 1 in Figure 5 (b).

If we accept the conclusions of Tajima and Claeys *et al.* to explain the present behavior of our nc-Si:H materials, then it follows that some of the crystallites contain dislocations. While we have yet to observe dislocations in our nc-Si:H films, the recent discovery of dislocations in 3 nm diameter nano-crystallites embedded in a-Si:H matrices increases the likeliness of this explanation [13]. In our experiments we expect that strain accompanying oxide formation on crystallite surfaces induces dislocations either during film growth or during the annealing treatments. If dislocations form during synthesis, hydrogen may passivate them initially but leave them electronically active as hydrogen evolves from the film. Alternatively, annealing could create dislocations as hydrogen evolves and oxygen reacts at dangling bonds as previously discussed. In either case, the deep traps we observed act as recombination centers in nc-Si:H films and we should expect related consequences in nc-Si:H based devices.

ACKNOWLEDGEMENTS

This research was partially supported by a DOE grant through United Solar Ovanic under the Solar America Initiative Program Contract, No. DE-FC36-07 GO 17053, by an NSF grant, DMR-0073004, and by an NSF cooperative agreement through the Renewable Energy MRSEC at Colorado School of Mines.

REFERENCES

1. C.S. Fuller, J.A. Ditzenberger, N.B. Hannay, and E. Buehler, *Acta Metallurgica*, **3**, 97 (1955).
2. A. Morimoto, M. Matsumoto, M. Yoshita, M. Kumeda, T. Shimizu, *Appl. Phys. Lett.* **59**, 2130 (1991).
3. T. Kamei, T. Wada, and A. Matsuda, *28th IEEE Photovoltaic Spec. Conf.* (2000) p.784
4. V. Dalal and P. Sharma, *Appl. Phys. Letts.* **86**, 103510 (2005)
5. J. Michel and L. Kimerling, *Oxygen in Silicon*, ed. F. Shimura, *Academic Press* (1994)
6. A. Borghesi, B. Pivac, A. Sassella, and A. Stella, *J. Appl. Phys.* **77**, 4169 (1995)
7. C. Claeys, E. Simoen, and J. Vanhellemont, *J. Phys. III France* **7**, 1469 (1997)
8. M. Tajima, *J. of Crystal Growth* **103**, 1 – 7 (1990)
9. K.G. Kiriluk, D.L. Williamson, D.C. Bobela, P.C. Taylor, B. Yan, J. Yang, S. Guha, A. Madan, F. Zhu, *MRS Symp. Proc.* (2010) published in present proceedings
10. J. D. Fields, P. C. Taylor, J. G. Radziszewski, D. A. Baker, G. Yue, and B. Yan, *MRS Symp. Proc.* **1153**, 3 (2009)
11. T. Merdzhanova, R. Carius, S. Klein, F. Finger, D. Dimova-Malinovska, *Thin Solid Films* **511-512**, 394 (2006)
12. S.S. Ostapenko, A.U. Savchuk, G. Nowak, J. Lagowski, L. Jastrzebski, *Materials Science Forum* **196-201**, 1897 (1995)
13. J. Kakelios, C.B. Carter, and C. Perrey, Private communication

Mater. Res. Soc. Symp. Proc. Vol. 1245 © 2010 Materials Research Society 1245-A13-02

A SAXS Study of Hydrogenated Nanocrystalline Silicon Thin Films

K.G. Kiriluk[1], D. L. Williamson[1], D. C. Bobela[2], P. C. Taylor[1], B. Yan[3], J. Yang[3], S. Guha[3], A. Madan[4], F. Zhu[4]

[1]Colorado School of Mines, Golden, CO 80401
[2]National Renewable Energy Laboratory, Golden, CO 80401
[3]United Solar Ovonic LLC, Troy, MI 48084
[4]MVSystems Inc. Golden, CO 80401

ABSTRACT

We have used small-angle x-ray scattering (SAXS) in conjunction with X-ray diffraction (XRD) to study the nanostructure of hydrogenated nanocrystalline silicon (nc-Si:H). The crystallite size in the growth direction, as deduced from XRD data, is 24 nm with a preferred [220] orientation in the growth direction of the film. Fitting the SAXS intensity shows that the scattering derives from electron density fluctuations of both voids in the amorphous phase and H-rich clusters in the film, probably at the crystallite interfaces. The SAXS results indicate ellipsoidal shaped crystallites about 6 nm in size perpendicular to the growth direction. We annealed the samples, stepwise, and then measured the SAXS and ESR. At temperatures below 350°C, we observe an overall increase in the size of the scattering centers on annealing but only a small change in the spin density, which suggests that bond reconstruction on the crystallite surfaces takes place with high efficacy.

INTRODUCTION

Hydrogenated nanocrystalline silicon (nc-Si:H) is starting to be used in multi-junction solar cells with thin film based materials. nc-Si:H has been shown to suffer less light induced degradation than a-Si:H [1], however like a-Si:H, the degradation mechanism is not currently understood. nc-Si:H is a complicated, inhomogeneous material and understanding its nanostructure will help to understand the degradation mechanism. During the nc-Si:H deposition process, silane gas is diluted with hydrogen to promote the growth of crystallites in an amorphous matrix. Varying the amount of hydrogen dilution during growth maintains near uniform density and size of the crystallites [2]. The hydrogen also serves to passivate Si dangling bonds found in the amorphous phase of the nc-Si:H film and on the surfaces of crystallites.

To understand the growth and size of the crystallites, X-ray diffraction (XRD) and small-angle x-ray scattering (SAXS) measurements were performed on several different nc-Si:H samples. For the as-deposited nc-Si:H samples, the average crystallite size in the direction perpendicular to the diffraction planes and parallel to the growth direction is related to the width of the diffraction peaks in XRD by the well-known Scherrer formula [3]. The average crystallite size in the direction perpendicular to the growth direction is obtained by analyzing the shape and intensity of the SAXS data [4].

The evolution of the nanostructure can be probed by isochronally annealing the samples, then measuring the resulting SAXS scattering intensity. When the anneal temperatures are sufficiently high to drive out hydrogen, the SAXS profile changes therefore indicating a change in the overall hydrogen microstructure and void distribution. Combining the SAXS scattering results with electron spin resonance (ESR) defect density measurements, we are able to understand how the paramagnetic defects in nc-Si:H correlate with the nanostructural scattering features.

EXPERIMENTAL DETAILS

For the first set of SAXS experiments, 1-μm thick nc-Si:H films were deposited by United Solar Ovonic (USO) on 10-μm thick, high purity aluminum foil. The recipe used for deposition is similar to that used in USO high efficiency solar cell material. Raman scattering measurements completed on a similar sample, taken with a laser of 532 nm^{-1}, show an initial crystalline volume fraction of approximately 31%. Typical amorphous silicon usually contains approximately 10-15 at. % hydrogen [5]. Hydrogen effusion results on the USO nc-Si:H material give a hydrogen content of 6 ± 3 at. %. 1.4-μm thick nc-Si:H samples used for the annealing experiments were made by MV Systems, Inc. in Golden, CO. One sample was deposited on approximately 80 μm thick c-Si wafer, while one sister sample was deposited on quartz glass with a very low ESR background signal. These samples were deposited at 170°C. Raman scattering measurements of the MV Systems samples show a crystalline volume fraction of approximately 52%. The aluminum foil substrate could not be used for annealing over 400°C, as SIMS showed the aluminum started to diffuse into the nc-Si:H.

The XRD experiments were performed using a Cu Target in a Siemens D-500 diffractometer. The first three nanocrystalline XRD peaks at $\theta = 28°$, 47°, and 56° were measured in the symmetric Bragg-Brentano geometry [3]. These peaks correspond to the (111), (220), and (311) planes respectively. Once the diffraction peaks were obtained, they were fit with a Pearson function to find the full width at half maximum and the average crystallite size, D_{XRD}, was obtained using the Scherrer equation [3, 6].

SAXS experiments were completed using an updated model of the Kratky system, called the SAXSess system. We performed the measurements with the sample placed perpendicular to a monochromatic Cu-Kα X-ray beam (transmission mode so that the momentum transfer vector, q, is in the plane of the sample) where $\lambda = 0.154$ nm, with an imaging plate as the detector. This geometry allows for fluctuations in the electron density parallel to q to be measured [4, 7]. In another geometry, the sample is tilted at 45° with respect to the X-ray beam. In this setup, q no longer lies in the plane of the sample, which allows scattering from non-spherical and oriented objects to be determined.

Hydrogen effusion experiments on the MV Systems nc-Si:H film were conducted at the National Renewable Energy Lab (NREL) under a high vacuum of 1×10^{-6} Torr. The H_2 aspect signal was taken using an Inficon Quadrex 200 mass spectrometer. The samples were isochronally annealed from 200°C to 600°C in approximately 50°C increments, for 20 minutes, on ESR substrates. After each anneal, paramagnetic defect concentrations at room temperature were measured via ESR using a Bruker Elexsys E5800 system operating in the X-band frequency. The spin concentration was calibrated against a strong pitch ESR standard.

RESULTS and DISCUSSION

Figure 1a shows the X-ray diffraction pattern for nc-Si:H made by USO. The peak near 45° is from the aluminum foil substrate. The peak around 47° is much larger in intensity than that around 28°. This is different from the powder diffraction pattern of crystalline silicon where the 28° peak is much more intense and indicates that there is a preferred [220] orientation of the crystallites in the growth direction of the film. The widths of the diffraction peaks give a crystallite size of 11.3 ± 1 for the 28° and 56° peak, and 24.4 ± 1 nm for the 47° peak respectively.

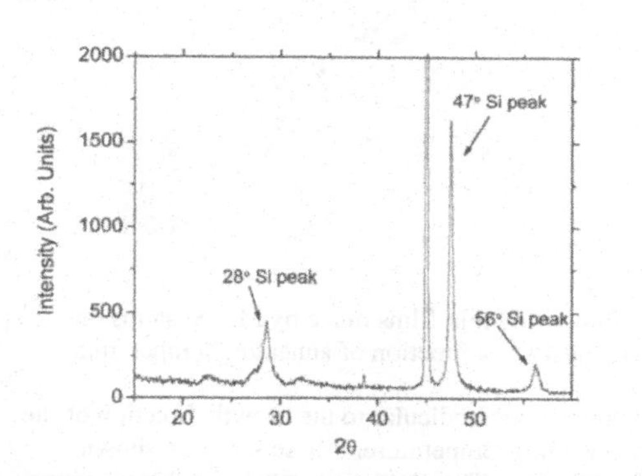

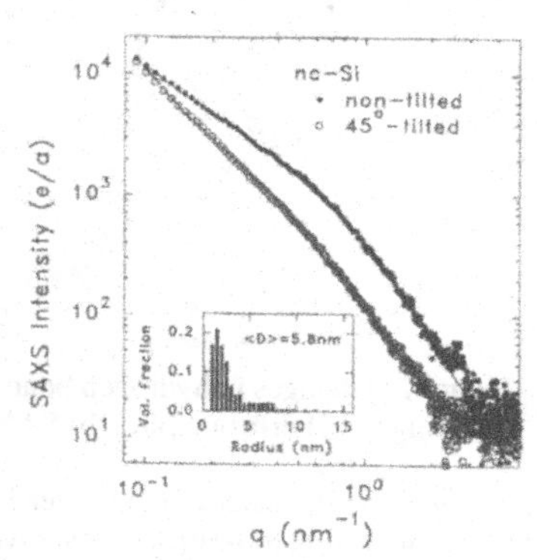

Figure 1. X-ray scattering from nc-Si:H thin films deposited by USO. a) XRD results b) SAXS results

The SAXS results for the nc-Si:H made by USO are shown in Figure 1b. The SAXS intensity can be described by

$$I(q) = I_L(q)+I_N(q)+I_D(q), \quad (1)$$

where $I_L(q)$ describes scattering from large-scale features such as surface roughness and corresponds to small q values, $I_N(q)$ describes scattering from nanostructural features and $I_D(q)$ describes diffuse scattering [7, 8]. The decreased intensity in the SAXS scattering between the tilted and non-tilted geometries reveals non-spherical and oriented nanostructural features in the film. An estimation of the size of the nanostructural features was obtained by fitting the SAXS results in Figure 1b, using the model based on Eq. 1 , and following the procedure documented in Ref [8]. The inset in Figure 1b is the distribution of sizes that contribute to the non-tilted SAXS intensity. The mean diameter, <D>, of these features is 5.8 nm ± 10%. Further modeling of the integrated intensity of Figure 1b suggests that three phases are needed to account for the amount of SAXS scattering seen in the nc-Si:H film: amorphous phase (a-Si), crystallites (c-Si), and voids or H-rich regions.

Figure 2a shows the H effused from the MV Systems film as a function of annealing temperature. H starts to move considerably above approximately 250°C. The first peak in the H effusion curve centered around 305°C is believed to originate from H moving from the crystalline grain boundaries. The second peak centered around 550°C is likely due to those H bonded deeper in the a-Si phase [9].

Figure 2b shows the SAXS intensity after the sample was annealed at the given temperature. For temperatures up to 310°C, there is little change in $I_N(q)$ from the as-grown intensity. Around 350°C, $I_N(q)$ shows a dramatic shape change and increase in intensity, indicating a change in the shape of the features contributing to the intensity. Above 350°C, the $I_N(q)$ decreases again although to a higher level than the original $I_N(q)$. Overall $I_D(q)$ increases with annealing temperature.

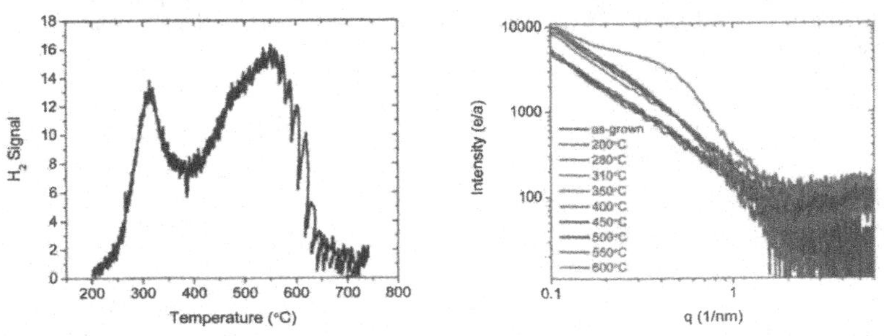

Figure 2. Changes in hydrogen bonding in nc-Si:H thin films made by MV Systems on annealing. a) H effusion plot. b) SAXS intensity as a function of annealing temperature.

The mean diameter of the scattering features, perpendicular to the growth direction of the film and contributing to $I_N(q)$, increases with annealing temperature. These sizes are shown in Figure 3 and are compared with the ESR defect density. The plot is separated into three regions designated by the region from where the H evolves; grain boundaries (black), grain boundaries and a-Si:H (light gray), and a-Si:H (dark gray). It is clear from Figure 3, that the initial release of H from crystallite boundaries results in an increase in the scattering size, yet the defect concentration remains almost the same. We suggest that the dehydrogenated crystalline surfaces undergo bond reconstruction, perhaps similar to the negative-U hydrogen pair release in vacancies proposed by Zafar and Schiff [10]. The hump that appears at 350 °C is from a sudden contrast in the electron density in the film, the cause of which is currently not understood. Above 400°C, we observe a large increase in defect density for a modest increase in the scattering size.

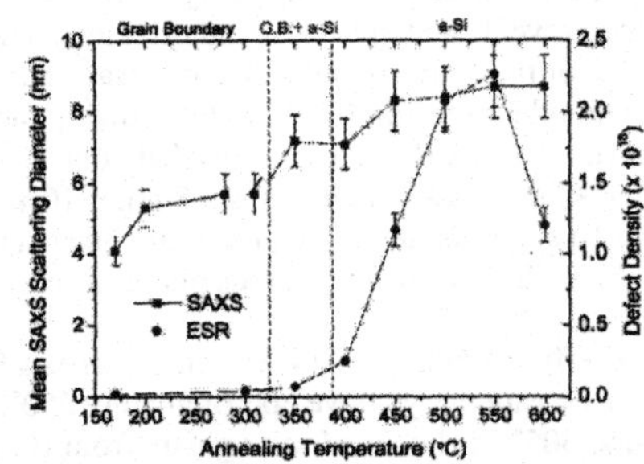

Figure 3. SAXS Scattering Feature Size (squares) with ESR Defect Density (circles) as a function of annealing temperature for MV Systems nc-Si:H thin films.

CONCLUSION

The XRD and SAXS results for the USO nc-Si:H thin films show elongated grains with a preferred [220] orientation in the growth direction of the film. Hydrogen effusion completed on the MV Systems nc-Si:H thin films shows that H starts rapidly evolving above 250°C, however, the SAXS intensity increases only once the H has moved out of the grain boundary region at 350°C. While the average scattering size increases with increasing annealing temperature, the defect density does not drastically increase until 400°C, which correlates to the a-Si regime of the hydrogen effusion curve. This result suggests that reconstruction is occurring at the crystallite interfaces between a-Si and c-Si for temperatures below 400°C.

ACKNOWLEDGEMENTS

The authors would like to thank the University of Colorado, Boulder for the use of their SAXSess setup. This work has been supported by the National Science Foundation (NSF) under grants DMR-0702351 and DMR-0073004, and by the NSF Renewable Energy Material Research Science and Engineering Center (REMRSEC) under grant DMR-08-20518. The work at CSM and USO has also been partially funded by the Department of Energy through the SAI program under subcontract (DE-FC36-07 GO 17053).

REFERENCES

1. J. Meier, R. Flückiger, H. Keppner, and A. Shah, Appl. Phys. Lett. 65, 860 (1994)
2. B. Yan, G. Yue, J. Yang, S. Guha, D. L. Williamson, D. Han, and C.-S. Jiang, Appl. Phys. Lett. **85**, 1955 (2004)
3. B. E. Warren, X-ray Diffraction (Dover, New York, 1990)
4. L. A. Feigen and D. I. Svergun, Structure Analysis by Small-Angle X-Ray and Neutron Scattering (Plenum Press, New York, 1987)
5. W. E. Carlos and P. C. Taylor, Phys. Rev. B 26, 3605 (1982)
6. D. L. Williamson, Mat. Res. Soc. Symp. Proc., Vol. 557, 251 (1999)
7. O. Kratky, Small Angle X-ray Scattering, edited by O. Glatter and O. Kratky (Academic Press, 1982).
8. D. L. Williamson, Mat. Res. Soc. Symp. Proc. 377, 251 (1995)
9. W. Beyer and H. Wagner, J. of Appl. Phys. 53(12) (1982)
10. S. Zafar and E. A. Schiff, Phys. Rev. B. 40 5235 (1989)

Mater. Res. Soc. Symp. Proc. Vol. 1245 © 2010 Materials Research Society 1245-A13-05

Assignment of High Wave-number Absorption and Raman Scattering Peaks in Microcrystalline Silicon

Erik.V. Johnson, Laurent Kroely, and Pere Roca i Cabarrocas
LPICM, CNRS, Ecole Polytechnique, 91128 Palaiseau, France

ABSTRACT

We present a detailed analysis of the narrow, twinned high-wave-number (2085 and 2100 cm^{-1}) infrared absorption and Raman scattering peaks observed in low-density hydrogenated microcrystalline (μc-Si:H) silicon. Peaks in this wave-number range originate from the stretching modes of the $Si-H_X$ bonds in the material, but the exact atomic configurations giving rise to these peaks is unclear. We attempt to elucidate the origins of the peaks through complementary experimental data on films grown by Matrix Distributed Electron Cyclotron Resonance (MDECR) PECVD. The different appearance and evolution of these peaks when using the two complementary measurement techniques mentioned above show that they have different origins, and cannot be attributed to a single, shifted peak. We additionally present data from secondary ion mass spectrometry (SIMS) measurements on the films to show the distribution of oxygen and carbon in the films after five months of air exposure. Finally, we provide X-ray diffraction (XRD) data and use the correlations between these measurements to propose a structural origin for the peaks.

INTRODUCTION

The high wave-number (HWN) (1800-2200 cm^{-1}) infrared (IR) absorption in hydrogenated silicon thin-films has long been used to characterize their quality. Absorption peaks at ~2000cm^{-1} have been assigned to isolated stretching Si-H bonds, with higher energy peaks variously assigned to clustered Si-H, platelet-like configurations, and the higher hydrides. This characterization tool has been extended to μc-Si:H thin films destined for use in photovoltaics. Smets, Matsui and Kondo [1] have recently made the case that the appearance of narrow, twin peaks at 2085 and 2100 cm^{-1} in the IR absorption spectrum, as measured by Fourier Transform Infrared (FTIR) spectroscopy, coincides with porous μc-Si:H material that will perform poorly in solar cells. Narrow absorption peaks in the HWN FTIR absorption spectra (2000-2200 cm^{-1}) of μc-Si:H thin films have been reported for many years [2,3,4] , but have only recently been directly correlated with the oxygen-related degradation of solar cells. In addition, the exact atomic configuration leading to these peaks is unclear, and has been attributed by some authors to Si-H bonds on two different silicon planes [5], to Si-H and $Si-H_2$ bonding configurations on the {111} and {100} surfaces of silicon crystallites [6], or to the natural Si-H resonance peak at 2083cm^{-1} being shifted to higher WN due to the bonds being located at the interface between silicon and void [7].

In a recent publication, we have underlined the analogous appearance of these twin peaks in Raman spectra measured on porous μc-Si:H [8] . The use of Raman scattering spectroscopy in place of FTIR eliminates the need for lightly-doped crystalline silicon (c-Si) wafers and the possibility of platelet formation near the wafer surface [9], and also reduces material quality uncertainties due to substrate-dependent growth [10] by allowing measurements to be done directly on films or devices on any type of substrate.

In this work, we apply HWN Raman scattering and FTIR absorption to the analysis of μc-Si:H. Using these techniques as well as additional X-ray diffraction (XRD) and Secondary Ion Mass Spectroscopy (SIMS) measurements, we aim to elucidate the origins of these twin peaks at 2085 and 2100cm^{-1}.

EXPERIMENT

All of the μc-Si:H samples studied in this work were deposited from SiH_4 using MDECR-PECVD. This deposition technology has been used to perform the deposition of μc-Si:H at rates up to 28Å/s [11], although the films for this study were deposited at lower rates (~5 Å/s). All the films were deposited from a gas mixture consisting of 8 sccm of SiH_4, 5 sccm of Ar and 75 sccm of H_2, and at a pressure of 5 mTorr. The substrate temperature during deposition was ~230°C, as determined by pyrometry, and a total microwave power of 1.5 kW was fed into the seven microwave antennas. The results presented herein are mostly shown for samples deposited on 10-20kΩcm, float-zone c-Si wafer substrates (100), as these samples provided the best opportunity to directly compare Raman and FTIR results, and because the twin peaks were always more pronounced for such substrates than for the Corning Glass 1737 and Eagle XG borosilicate substrates on which they were co-deposited. Three sets of samples, made under identical conditions except for deposition time, are discussed in this work Sample Set 1 (#90514, ~1μm), Sample Set 2 (#100208, 800nm), and Sample Set 3 (#100315, ~1μm).

The films were characterized by FTIR using a Nicolet 6700 spectrometer. Raman spectra for the three sets of films were acquired using a backscattering configuration and a HeNe laser to provide optical excitation at 632 nm. The optical absorption length at this wavelength is α^{-1} = 0.5-1 μm for these materials, meaning these samples are quite homogenously sampled. XRD spectroscopy was performed on the samples of Set 3 using a Philips X'Pert X-ray diffractometer using the Cu-$K_{\alpha 1}$ and Cu-$K_{\alpha 2}$ lines. SIMS was performed on the samples of Set 1.

RESULTS

Figures 1a and 1b present HWN Raman scattering and FTIR absorption spectra, respectively, measured for a μc-Si:H sample from Set 1 at times from directly after deposition until 82 days later. In Figure 1a, we present the evolution of the Raman spectra peaks. Although their intensity appears to decrease, little can be said about the absolute values of the peak intensities, as the Raman data cannot be quantified (and changes may be due to alignment, etc). One may note, however, that relative to the height of the twin Si-H peaks, the peak at 2250 cm^{-1} does not grow significantly (just as was previously observed [5]). Even more importantly, the *relative* height of the two Si-H peaks stays the same (i.e. the amplitude of the peak at 2100cm^{-1} remains greater than that of the one at 2083cm^{-1}).

Figure 1b displays FTIR measurements taken on the same films as Fig.1a, and shows the analogous evolution in the absorption peaks. Two important differences should be noted between these two sets of measurements. Firstly, the peak at 2250cm^{-1} (attributed to Si-H bonds with a back-bonded O, O_y-Si-H_x [1,12]) becomes much more evident than in the Raman spectrum, indicating that if this assignment is correct, its symmetry should make it FTIR-active and not Raman-active. In general, the assignment of the 2250cm^{-1} peak to an oxygen-related structure is supported by the similar, slowed evolution of the Si-O-Si peak at 1050cm^{-1} (not shown). Secondly, in contrast to the Raman spectra, the *relative height* of the two FTIR absorption peaks changes during the exposure-related evolution. Initially, the intensity of the peak at 2100 cm^{-1} (I_{2100}) is stronger than that of the peak at 2083cm^{-1}(I_{2083}), but after air exposure, the two peak heights rapidly become equal.

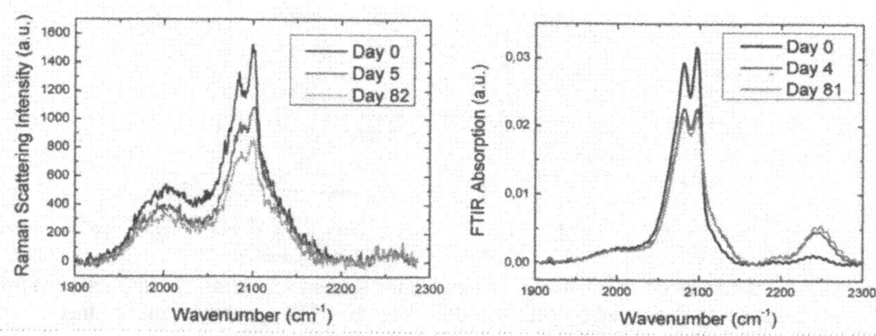

Figure 1. Time evolution of (a) Raman scattering and (b) FTIR absorption spectra obtained for a µc-Si:H thin film from Set 1 (Sample B) deposited by MDECR-PECVD on c-Si.

As the difference in peak evolution between Raman and FTIR may hold keys to the origin of these peaks, the relative peak height (I_{2100} / I_{2083}) is graphed as a function of time in Figure 2. It can be seen that for FTIR, the 2100cm^{-1} peak height is initially greater, but then equalizes with the low energy peak (ratio of unity). However, this phenomenon is not consistently seen in the Raman scattering spectra, where the relative heights remain almost constant (within the noise of the measurement), the high energy peak remains stronger, and I_{2100} / I_{2083} retains a value greater than unity.

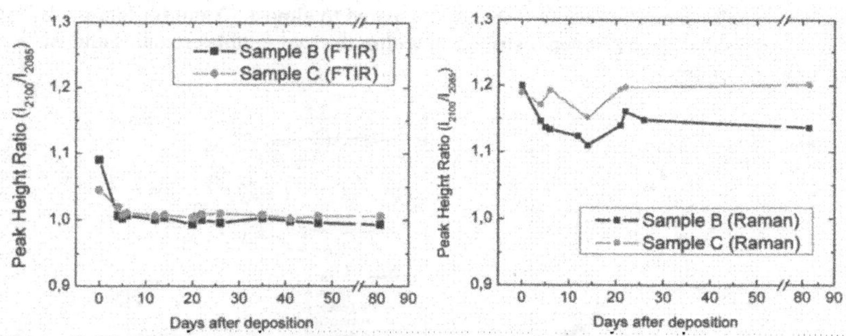

Figure 2. Time dependence of relative heights (I_{2100} / I_{2083}) of HWN peaks as measured by (a) FTIR absorption and (b) Raman scattering. Spectra obtained on a µc-Si:H thin film from Set 1.

This phenomenon is consistently observed in all the MDECR samples of this study. In Figure 3a, the relative peak height evolution is shown for a sample from Set 2. This set of samples is significantly thinner than those of Set 1 (Figures 1-2) but the same pattern in evolution is observed. In addition, an FTIR spectrum measured on a c-Si wafer from which the sample peeled off after deposition is presented in Figure 3b alongside spectra from two other co-deposited samples from Set 2. The complete lack of narrow peaks in this "peeled" sample underlines the fact that both peaks originate in the deposited film and not the underlying wafer.

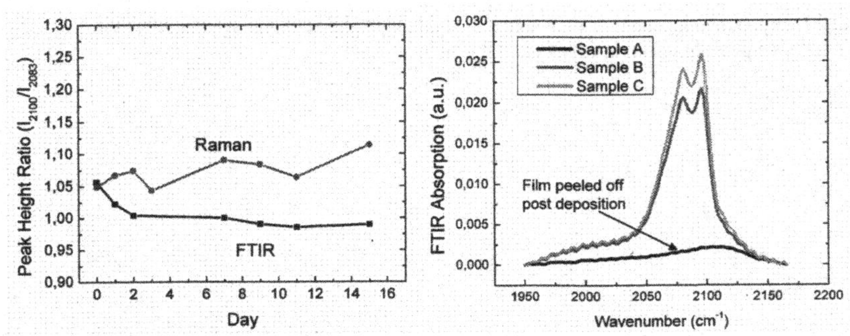

Figure 3. (a) Time dependence of relative peak heights for Raman scattering spectra obtained for a μc-Si:H thin film (Set 2) deposited by MDECR-PECVD. (b) FTIR spectra for these films including for a substrate from which the film peeled off.

In Figure 4, we present SIMS results from two co-deposited samples of Set 1 measured 160 days after deposition. The as-deposited Raman spectra of these two samples (Fig 4a) show different features due to different quality of clamping (and thermal contact) to the substrate holder, with Sample B showing much more prominent twin peaks. The SIMS data from these two samples (Fig 4b) bears out the prediction that Sample B, showing prominent twin peaks in Raman, should undergo a rapid and deep oxidation. Sample A, however, shows an oxygen tail originating from the wafer, and only very little penetrating from the surface. This "substrate" tail is surprising as a load-lock was used, and such a tail was only observed under such deposition conditions. Both samples show penetration of carbon from the surface, although Sample A shows only a very small penetration depth (100nm) compared to almost 500nm for Sample B. Finally, the data shows that the hydrogen distribution within the two samples remains almost constant as a function of depth.

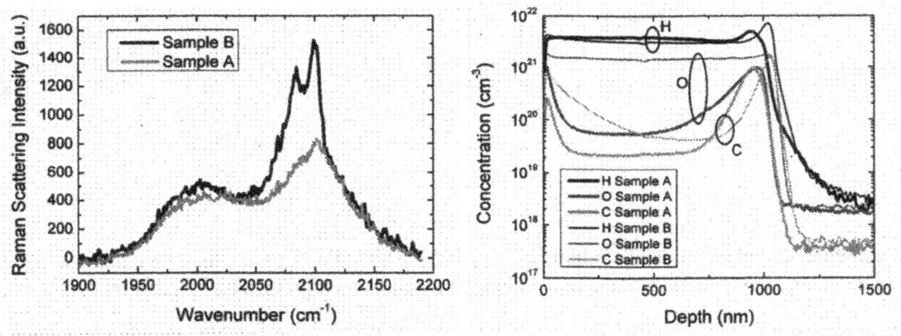

Figure 4. Data for co-deposited samples A and B from Sample Set 1. (a) High WN Raman spectra, and (b) SIMS data for both samples, following H, C, and O.

Finally, XRD data from three co-deposited samples from Set 3 are presented in Figure 5a : Samples A and B are deposited on c-Si, and to underline the substrate selectivity, data from

Sample E (deposited on Corning Glass) are also shown. Variations in substrate position lead to Sample B having a more dominant [111] XRD peak (compared to [220]) than Sample A. The Raman spectra of the two c-Si samples are shown in Fig 5b. Such a correlation of oxidation with [111] growth has been shown in larger studies as well [13].

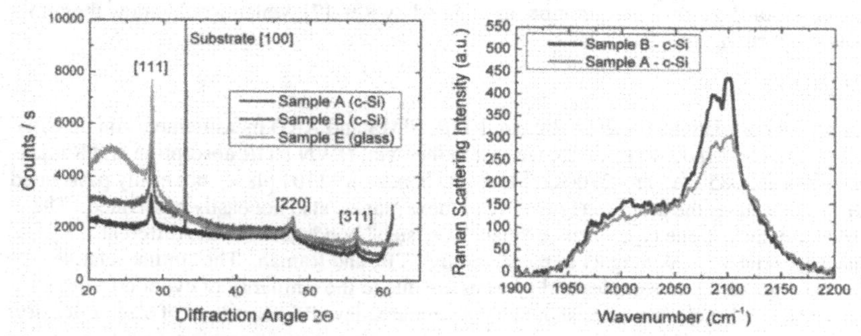

Figure 5. (a) XRD data for samples A and B on c-Si and sample E on Corning Glass, and (b) Raman scattering spectra for samples A and B, all from Sample Set 3. The I_{220}/I_{111} ratios are 0.52 for Sample A and 0.24 for Sample B (random is 0.5).

DISCUSSION

The data presented above allow for some conclusions to be drawn about the origins of the HWN peaks at 2085, 2100, and 2250cm^{-1}. Firstly, we observe that the peak at 2250cm^{-1} (attributed to O-Si-H_y bonds [12]) is FTIR-active but not Raman-active. If this peak has been correctly assigned, the symmetry of this structure should support this, and given the selectivity of the measurement, a very specific configuration should be responsible. Secondly, the relative height of the peaks at 2100 and 2085cm^{-1} evolve differently in Raman and in FTIR (Figs 1 and 2). As hydrogen content remains roughly constant with depth (SIMS results, Fig.4), this phenomenon cannot be due to the slightly different probe depths (Raman slightly favouring the surface and FTIR probing all film depths evenly). Furthermore, measurements on substrates from which the films peeled off show that the peaks are not due to platelets in the substrate. The selectivity of the Raman versus FTIR measurements (i.e. the different evolutions in relative peak heights) therefore means that the peak at 2100cm^{-1} must represent *at least two different Si-H_x configurations*,

Together with the XRD data (Fig. 5a), these measurements allow one to evaluate the three theories presented for the peaks at 2085 and 2100 cm^{-1}: namely (1) Si-H on two different high density Si planes [5], (2) Si-H vs Si-H_2 bonding configurations (on the {111} or {100} surfaces of crystallites) [6], or (3) to the natural resonance of the Si-H peak at 2085cm^{-1} being up-shifted due to being at an interface between silicon and void [7]. The XRD data of Fig 5 shows that the HWN peaks at 2085 and 2100 cm^{-1} correlate with {111} dominant growth (as has been previously shown by other authors [13]). Assuming a cylindrical growth model, a {220} growth surface allows for a mixture of {100}, {110}, and {111} crystalline cylinder sidewalls, whereas a cylinder with a {111} top surface can *only* have {110} sidewalls, therefore weakening the argument of Refs [5] and [6]. We suggest that tightly packed {100} and {111} sidewalls would provide a better template for a-Si:H "tissue" to passivate the cylinders (rather than H, causing the narrow HWN peaks) whereas the odd atomic spacing on the {110} planes will result in more passivation by H bonds, more voids, and therefore a greater number of vertical diffusion paths. Finally, as the peaks always occur in approximately the same proportions in both Raman and FTIR, (for example, the 2085 cm^{-1} peak is never stronger than the 2100 cm^{-1} peak), they cannot be shifted values of the same configuration, as claimed by Ref. [7], which would allow them to exist in any proportion.

CONCLUSIONS

Applying the correlations between Raman, FTIR, SIMS and XRD measurements on μc-Si:H samples deposited by MDECR-PECVD, we assign the twin, HWN FTIR absorption and Raman scattering peaks at 2085 cm^{-1} and 2100cm^{-1} to Si-H_x bonds on {110} planes not easily passivated by a-Si:H tissue, and on the atomic steps between these planes, also not easily passivated. The preferential oxidation of one type of these "plane" or "step" bonds leads to the different evolution of the relative peak heights as measured by FTIR and Raman. The similar relative peak sizes amongst different samples and systems are due to the similarity of cylinders with a {111} top surface; all such structures should have similar sidewalls, regardless of size or density.

ACKNOWLEDGMENTS

The authors would like to thank P. Bulkin for valuable discussions. The work of this study was partially funded by the EU Project SE Powerfoil (Project number 038885 SES6).

REFERENCES

1. A. H. M. Smets, T. Matsui and M. Kondo, *J. Appl. Phys* **104**, 034508 (2008).
2. S. Veprek, Z. Iqbal, H.R. Oswald and A.P. Webb, *J. Phys. C: Solid State Phys.* **14**, 295 (1981).
3. T.Imura, K. Mogi, A. Hiraki, S. Nakashima and A. Mitsuishi, *Solid State Commun.* **40**,161 (1981).
4. D.C. Marra, E.A. Edelberg, R.L. Naone and E.S. Aydil, *J. Vac. Sci. Technol. A* **16**, 3199 (1998)
5. T. Satoh and H. Hiraki, *Jpn. J.Appl.Phys.* **24**, L491–L494 (1985).
6. U.Kroll, J.Meyer, A.Shah, S.Mikhailov, and J.Weber, *J. Appl. Phys.* **80**, 4971–4975 (1996).
7. D. Stryahilev, F. Diehl, B. Schröder, M. Scheib and A.I. Belogorokhov, *Phil. Mag. B* **80**, 1799 (2000).
8. E.V. Johnson, L. Kroely, and P.Roca i Cabarrocas *Solar Energy Mater. Solar Cells* **93**, 1904 (2009).
9. J.N. Heyman, J.W. Ager III, E.E. Haller, N.M. Johnson, J. Walker, and C.M. Doland, *Phys.Rev.B* **45**, 13363 (1992).
10. P. Roca i Cabarrocas, N. Layadi, T. Heitz, B. Drévillon and I. Solomon, *Appl. Phys. Lett.* **66**, 3609 (1995).
11. P. Roca i Cabarrocas *et al*, *Thin Solid Films* **516**, 6834 (2008).
12. M. Niwano, J. Kageyama, K. Kurita, K. Kinashi, I. Takahashi, and N. Miyamoto, *J. Appl. Phys.* **76**, 2157 (1994).
13. S. Nunomura and M. Kondo, *J. Phys. D: Appl. Phys.* **42**, 185210 (2009).

Mater. Res. Soc. Symp. Proc. Vol. 1245 © 2010 Materials Research Society 1245-A13-06

Standard Characterization of Multi-junction Thin-film Photovoltaic Modules: Spectral Mismatch Correction to Standard Test Conditions and Comparison with Outdoor Measurements

Mauro Pravettoni[1,2], Georgios Tzamalis[1], Komlan Anika[1], Davide Polverini[1], and Harald Müllejans[1]
[1] European Commission, DG JRC, IE, Renewable Energy Unit, 21027 Ispra (VA), Italy
[2] Imperial College London, Blackett Laboratory, London, SW7 2BW, United Kingdom

ABSTRACT

Multi-junction thin-film devices have emerged as very promising PV materials due to reduced cost, manufacturing ease, efficiency and long term performance. The consequent growing interest of the PV community has lead to the development of new methods for the correction of indoor measurements to standard test conditions (STC), as presented in this paper. The experimental setup for spectral response measurement of multi-junction large-area thin-film modules is presented. A method for reliable corrections of indoor current-voltage characterization to STC is presented: results are compared with outdoor measurements where irradiance conditions are close to standard ones, highlighting ongoing challenges in standard characterization of such devices.

INTRODUCTION

Multi-junction photovoltaic (PV) devices consist of a stack of two or more semi-conductive layers ("junctions"), each with different characteristic band gap. The layer with the widest band gap is the top junction, absorbing the short wavelength radiation. Unabsorbed rays with longer wavelengths enter the following junction with smaller band gap, resulting in a more efficient use of the solar spectrum. Broadening the absorption spectrum by means of multi-junction structures has shown to increase the efficiency of thin-film modules above 10% [1], thus helping to strengthen the importance of thin-film modules in the PV market.

As a consequence, researchers at the European Solar Test Installation (ESTI) laboratories have started to investigate long-term stability and indoor-outdoor characterization of multi-junction a-Si thin-film modules. In view of the evaluation of spectral mismatch correction of indoor current-voltage (IV) characterization to standard test conditions (STC) an experimental setup for spectral response (SR) measurement of large area multi-junction modules has been developed [2]. In this work we present results of SR measurements and indoor characterization of multi-junction a-Si thin-film modules of two different technologies. Methods for a reliable correction to STC are also analyzed. Data are then compared with results of outdoor characterization, where spectral irradiance is close to the standard.

Part of the work presented has been performed within the Integrated Project (IP) PERFORMANCE [3]: the 4-year project started in January 2006 and has involved 27 European partners, representing the most experienced companies and research centres in Europe. It aimed to provide the PV community worldwide with tools to measure the quality of PV products, devices, systems and services, ensuring their usefulness and reliability.

An Experimental Setup for SR Measurements of Multi-junction PV Modules

SR measurements of multi-junction PV devices in general have been widely studied in the literature since the late 1980s [4] and require the saturation of the junctions not being measured by means of proper light and voltage biasing. Additional challenges occur for large area modules SR measurements [5]. A description of the standard procedure for non-concentrating PV cells and modules is given in the ASTM standard test method E2236-05 [6].

Figure 1 shows a scheme of the experimental setup developed at ESTI. A 300 W xenon lamp provides the source light, filtered by up to 64 bandpass interference filters available (8 to 20 nm width; range from 300 to 1200 nm). The monochromatic beam obtained is chopped, collimated by a lens system and then driven to the testing device and the reference cell by means of an off-axis reflector with broad spectral reflectivity. The module can be moved horizontally and vertically, so that the monochromatic spot can be focused in various areas of the test module, in the frequent case where the spot area is smaller than module area.

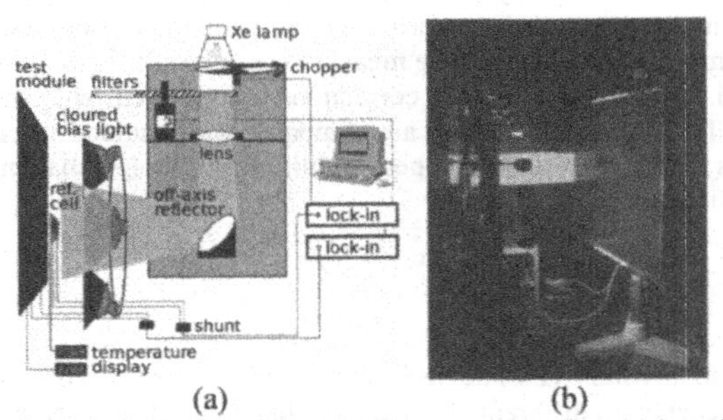

Figure 1. ESTI new experimental setup for SR measurements of multi-junction large area modules: (a) a schematic diagram; (b) the bias light system.

Bias light is provided by up to 24 QTH bulbs and various sets of 6 high power LEDs by ENFIS ltd. (available peak wavelengths: 405, 520, 630 and 870 nm) for the saturation of junctions not to be measured: due to the series connection of component junctions, the junction under examination limits the current of the entire device and is responsible of the AC signal at the frequency of the chopper. The bias light system is able to fully illuminate the test module to avoid current limitation by component cells eventually in the dark and to put the whole device at desired total irradiance conditions.

Short-circuit current (I_{SC}) AC signals are filtered out over a shunt resistor by means of lock-in technique. The SR of the *i*-th junction of the testing device, $SR_i(\lambda)$ (in AW^{-1}), is calculated as the ratio between short-circuit current values measured from the *i*-th junction and from the reference cell respectively at wavelength λ, times the absolute spectral response of the calibrated reference cell in use at that wavelength, $ASR_{ref}(\lambda)$ (in AW^{-1}).

Photon fluxes reaching the testing device and the reference cell respectively are generally different, since the monochromatic spot is not large enough to cover the whole module area and the reference cell and the test module do not lie on the same testing plane: the measured SR is therefore only "relative" to the area covered by the photon flux, but is assumed to be proportional to the absolute spectral response of the whole module and is enough for spectral mismatch correction, as shown in the next section. The absolute SR of the testing module can be calculated by applying a scaling factor to each junction (see the cited work [2]).

Spectral Mismatch Correction

The spectral mismatch between the standard air mass 1.5 global spectrum (AM1.5g, IEC 60904-3) and the spectral irradiance of the solar simulator in use determines a correction to be applied to the measured short-circuit current value, before reporting it as a STC value. The correcting factor can be close to unity when indoor electrical characterization may be performed on a spectrally-adjustable solar simulator, but the procedure to be followed for proper spectrum adjustment typically differs from module to module and may be time-costing, if not prohibitive. To minimize the uncertainties in the determination of short-circuit current at STC, a procedure derived from the international standard IEC 60904-7 and described in [2] is suggested, involving the calculation of the following spectral mismatch adimensional factor (MMF) to the i-th junction

$$MMF_i = \frac{\int SR_i(\lambda) E_{AM1.5g}(\lambda)\, d\lambda}{\int SR_i(\lambda) E_{SS}(\lambda)\, d\lambda} \cdot \frac{\int ASR_{ref}(\lambda) E_{SS}(\lambda)\, d\lambda}{\int ASR_{ref}(\lambda) E_{AM1.5g}(\lambda)\, d\lambda} \qquad (1)$$

where $E_{AM1.5g}(\lambda)$ and $E_{SS}(\lambda)$ (in $Wm^{-2}nm^{-1}$) are AM1.5g spectral irradiance and the spectral irradiance of the solar simulator in use; $SR_i(\lambda)$ is the SR of the i-th junction. A constant scaling factor in $SR_i(\lambda)$ cancels in eq. (1) and the absolute SR is not necessary for MMF_i calculation.

The spectral mismatch factor in eq. (1) is calculated for each junction of the multi-junction device. The current-limiting junction can be individuated, finding the j-th junction in correspondence of which the following integral is minimum (representing the current contribution of the j-th junction in short-circuit conditions on the solar simulator in use)

$$I_{SC,j} = A \int SR_j(\lambda) E_{SS}(\lambda)\, d\lambda\,, \qquad (2)$$

where A (in m^2) is the test device cell area.

To perform spectral mismatch correction of the short-circuit current, ideally the current-limiting junction on the solar simulator should be the same junction that limits the current of the device at STC: in this case, the mismatch factor of that junction should be considered. When this is not the case, the consequent current imbalance may introduce high uncertainties in the I_{SC}, in the fill factor (FF) and therefore in the maximum power P_{MAX}: the method described here is therefore no more applicable. An example where current imbalance is shown to introduce uncertainties in I_{SC} is given below.

Outdoor Measurement Setup

Test devices can be measured outdoor on a sun tracker in clear-sky conditions and at air mass values close to AM1.5. As a result, in eq. (1) $E_{SS}(\lambda) \sim E_{AM1.5g}(\lambda)$ and therefore $MMF_i \sim 1$, thus reducing mismatch correction uncertainties: indoor measurements and spectral mismatch correction procedures can therefore be cross-checked by comparison with outdoor measurements.

When outdoor measurements are performed, the total irradiance is monitored by means of a c-Si reference cell and correction to 1000 Wm^{-2} is performed. The temperature of the module is also monitored with a Pt-100 temperature sensor attached to the rear of the module.

RESULTS AND DISCUSSION

Spectral Response Measurements and Spectral Mismatch Correction

Results from the following two tandem thin-film modules are presented: MM705 (a-Si:H/a-Si:Ge; 14 cells in series, 4 cells in parallel; cell area: 242.25 cm^2; module area: 1.45 m^2); MM707 (a-Si/μc-Si; 48 cells in series, 2 cells in parallel; cell area: 99.00 cm^2; module area: 1.05 m^2).

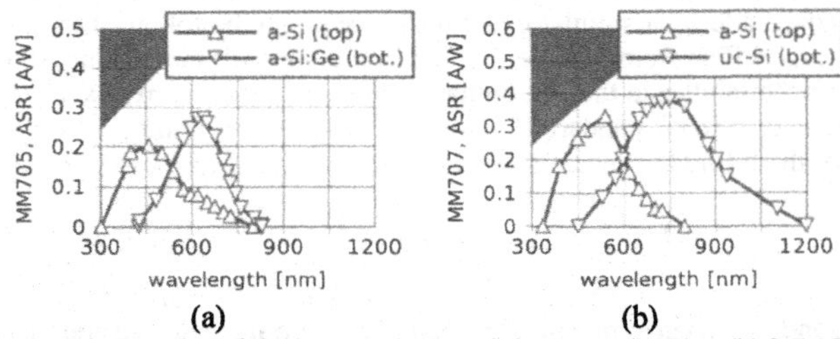

(a) (b)

Figure 2. Absolute SR. Results of the two testing modules: (a) MM705; (b) MM707. Quantum efficiency limit (*QE*=1) is shown.

Table I. Spectral mismatch calculation of indoor measurements (LAPSS) and outdoor measurements, performed the 3rd November 2009 (MM705: 11.48AM, MM707: 12.03PM). $I_{SC,STC}$ is calculated in correspondence of the current-limiting junction at STC.

ESTI code	LAPSS /outdoor	$I_{SC,i}$ calc. from eq. (2)		$I_{SC,meas}$ [A]	MMF_i from eq. (1)		$I_{SC,STC}$ [A]
		top [A]	bottom [A]		top [a.u.]	bottom [a.u.]	
MM705	LAPSS	6.225	6.283	6.225	0.9681	1.1586	6.03±0.15
	outdoor	6.174	7.578	6.174	1.0281	1.0119	6.35±0.15
MM707	LAPSS	2.055	2.720	2.055	1.0153	1.0679	2.09±0.05
	outdoor	1.972	2.798	1.972	1.0192	1.0003	2.01±0.05

The absolute SR of both devices has been measured: results are shown in Figure 2, where the external quantum efficiency limit *QE*=1 (i.e., one electron-hole pair per incident photon) is also shown. Table I shows results of spectral mismatch corrections to indoor measurements performed on ESTI large area Class AAA pulsed solar simulator (LAPSS): both devices are current-limited by the top junction, both at STC and on LAPSS. Spectral correction of outdoor measurements performed on a tracker at ESTI solar field is also shown, where devices are current-limited by the top junction as well.

The STC corrected values overlap within measurements uncertainties on MM707 (a-Si/μc-Si) case, where the top junction mismatch factor is below 2% both on LAPSS and outdoor. A difference in STC values is shown on MM705 (a-Si:H/a-Si:Ge) instead, where higher uncertainty is expected due to the high mismatch factor values: bottom junction MMF on LAPSS is over 15%, resulting in a strong current imbalance that may affect all electrical parameters. More confidence is given on outdoor measurements on that device, showing a discrete current balance.

Comparison between Indoor and Outdoor IV Characterization

Table II shows MM707 electrical parameters after applying the spectral mismatch correction. Two methods have been used. "Method 1" simply multiplies the average I_{SC} and I_{MP} values (that have already been corrected to 1000 Wm^{-2}) by the MMF of the current limiting junction at STC, while V_{OC} and the FF are not affected. With "Method 2" each measured IV curve is readjusted to $1000 \times MMF_i$ Wm^{-2} (where i refers to the current-limiting junction at testing conditions) resulting in corrections to all the electrical parameters of the module and transforming the whole IV curve rather than simply the current values. "Method 2" leads to more accurate corrections to STC conditions and can be used in combination to the correction procedures of the IEC 60891 for a definite correction of the measured data to STC conditions.

Table II. IV characterization: correction to STC.

MM707	I_{SC} [A]	V_{OC} [V]	I_{MP} [A]	V_{MP} [V]	P_{MAX} [W]	FF [%]
Method 1	2.01	64.89	1.714	49.80	85.35	65.29
Method 2	2.01	65.08	1.717	49.97	85.79	65.70
% diff.	<0.01	0.29	0.17	0.34	0.51	0.62

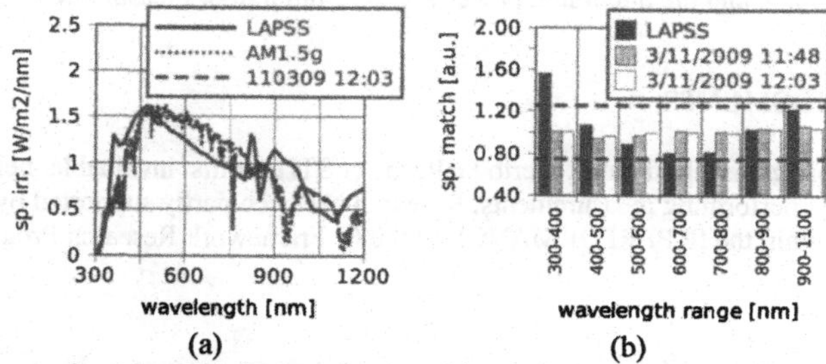

(a) (b)

Figure 3. Indoor and outdoor IV characterization: (a) comparison between AM1.5g, Class A ESTI solar simulator (LAPSS) and outdoor spectral irradiances; (b) spectral match (limit for Class A standard classification of solar simulators IEC 60904-9 is shown).

Figure 3a shows a comparison between LAPPS, AM1.5g and recorded outdoor spectral irradiances at time and day when outdoor measurements were performed. The spectral match (i.e. the ratio of percent spectral distributions) of LAPSS and outdoor spectra with respect of standard AM1.5g spectrum can be calculated as specified in the IEC 60904-9 (Figure 3b): the Class A spectral irradiance requirement is met when the spectral match value is between 0.75 and 1.25 in the six specified wavelength intervals from 400 to 1100 nm. The spectral match value of LAPSS in 300-400 nm range exceeds the limit for Class A requirement, but this does not affect its standard classification, since no specification is given in IEC 60904-9 for wavelengths below 400 nm.

LAPSS poor spectral match both between 300 and 400 nm and between 600 and 800 nm results in high MMF values shown in Table I (bottom junctions of both devices and top junction of MM705). On the other hand, a richer spectral percentage in 900-1100 nm range (due to Xe peaks) helps to reduce the MMF of the μc-Si junction (bottom) of MM707.

Outdoor measurements reduce the spectral mismatch to less than 3% and may therefore provide an electrical characterization of higher quality for many multi-junction a-Si devices.

CONCLUSIONS

Measurements of the spectral response of a multi-junction thin-film module are of pivotal importance to predict the short-circuit current at STC and the current balance. The recent development of a new experimental setup for the spectral response measurement of multi-junction large area thin-film modules has provided an important tool to ESTI laboratories: the determination of spectral mismatch correction to be applied to indoor and outdoor current-voltage characterization on multi-junction thin-film modules can now be performed properly, if the current limiting junction under the solar simulator remains the same as under the reference spectral irradiance. Results show that a high spectral mismatch may occur even with measurements on a Class A solar simulator. Nevertheless, the correction procedure has been shown to give short-circuit current values at STC which can be close to the outdoor measurements, where spectral irradiance is close to AM1.5g. Further work needs to be performed to reduce the uncertainty on the FF related to a strong current imbalance and to predict the FF value and the maximum power at STC from indoor measurements.

ACKNOWLEDGMENTS

The authors gratefully thank Roberto Galleano (ESTI) for his invaluable help in setting up instruments and performing measurements. Research has been partly supported by the European Commission within the IP PERFORMANCE of the 6th Framework Research Programme.

REFERENCES

1. M. A. Green et al., Prog. Photovolt: Res. Appl. **17**(1), 85-94 (2009).
2. M. Pravettoni et al., in *Proceedings of the 24th European Photovoltaic Solar Energy Conference*, 3338-3342, Hamburg (2009).
3. C. Reise et al., in *Proceedings of the 21st European Photovoltaic Solar Energy Conference*, Barcelona (2006).
4. J. Burdick and T. Glatfelter, Solar Cells **18**, 301-314 (1986).
5. Y. Tsuno et al., in *Proceedings of the 23rd European Photovoltaic Solar Energy Conference*, 2723-2727, Valencia (2008).
6. ASTM E 2236-05 "Standard Test Method for Measurement of Electrical Performance and Spectral Response of Nonconcentrator Multijunction Photovoltaic Cells and Modules", in *Annual Book of ASTM Standards*, vol. 12, 868-872 (2005).

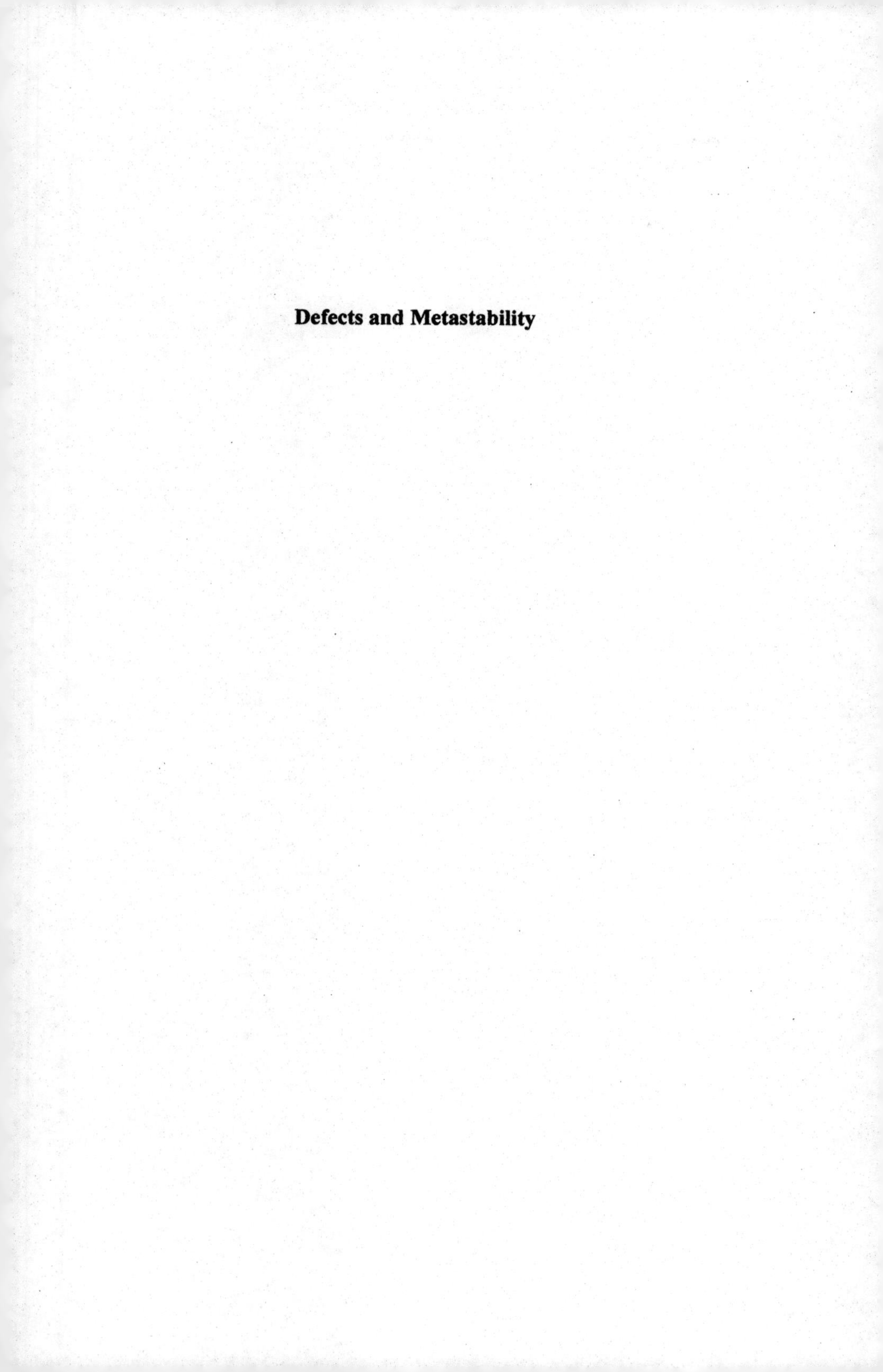

Defects and Metastability

Mater. Res. Soc. Symp. Proc. Vol. 1245 © 2010 Materials Research Society 1245-A14-01

A New Perspective on an old Problem: The Staebler-Wronski Effect

Hellmut Fritzsche

Department of Physics, The University of Chicago, Emeritus, Tucson, AZ

ABSTRACT

Photo-induced structural changes and defect creation are common phenomena in a large variety of polymeric and non-crystalline semiconductors. The photo-induced degradation of a-Si:H and its alloys, discovered by Staebler and Wronski in 1977, belongs to a special category with quite unique features, which so far has resisted an explanation. Part of the problem is that the near 4-fold coordinated network does not naturally form an amorphous material. It is over-constrained and forms a stress relief void structure. While reviewing the experimental evidence it will be argued that some of our commonly held views regarding the underlying mechanisms of the Staebler-Wronski effect (SWE) may have to be abandoned. First, the internal void surfaces seem to be the principal locations of the photo-structural changes. Second, non-radiative bimolecular recombinations of photo carriers do not seem to be the driving force of defect creation at helium temperatures. Alternative pathways for the photo-induced processes will be suggested.

INTRODUCTION

This paper explores why, after 33 years of active research, the detailed mechanisms causing the light-induced creation of dangling bond defects in a-Si:H and related materials [1], have escaped a satisfactory explanation [2,3]. One of the reasons appears to be the heterogeneous microstructure of the material which suggests the presence of not one but several different environments for light-induced defect creation. This will be linked to the recent discovery of groups of light-induced defects (LIDs) which have distinctly different anneal energies and electron capture cross sections [3]. The relation of some groups of LIDs with certain characteristics of the material's microstructure becomes evident by studying the significant improvements of the stability of solar cells which were achieved by deposition conditions that change the microstructure [4-6]. Finally, a generally accepted cause of the Staebler-Wronski effect, that is non-radiative recombination, will be re-examined in view of experimental evidence and recent theoretical work. It will be suggested that the creation mechanism of LIDs at helium temperatures is charge driven and occurs prior to recombination.

STRESS AND MICROSTRUCTURE IN a-Si:H

Despite the hydrogen content of about 8 at%, the material is essentially 4-fold coordinated and a highly over-constrained amorphous network [7]. In order to relieve the strain the films develop micro voids of different sizes and connectivity. The internal surfaces of the voids contain bonded hydrogen, about 70% of the film's hydrogen content is clustered [8] and about 10% exists as H_2 molecules in voids [9]. The remaining 20% is atomically dispersed.

The best a-Si:H material which exhibits the smallest SWE has the smallest micro-void volume and hence the highest density as shown in Fig.1 [10]. In the films prepared by the hot wire method (HW), hydrogen dilution of the silane processing gas decreases the density. The opposite happens in films grown by plasma enhanced chemical vapor deposition (PECVD) where the best film are obtained at large hydrogen dilution ratios just before the films become nano-crystalline. A consequence of reducing stress relieving voids is the accumulation of large compressive stresses in the films as shown by the shaded area in Fig. 2 [11]. The maximum stress coincides with the highest photovoltaic quality of the material at a hydrogen dilution ratio $R=H_2/SiH_4=9$ in this case [11].

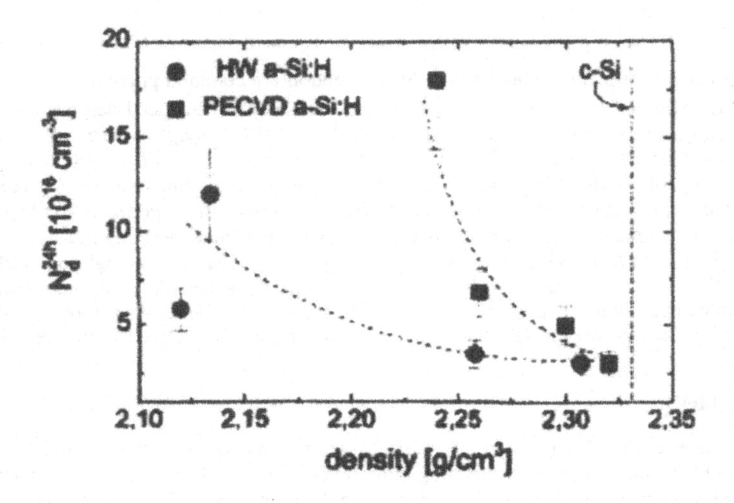

Figure 1. Defect concentration after 24h flash light soaking vs. ellipsometrically determined film density [10].

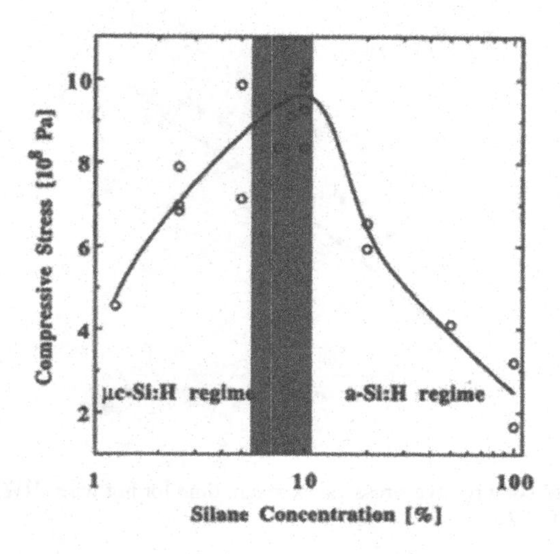

Figure 2. Compressive stress vs. hydrogen dilution [11], where the shaded area indicates the accumulation of large compressive stresses in the films.

The large stress further grows with light exposure [12,13], see Fig.3.These stress values were calculated from the light-induced bending of silicon micro-cantilevers. The largest light-induced stresses occur in the films of highest photovoltaic quality, which have the smallest SWE. They correspond to a volume change in the a-Si:H film as large as 0.1%.[12,13]

All stress disappears after annealing for several hours at 425^{0}C and no stress could be induced by light after that [12]. The anneal treatment reduced the hydrogen concentration by less than 10%. A reconstruction of the Si network must have occurred at this temperature presumably with the development of a stress relief void structure. Evidence for this was reported by Vanecek et al.[14]. They discovered that one of their best HW deposited films, which had a very small SWE, became a film with a normal, that is, large SWE after annealing at 450^{0}C.

Essentially all significant improvements of the stability of solar cells involved the preparation of materials of higher density and less micro-void volume. One is therefore tempted to claim that the internal surfaces of the micro-voids are the principal locations of the Staebler-Wronski effect. However, the situation is not that clear, unfortunately. The bulk of PECVD material also changes with an increase in hydrogen dilution. High quality material exhibits evidence of intermediate range order in the bulk [15,16], TEM images reveal the presence of highly ordered chain-like objects imbedded in the amorphous matrix. Therefore, the improvements observed by using hydrogen dilution may be an effect of greater structural order on the LIDs in the bulk. Moreover, high temperature annealing of the SWE is related to the hydrogen diffusion constant in the bulk which would not be the case if the SWE were confined to voids.

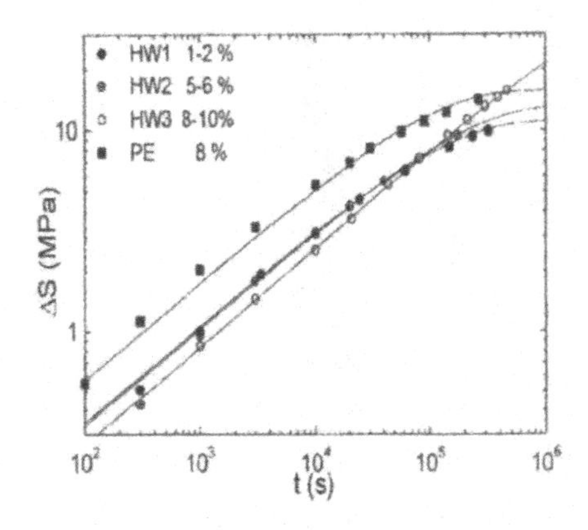

Figure 3. Light induced compressive stress vs. exposure time for hot wire (HW) and RF PECVD samples (PE). After ref. [12].

The importance of the microstructure of a-Si:H for understanding the SWE has been pointed out by many authors and several techniques have been employed to elucidate its details. These include studies of the hydrogen evolution rate as a function of temperature or the in- and out- diffusion of inert gas atoms of different sizes [17]. The evolution of the microstructure can be followed by real time spectroscopic ellipsometry [18] and the growth mechanism of a-Si:H has been studied in detail with the goal of avoiding the conditions that yield detrimental microstructure and SiH_2 bonding configurations [5]. Small angle X-ray scattering (SAXS), neutron scattering [19] reveal the number, sizes and shapes of the voids.[19-21] In addition, the most advanced resonance techniques are being used to identify the immediate environments of the dangling bond defects and of the hydrogen. [22-25]. The internal surfaces of the micro-voids have been found to provide special bonding sites for hydrogen and pairs of hydrogen with low binding energies, making them likely candidates for playing a role in the SWE [26]. Indeed, a-Si:H has become one of the most thoroughly analyzed materials thanks to our need to understand the SWE.

NATIVE AND LIGHT-INDUCED DEFECTS

The heterogeneous structure of a-Si:H provides a large spectrum of local environments for the dangling bond defects. Even though they have the same g- factor g=2.005, the native and the light-induced defects behave differently and even the latter need further differentiation.

A more voided structure is obtained by increasing the RF PECVD deposition rate. Figure 4 shows that both the initial and the degraded efficiency of solar cells decrease [21] which usually is interpreted to mean that the concentrations of both the native and the light-induced defects increase with the deposition rate. Figure 5 gives a different picture of what is truly happening [27]. Here one sees that the major effect of the increase in deposition rate is a shift of the distribution of annealing energies to higher values. The change in the microstructure has affected the ease at which the LIDs anneal. Even though each distribution is continuous it is instructive to call defects 'soft' or 'hard' depending on whether their anneal energies are in the lower or upper range of such distribution [3].

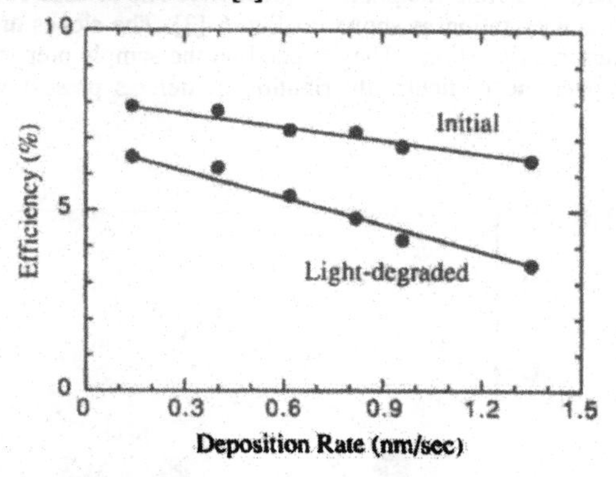

Figure 4. Initial and light-degraded efficiencies of a-Si:H solar cells as a function of deposition rate. After ref. [21].

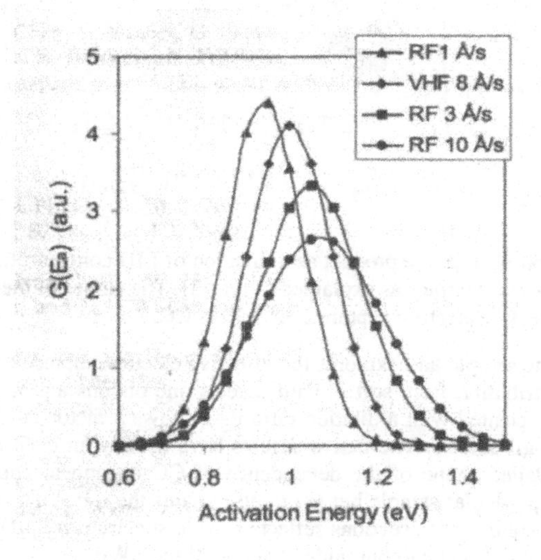

Figure 5. Annealing activation energy distribution of a-Si:H alloy solar cells deposited at different deposition rates. After ref. [27].

An increase in the defect concentration decreases the mobility-lifetime product of the electrons which is determined from the photo conductivity. The inverse of this product depends linearly on the defect concentration as shown in Fig. 6 [3]. The slopes are proportional to the average electron capture cross sections. They depend on the sample preparation parameters and the average is taken over the particular distribution of defects present with different anneal energies.

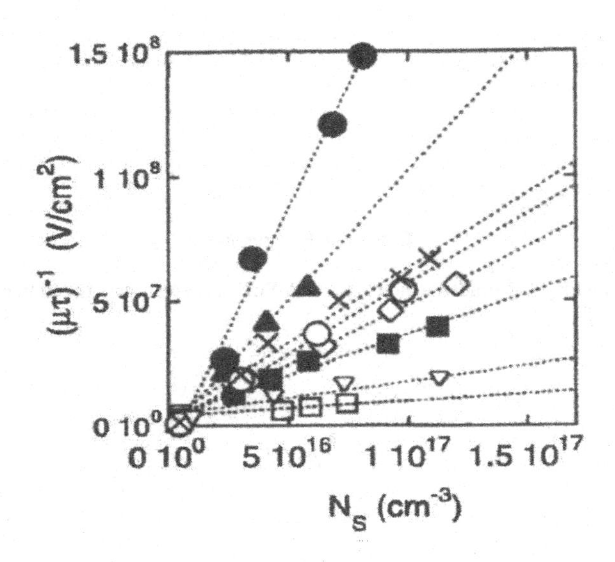

Figure 6. Inverse mobility- lifetime product as a function of LID concentration for different exposure temperatures and samples as explained in ref. [3]. The slopes of the curves are proportional to the effective electron capture coefficients CC.

If one takes one sample and explores the effective electron capture cross section CC over the anneal energy distribution from soft to hard defects one obtains a plot like Fig.7 [28]. This RF PECVD sample deposited with a dilution ratio R=5 shows a factor 20 change of CC as one moves from soft to hard defects. The native defects have usually an even smaller CC than the hard defects. The detailed shape of the dependence of CC on anneal energy depends on the microstructure of the particular sample but the trend remains the same. The rather wide range of anneal energies and capture cross sections reflects a wide spectrum of LID environments. This must form an important challenge to our understanding of the SWE.

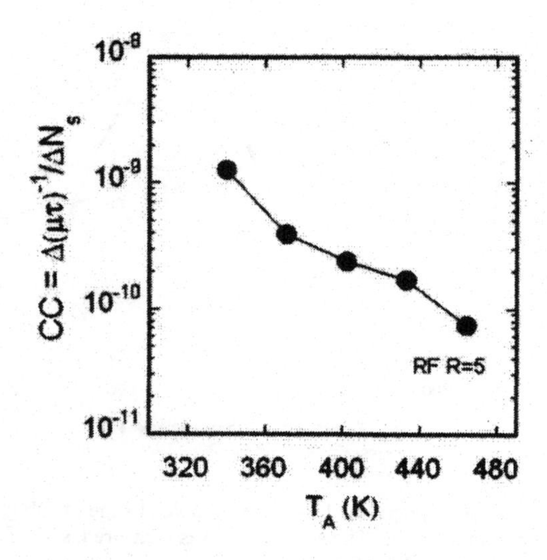

Figure 7. Effective electron capture cross section CC as a function of annealing temperature after light soaking at room temperature. The CC values are characteristic of only the annealed defect group during step wise annealing the sample. After ref. [28].

A FULL REPRESENTATION OF LIGHT-INDUCED DEFECTS

The creation kinetics of LIDs is remarkably independent of temperature and sample preparation [29]. We therefore can plot the concentration of LIDs or the resultant change of the sub gap absorption as a function of exposure temperature for a fixed light exposure. This is shown in Fig.8. Between 1 and 20K one can create LIDs with nearly the same efficiency as at room temperature [30]. These low T defects are different from the soft and hard defects. They have very small annealing energies, and their capture cross sections exceed even those of soft defects by at least a factor 10. The dip in the curve of Fig. 8 at intermediate temperatures presumably arises from two factors. The creation efficiency is low for low T defects because they already start to anneal at these intermediate temperatures, and the soft and hard defects appear to require a certain activation energy for their creation [31]. The low T defects can be stabilized only by a very local bonding re-arrangement since hydrogen diffusion is ruled out at these low temperatures. Their creation mechanism is probably different from that of the normal SWE defects. For that matter it appears to be important to remember the rather pronounced decrease of the low T creation efficiency in strong electric fields shown in Fig.9 [30]. Above room temperature electric fields were found to enhance the annealing of the SWE in darkness and under illumination [32].

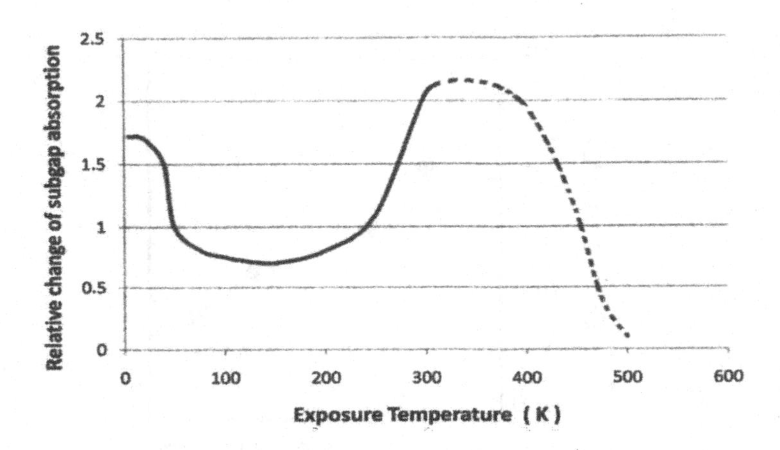

Figure 8. LID concentration as determined from the relative change of sub gap absorption after 2h exposure at different temperatures. The solid curve was measured, see ref. [30]. The dashed curve is conjectured from various literature sources.

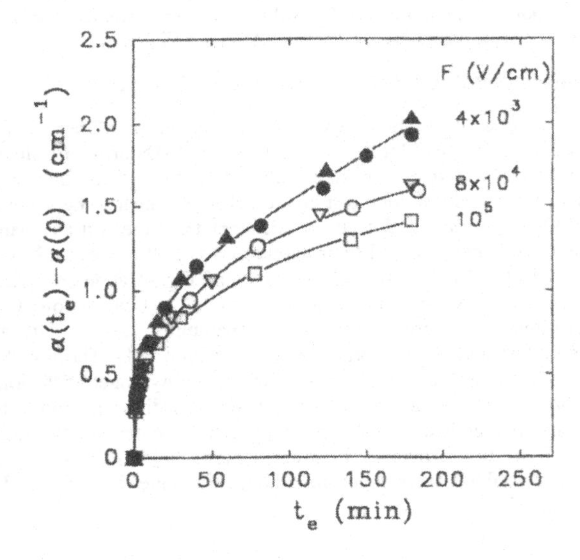

Figure 9. Increase in sub gap absorption or LID concentration as a function of exposure time at 4.2K of a RF PECVD sample for different electric fields F. After ref. [30].

DISCUSSION AND RESULTS OF RECENT COMPUTER SIMULATIONS

Three major issues will be discussed in the following, the relation of the SWE with stress, the spectrum of defects having different anneal energies and electron capture cross sections, and the apparently special nature of the low T defects.

The large compressive stress in high quality a-Si:H films is a consequence of the correlation of quality with high density and a small void volume. Drabold's computer simulations revealed that the valence band tail states, which are held responsible for producing the SWE defects, are shorter than the normal Si-Si bonds [33]. Hence the compressive stress should be detrimental in contrast to observation. Perhaps this is an additional argument for supposing that the LIDs are primarily at the void surfaces and not in the bulk. The light-induced stress is also largest for the best films, presumably because of the lack of stress relief voids. There does not seem to be a causal relation between the SWE and the light-induced stress [3]. The former saturates much earlier than the latter and they follow different creation kinetics [3]. The presence of a light-induced stress which can amount to a volume change of 0.1% shows that light exposure causes structural changes in the material which are much larger than the SWE which affects at most 1 in 10^5 bonds. Photo-structural changes involving the larger Si network have also been observed in the infrared spectra [34] and as an irradiation-induced shift to lower binding energies of the Si2p and Si2s electrons detected by X-ray photoemission spectroscopy [35].

In the past, the anneal temperatures of SWE defects, different for n-type, p-type and intrinsic a-Si:H [36] were successfully related to the doping dependent hydrogen diffusion coefficients. One now finds that the anneal process involves a spectrum of energies and temperatures in the same sample and that it depends on the sample density and void volume. In contrast, despite the heterogeneous hydrogen bonding structure, one never observed more than one hydrogen diffusion process. It appears that part of the anneal energy is associated with the local bonding environment of the light-induced defect and the position of the Fermi level.

The large electron capture cross sections which are related to the defect anneal energies are difficult to explain. The low T defects have even larger capture cross sections than the soft and hard ones. If the defects are located at internal surfaces such as void surfaces then an electric field directed toward the surface can enhance the effective capture cross sections. Indeed it is unlikely that the local potentials remain unaffected by the presence of voids. Differences between native and LIDs have also been explained by the material's heterogeneity in which photo carriers drift to regions where the mobility gap is narrower, producing LIDs there [37].

At low temperatures, between 1-20K, the defects creation mechanism is probably different from that at the higher temperatures. Once the carriers remain trapped for longer than 10^{-8} seconds plus the inter- pair tunneling time the radiative recombination rate greatly exceeds the non-radiative rate. This gives rise to the efficient photo luminescence at low temperatures. It is likely that defect creation at low temperatures is essentially charge driven similar to the processes that lead to photo-structural changes in chalcogenide glasses [38]. There the bonding change responds to a local change in valency because of the trapped charge before recombination takes place. As a result, the photo-induced changes in chalcogenide glasses do not depend on the details of the recombination process. There are of course major differences between tetrahedral and chalcogenide materials. The covalent connectivity of the latter is much lower which provides a larger free volume around each atom and consequently a large electron phonon coupling. We find, however, circumstances in a-Si:H which favor a similar process. Drabold found by simulation that the vibrational behavior of the disordered Si lattice is greatly enhanced by a positive charge, i.e., the electron phonon interaction is enhanced [33]. The simulation dealt with the bulk of the Si network. At the internal void surfaces, on the other hand, the atoms find an even larger degree of freedom of motion than in the bulk, hence the local displacement in response to a change in charge state, the electron phonon interaction, should be correspondingly larger there. Very interesting and relevant is a calculation and computer simulation by Wagner and Grossman [39].They found that an electron-hole pair near a trapped hole leads to the breaking of a weak bond without an energy barrier. The electron-hole recombination proceeds then afterwards in agreement with our assumption. The low T defects have a small anneal energy presumably because the stabilizing mechanism involves only a very local bonding change.

FINAL COMMENTS

Some experiments should be mentioned which could elucidate and fill gaps in our understanding of the SWE. The generation rate of room temperature LIDs is much larger in doped a-Si:H, particularly in p-type films, almost mimicking the dependence of H diffusion on the position of the Fermi level [36]. The generation rate and the final concentration of LIDs are smallest for compensated films [40]. How do these trends depend on the micro-void density and can one distinguish soft and hard defects in these materials? The native defects reside most likely in the bulk and not at void surfaces. The electron capture cross section of the native defects is even smaller than that of the hard LIDs. This can be determined by changing the native defect concentration by quenching films from say 350^0C.

One hypothesis for explaining the small SWE in compensated films is that recombination proceeds via the charged dopants rather than via the tail states in bimolecular fashion. If recombination occurs after defect formation at low temperatures, this alternative recombination channel should not influence the creation of low T defects. The dopant dependence of the creation rate is quite pronounced for the high T defects as mentioned earlier. Does the same hold for low T defects? Experiments can answer this question. The creation of LIDs at low temperatures was studied with PECVD samples prepared without H-dilution. In order to see the effect of the void structure one should repeat these experiments with high quality samples of high density.

A major breakthrough in our research of the SWE was reported by Bobela et al. at this meeting.[41] This group found that annealing at 375°C greatly diminishes the creation of LIDs. Only a small fraction of loosely bonded hydrogen is effused at this temperature. This discovery may well be a key to the obstinate puzzle of the SWE.

Finally one might ask whether the results reviewed here favor one of the models which were proposed to explain the SWE: the weak bond breaking model of Stutzmann et al.[31] and the hydrogen collision model of Branz. [42] Both models apply to the creation of LIDs either in the bulk or at void surfaces. Defect creation at low temperatures obviously falls out of the realm of the hydrogen collision model because hydrogen does not diffuse at these low temperatures. The weak bond breaking model suffered from the fact that the predicted hyperfine interaction between the dangling bond and a nearest hydrogen atom was never detected for LIDs created at room temperature. The hyperfine interaction might become detectable however when the LIDs are created at low temperatures where the bond rearrangement remains very local.

ACKNOWLEDGMENTS

I am grateful to Klaus Lips for organizing the very informative and stimulating International Workshop on the Staebler-Wronski Effect in Berlin, April 2009. I also profited greatly from discussions with Paul Stradins and my colleagues at United Solar Ovonic LLC.

REFERENCES

1. D. L. Staebler and C. R. Wronski, *Appl. Phys. Lett.* **31,** 292 (1977).
2. H. Fritzsche, *Ann. Rev. Mater. Res.* **31,** 47 (2002).
3. P. Stradins, *Sol. Energy Mater. Sol. Cells* **78,** 347 (2003).
4. S. Guha, J. Yang, S. J. Jones, Yan Chen, and D. L. Williamson, *Appl. Phys. Lett.* **61,** 1444 (1992).
5. A. Matsuda, M. Takai, T. Nishimoto, and M. Kondo, *Sol. Energy Mater. Sol. Cells* **78,** 3 (2003).
6. S. Guha, J. Yang, A. Banerjee, B. Yan, and K. Lord, *Sol. Energy Mater. Sol. Cells* **78,** 329 (2003).
7. R. A. Street, *Hydrogenated amorphous silicon*, (Cambridge University Press, 1991).
8. K. K. Gleason, M. A. Petrich, and J. A. Reimer, *Phys. Rev.* **B36,** 3259 (1987).
9. T. Su, P. C. Taylor, S. Chen, R. S. Crandall, and A. H. Mahan, *J. Non-Cryst. Solids* **266-69,** 195 (2000).
10. S. Bauer, B. Schroeder, and H. Oechsner, *J. Non-Cryst. Solids* **227-230,** 34 (1998).
11. U. Kroll, J Meier, A. Shah, S. Makhailov, and J. Weber, *J. Appl. Phys.* **80,** 4971 (1996).
12. E. Spanakis, *Ph.D. thesis*, Univ. of Crete (2001).
13. P. Tzanetakis, *Sol. Energy Mater. Sol. Cells* **78,** 369 (2003).
14. M. Vanecek, J. Fric, A. Poruba, A. H. Mahan, and R. S. Crandall, *J. Non-Cryst. Solids* **198-200,** 478 (1996).
15. D. V. Tsu, B. S. Chao, S. R. Ovshinsky, J. Yang, and S. Guha, *Appl. Phys Lett.* **71,** 1317 (1997).
16. D. V. Tsu, B. S. Chao, and S. J. Jones, *Sol. Energy Mater. Sol. Cells* **78,** 115 (2003).
17. W. Beyer, *Sol. Energy Mater. Sol. Cells* **78,** 235 (2003); *phys. stat. solidi* (c) **1,** 1144 (2004).
18. R. W. Collins, A. S. Ferlauto, G. M. Ferreira, Chi Chen, Joohyun Koh, R. J. Koval, Yeeheng Lee, J. M. Pearce, and C. R. Wronski, *Sol. Energy Mater. Sol. Cells* **78,** 143 (2003).
19. D. L. Williamson, *Sol. Energy Mater. Sol. Cells* **78,** 41 (2003).
20. D. L. Williamson, *Mater. Res. Soc. Symp. Proc.* **377,** 251 (1995).
21. S. Guha, J. Yang, S. J. Jones, Yan Chen, and D. L. Williamson, *Appl. Phys. Lett.* **61,** 1444 (1992).
22. T. Su and P. C. Taylor, *Sol. Energy Mater. Sol. Cells* **78,** 269 (2003).
23. M. Fehr, A. Schnegg, C. Teutloff, R. Bittl, O. Astakhov, F. Finger, B. Rech, and K. Lips, *phys. stat. solidi* (a) **207,** 552 (2010).
24. S. Yamasaki, T. Umeda, J. Isoya, J. H. Zhou, and K Tanaka, *J. Non-Cryst. Solids* **227-230,** 332,353 (1998).
25. T. Umeda, S. Yamasaki, J, Isoya, and K. Tanaka, *Phys. Rev.* **B62,** 15702 (2000).
26. S. B. Zhang and H. M. Branz, *Phys. Rev. Lett.* **87,** 105503 (2001).

27. B. Yan, J. Yang, K. Lord and S. Guha, *Mater. Res. Soc. Symp. Proc.* **664,** A25.2.1 (2001).
28. P. Stradins, private communication, to be published.
29. P. Stradins, H. Fritzsche, and M. Q. Tran, *Mater. Res. Soc. Symp. Proc.* **336,** 227 (1994).
30. P. Stradins and H. Fritzsche, *J. Non- Cryst. Solids* **200,** 432 (1996).
31. M. Stutzmann, W. B. Jackson, and C. C. Tsai, *Phys. Rev.* **B32,** 23 (1985).
32. D. E. Carlson and K. Rajan, *J. Appl. Phys.* **83,** 1726 (1998).
33. D. A. Drabold, *Eur. Phys. J.* **B68,** 1 (2009).
34. G. Kong, D. Zhang, G. Yue, Y. Wang, and X. Liao, *Mater. Res. Soc. Symp. Proc.* **507,** 697 (1998).
35. A. Yelon, A. Rocheford, S. Sheng, and E. Sacher, *Sol. Energy Mater. Sol. Cells* **78,** 391 (2003).
36. A. Hamed and H. Fritzsche, *J. Non-Cryst. Solids* **114,** 717 (1989).
37. N. Hata and A. Matsuda, *J. Non-Cryst. Solids* **164-166,** 187 (1993).
38. H. Fritzsche, *Phil. Mag.* **B68,** 561 (1993).
39. L. K. Wagner and J. C. Grossman, *Phys. Rev. Lett.* **101,** 265501 (2008).
40. P. Tzánetakis, N. Kopidakis, M. Androulidaki, C. Kalpouzos, P. Stradins and H. Fritzsche, *J. Non-Cryst. Solids* **198-200,** 458 (1996).
41. D. C. Bobela, H. Branz, P. Stradins, B. Yan, and X. Xu, *Mater. Res. Soc. Symp. Proc.* **1245,** A14.4 (2010).
42. H. Branz, *Phys. Rev.* **B59,** 5498 (1999).

Mater. Res. Soc. Symp. Proc. Vol. 1245 © 2010 Materials Research Society 1245-A14-02

The Staebler-Wronski effect: new physical approaches and insights as a route to reveal its origin

A.H.M. Smets[1,2,4], C.R. Wronski[3], M. Zeman[4] and M.C.M. van de Sanden[1]

[1]Eindhoven University of Technology, the Netherlands, [2]National Institute of Advanced Industrial Science and Technology, Japan, [3]Pennsylvania State University, USA, [4]Delft University of Technology, the Netherlands

Abstract

In the recent years more and more theoretical and experimental evidence have been found that the hydrogen bonded to silicon in dense hydrogenated amorphous silicon (a-Si:H) predominantly resides in hydrogenated divacancies. In this contribution we will philosophize about the option that the small fraction of divacancies, missing at least one of its bonded hydrogen, may correspond to some of the native and metastable defect states of a-Si:H. We will discuss that such defect entities are an interesting basis for new and alternative views on the origin of the SWE.

Introduction

Recent experimental studies have revealed two crucial features of the Staebler-Wronski effect (SWE) which up to now did not receive any attention in the models proposed to explain its mechanism. First, using charge deep-level transient spectroscopy (Q-DLTS) [1] at Delft University of Technology and dual beam photoconductivity (DBP) [2] analysis at Penn State, it has been shown that hydrogenated amorphous silicon (a-Si:H) has at least three native gap states related to defects. Under light soaking the three distributions increase with distinctly different kinetics for the creation of the metastable states as well as annealing characteristics. Both groups have identified three defects states with a broad density distribution, here for clarity denominated state A, B and C. The peak energies of the Gaussian distributions from DBP are [2]: for the A states 0.05 eV above midgap; for the B states 0.095 eV below midgap; and for the C states 0.39 eV below midgap. It was found that that state A and B are efficient electron recombination centers which dominate the photoconductivity in the protocrystalline a-Si:H films [3]. The C states positioned further from midgap on the other hand are inefficient electron, but very efficient hole recombination centers [3] and are the metastable defects which dominate the degradation in the solar cells under 1 sun illumination. Secondly, Wronski and co-workers [4,5] showed that it is far from straightforward to reveal the correct kinetics of the SWE based on the evolution of defects during light-soaking. Their detailed study on films and cells showed that the contribution of the initial defect density obscures the real appearance of the SWE during the time scales of typical light-soaking experiments. After correction for the contribution of the initial defects, the evolution of meta-stable defect states A and B for protocrystalline silicon show a scaling of $\sim Gt^{1/2}$ under one sun illumination (with G being the generation rate and t the time) during the initial two hours of light soaking in contrast to the commonly reported $\sim G^{2/3}t^{1/3}$.

In view of these two important experimental results, we have to conclude that even the most advanced models describing the SWE [6,7] do not reveal the correct kinetics for the fast A and B states or even excludes the presence of more than one metastable defect state. One of the reasons for this is that the micro-structural view on the a-Si:H network has been limited to a treatment as a continuously random network (CRN) in which isolated dangling bonds and the randomly distributed isolated hydrogens are the only entities considered as defect sites and hydrogen emission sites, respectively. This is despite the fact that many studies have shown that hydrogen passivated divacancies are the dominant entities in dense a-Si:H [8-17].

Since it is believed that the origin of the SWE is related to a redistribution of a small fraction of the bonded hydrogen in the material, a not fully hydrogenated divacancy in a disordered network is an interesting configuration that should be considered as possible native or metastable defect site. In this paper we discuss that the relative positions between the defect states favors the approach of describing some of the native and metastable defects as not-fully-hydrogen-passivated divacancies.

Recent progress: self-consistent model for H incorporation in a-Si:H

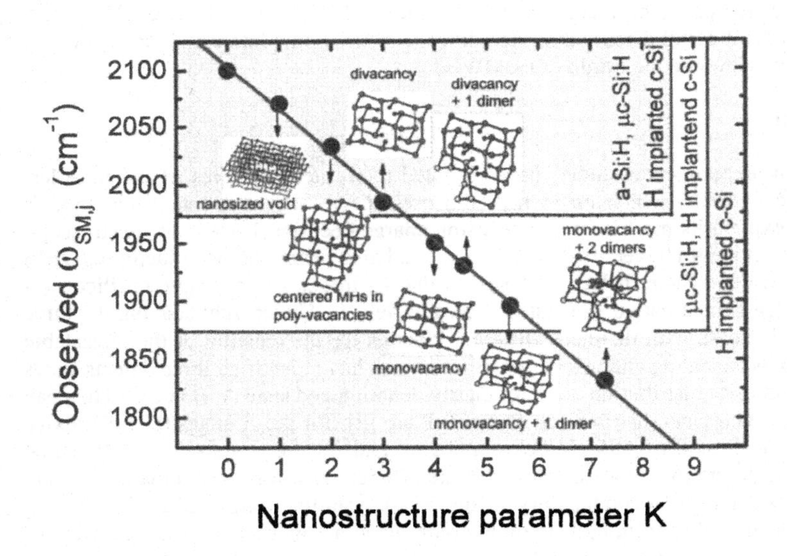

Figure 1. *The relation between the frequency positions of the stretching mode of monohydrides incorporated in various Si:H networks from Ref.[10]. Only data for Si:H materials with the same screening term Ω~1.43±0.02 have been taken.*

An interesting development which leads to an alternative view on the origin of the defect entities related to SWE is the recent progress made in the insight into the microstructure of a-Si:H films. Recently, we have reported on a self-consistent model for hydrogen incorporation in a-Si:H based on infrared absorption spectroscopy [8-10]. In these papers the point was made that a-Si:H described by a conventional multiple Lorentz-Lorenz (LL) dielectric of Si-H and Si-Si dipoles can only represent homogeneously and randomly distributed Si-H bonds in the network. The fact that such a model is not able to explain the experimental IR data of hydride stretching modes (SMs) in the bulk, is already a good indication that the hydrogen distribution in the a-Si:H network, unlike in CRN, is not random. This problem has been overcome by the description of anisotropic LL dielectrics by the introduction of an effective medium approximation of LL dielectrics; one (ε_{Si}) for the Si matrix and the others (ε_j, $j \geq 1$) for the network configurations in which the hydrogen resides. Employing this approach on the SMs of monohydrides (MHs) in the infrared, it was possible to self-consistently explain: the configurations of incorporated hydrogen responsible for the SMs; the relation of its density to the mass deficiency of a-Si:H network; the oscillator strength of dipole vibrations of the hydrides; the frequency shift of hydride vibrations in the bulk relative to their unscreened frequencies in molecules; and the screening mechanism of the hydrides by their environment using only THREE parameters [9,10]. The first parameter is the nanostructure parameter K defining the local hydride density in the material, i.e. the number of hydrides in the volume equal to that of a missing silicon atom in the network (V_{Si}~2×10^{23} cm^{-3}). For example a mono- and divacancy have K=4 and 3, respectively. The second parameter is the unscreened effective dynamical charge of a MH $q_{0,MH}$ and the third parameter is the term for the Silsbee screening of the dipole Ω as a function of the surrounding dielectric function of the Si matrix ε_{Si} like $\Omega = 3\varepsilon_{Si}/(2\varepsilon_{Si}+1)$. In this model the frequency shift of a MH $\Delta\omega_{MH,j}$ relative to its unscreened frequency position $\omega_{0,MH} = (2099\pm2)$ cm^{-1} and residing in configuration type j with nanostructure parameter K_j is given by:

$$\Delta\omega_{MH,j} = -\frac{10^{-4}}{24\pi^2 c^2 m_{MH}\omega_{0,MH}\varepsilon_0}\frac{K_j q_{0,MH}{}^2\Omega^2}{V_{Si}} \quad (1)$$

with c the velocity of light and m_{MH} the effective mass of the MH oscillation. If the screening of the hydrides by ε_{Si} is the same for every MH (Ω~constant), the frequency shift becomes only dependent on the configuration of incorporated H via K. In other words, the frequency shift becomes a measure of the MHs environment within volumes as small as that of a Si atom in the network! Here we will generalize this approach by considering the K-values for all possible configurations of incorporated MHs, like nanosized voids, mono-, di- and mutli-vacancies without and with dimer reconstructions (see Ref. [10]). The surprising result is that this approach enables us to identify all the frequency positions of the MHs relative to $\omega_{0,MH}$ as observed in a-Si:H, μc-Si:H [18,19] and H implanted c-Si [20] (with Ω~1.43±0.02) as shown in Fig. 1. All frequency positions observed are described by equation (1). This shows that even in disordered material as a-Si:H the configuration of incorporated hydrogen are all vacancy-like!

Based on these assignments, it was found that the fully hydrogen passivated divacancy is the dominant hydrogen complex in dense (current device-grade) a-Si:H material. The results are in agreement with studies based on nuclear magnetic resonance (NMR) [11,12], positron annihilation (PA) [13-15] and film density analysis [8,16,17]. Furthermore, Smets *et al.* [21] made the case that an amorphous network controlled by hydrogenated vacancies leads to a more consistent view on the properties of plasma processed hydrogenated amorphous silicon (a-Si:H), compared to the extensively held views based on CRN exhibiting randomly and homogenously distributed silicon and hydrogen atoms. A network predominantly exhibiting fully hydrogen passivated divacancies self-consistently reveals the amorphous nature of a-Si:H, its phase transition from amorphous-to-microcrystalline, and the characteristic details of its electronic structure like the band gap, the a-symmetric widths of the gap tails and the various defect states in the gap. The latter issue will be discussed in more detail.

Defect states of not fully hydrogenated divacancies

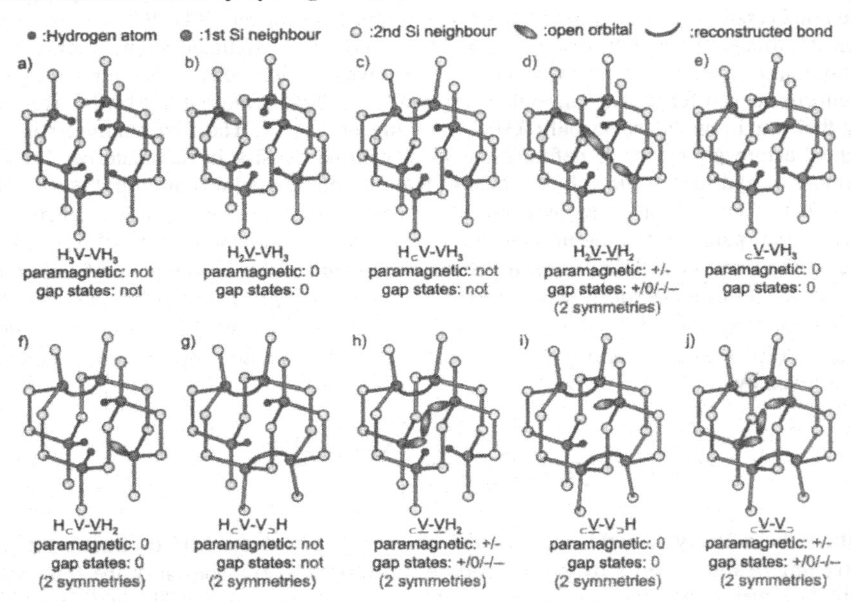

Figure 2. *Representation of all theoretical possible configurations of divacancies, from fully hydrogenated to hydrogen free ones. Note, that in a-Si:H the divacancy structure will be more relaxed and is one of the sources for disorder in the network.*

If a redistribution of very small fraction of the bonded hydrogen plays a role in the SWE, the entity of the 'divacancy' offers an alternative to the extensively held view that all gap states represent solely the isolated dangling bond. If a fraction of 10^{-4} to 10^{-5} of the divacancies is not fully passivated, it will certainly lead to states in the gap. This is in line with the results of analysis based on PA and constant photocurrent method (CPM) by Gordo *et al.* [15], which show that vacancies are responsible for defect states above the Fermi-level.

In figure 2 all theoretical possible configurations of divacancies, from fully hydrogen passivated ones to hydrogen free ones are illustrated. Unfortunately, no modeling work has been carried out to calculate the energy positions of the defect states in the gap of these configurations in disordered networks [22-26]. Only for two hydrogen poor vacancies in crystalline Si network, i.e. the configurations $_{\subset}\underline{V}\text{-}V_{\supset}H$ and $H_{\subset}V\text{-}V_{\supset}H$, the gap states have been estimated using ab initio modeling [22]. (Note, the underlined $\underline{V}$ reflects that one dangling bond is residing at the vacancy site and both subscripts $_{\subset}$ and $_{\supset}$ correspond to a reconstructed bond at the vacancy site between two silicons missing both a H atom). For the other configurations of not-fully passivated divacancies we are currently limited to speculations based on similar configurations known for c-Si [22-26]. In $H_2\underline{V}\text{-}\underline{V}H_2$ the two hydrogen free bonds will form an orbital similar to that of the hydrogen free divacancy in crystalline silicon $_{\subset}\underline{V\text{-}V}_{\supset}$ with two bond reconstructions. In line with the $_{\subset}\underline{V\text{-}V}_{\supset}$ in c-Si, this state would be expected to have four gap states depending on the charged state of the entire defect entity, i.e. positively charged (+), neutrally charged (0), and negatively charged (-) or (--). The energy position of the defects states of the $_{\subset}\underline{V\text{-}V}_{\supset}$ in c-Si are given in Table I. The relative positions between the states of the charged states of $_{\subset}\underline{V\text{-}V}_{\supset}{}^{+/0/-/--}$ are almost similar to the ones observed for a-Si:H. (Since the band gap differs between these samples, it is better to use the energy difference between the states as a comparison). This makes the possible charged states of $H_2\underline{V}\text{-}\underline{V}H_2{}^{+/0/-/--}$ (having states similar to $_{\subset}\underline{V\text{-}V}_{\supset}{}^{+/0/-/--}$) a plausible candidate to correspond to some of the native and/or metastable defects observed. Furthermore, a redistribution of hydrogen bonds in a divacancy having initially no gap states to one with gap states, like $H_3V\text{-}V_{\supset}H \rightarrow H_2\underline{V}\text{-}\underline{V}H_2$, could result in the creation of additional defect states in the gap [27] and might be the origin of some of the metastable defects states observed in a-Si:H. The possibility that hydrogen redistribution like $H_3V\text{-}V_{\supset}H \rightarrow H_2\underline{V}\text{-}\underline{V}H_2$ occurs through a proton motion within the divacancy has already been suggested by Carlson *et al.* [27] as an explanation for the fact that the SWE can be slowed down by employing an electric field over the film.

Table I *Overview of the measured gap states in a-Si:H as reported in literature 1) using Q-DLTS [1] and DBP [2] and 2) as measured or calculated for the hydrogen free divacancies in c-Si [22-26]. The red numbers between the brackets are the energy differences between the two following gap states.*

	a-Si:H Q-DLTS [1] (eV)	a-Si:H DBP [2] (eV)	$_{\subset}\underline{V\text{-}V}_{\supset}$ in c-Si IR,DLTS[22-26] (eV)
			E_c-0.36
			(0.03)
A	E_c-0.63	E_c-0.85	E_c-0.39
	(0.19)	(0.14)	(0.15)
B	E_c-0.82	E_c-0.99	E_c-0.54
	(0.43)	(0.40)	(0.38)
C	E_c-1.25	E_c-1.39	E_c-0.92

Conclusion

We have made the point that the entity of the 'divacancy' offers an alternative to the extensively held view that all gap states represent solely the isolated dangling bond. We have demonstrated that hydrogen in dense a-Si:H predominantly resides in divacancies. Since the origin of the SWE is considered as a small redistribution of bonded hydrogen, the not fully hydrogenated divacancy is a potential candidate for hydrogen reconstruct-tions induced by charge carrier recombination. We have discussed the resemblance of the energy positions between the defect states of the $\underline{\text{V-V}}$ configuration in c-Si and those observed for the stable and meta-stable defect states in protocrystalline a-Si:H. Modeling activities on vacancies like $H_2\underline{\text{V-V}}H_2$ in disordered silicon networks are badly needed to identify the defect entities corresponding to native or metastable defects.

References

1. V. Nadazdy and M. Zeman, Phys. Rev. B **69**, 165213 (2004).
2. X. Niu, PhD Thesis, *"Nature and evolution of light induced defects in hydrogenated amorphous silicon"*, Pennsylvania state University (2006).
3. C.Wronski *et al.*, to be submitted to proc. of 35th IEEE PSC (2010).
4. J. Deng, *et al.*, Proceedings of the 4th WCPE, p. 1576 (IEEE NY 2006).
5. J. Deng, *et al.*, Mater. Res. Soc. Symp. Proc. **910**, A02-02 (2006).
6. M. Stutzmann, W. B. Jackson, and C. C. Tsai, Phys. Rev. B **32,** 23 (1985).
7. H.M. Branz, Phys. Rev. B **59**, 5498 (1999).
8. A.H.M. Smets, *et al.*, Appl. Phys. Lett. **82**, 1547 (2003).
9. A.H.M. Smets and M.C.M. van de Sanden, Phys. Rev. B **76**, 073202 (2007).
10. A.H.M. Smets, *et al.,* submitted for publication (2010).
11. J. Baum, *et al.*, Phys. Rev. Lett. **56**, 1377 (1986).
12. J.T. Stephen, *et al.*, Mat. Res. Soc. Symp. Proc. Vol. **467**, 159 (1997).
13. V.G. Bhide, *et al.*, J. Appl. Phys. 62, 108 (1987).
14. R. Suzuki, *et al.,* Jap. J. of Appl. Phys. **30**, 2438 (1991).
15. P.M. Gordo, *et al.*, Rad. Phys. Chem. **76**, 220 (2007).
16. Z. Remes, *et al.*, Phys. Rev. B **56**, 12710 (1997).
17. Z. Remes, *et al.*, J. Non-Cryst. Solids **227-230**, 876 (1998).
18. A.H.M. Smets, T. Matsui, and M. Kondo, Appl. Phys. Lett. **92**, 033506 (2008).
19. A.H.M. Smets, T. Matsui, and M. Kondo, J. Appl. Phys. **104**, 034508 (2008).
20. Y.J. Chabal, *et al.*, Physica B 273-274, **152** (1999).
21. A.H.M. Smets and M.C.M. van de Sanden, to be published.
22. J. Coutinho, *et al.*, J. Phys.: Condens. Matter **15**, S2809 (2003).
23. G.D. Watkins and J.W. Corbett, Phys. Rev. **138**, A543 (1963).
24. A.H. Kalma and J.C. Corelli, Phys. Rev. **173**, 734 (1968).
25. R.C. Young, and J.C. Corelli, Phys. Rev. B **5**, 1455 (1977).
26. U. Lindefelt, and W. Yong-Liang, Phys. Rev. B **38**, 4107 (1988).
27. D.E. Carlson and K. Rajan, J. Appl. Phys. **83**, 1726 (1998).
28. C. Herring, *et al.*, Phys. Rev. B **64**, 125209 (2001).

Carrier Transport

Mater. Res. Soc. Symp. Proc. Vol. 1245 © 2010 Materials Research Society 1245-A15-04

FDTD simulation of light propagation inside a-Si:H structures

A.Fantoni[1,2] and P. Pinho[1,3]
[1]Instituto Superior de Engenharia de Lisboa (ISEL), Departamento de Engenharia Electrónica e Telecomunicações e de Computadores (DEETC), Lisboa, Portugal
[2] CTS, Uninova, Departamento de Engenharia Electrotécnica, Faculdade de Ciências e Tecnologia, FCT, Universidade Nova de Lisboa, 2829-516 Caparica, Portugal
[3]Instituto de Telecomunicações, Campus Universitário de Santiago, Aveiro, Portugal

ABSTRACT

We have developed a computer program based on the Finite Difference Time Domain (FDTD) algorithm able to simulate the propagation of electromagnetic waves with wavelengths in the range of the visible spectrum within a-Si:H p-i-n structures. Understanding of light transmission, reflection and propagation inside semiconductor structures is crucial for development of photovoltaic devices. Permitting 1D analysis of light propagation over time evolution, our software produces results in well agreement with experimental values of the absorption coefficient. It shows the light absorption process together with light reflection effects at the incident surface as well as at the semiconductor interfaces. While the effects of surface reflections are easily taken into account by the algorithm, light absorption represents a more critical point, because of its non-linear dependence from conductivity. Doping density, density of states and photoconductivity calculation are therefore crucial parameters for a correct description of the light absorption-transmission phenomena through a light propagation model.

The results presented in this paper demonstrate that is possible to describe the effect of the light-semiconductor interaction through the application of the FDTD model to a a-Si:H solar cell. A more general application of the model to 2D geometries will permit the analysis of the influence of surface and interface roughness on the device photovoltaic efficiency.

INTRODUCTION

Finite Difference Time Domain (FDTD) is a computational algorithm [1,2,3,4] based on the differential formulation of the Maxwell equations, widely used for simulation of electromagnetic waves propagation. Application of this simulation technique has been mainly in the telecommunication domain for simulations of radio-wave propagation [5,6] and for simulation of radar systems [7,8]. More recently, the FDTD algorithm has been applied in different areas for modeling electromagnetic fields in multilayer inhomogeneous objects such as biological tissues for clinical application as breast tumor imaging [9], for generation, recombination, and transport process in semiconductors [10], for simulating semiconductor gain medium and lasing dynamics of a four-level two-electron atomic system [11]. As in thin film research area, it has been recently presented a study about the influence of the film nanostructure on the conversion efficiency of thin-film silicon solar cells [12], and a work on photoconductivity in thin film semiconductors [13], showing the importance of extending the light/ matter interaction from material properties to a device engineering point of view. Antireflective coating, textured surface, device area, layer thickness, interface quality, are all examples of factors that, in addition to material structure and properties, determine the performance of a thin film device and that can be handed by the FDTD algorithm. As a general solution of the Maxwell equations, the FDTD algorithm, in its full implementation, would permit a complete analysis of light propagation within a thin film device, able to take into account all the factors that determine, absorption, internal interface reflection, local scattering due to inhomogeneous materials or defect distribution. Finally, such a result can be used to define the carrier photogeneration profile (in space and time) and related to the device electrical properties to foresee device performance. We present in this communication results obtained during our first steps on the way to develop a complete optical model for thin film devices based on the FDTD algorithm. Such a model is integrated into the ASCA simulation program [14] for analysis of multilayer photo-devices.

THEORY

The FDTD model

The FDTD method utilizes the central difference approximation to discretize the two Maxwell's curl equations, namely, Faraday's and Ampere's laws, in both the time and spatial domains, and then solves the resulting equations numerically to derive the electric and magnetic field distributions at each time step using an explicit leapfrog scheme. If a medium is dispersive, then the propagation velocities of electromagnetic waves will vary with frequency in such a medium. The most commonly used approach to simulating dispersive materials is the recursive convolution method, which can be applied to the cold plasma, as well as Debye and Lorentz materials [15] which was later improved to the piecewise linear recursive convolution (PLRC) method [16].In addition, the transform method [17] and the auxiliary differential equation (ADE) method [18], were developed for modeling frequency-dependent media using total-field FDTD. The ADE method leads in general to the most efficient numerical implementation both in terms of memory and operational count. In this work uses a FDTD formulation for modeling Lorentz media, based on the ADE technique with a 5-points finite difference approximation to spatial derivatives.

Optical functions of amorphous material

In order to model a-Si:H as a dispersive medium in a correct implementation of the FDTD algorithm, analytical expression for the real and imaginary part of the dielectric function (ε_1,ε_2) as a function of photon energy(E_{ph}) are necessary. The Tauc–Lorentz analytical expression [19,20] is currently the most widely-used for describing the optical functions of amorphous semiconductors and it is used as a parameterization for optical measurement interpretation. A generalization of this model directed to improve simulations of amorphous silicon optoelectronic devices such as solar cells and photodetectors has been recently presented [21]. While ε_1 is defined through Kramers–Kronig consistence, ε_2 is modeled by three different regions described by an Urbach tail (sub-bandgap absorption) a Tauc function (band to band transition occurring between parabolic bands with constant dipole matrix element) and an above-bandgap region where a Lorentz oscillator model is applicable (Urbach-Tauc-Lorentz model: UTL). An explicit function, built on fitting parameters extracted from experimental data, determines the value of ε_2 as a function of the optical gap on a large interval of photon energies centered in the visible spectrum. Unfortunately such a model is not directly usable in the FDTD scheme, which requires a complex function $C^{\infty}(\mathbf{R}^{+} \Rightarrow \mathbf{C})$ to define the electric permittivity. We propose here a model for $\varepsilon(E_{ph})$ based on the superposition of two Lorentz oscillators (LL model):

$$\varepsilon(E_{ph}) = \varepsilon_\infty + \sum_{k=1}^{2} \frac{\Delta\varepsilon_k E_k^2}{E_k^2 + 2j\delta_k E_{ph} - E_{ph}^2} \quad (1)$$

Where $\varepsilon_\infty$is is the permittivity as energy approaches infinity and for each Lorentz pole k, E_k is the resonant energy of the medium, $\Delta\varepsilon_k$ is the difference between the permittivity at zero energy and ε_∞, δ_k is a damping constant. We have calculated the parameters needed in equation (1) by fitting the data obtained with the UTL model. In Figure 1a is depicted the plot of ε_2 obtained by the UTL model (optimum a-Si:H, optical gap 1.8 eV) compared with the values obtained by the LL model. The two plots are almost identical and a measure of the goodness of fit can be deducted for Figure 1b, where are reported the fit residuals.

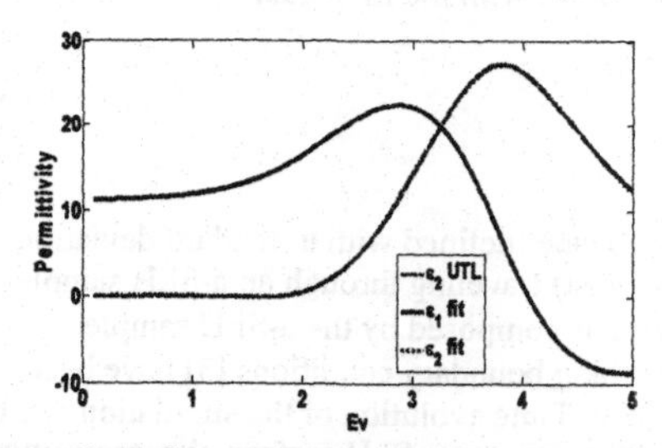

a)

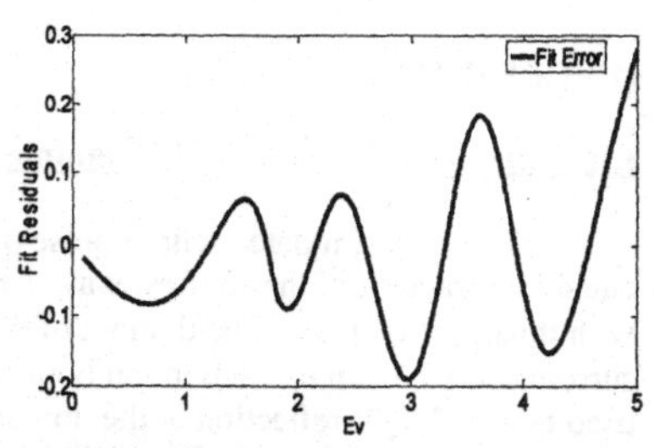

b)

Figure 1: a) Comparison between the imaginary part of the dielectric function in a-Si:H obtained with the UTL and LL models. The real part obtained by the LL model is also presented for completeness. b) Fit Residuals

The parameters used in the LL model are reported in Table I. The two Lorentz poles have different polarity; the parameter $\Delta\varepsilon_k$ in pole 2 has a negative value which determines a negative value of ε_2. For sub-bandgap energies the effect caused by the two poles mutually compensate each other, leading to a value of ε_2 close to zero. Increasing the photon energy the first pole becomes dominant and overcomes the effect of the second one. The imaginary parts of the two Lorentz poles are reported in Figure 2, while in Figure 3 is depicted the absorption coefficient calculated from the values of ε_1 and ε_2 calculated with the LL model.

The same approach has been used for fitting the optical functions of other materials presented in [21]: non-optimized a-Si:H (higher density of states) and a-SiC:H. The results obtained showed the same degree of approximation.

The LL model is merely a mathematical construction which permits the straight application of the FDTD method to thin film semiconductors. Anyway, it should be remarked that a Lorentz oscillator with a negative strength is generally associated to materials with optical gain. From this point of view the interaction of light with sub-

Table 1. The parameters used in the LL model for a-Si:H dielectric function

Parameter	*E_k (eV)*	*$\Delta\varepsilon_k$*	*δ_k (eV)*
Lorentz Pole 1	3.928	16.52	1.102
Lorentz Pole 2	2.588	-6.351	1.145

bandgap photon energies is described by the LL model as a mutual compensation between light absorption and optical gain.

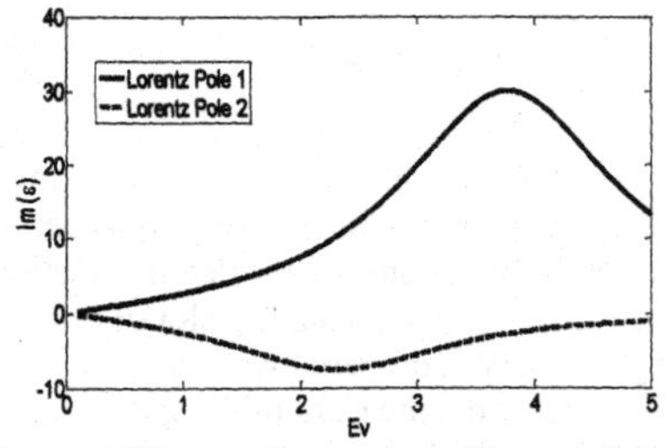

Figure 2. The two Lorentz oscillators of the LL model

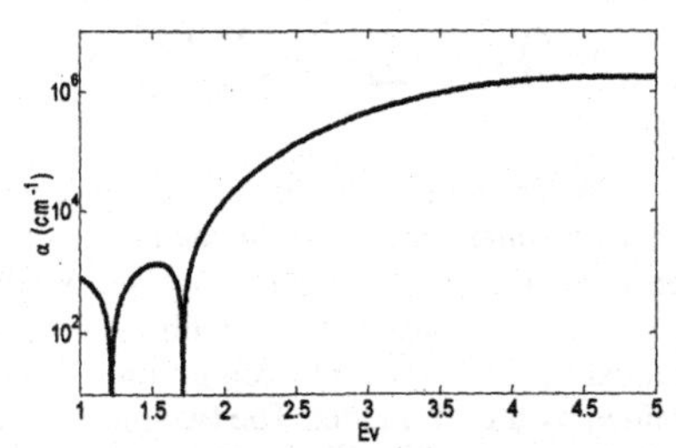

Figure 3: Absorption coefficient of a-Si:H obtained with the LL model

DISCUSSION

Light propagation in an a-Si:H sample

We have simulated a light pulse of Gaussian shape (defined with a standard deviation equal to a fraction of the shortest wave period of interest) traveling through an a-Si:H sample with thickness of 1 μm. The domain of the simulation is composed by the a-Si:H sample surrounded by 1 μm of vacuum on both sides. Absorbing boundary conditions [3] have been used to avoid light reflection at the domain boundaries. Time evolution of the simulation can be observed in Figure 4. Part of the light is reflected at the vacuum/a-Si:H surface, the transmitted light travel along the a-Si:H sample which is acting as a dispersive media, absorbing part of the light. Arriving to the back surface of the sample, the light pulse suffers a second internal reflection. A small portion of it is transmitted to the external vacuum space. The Gaussian pulse is narrow enough to be considered a good approximation to an ideal impulse, so it contains information about a large spectrum of frequencies (i.e of photon energies). By calculating the Fourier transform it is possible to retrieve information about any frequency of interest. We have performed the Fourier analysis by applying the Discrete Fourier Transform to the Gaussian pulse propagation over an interval of time corresponding to the complete damping of the pulse intensity. Transferring the analysis to the frequency domain we obtain as a final result the representation of a stationary wave for any frequency (photon energy) of interest.
An example of the results produced by the simulation is reported in Figure 5 for two different wavelengths. The effect of the reflection can be observed from the pattern formed by the interaction between the original incident wave and the reflected one. The 550 nm wavelength is absorbed in the first 0.5 μm of the a-Si:H film, while the 650 nm wave falls in a low absorption regime, and suffers internal reflection at the back surface. From this result is possible to calculate the Poynting vector and the associated photon flux. Combining it with the data on the absorption coefficient reported in Figure 3, it is possible to determine a photogeneration profile which takes into account for reflection effects.

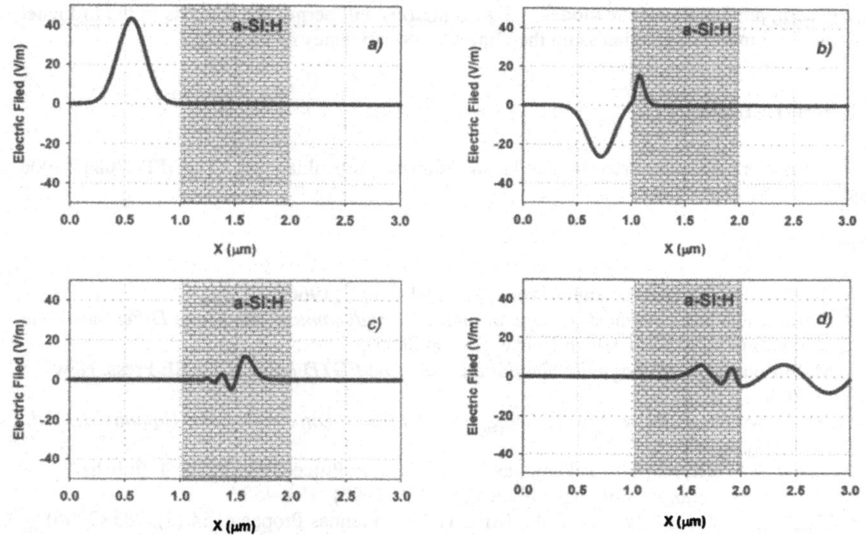

Figure 4. Electric field component of the EM wave traveling across an a-Si:H sample with thicknes of 1 µm. a)Iimpulse approaching the a-Si:H sample, b) Impulse is in part reflected and in part transmitted. c)Impulse travels along the a-Si:H film. d)Partial reflection at the a-Si:H/vacuum interface

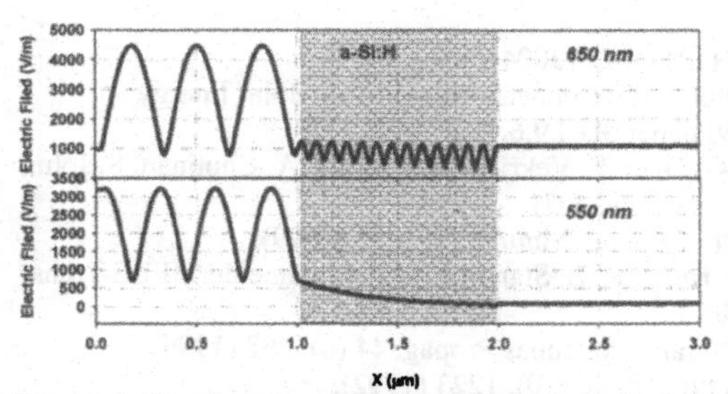

Figure 5. Steady state wave obtained through the application of the discrete Fourier transform for two different wavelengths

CONCLUSIONS

It has been presented a model for simulation of light propagation within a-Si:H films. The simulation is based on the FDTD algorithm for Lorentz dispersive materials with multiple poles.

A simplified model for the optical function of a-Si:H, suitable to be used with the FDTD algorithm has been developed. Examples of some result obtained by this simulation scheme show how this model can be applied as an optical model for simulation of thin film photodevices taking into account for multiple reflection effects, as well as interference effects of internal reflected light. The results presented in this paper demonstrate that is possible to describe the effect of the light-semiconductor interaction through the application of the FDTD model to a a-Si:H solar cell. A more general application of the model to 2D geometries will permit the analysis of the influence of surface and interface roughness on the photovoltaic efficiency of the device.

ACKNOWLEDGMENTS

This work was supported by Fundação Calouste Gulbenkian and PTDC/FIS/70843/2006 project.

REFERENCES

1. K. S. Yee, IEEE Trans. Antennas Propagat. AP-14, 302, (1966).
2. A. Taflove and S. C. Hagness, *Computational Electrodynamics: The Finite Difference Time Domain Method,* 3rd ed (Artech House, Boston 2005).
3. D.M. Sullivan, *Electromagnetic Simulation Using the FDTD Method.* (IEEE Press, New York 2000).
4. W.Yu, R. Mittra, T. Su, Y. Liu, X. Yang. *Parallel Finite-Difference Time-Domain Method.* (Artech House, 2006).
5. A. Lauer, I. Wolff, A. Bahr, J. Pamp, and J. Kunisch, in Proceedings of the 45th IEEE Vehicular Technology Conference (Chicago, USA, 1995), 454–458
6. G. Rodriguez, Y. Miyazaki, N. Goto, IEEE Trans. Antennas Propagat. 54 (3), 785 (2006)
7. J. M. Bourgeois, G. S. Smith.. IEEE Trans. Geosci. Remote Sensing. 34 (1), 36 (1996)
8. U. Oguz, L. Gurel, IEEE Trans. Geosci. Remote Sensing 40, 1385 (2002)
9. A. Sabouni, S. Noghanian, S. Pistorius, IEEE AP-S International Symposium, (New Mexico, USA, 2006).
10. S. M. El-Ghazaly, R. P. Joshi, R. O. Grondin, IEEE Trans. Microw. Theory Tech. 38 (5), 629 (1990).
11. S. H. Chang, A. Taflove, Opt. Express (12), 3827 (2004).
12. C. Pflaum, C. Haase, H. Stiebig, C. Jandl, 24th European Photovoltaic Solar Energy Conference, (Hamburg, Germany 2009) paper 3BO.9.6.
13. P.O'Brien, N.Kherani, S. Zukotynski, G. Ozin, E. Vekris, N. Tetreault, A. Chutinan, S. John, A.Mihi, and H. Míguez, Adv. Mater. 19, 4177 (2007).
14. A. Fantoni, M. Vieira, R. Martins. Math. Comput. Simulat. 49, 381 (1999).
15. R. J. Luebbers, F. P. Hunsberger, K. S. Kunz, R. B. Standler, and M. Schneider, IEEE Trans. Electromag. Compat. 32 (3), 222 (1990).
16. D. K. Kelley and R. J. Luebbers, IEEE Trans. Antennas Propag. 44 (6), 792 (1996)
17. D. M. Sullivan, IEEE Trans. Antennas Propag. 40 (10), 1223 (1992).
18. R. M. Joseph, S. C. Hagness, and A. Taflove, Optics Lett. 16, 1412 (1991).
19. G.E. Jellison Jr., F.A.Modine, Appl.Phys.Lett. 69, 371 (1996).
20. G.E. Jellison Jr., F.A.Modine, Appl.Phys.Lett. 69, 2137 (1996)
21. A.S. Ferlauto , G.M. Ferreira , J.M. Pearce, C.R. Wronski, R.W. Collins, X. Deng, G. Ganguly, Thin Solid Films 455–456 , 388 (2004)

Poster Session: Films and Growth

Mater. Res. Soc. Symp. Proc. Vol. 1245 © 2010 Materials Research Society 1245-A16-02

A thermodynamic model for the laser fluence ablation threshold of PECVD SiO_2 on thin a-Si:H films deposited on crystalline silicon

Krister Mangersnes and Sean E. Foss
Institute for Energy Technology, Solar Energy Department, Instituttveien 18, 2027 Kjeller, Norway.

ABSTRACT

We have developed a thermodynamic model that predicts the heat distribution in a stack of PECVD SiO_2 and a-Si:H on crystalline Si after laser irradiation. The model is based on solving the total enthalpy heat equation with a finite difference scheme. The laser used in the model is a frequency doubled Nd:YVO_4 green laser with pulse duration in the nanosecond range. The modeling was done with the aim of getting a better understanding of our newly developed laser ablation process for making local contacts on back-junction silicon solar cells. Lasers with pulse duration within the nanosecond range are usually believed to induce too much thermal damage into the underlying silicon to make them suitable for high efficiency solar cells. In our case, insertion of a thin layer of a-Si:H between the SiO_2 and the Si absorbs much of the laser irradiation both optically and thermally. This makes it possible to form local contacts to Si in a damage-free way. In addition, the residual a-Si:H serves as an excellent surface passivation layer for the Si substrate. We have also developed a simple static model to determine the onset of SiO_2 ablation on a-Si:H layers of varying thickness. The models, both the static and the dynamic, are in good agreement with experimental data.

INTRODUCTION

Lasers have in the recent years been shown to be a very promising tool for making local contacts on silicon solar cells through a dielectric layer[1-7]. The local ablation is done to reduce the total metal semiconductor area of the solar cells, and thus improve the efficiency of the device. Lasers in the nanosecond (ns) range are usually believed to induce too much thermal damage into the underlying silicon lattice to make them suitable for high efficiency solar cell concepts[4, 5, 8]. In the case of a frequency doubled Nd:YVO_4 green laser, SiO_2 is transparent to the laser light. Ablation of SiO_2 on Si takes place through an indirect process where the oxide is lifted off from expansion of molten or vaporized silicon. We have recently shown that insertion of a buffer layer of amorphous silicon, a-Si:H (a-Si for simplicity), between the SiO_2 and the Si will absorb much of the laser irradiation[7]. The high optical absorption and low thermal conductivity of a-Si compared to Si confines the laser energy to a much smaller volume. This makes it possible to ablate SiO_2 in a damage free way with a laser fluence five times lower than that needed to ablate SiO_2 on crystalline silicon. In addition, the residual a-Si serves as an excellent surface passivation layer for the Si substrate[9-11]. a-Si is not a very well defined material as most of the material parameters are very sensitive to the deposition technique and parameters. Especially the absorption coefficient and the thermal conductivity vary over a broad range according to different references[12-19]. Previous work on laser interaction with a-Si has mostly focused on melting and recrystallization of a-Si, both experimentally[15, 20-22] and numerically[21, 23-25]. The goal of this paper is to develop a thermodynamic model that predicts the laser fluence ablation threshold of SiO_2 on a-Si layers of varying thickness, and that also describes the heat distribution in the SiO_2 - a-Si - Si stack after laser irradiation.

THEORY

To find an expression for the laser fluence ablation threshold we make some simple assumptions. We assume that the onset of melting or vaporization is given by a critical energy density $E_T = F_T/L_{th,eff}$, where F_T is the threshold fluence. $L_{th,eff}$ is the effective thermal diffusion length defined by $L_{th} = \sqrt{2D_{eff}\tau_p}$, where D_{eff} is the effective thermal diffusivity and τ_p is the pulse duration of the laser at full width half maximum. D_{eff} is the ratio between the effective thermal conductivity, k_{eff}, and the volumetric heat capacity. k_{eff} is weighted between the thermal conductivity of a-Si and Si, also including a thermal resistance at the interface between the two. A version of this model, not including interface resistance, was proposed by Matthias et al., when investigating laser ablation of metal films on quartz[26]. Herein we also need to account for an optically absorbing substrate under the thin film. If we assume a uniform temperature rise throughout the volume defined by $L_{th,eff}$, it can be shown that the threshold fluence for melting is given by

$$F_{TM} = \frac{\Delta T_M}{(1-R)\left[(1-e^{-\alpha d})+(1-e^{-\beta d_s})e^{-\alpha d}\right]} C_{P,eff} L_{th,eff} \tag{1}$$

ΔT_M is the needed temperature increment for melting, d and d_s are the thicknesses of the a-Si layer and the Si substrate, respectively. We assume that the transparent SiO_2 constitutes a thermal barrier and only influences the reflection, R, at the surface. α and β are the optical absorption coefficients of a-Si and Si, respectively. $C_{p,eff}$ is the effective volumetric heat capacity; linearly weighted between a-Si and Si. The corresponding model for the onset of surface vaporization is the same as for melting but with ΔT_M replaced by $\Delta Tv + \Delta H_m/C_{p.a\text{-}Si}$, where ΔTv is the needed temperature rise for vaporization, ΔH_m is the latent heat of melting, and $C_{p.a\text{-}Si}$ is the volumetric heat capacity of a-Si. The parameters used in the model are taken from Table I. We used the respective values at 1000 K for the temperature dependent parameters, and a pulse duration of 142 ns. Our experimental data from reference 7 is plotted together with the static thermodynamic models for surface melting and vaporization in Figure 1. There is a good fit between the experimental data and the static model for surface evaporation even without temperature dependent parameters, and with the assumption of a uniform temperature rise. Still, this model only gives an indication of the onset of evaporation and does not include any information about the actual temperature distribution within the film and the substrate. Neither have we taken into account that the pulse length of our laser varies with fluence as given in Table I. A more dynamic description that also includes phase transitions will require a solution of the total enthalpy version of the heat equation. The diameter of our laser spot is much wider than the thermal diffusion length of a-Si, and it is a good approximation to solve the one-dimensional version of the equation[27].

$$\frac{\partial \Delta H(x,t)}{\partial t} = \frac{\partial}{\partial x}\kappa(T)\left(\frac{\partial}{\partial x}T(x,t)\right) + S(x,t) \tag{2}$$

$\kappa(T)$ is the temperature dependent thermal conductivity, x is the depth from the top of the a-Si layer, and $S(x,t)$ is the laser source given by

$$S(x,t) = \sqrt{\frac{4\ln 2}{\pi}} \frac{(1-R)F\cdot\alpha}{\tau_p} \exp\left(-\alpha x - 4\ln 2 - \frac{(t-t_{peak})^2}{\tau_p^2}\right) \tag{3}$$

F is the fluence at the peak of the Gaussian pulse, and t_{peak} is the time for the peak fluence. The total enthalpy, ΔH, is given by[27]

$$\Delta H(T) = \int_{T_0}^{T} \rho(T')c_p(T')dT' + \eta(T-T_M)\Delta H_M + \eta(T-T_V)\Delta H_V \tag{4}$$

T_0 is the ambient temperature, and T_M and T_V are the melting and vaporization temperatures, respectively. η is the Heaviside function (1 or 0 if the argument is positive or negative, respectively) and ΔH_m and ΔH_v are the latent heat of melting and vaporization, respectively.

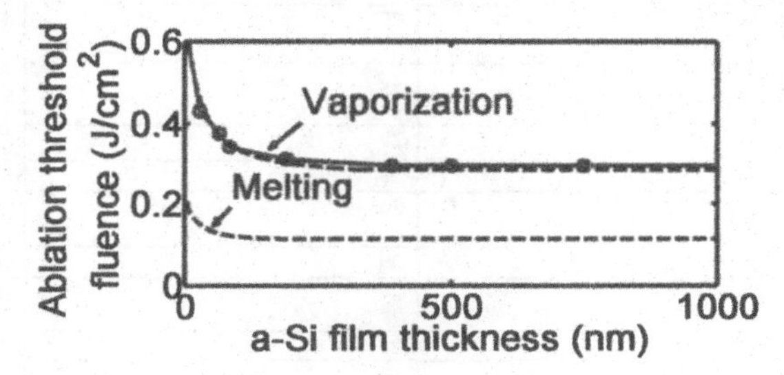

Figure 1: The fluence threshold for laser ablation as a function of the thickness of the a-Si film. Our experimental data from reference 7 (solid line with circles) is shown together with the modeled data for surface melting and vaporization.

To omit numerical instabilities we have assumed that the phase transitions occur over a temperature interval of $\Delta T = 5$ K. The SiO_2 is assumed to be optically transparent and thermally insulating, and is only influencing the surface reflection. The structure is at ambient temperature before the laser heating starts. We have used the following initial and boundary conditions:

$$T(x,t) = T_0\big|_{t=0} \qquad T(x,t) = T_0\big|_{x\to\infty} \qquad \frac{\partial T(x,t)}{\partial x} = 0\big|_{x=0} \tag{5}$$

A finite difference scheme was used to solve the total enthalpy heat equation. The simulation parameters are listed in Table I. a, c, and l in the table refer to amorphous Si, crystalline Si, and liquid Si, respectively

DISCUSSION

In Figure 2 we show the temperature distribution in a SiO_2 covered silicon wafer at different depths after laser irradiation with a single laser pulse with a fluence of 0.32 J/cm^2. In Figure 3 we show the temperature distribution in a similar sample, but now with a 300 nm buffer layer of a-Si. We observe that the a-Si obtains a much higher surface temperature, and that 0.32 J/cm^2 is enough to reach the onset of surface vaporization. This result, and also results for a-Si layers of different thicknesses (not shown) are in good agreement with our experimental data.

TABLE I

Parameters	Value/Expression	Reference
Absorption coefficient (cm^{-1})	a: 1.9×10^5	28
	c: $5.02\times10^3 e^{T/430}$	20
	l: 1×10^6	23
Density (g/cm^3)	a: 2.20	23
	c: 2.32	27
	l: 2.52	27
Thermal Conductivity (W/cm K)	a: $1.3\times10^{-11}\times(T-900)^3 + 1.3\times10^{-9}\times(T-900)^2 + 1\times10^{-6}\times(T-900) + 1\times10^{-2}$	20
	c: $1521/T^{1.226}$ (T<1200 K) $8.96/T^{0.502}$ (1200 K <= T <1690 K)	20
	l: 0.62	29
Heat capacity (J/cm^3 K)	a: $2.2\times(0.952 + 0.171\times T/685)$	20
	c: $2.32\times(0.711 + 0.255\times(T^{1.85}-1)/(T^{1.85}-0.255/0.711)$	30
	l: 1.0	29
Melting temp (K)	a: 1420	27
	c: 1690	27
Vaporization temp. (K)	l: 2680	27
Interface thermal resistance (cm^2 K/W)	0.0054	16
Latent heat of melting (J/K)	a: 1250	27
	c: 1780	27
Latent heat of Vaporization (J/K)	l: 15000	27
Reflection	a: 0.18	Measured
	c: 0.18	Measured
	l: 0.07	Calculated with data from ref 31
Pulse duration, full width at half maximum (ns)	$-37\times F + 154$	Measured. F is the peak fluence in the Gaussian beam profile (J/cm^2)
Pulse peak (ns)	200	Set

Interface thermal resistance

Kuo et al. measured the thermal resistance of the interface between a-Si films and crystalline Si[16]. The value we have used in our simulations, 0.54 mm^2 K/ W, is the resistance they measured for a structure with a native oxide between the a-Si and the Si. This value was found to give the best fit between the modeling and our experiments. A clean surface, prepared by pre-sputtering in a vacuum chamber, showed an interface resistance of 0.15 mm^2 K/ W[16]. A lower interface resistance increases the laser fluence ablation threshold in our model, as less heat is contained in the film. The increased thermal resistance at the interface between a-Si and the Si (presumably provided by an oxide) is therefore a necessary element of our damage-free laser-ablation process.

CONCLUSIONS

We have developed a static thermodynamic model that describes the fluence threshold for pulsed laser ablation of SiO_2 on Si with a buffer layer of a-Si of varying thickness. We have also solved the heat equation for the SiO_2 – a-Si –Si stack to predict the temperature distribution within the a-Si film and the Si substrate. We found that an interface thermal resistance between a-Si and Si is important for lowering the laser fluence ablation threshold and for protecting the underlying silicon lattice. The results are in good agreement with our experimental data, when we include in the model the expected contribution from the native oxide layer to the thermal resistance of the interface.

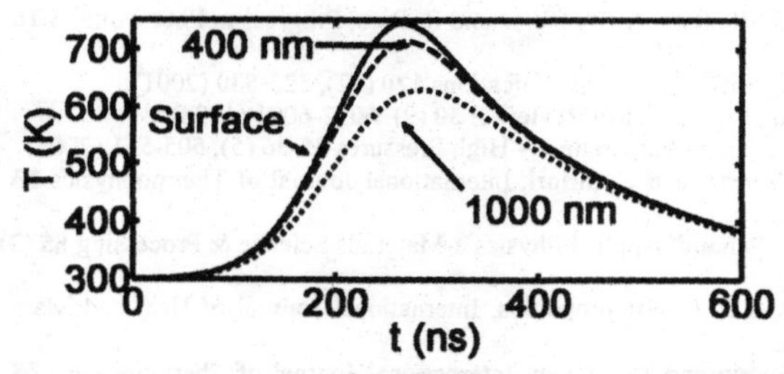

Figure 2: Heat distribution in Si as a function of time at different depths after single pulse laser irradiation with a fluence of 0.32 J/cm^2 and a pulse duration of 142 ns. $x = 0$ (surface) corresponds to the top of the Si substrate.

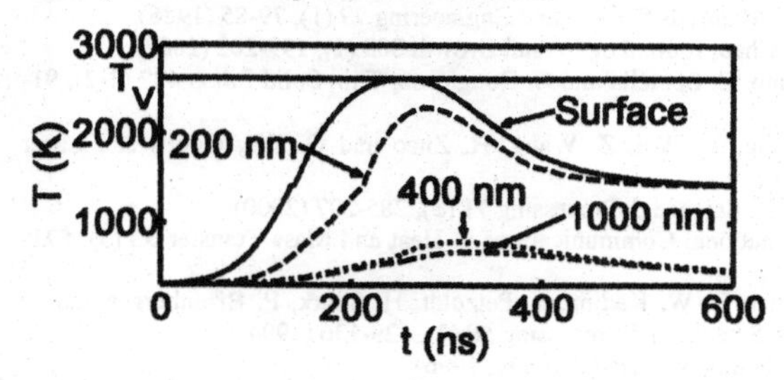

Figure 3: Heat distribution as a function of time at different depths in a stack of 300 nm a-Si on Si after single pulse laser irradiation with a fluence of 0.32 J/cm^2 and a pulse duration of 142 ns. $x = 0$ (surface) corresponds to the top of the a-Si film.

REFERENCES

1. P. Engelhart, N. P. Harder, T. Horstmann, R. Grischke, R. Meyer and R. Brendel, Conference Record of the 2006 IEEE 4th World Conference on Photovoltaic Energy Conversion (IEEE Cat. No. 06CH37747), 4 pp.|CD-ROM (2006).
2. A. Grohe, C. Harmel, A. Knorz, S. W. Glunz, R. Preu, G. P. Willeke, presented at the 4th IEEE World Conference on Photovoltaic Energy Conversion, Vols 1 and 2, 2006.
3. P. Engelhart, N. P. Harder, R. Grischke, A. Merkle, R. Meyer and R. Brendel, Progress in Photovoltaics **15** (3), 237-243 (2007).
4. P. Engelhart, S. Hermann, T. Neubere, H. Plagwitz, R. Grischke, R. Meyd, U. Klug, A. Schoonderbeek, U. Stute and R. Brendel, Progress in Photovoltaics **15** (6), 521-527 (2007).
5. S. Hermann, T. Neubert, B. Wolpensinger, N.-P. Harder and R. Brendel, presented at the 23rd European Photovoltaic Solar Energy Conference, Valencia, Spain, 2008.
6. A. Knorz, M. Peters, A. Grohe, C. Harmel and R. Prett, Progress in Photovoltaics **17** (2), 127-136 (2009).
7. K. Mangersnes, S. E. Foss and A. Thøgersen, Journal of Applied Physics **107** (4) (2010).
8. K. Mangersnes and S. E. Foss, presented at the 24th European PVSEC, Hamburg, Germany, 2009.
9. A. G. Aberle, Progress in Photovoltaics **8** (5), 473-487 (2000).
10. A. Bentzen, A. Ulyashin, E. Sauar, D. Grambole, D. N. Wright, E. S. Marstein, B. G. Svensson and A. Holt, presented at the 15th international Photovoltaic Science and Engineering Conference, Shanghai, China, 2005.
11. M. Hofmann, C. Schmidt, N. Kohn, J. Rentsch, S. W. Glunz and R. Preu, Progress in Photovoltaics **16** (6), 509-518 (2008).
12. N. Attaf, M. S. Aida and L. Hadjeris, Solid State Communications **120** (12), 525-530 (2001).
13. D. G. Cahill, M. Katiyar and J. R. Abelson, Physical Review B **50** (9), 6077-6081 (1994).
14. R. K. Endo, Y. Fujihara and M. Susa, High Temperatures - High Pressures **35-36** (5), 505-511 (2003).
15. M. G. Grimaldi, P. Baeri, M. A. Malvezzi and C. Sirtori, International Journal of Thermophysics **13** (1), 141-151 (1992).
16. B. S. W. Kuo, J. C. M. Li and A. W. Schmid, Applied Physics a-Materials Science & Processing **55** (3), 289-296 (1992).
17. S. Moon, M. Hatano, M. H. Lee and C. P. Grigoropoulos, International Journal of Heat and Mass Transfer **45** (12), 2439-2447 (2002).
18. S. Volz, X. Feng, C. Fuentes, P. Guerin and M. Jaouen, International Journal of Thermophysics **23** (6), 1645-1657 (2002).
19. H. Wada and T. Kamijoh, Japanese Journal of Applied Physics Part 2-Letters **35** (5B), L648-L650 (1996).
20. C. K. Ong, H. S. Tan and E. H. Sin, Materials Science and Engineering **79** (1), 79-85 (1986).
21. Y. R. Chen, C. H. Chang and L. S. Chao, Journal of Crystal Growth **303** (1), 199-202 (2007).
22. L. Mariucci, A. Pecora, G. Fortunato, C. Spinella and C. Bongiorno, Thin Solid Films **427** (1-2), 91-95 (2003).
23. Z. Yuan, Q. Lou, J. Zhou, J. Dong, Y. Wei, Z. Wang, H. Zhao and G. Wu, Optics and Laser Technology **41** (4), 380-383 (2009).
24. S. Tosto, Applied Physics a-Materials Science & Processing **71** (3), 285-297 (2000).
25. C. H. Chang and L. S. Chao, International Communications in Heat and Mass Transfer **35** (5), 571-576 (2008).
26. E. Matthias, M. Reichling, J. Siegel, O. W. Kading, S. Petzoldt, H. Skurk, P. Bizenberger and E. Neske, Applied Physics a-Materials Science and Processing **58** (2), 129-136 (1994).
27. D. Bauerle, Laser Processing and Chemistry, 2 ed. (Springer, 1996).
28. E. D. Palik, Handbook of Optical Constants of Solids. (pp: 571-586), Elsevier.
29. H. Kobatake, H. Fukuyama, I. Minato, T. Tsukada and S. Awaji, Applied Physics Letters **90** (9) (2007).
30. V. Palankovski, R. Schultheis and S. Selberherr, IEEE Transactions on Electron Devices **48** (6), 1264-1269 (2001).
31. G. E. Jellison and D. H. Lowndes, Applied Physics Letters **51** (5), 352-354 (1987)

Mater. Res. Soc. Symp. Proc. Vol. 1245 © 2010 Materials Research Society 1245-A16-07

The Grain Size Distribution in Crystallization Processes with Anisotropic Growth Rate

Kimberly S. Lokovic[1], Ralf B. Bergmann[2] and Andreas Bill[1]
[1] California State University Long Beach, Department of Physics & Astronomy, 1250 Bellflower Blvd., Long Beach, CA 90840, U.S.A.
[2]Bremen Institute for Applied Beam Technology (BIAS), Klagenfurter Str. 2, 28359 Bremen, Germany.

ABSTRACT

The grain size distribution allows characterizing quantitatively the microstructure at different stages of crystallization of an amorphous solid. We propose a generalization of the theory we established for spherical grains to the case of grains with ellipsoidal shape. We discuss different anisotropic growth mechanisms of the grains in thin films. An analytical expression of the grain size distribution is obtained for the case where grains grow through a change of volume while keeping their shape invariant. The resulting normalized grain size distribution is shown to be affected by anisotropy through the time-decay of the effective growth rate.

INTRODUCTION

Recently, we developed a theory of the grain size distribution (GSD) $N(g,t)$ for the crystallization of an amorphous solid [1-3]. N is the normalized count of the number of grains of a certain size g found at time t in a sample. The quantity g is, for example, the diameter of the grain, the number of atoms contained in the grain, its volume, etc. As discussed below, g can also be replaced by a vector quantity. The interest in determining the GSD lies in the fact that physical properties of a solid such as their electrical conductivity, magnetization or optical absorption may be substantially affected by the degree of crystallization and the microstructure of a material. Thus, it is important to determine the grain size distribution during the crystallization of a sample. In previous work [1,3], we obtained a closed analytical form of the grain size distribution for spherical grains. The expression was derived for any dimension d of the crystallization process and can easily be used for the description of experimental data on bulk or thin film systems, as shown in the case of solid-phase crystallization of silicon in Ref. [2].

The theory is developed for constant *microscopic* nucleation and growth rates I_0 and v_0 and incorporates the well-known Kolmogorov-Avrami-Mehl-Johnson (KAMJ) expression for the fraction of crystallized material during a random nucleation and growth (RNG) crystallization process [4-6]. This leads to a time-dependent decay of the effective nucleation rate $I(t)$ [see below]. In addition, we introduced a new *effective* time-dependent growth rate $\bar{v}(t)$ that accounts for the fact that a specific grain stops growing once a sufficient number of other grains impinge on it [3]. Coalescence is not considered in the theory. The aim of the present work is to generalize the theory of Refs. [1-3] beyond the spherical grains assumption. We thus allow for the existence of grains with different shapes and sizes in the crystallization process.

It is worth pointing out that the derivation of the GSD for spherical grains is not of pure academic relevance. Indeed, when displaying the GSD resulting from experimental data it is usual to replace the grains' great variety of shapes by spheres with a radius chosen in such a way that the volume of the sphere equals the volume of the actual measured grain. Nevertheless, it is of interest to extend the analysis because growth inhibition of a grain due to impingement is expected to be affected by the shape of individual grains and might, therefore, modify the GSD. The aim of the paper is to derive an expression for the GSD that accounts for the presence of anisotropic growth rates and non-spherical grains.

In materials such as silicon two main mechanisms determine the growth and shape of grains during homogeneous nucleation at moderate crystallization temperatures of around 600°C. Non-spherical, faceted growth during primary grain growth results from preferred directions of growth within the film according to the crystallographic structure of the diamond lattice. During secondary grain growth at elevated temperatures around 900°C, the orientation of the surface of the grains is surface energy driven. Grains oriented with a favorable surface energy grow at the expense of unfavorably oriented grains leading to a textured film with random orientation of the crystalline structure in the plane of the film.

Another mechanism that may lead to a surface texture is heterogeneous nucleation, which is the most common situation encountered in nature. This generally results from imperfections of the substrate and is random over the surface of the film. In addition, preferred sites or a texture of the substrate may determine the location of nucleation centers. As shown in Ref. [7,8] it is possible to generate an artificial periodic lattice of nucleation sites that results in a periodic arrangement of grains. As for the homogeneous grain growth mechanism, the in-plane orientation of grains remains random. Many studies have been made that attempt at predetermining the latter orientation as well and a summary of that work can be found in Ref. [9].

Finally, the formation of grains also occurs in systems made of molecules where the interaction between the entities is anisotropic, that is where the intermolecular interactions depends on the relative orientation of the molecules. The growth will occur predominantly in the direction of strongest binding. Depending on the particular thermodynamic conditions of the crystallization process these interactions may lead to grains with spherical, ellipsoidal or even tubular shapes [10].

In the present paper we are interested in the case where the grain growth in the plane of the film is anisotropic, but the orientation of the crystalline structure is random.

THEORY

We introduced in Refs. [1] and [3] a partial differential equation for the grain size distribution $N(\vec{r},t)$ that includes a source term and a growth rate

$$\frac{\partial N(\vec{r},t)}{\partial t} + \vec{\nabla}_{\vec{r}} \cdot \left[N(\vec{r},t)\vec{v}(\vec{r},t) \right] = I(t)\,D(\vec{r}) \tag{1}$$

It is important to note that the vector $\vec{r}$ is not a vector in regular space, but is a vector in the space of quantities that define the size and shape of the grain. Below we will consider ellipsoidal shapes of grains and the components of $\vec{r}$ will be the semi-axes of the ellipsoid. We assume a separable source term for nuclei on the right hand side of the equation. The function $D(\vec{r})$ can have any form *a priori*, but we have considered the case $D(\vec{r}) = \delta(\Omega - \Omega_c)$, where $\delta(\cdot)$ is the

Dirac distribution, Ω is the volume of a grain and Ω_c the critical volume of the nucleus, the smallest stable grain in the system. Other forms such as a Gaussian distribution have also been considered, but we will limit the present study to the Dirac distribution.

In principle, the various quantities appearing in Eq.(1) depend on temperature, but we will assume that the crystallization process occurs at fixed thermodynamic conditions.

Equation (1) was written in its most general form in Refs. [1,3], but has been solved analytically and discussed in detail for the case of spherical grains only. One important outcome of our previous work is that for a nucleation rate $I(t)$ complying with the KAMJ model and a growth rate $\vec{v}(t)$ of similar functional form, the resulting distribution is lognormal in the asymptotic limit of large times (*i.e.* at full crystallization). Importantly, this remarkable result requires specific dynamics of nucleation and growth processes such as those used in Refs. [1-3].

Equation (1) can be solved by quadrature for well-behaved functions for nucleation and growth rates. However, we are interested in obtaining a *closed analytical* form of the solution for RNG processes and non-spherical grains. The main assumption allowing solving the problem analytically is to consider that $\vec{v}(\vec{r},t)$ is independent of $\vec{r}$. This assumption is valid whenever the diffusion of atoms contributing to the growth of a grain can be neglected. This is, for example, the case of solid-phase crystallization of Si [1,2]. With this assumption Eq. (1) can be solved analytically.

We introduce generic time-dependent effective nucleation and growth rates

$$I(t) = I_0 f(t)\Theta(t - t_0), \quad \vec{v}(t) = \vec{v}_0 g(t)\Theta(t - t_0), \tag{2}$$

where I_0 and $\vec{v}_0$ are the constant microscopic growth and nucleation rates, respectively, $f(t)$ and $g(t)$ describe the effective time decays of nucleation and growth rates, and $\Theta(t - t_0)$ is the Heaviside function containing the incubation time t_0. The latter describes the time at which the first nucleus is created. As discussed in Refs. [1-3], the time dependence of nucleation and growth rates has different origins. The nucleation rate decays in time because the volume fraction available for the creation of new nuclei decreases as crystallization takes place. Avrami and Kolmogorov derived the analytical expression for $t \geq t_0$

$$f(t) = \exp\left[-\left(\frac{t - t_0}{t_{cl}}\right)^{d+1}\right]. \tag{3}$$

This expression depends on the critical time decay t_{cl}, on the incubation time and on the dimensionality of the crystallization process [4,5]. It should be noted that Avrami's expression for $f(t)$ is very general and does, in particular, not depend on the shape of the grains. Therefore, it was applicable to the crystallization process described in Refs. [1-3] and is also applicable to the case studied in the present paper. The growth rate, on the other hand, decays in time because grains impinge on each other and prohibit further growth in the direction perpendicular to the interface of the impinging grains. This rate may depend on the shape of the grains.

To demonstrate how the grain size distribution is modified when considering non-spherical grains we limit ourselves to the two-dimensional (thin film) case for convenience. The simplest extension beyond the spherical shape of grains is to consider ellipses (see Fig.1). The variable $\vec{r}$ defining the shape and size of the grain is then composed of the two semi-axes of the ellipse $\vec{r}=(r_1,r_2)$. We define the critical semi-axes of the nuclei by $\vec{r}_c=(r_{c1},r_{c2})$. Several cases need to be considered.

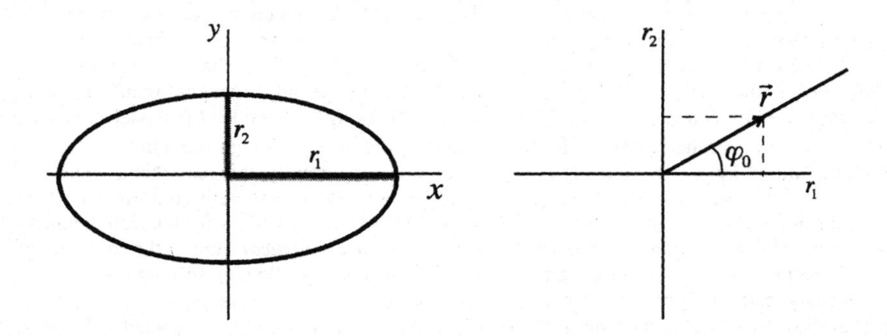

Figure 1. Left: Crystalline grain modeled by an ellipse (case for thin films). A vector $\vec{x}=(x,y)$ denotes a point in the ellipsoidal grain. Right: Graphical representation of the vector $\vec{r}$ for two dimensions. A vector $\vec{r}=(r_1,r_2)$ in that space denotes a specific ellipse with values of the semi-axes given by r_1 and r_2.

Spherical nuclei and grains

In this case the nuclei have radius $r_c=r_{c,1}=r_{c,2}$ and the grain size distribution has been derived in Refs. [1-3]. Worth noting is that the GSD is of the lognormal type at full crystallization when the calculation is done within the KAMJ model and the growth rate has the exponential form $g(t)=\exp\left[-(t-t_0)/t_{cv}\right]$.

Shape invariant ellipsoidal grain growth

In this situation the nuclei are ellipses and the growth into grains results in the increase of the grain's volume while the shape remains invariant, that is $r_2/r_1=r_{c,2}/r_{c,1}$. Introducing polar coordinates $\vec{r}=(r,\varphi)$ in the space of the ellipses semi-axes (r_1,r_2) and defining $\tan\varphi_0=r_{c,2}/r_{c,1}$ (see Fig. 1, right) we can perform calculations similar to those of Ref. [3] and obtain the following expression for the time-dependent GSD

$$N(r,t)=\frac{I_0}{v_0}\frac{r_\infty}{2A_\infty}\left(\frac{r_\infty}{r}\right)\sum_i\frac{f(\sigma_i)}{g(\sigma_i)}\times\left[\Theta\left(\frac{r-r_c}{r_\infty-r_c}\right)-\Theta\left(\frac{r-r_{\max}(t)}{r_\infty-r_c}\right)\right], \qquad (4)$$

where $t_r^2 = t_{cv}/t_{cl}$ and $A_\infty = \pi r_{1,\infty} r_{2,\infty}$ is the area of the largest grain (ellipse) found at full crystallization with semi-axes $r_{1,\infty} = r_\infty \cos\varphi_0$, $r_{2,\infty} = r_\infty \sin\varphi_0$. In addition, the functions σ_i $(i = 1,2,\ldots)$ are solutions of the equation

$$r(t) = r_c + \int_{\sigma_i}^{t} v(t')dt'. \tag{5}$$

When σ_i in the integral's lower limit is replaced by t_0, Eq.(5) defines the quantity $r_{\max}(t)$ appearing in the Heaviside function. Similarly to the spherical case we expect that Eq.(5) has only one solution ($i = 1$) for physically relevant time-dependent RNG rates. The time-dependent $r_{\max}(t)$ is related to the largest ellipsoidal grain found at time t in the sample, which has semi-axes $r_{1,\max}(t) = r_{\max}(t)\cos\varphi_0$ and $r_{2,\max}(t) = r_{\max}(t)\sin\varphi_0$. This result is only valid for the present case where the shape of the grain is invariant during its growth.

The result presented in Eqs. (4,5) has several interesting features. First, it shows that the GSD is essentially determined by the ratio of the effective nucleation rate to the effective growth rate, except for the important fact that the variable appearing in f and g is not simply time but a more complicated function $\sigma_i(t)$, solution of Eq. (5). Second, the result obtained for ellipsoidal grains transforms into the expression found previously for spherical grains when replacing $f(\sigma_i)$ by the exponential term derived by Avrami [3,5]. Remarkably, only the radial coordinate $r = \sqrt{r_1^2 + r_2^2}$ appears in the expression, except in the prefactor, which contains A_∞ and therefore φ_0 that is directly related to the ratio of semi-axes of the nucleus' ellipsoidal shape. Hence, introducing the normalized grain size distribution $N(r,t)/N(t)$ with $N(t) = \int_0^\infty N(r,t)\,dr$, as usually done to compare theory with experiment (see, *e.g.*, Refs. [2,3]), this prefactor cancels out and the normalized GSD takes the same functional form for spherical and ellipsoidal grains. It is important to point out, however, that the shape of the grain is determined by $g(t)$ and leads thus to different forms of $N(r,t)$ for spheres or ellipsoids. The distinction resides in the time dependence of the effective growth rate $\bar{v}(t)$. As mentioned above, the growth rate is expected to depend on the shape of the crystallized grains.

Ellipsoidal nuclei ending as spherical grains

This case requires a different approach to solve Eq.(1) than the one presented in Ref. [3] and used above. The obtained expression is more involved and depends on the particular functional form of the effective growth rate $\bar{v}(t)$. This case will be discussed elsewhere. It is worth pointing out that even though the microscopic growth rate $\bar{v}_0$ leads free grains to grow into spheres, the sample at full crystallization will display grains with a distribution of shapes ranging from the original critical ellipse to spheres because impingement prematurely stops the growth of grains at different stages.

SUMMARY

We analyzed the changes occurring in the form of the grain size distribution $N(r,t)$ when replacing spherical grains by ellipsoidal grains. Using the two-dimensional (thin film) case as an example, we demonstrated that the general functional form of the normalized GSD expressed in terms of the generic effective nucleation and growth rates $I(t), \vec{v}(t)$ remains unaffected by the shape of the grains. The major difference between the two cases resides in the different time dependences of the growth rate. What functional form these rates take is presently under investigation.

ACKNOWLEDGMENTS

We gratefully acknowledge the support of the Research Corporation, the DAAD and SCAC at CSU Long Beach.

REFERENCES

1. R.B. Bergmann and A. Bill, *J. Cryst. Growth* **310**, 3135 (2008).
2. A.V. Teran, R.B. Bergmann and A. Bill, *Mater. Res. Soc. Symp. Proc.* **1153**, A05-03 (2009).
3. A.V. Teran, R.B. Bergmann and A. Bill, *Phys. Rev. B* **81**, 075319 (2010).
4. A.N. Kolmogorov, Akad. Nauk SSSR, Izv. Ser. Matem. **1**, 355 (1937).
5. M. Avrami, J. Chem. Phys. **7**, 1103 (1939); *ibid.*, **8**, 212 (1940).
6. W. Johnson and R. Mehl, Trans AIME **135**, 416 (1939); W. Anderson and R. Mehl, *ibid.*, **161**, 140 (1945).
7. H. Kumomi and T.Yonehara, *Jpn. J. Appl. Phys.* **36**, 1383 (1997)
8. H. Kumomi, in *Growth, Characterization and Electronic Applications of Si-based Thin Films*, edited by R. B. Bergmann (Research Signpost, Trivandrum, India, 2002).
9. *Oriented Crystallization on Amorphous Substrates* E.I. Givargizov (Springer Verlag, 1991).
10. K.P. Gentry, T. Gredig and I.K. Schuller, *Phys. Rev. B* **80**, 174118 (2009).

Mater. Res. Soc. Symp. Proc. Vol. 1245 © 2010 Materials Research Society 1245-A16-09

Low Temperature Dopant Activation Using Variable Frequency Microwave Annealing

T. L. Alford, K. Sivaramakrishnan, and A. Indluru
School of Mechanical, Chemical, and Materials, Arizona State University, Tempe, AZ 85287
School of Electrical Engineering, Arizona State University, Tempe, AZ 85287

Iftikhar Ahmad and R. Hubbard
Lambda Technologies, Morrisville, NC 27560

N.D. Theodore
Freescale Semiconductor Inc., 2100 East Elliot Rd., Tempe, AZ 85284

ABSTRACT

Variable frequency microwaves (VFM) and rapid thermal annealing (RTA) were used to activate ion implanted dopants and re-grow implant-damaged silicon. Four-point-probe measurements were used to determine the extent of dopant activation and revealed comparable resistivities for 30 seconds of RTA annealing at 900 °C and 6-9 minutes of VFM annealing at 540 °C. Ion channeling analysis spectra revealed that microwave heating removes the Si damage that results from arsenic ion implantation to an extent comparable to RTA. Cross-section transmission electron microscopy demonstrates that the silicon lattice regains nearly all of its crystallinity after microwave processing of arsenic implanted silicon. Secondary ion mass spectroscopy reveals limited diffusion of dopants in VFM processed samples when compared to rapid thermal annealing. Our results establish that VFM is an effective means of low-temperature dopant activation in ion-implanted Si.

INTRODUCTION

Goals established in the current International Technology Roadmap for Semiconductors (ITRS) anticipate physical gate lengths of silicon-based devices at 13 nm by 2013 [1]. These smaller feature sizes are instrumental in achieving faster devices. One major undesirable effect of reduced junction depths is increased contact resistance. To counteract this effect, shallow implants are done to elevate the concentrations (10^{20} cm^{-3}) of electrically active dopants; thus, reducing the contact resistance and compensating for the effects of the reduction of device feature sizes [2].

Another adverse effect of high dose, low energy ion implantation is the increased amount of nuclear energy loss deposited into the near surface region of the silicon substrate [2]. This large amount of lattice damage results in increased sheet resistance. To repair the damage and activate the dopants, high temperature annealing is used. During a high temperature anneal, thermally activated processes (*e.g.*, diffusion) can take place [2]. To lessen the extent of dopant diffusion during heating, rapid thermal annealing (RTA) is done with the use of lasers and/or lamps [3]. A shortcoming of both of these techniques is that they create uneven heating due to emissivity differences in near-surface device materials and because the photons used in lamp and laser heating cannot penetrate past the near surface region of silicon [4].

Earlier work shows that single frequency microwave annealing can activate dopants and remove radiation induced damage in boron implanted Si [5]. Alford *et al.* have shown that microwave annealing can repair lattice damage and activate As dopants at temperatures as low as 500 °C [6]. Conventional single-frequency microwave (SFM) annealing generates standing waves (*i.e.*, nodes and antinodes) throughout the processing volume and results in non-uniform heating. As a means to thwart this adverse effect of SFMs, variable frequency microwave (VFM) annealing is used. This technique sweeps through a bandwidth of 4096 frequencies over a short time period (*e.g.*, 0.1 second). During VFM, any standing waves generated have typical resident time on the order of 25 μs. This allows for VFM to result in very uniform heating when compared to SFM and also does not allow any charge buildup on metal circuitry, thereby eliminating arcing and damage.

This paper shows that VFM is an effective annealing tool for removing implantation damage and activating dopants, comparable to RTA in terms of effectiveness, but reducing diffusion of dopants significantly.

EXPERIMENTAL

Single crystal *p*-type boron doped (001) Si wafers were cleaned using the Radio Corporation of America procedure. Wafers were mounted into the implanter using back-side thermal conductive paste in order to minimize wafer heating during ion implantation. The actual ion implants took place at room temperature with 30 keV arsenic ions with doses ranging from 0.5-5×10^{15} ions-cm^{-2}. During the implantation process, the normal of each wafer was tilted 7° from the incident beam and given a 45° plane twist in the azimuth.

All of the samples were loaded into a sample chamber with the ability to be vacuum pumped and N_2 backfilled (O_2 < 20 ppm). VFM anneals are done using a Lambda Technologies VFM System that can do frequency sweeps from 5.85–6.65 GHz while transferring 1600 Watts of power into the sample chamber. The temperature of the silicon samples is monitored as a function of time within the reactor using a calibrated IR pyrometer. The VFM anneals are able to achieve temperatures in the range of 500-600 °C. The anneal times are defined as the time period extending from the time the VFM is switched on until the time it is switched off.

To monitor the extent of dopant activation, sheet resistance measurements and Hall measurements were done using an in-line four-point-probe equipped with a 100 mA Keithley 2700 digital multimeter and an Ecopia HMS 3000 Hall measurement system, respectively. Several secondary in mass spectroscopy (SIMS) profiles were obtained in order to monitor the extent of As diffusion during VFM and RTA processing [7]. The extent of implant damage was quantified by ion channeling analysis using a 2.0 MeV He^+ analyzing beam [7]. Samples were analyzed in both random and [001] channeled orientations. Inspection of the microstructure and defects was done utilizing cross-section transmission electron microscopy (XTEM) on a Philips CM200-FEG TEM. The operating voltage was 200 kV during both bright-field and dark-field imaging.

RESULTS AND DISCUSSION

In order to ascertain the extent of As^+ implantation damage prior to and after VFM anneals, samples were monitored using ion channeling analysis. Figure 1 shows Si [001] channeling spectra. Spectrum ***d*** corresponds to the [001] channeled spectrum for un-implanted, virgin silicon. Spectrum ***b***, demonstrates that a highly damaged silicon layer exists near the surface of the silicon. Comparison of the normalized yield of spectrum ***b*** with the normalized yield of spectrum ***a*** demonstrates that the as-implanted sample contains a layer of disordered silicon. Spectrum ***c*** demonstrates that annealing for 9 minutes in VFM results in a significant reduction in the lattice damage incurred during high dose ion implantation. Comparison of the normalized yield of spectrum ***c*** with that of un-implanted silicon in ***d*** demonstrates that VFM processing results in almost complete repair of the ion implantation damage. Normalized yield comparison (χ_{min} = $yield_{channeled}$/ $yield_{ramdon}$) of spectrum ***c*** and ***d*** have values of 0.30 and 0.28, respectively. These results suggest solid phase epitaxial regrowth.

The change in sheet resistance (R_s) and carrier concentration with anneal time is shown in Fig. 2 for 35 keV, 5×10^{15} cm^{-2} As^+ implanted samples. Examination reveals that the R_s of the implanted silicon reduces dramatically after 1 minute of VFM annealing. This drop in R_s is indicative of activation of arsenic dopants. For microwave times greater than 6 minutes, the value of R_s is essentially constant. For comparison, the filled triangle (in Fig. 3) represents the data from a 30s, 900 °C conventional RTA. These findings confirm that VFM processing achieves dopant activation levels comparable to RTA.

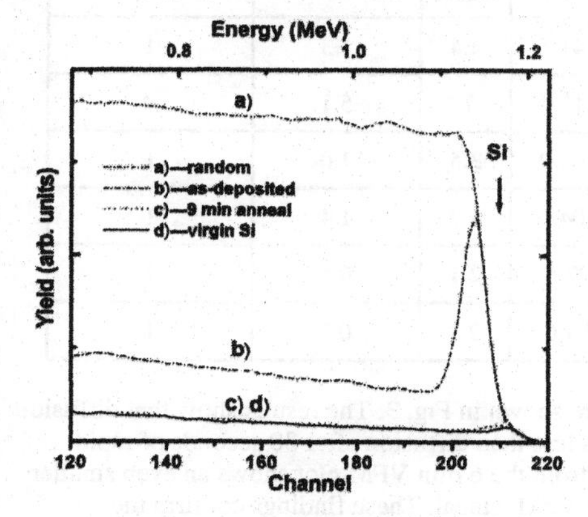

Figure 1 Backscattering spectra of 5×10^{15} As^+ cm^{-2} (focusing on Si signal): (a) as-implanted Si in the random orientation, (b) in a [001] channeled direction, (c) post 9 minute anneal in a [001] channeled direction, and (d) virgin silicon in a [001] channeled direction.

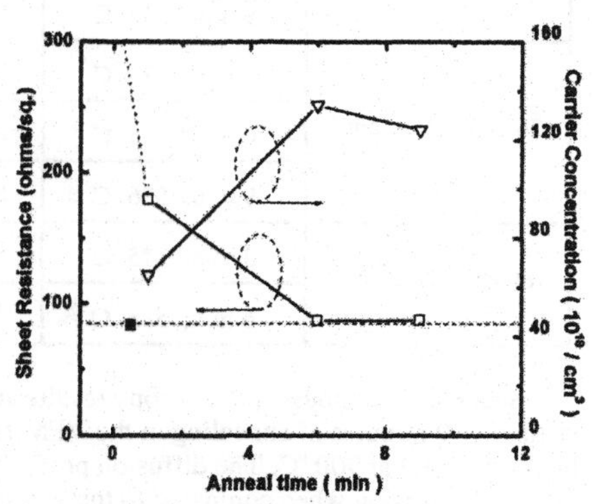

Figure 2 Sheet resistance R_s and carrier concentration values as a function of microwave anneal time. The filled triangle represents the sheet resistance of 30s, 900 °C conventional RTA sample.

Table 1 shows the typical temperature reached for each implant and the uniformity of the sheet resistance across the wafer. The standard deviation of the sheet resistance is seen to be small in all of the samples, independent of the dopant concentration. This suggests that VFMs provide uniform heating of the samples. The carrier concentration values obtained from Hall measurements show improvement with increasing VFM annealing. The resulting values after 6 minutes are consistent with conventional RTP values.

Table 1 Temperatures reached for each implant and the uniformity of the R_s across the wafer

Implantation (energy, dose)	Time, Temperature	Sheet Resistance (Ω/sq.)			
		avg	stdev	std error (±)	std error (± %)
30 keV, 5E14 cm^{-2}	3 min; 570 °C	235.8	8.9	2.6	1
	6 min; 586 °C	222.3	5.5	1.7	1
	9 min; 552 °C	222.5	6.5	1.8	1
30 keV, 1E15 cm^{-2}	3 min; 569 °C	142.1	3.0	0.9	1
	6 min; 584 °C	139.5	3.7	1.2	1
	9 min; 585 °C	141.1	4.4	1.3	1
30 keV, 5E15 cm^{-2}	3 min; 573 °C	115.7	17.1	5.1	4
	6 min; 585 °C	101.9	3.5	1.0	1
	9 min; 586 °C	104.6	4.7	1.4	1
180 keV, 1E15 cm^{-2}	3 min; 575 °C	97.6	2.1	0.6	1
	6 min; 585 °C	94.6	2.3	0.7	1

Secondary ion mass spectroscopy results are shown in Fig. 3. The results show that diffusion even after 9 minutes of annealing in the VFM is less than diffusion after 30 seconds of rapid thermal anneal at 900 °C. The diffusion profile from the 6 min VFM plot shows an even smaller extent of diffusion when compared to the 9 min VFM anneal. These findings confirm the inference from sheet resistance measurements that the VFM anneal is comparable to a 900 °C, 30s RTP anneal; however it occurs with less diffusion.

The extent of microwave-induced recrystallization is monitored by XTEM (as shown in Fig. 4). Micrographs *a*) and *b*) depict the as-implanted sample and 9 min VFM annealed sample, respectively. The first micrograph shows diffuse rings in the selected area diffraction (SAD) pattern suggesting an amorphous surface layer. The second XTEM micrograph reveals that the 9 min VFM recrystallizes the previously amorphous Si. Dopants are activated in the samples due to heating of the extrinsic Si by absorption of microwaves due to conduction and polarization losses. The high dopant concentrations lead to high electronic conduction losses while the presence of the dopant ions can lead to ionic polarization losses. The two mechanisms combine to produce a rapid heating effect. Moreover, as the activation occurs, the carriers available for conduction increase, thus increasing the rate of microwave absorption and the presence of arsenic ions in the Si matrix. This leads to a spiraling effect and very rapid activation once the initial barrier to activation is overcome. Once most of the dopants are activated, the rate of heating tapers off as most of the polarization loss mechanisms are now active and the saturation temperature is limited by the number of carriers, and ions.

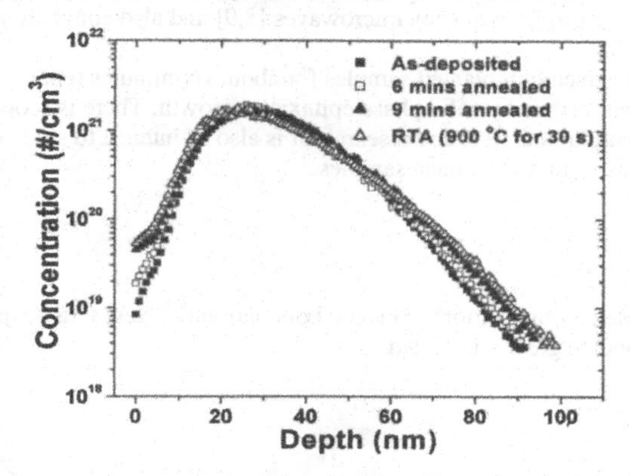

Figure 3 Secondary ion mass spectroscopy of as-implanted As^+ (filled squares), 6 minute anneal (open squares), 9 minute anneal (filled triangles), and RTA (open triangles) (900 °C for 30 seconds).

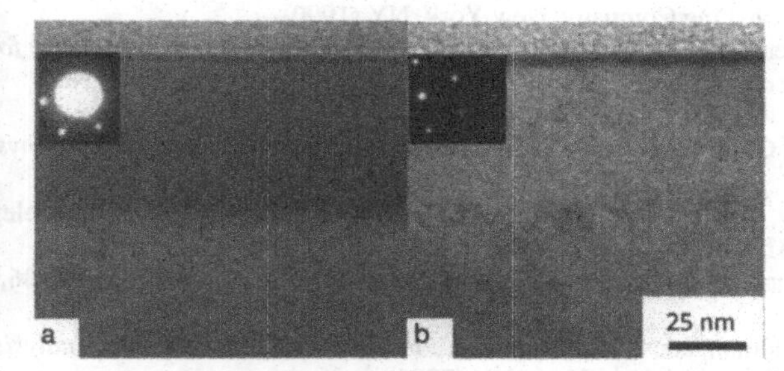

Figure 4 Cross-section TEM of (a) as-implanted sample, and (b) post 9 minute VFM anneal.

SUMMARY

Microwave processing of arsenic implanted silicon for approximately 9 minutes, completely repairs the radiation damage as monitored by XTEM and ion channeling analyses. This work establishes that VFMs can be used for repairing implantation-induced damage and to activate dopants in extrinsic silicon. Samples implanted with 5×10^{15} cm^{-2} arsenic ions were microwave processed at temperatures up to 550 °C to induce solid phase epitaxial regrowth. The results are in accord with previous research using single frequency microwaves [8,9] and also conventional rapid thermal processing. [8]

VFM annealing treatment of the arsenic implanted samples for about six minutes repairs radiation-induced damage. This repair occurs by solid phase epitaxial regrowth. There is scope for examining growth rate and activation rate, in future research. It is also of interest to study the effects of VFM annealing on low dose and low damage samples.

ACKNOWLEDGEMENTS

This work was partially supported by the National Science Foundation (L. Hess Grant No. DMR-0902277) to whom the authors are greatly indebted.

REFERENCES

1. *Technology Roadmap for Semiconductors* (Semiconductor Industry Association) (http://www.itrs.net)
2. J.W. Mayer and S.S. Lau, *Electronic Materials Science: For Integrated Circuits in Si and GaAs*, Macmillan Publishing Company, New York, NY (1990).
3. K.-N. Tu, J.W. Mayer, L.C. Feldman, and J.W. Mayer, *Electronic Thin Film Science for Electrical Engineers and Materials Scientist*, Macmillan Publishing Company, New York, NY (1992).
4. A.V. Rzhanov, N.N. Gerasimenko, S.V. Vasil'ev, and V.I. Obodnikov, *Sov. Tech. Phys. Lett.* **7**, 521 (1981).
5. P. Kohli, S. Ganguly, T. Kirichenko, H.-J. Lee, S. Banerjee, E. Graetz, and M. Shevelev, *J. Electron. Mater.* **31**, 214 (2004).
6. T.L. Alford, D.C. Thompson, J.W. Mayer, and N.D. Theodore, *J. Applied Physics* **106**, 114902 (2009).
7. T.L. Alford, L.C. Feldman, and J.W. Mayer, *Fundamentals of Nanoscale Film Analysis*, Springer Publishing Company, New York, NY (2008).
8. D.C. Thompson, J. Decker, T.L. Alford, J.W. Mayer, N.D. Theodore, *Materials Research Society Proceedings*, **0989** A06-18 (2007).
9. Y.-J. Lee, F.-K. Hsueh, S.-C. Huang, J.M. Kowalski, J.E. Kowalski, A.T. Y. Cheng, A. Koo, G.-L Luo, and C.-Y. Wu, *IEEE Electron Device Lett.* **30**, 123 (2009).

Mater. Res. Soc. Symp. Proc. Vol. 1245 © 2010 Materials Research Society 1245-A16-11

Measure of carrier lifetime in nanocrystalline silicon thin films using transmission modulated photoconductive decay

Brian J. Simonds[1,2], Baojie Yan[4], Guozhen Yue[4], Donald J. Dunlavy[3], Richard K. Ahrenkiel[2,3], P. Craig Taylor[1,2]
[1] Department of Physics, Colorado School of Mines, Golden CO 80401
[2] Renewable Energy Materials Research Science and Engineering Center, Colorado School of Mines
[3] Department of Metallurgical and Materials Engineering, Colorado School of Mines
[4] United Solar Ovonic, LLC, Troy, Michigan 48084

ABSTRACT

We present results of extremely short carrier lifetime measurements on a series of hydrogenated nanocrystalline silicon (nc-Si:H) thin films by a novel, non-destructive, non-contact method. Transmission modulated photoconductive decay (TMPCD) is a newly developed technique which appears to have high enough sensitivity and time resolution to measure the extremely short carrier lifetimes on the order of a nanosecond. As a proof of this, we measure various nc-Si:H samples of varying crystalline volume fraction as well as a fully amorphous sample. To ascribe an effective lifetime to the materials, we use a simple model which assumes a single exponential decay. By using this model, effective lifetimes can be deconvoluted from our pump beam giving nanosecond lifetimes. Lifetimes of between 1.9 and 0.9 nanoseconds are reported and trend to decreasing lifetimes as crystalline volume fraction is increased.

INTRODUCTION

Minority carrier lifetime has long been an important measure in discerning the quality of solar cell absorber materials. As these materials have shifted from bulk silicon which has a high mobility and long carrier decay lifetimes to thin film materials, which enjoy neither of these traits, it has become more difficult to measure carrier lifetime. Typical methods of measuring photoconductive carrier kinetics in the past include time-of-flight experiments [1-3] as well as several other transient photocurrent measurements. Both of these experiments require that contact is made to the material through Schottky contacts and that the sample is biased. Other techniques have been developed which allow for transient photoconductivity to be measured without contacts, such as resonant coupled photoconductive decay (RCPCD)[4] and microwave photoconductive decay (μ-PCD) [5]. Both of these measurements are available in our lab and we have seen that transmission modulated photoconductive decay (TMPCD) appears to combine the faster time response of μ-PCD with the sensitivity of RCPCD.

Nanocrystalline silicon is a thin film material of interest to the solar cell community as a possible absorber layer in a thin film solar cell. It is deposited using plasma enhanced chemical vapor deposition, a technique long familiar to the amorphous thin film community. It is a promising material as it appears to combine strengths of both amorphous and crystalline silicon and is less sensitive to the light induced degradation seen in amorphous silicon. Recently United Solar Ovonics, Inc. has reported a current world record nc-Si:H solar cell with a stable efficiency of 13.5%. [6]

EXPERIMENT

TMPCD is a novel technique developed for measuring minority carrier lifetimes in bulk semiconductors. Improvements have been made in TMPCD specifically aimed at improving time resolution in order to measure and study the carrier decay in nc-Si:H as well as other nanostructured and thin film semiconductor materials. The experimental setup is shown in figure 1 and is discussed in more detail in reference [7]. The basic mechanism of detection is based on

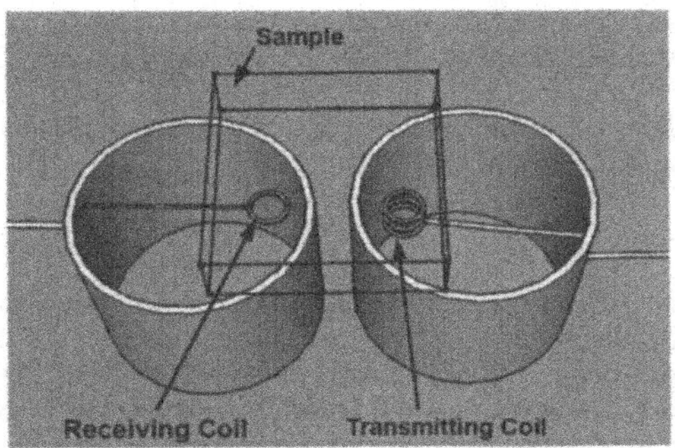

Figure 1: Schematic of active area of TMPCD with sample bridging transmission and receiving coil.

the fact that the sample acts as a conductive bridge for an alternating RF field (~500MHz) produced by the transmitting coil. The RF conductivity is dependent upon the number of free carriers in the sample, which are produced by the pump beam (Q-switched doubled YAG laser at 532nm, pulse width 5.4ns). In this study, we are using an energy density on the order of 10 mW/cm^2 which gives very high generation rates (~10^{28} cm^{-3} s^{-1}). The key to the fast response of the TMPCD experiment is the use of a second, low-Q detection coil which is shielded from the transmission coil. The receiving coil is a single coil inductor with a 43 Ω series resistor. This receiving circuit has a measured inductance of 14 nH and a measured impedance of 73 Ω which gives a circuit response time of about 200 picoseconds.

The nc-Si:H samples were provided by United Solar Ovonics, Inc. and were prepared by PECVD at varied hydrogen dilution ratios. Through a combination of small angle x-ray scattering and x-ray diffraction techniques, it has been determined that these films contain small, elongated crystallites that are approximately 6 nm by 20 nm in size, which are dispersed approximately uniformly throughout the sample [8]. The crystalline volume fraction of our samples was determined by Raman spectroscopy using a commercial confocal micro-Raman apparatus made by WiTec Inc. Raman data on the samples is shown in figure 2. In using a confocal geometry care was needed to ensure that no further crystallization of the samples occurred with the laser beam focused on the sample. An amorphous silicon sample was used to determine a threshold intensity above which the laser would begin to crystallize the sample as determined by the appearance of a Raman peak at 520 cm^{-1}. To ascribe a crystalline volume fraction to the samples we used the standard method of fitting the data between about 450 cm^{-1} and 550 cm^{-1} with three Gaussian functions [9]: one centered at 480 cm^{-1} representing the amorphous phase, one around 510 cm^{-1} for the crystalline phase, and another near 500 cm^{-1}, which is commonly referred to as a "grain boundary" phase. We determined the crystalline volume fraction using the common method of a ratio of the integrated areas of these peaks. There is widespread debate concerning the accuracy of this fitting technique, as well as others, for determining crystalline volume fraction in nanocrystalline materials [10]. However, for the present work this procedure is sufficient to determine the correct trend and, to a satisfactory degree of accuracy, an estimate of the crystalline volume fraction. The spectra in figure 2 corroborate this assertion as one can easily see that the intensity of the crystalline peak grows with the crystalline volume fraction. Throughout this work we refer to the samples by their crystalline volume fraction which is 0% or completely amorphous, 55%, 75%, and 77%.

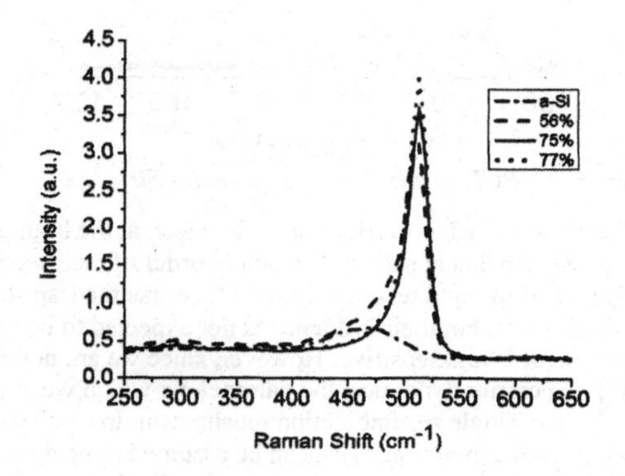

Figure 2: Raman results for 3 nc-Si:H materials with a-Si:H sample for reference.

RESULTS AND DISCUSSION

The results of our TMPCD measurements are shown in figure 3. The signal is a transient response of the photoconductivity decay following a short (~5.4 ns) pulse. As a result of the pump pulse duration being of significant duration compared to the expected photoconductive decay time in these samples it is necessary to deconvolve this signal. To do so, the pulse profile was measured by a fast *pin* photodiode (Hamamatsu model S5973). The beam was measured to be approximately Gaussian with a 5.4 ns FWHM. This fit was then used in the data analysis. As a simple approach, we model the data with a single exponential decay of photoexcited carriers convolved with a Gaussian pump pulse:

$$\Delta\sigma(t) \propto e^{\frac{-t^2}{c^2}} * e^{\frac{-t}{\tau}} \quad (1)$$

The exponential decay is parameterized by a single time constant τ which is related to the characteristic time a photoexcited carrier exists before recombination. We admit that this approach is an oversimplification of the very complex recombination that occurs in these materials. This complexity is due to the nature of their density of states that includes a substantial

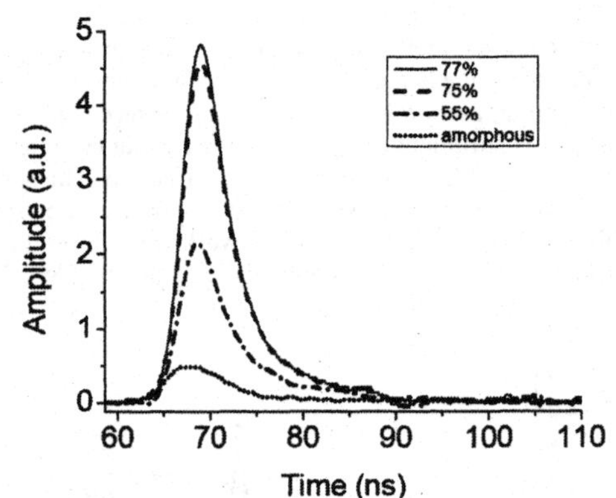

Figure 3: TMPCD results for nc-Si:H and a-Si:H.

distribution of localized states, which act as carrier traps. However, at the high generation rates used here (~10^{28} cm^{-3} s^{-1}) we assume that at short times on the order of nanoseconds we are seeing mainly non-geminate, band-to-band recombination. Of course the trap states most certainly play a role in recombination but their influence is not expected to be dominate until later times when our measurement is less sensitive. However, since we are not explicitly eliminating the effect of trapped carrier dynamics, the parameter τ which we extract cannot be viewed as a lifetime related to any single recombination mechanism. Instead, it should be viewed as an effective lifetime that, at high generation rates, can be assumed to be dominated by direct recombination but still influenced by longer trap-and-release mechanisms.

As a consequence of the convolution theorem it is most natural for our data to be analyzed in frequency space which turns a convolution into simple multiplication. Thus, we actually fit the discrete Fourier transform (FT) of our data to the Fourier transform of equation (1) which is given below in equation (2). The FT of our Gaussian pulse gives another Gaussian and the FT of the exponential decay gives a Lorentzian.

$$\Delta\sigma(\omega) \propto e^{\frac{-\omega^2}{c'^2}} \bullet \frac{2\tau'}{4\pi\omega^2 + \tau'^2} \qquad (2)$$

The parameter c' is related to the FWHM of our measured pulse and is held fixed. The parameter τ' is actually the inverse of the time constant in the time domain and is used as a free parameter in a non-linear least squares fitting routine.

Examples of the discrete FT of our data and the results of least squares fitting are shown in figures 4a and 4b. Part b shows the data from a-Si:H which has a computed lifetime τ of 1.94 ns. In figure 4a we show the fit to the nc-Si:H with 77% crystalline volume fraction. In comparing the two data sets one notices that although both fits are reasonable, in the nc-Si:H sample there are higher frequency 'wings' which are not accounted for by our simple, single exponential model. We see this trend in all nc-Si:H samples. This behavior is an indication that the recombination process is not accurately accounted for by a single exponential decay. This

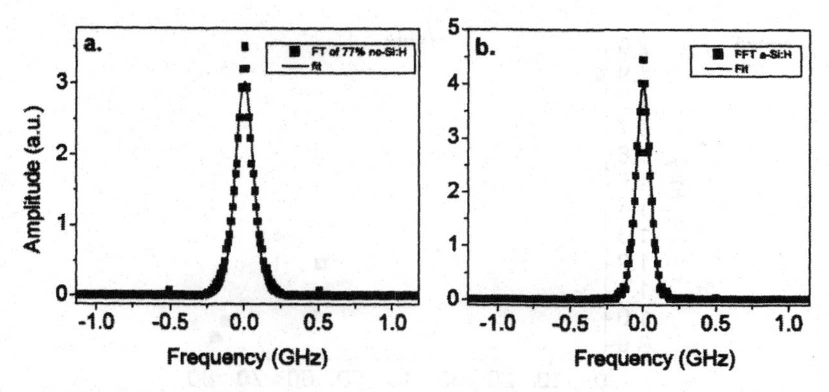

Figure 4: Data fits to FT of data using equation (1). Panel a is 77% nc-Si:H and b is a-Si:H.

result is not entirely surprising as the complex structure of nc-Si:H will lead to a complicated density of states with a series of recombination mechanisms.

Figure 5 shows the lifetime result for all 4 materials plotted versus crystalline volume fraction. In going from completely amorphous to 77% crystalline nc-Si:H there is about a factor of 2 decrease in the carrier lifetime calculated from TMPCD. This trend supports the assertion that the complex structure introduced by adding more crystallites strongly influences the density of states that control the recombination in these materials. The fact that the lifetimes decrease with increasing crystallinity suggests that it is the surface defects at the grain boundary regions, which act as the dominate recombination sites in these materials. Recent work in time-resolved terahertz spectroscopy (TRTS) is consistent with these results [11]. Although TRTS has the capability of resolving much faster decay processes, on the order of picoseconds, there is a range of overlapping information between the two techniques in the nanosecond regime. In addition, although there is only about a 2% change in crystallinity from the highest two nc-Si:H samples, there is about a 7% change in the carrier lifetime. This result suggests that the carrier lifetime could be more sensitive to subtle changes in the microstructure of these films than is Raman spectroscopy.

CONCLUSIONS

It has been shown that transmission modulated photoconductive decay is an effective technique to measure the short transient carrier decay in nc-Si:H thin films. We have modeled this decay by a single exponential, which we parameterize by a time constant τ. For these films, τ is on the order of a nanosecond. As the films increase in crystalline volume fraction the carrier lifetimes decrease by a factor of about 2, which we attribute to an increase in recombination sites located at grain boundaries. The carrier lifetimes are more sensitive to subtle changes of microstructure than is crystalline volume fraction as determined by Raman spectroscopy. Further work is needed to better characterize the carrier dynamics in these films with TMPCD because the simple model of a single exponential does not accurately account for all of the details of the experiment.

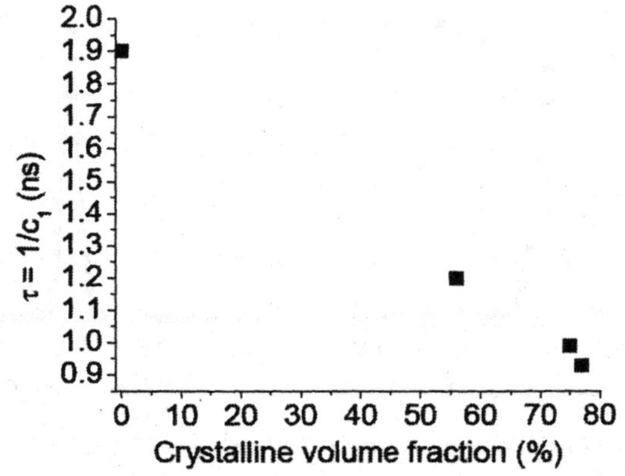

Figure 5: Extracted lifetime parameter from data fits based on equation 1 as a function of crystalline volume fraction.

ACKNOWLEDGEMENTS

This research is supported by the Solar America Initiative of DOE (sub-contract #DE-FC36-07G017053) through United Solar Ovonics, LLC as well as through the Renewable Energy Materials Research Science and Engineering Center at The Colorado School of Mines funded by NSF (DMR0820518).

REFERENCES

1. M. Brinza, Guy J. Adiaenssens, K. Iakoubovskii, A. Stesmans, W.M.M. Kessels, A.H.M. Smets and M.C.M. Van de Sanden, Journal of Non-Crystalline Solids, v 299-302, p 420-424 (2002).
2. T. Dylla, S. Reynolds, R. Carius and F. Finger, Journal of Non-Crystalline Solids, v 352, p 9-20 (2006).
3. R.I. Devlen, J. Tauc and E.A. Schiff, Journal of Non-Crystalline Solids, v 114, n Pt2, p 567-569 (1989).
4. R.K. Ahrenkiel and S.W. Johnston, Mater. Sci. Eng. B102, 161-172 (2003).
5. C. Swiatkowski, M. Kunst, Appl. Phys. A 61, 623-629 (1995).
6. G. Yue, L. Sivec, B. Yan, J. Yang, and S. Guha, Mater. Res. Soc. Symp. Proc. 1245, A21.1 (2010).
7. Richard K. Ahrenkiel and Donald J. Dunlavy, Sol. Energy Mater. Sol. Cells (2010).
8. Kristin Kiriluk, Don Williamson, David Bobela, P. Craig Taylor, Baojie Yan, Jeffrey Yang and Subhendu Guha, Mater. Res. Soc. Symp. Proc. 1245, 13.2 (2010).
9. E. Bustarret, M. Hachicha and M. Brunel, Appl. Phys. Lett. 52 (1988).
10. Ch. Ossadnik, S. Veprek and I. Gregora, Thin Film Solids 337 (1999).
11. L. Fekete, P. Kuzel, H. Nemec, F. Kadlec, A. Dejneka, J. Stuchlik and A. Fejfar, Phys. Rev. B 79 (2009).

Mater. Res. Soc. Symp. Proc. Vol. 1245 © 2010 Materials Research Society 1245-A16-13

Molecular Dynamics Modeling of Stress and Orientation Dependent Solid Phase Epitaxial Regrowth

Haoyu Lai[0], Stephen M. Cea[0], Harold Kennel[0] and Scott T. Dunham[0]
[1]Electrical Engineering, University of Washington, Seattle, Washington, USA.
[2]Design and Technology Solutions, Intel Corporation, Hillsboro, Oregon, USA.

ABSTRACT

Solid Phase Epitaxial Regrowth (SPER) is of great technological importance in semiconductor device fabrication. A better understanding and accurately modeling of its behavior are vital to the design of fabrication processes and the improvement of the device performance. In this paper, SPER was modeled by Molecular Dynamics (MD) with Tersoff potential. Extensive MD simulations were conducted to study the dependence of SPER rate on growth orientation and uniaxial stress. The results were compared with experimental data. It was concluded that MD with Tersoff potential can qualitively describe the SPER process. For a more quantitatively accurate model, a better interatomic potential are needed.

INTRODUCTION

Solid Phase Epitaxial Regrowth (SPER), the epitaxial recrystallization of amorphous Si in solid phase, is used extensively in semiconductor device fabrication to repair the crystal damage caused by ion implantation as well as incorporate and activate the implanted dopants. In many cases, it can even be used to obtain solid solution beyond the dopant's solubility. However, it is well known that this process also generates defects such as stacking faults, dislocation loops and micro twins [1]. These defects can act as scattering centers or electron-hole recombination centers and severely degrade device performance [2, 3]. In addition, dopants diffuse differently in crystalline Si, amorphous Si and at the amorphous/crystal (a/c) interface. As a result, the final dopant profile, which is very critical to nanoscale device structures such as ultra-shallow junctions, can also be altered. Therefore, a better understanding of SPER and accurately modeling of its behavior are vital to the design of fabrication processes and the improvement of the device performance.

It has been demonstrated by experiments that the rate of SPER depends on temperature [4], orientation [5], uniaxial stress [6, 7] and pressure [8]. Combinations of these factors may result in many critical phenomena that affect the device performance. For example, the orientation dependence leads to mask edge defect formation [9, 10], augmented by the dependence on local curvature of the a/c interface which may originate from the stress dependence.

In this paper, Molecular Dynamics (MD) is used to model and study two of the four important aspects of SPER: orientation and uniaxial stress dependence. MD is a powerful computer simulation tool. With the proper interatomic potential, it can be used to study the dynamics of the SPER in atomistic detail and provide physical insight. It also can be used to extract parameters for higher level modeling.

INTERATOMIC POTENTIAL USED IN MD

There are three types of interatomic potentials for MD simulation: *ab-initio*, semi-empirical and empirical. They provide quite different levels of accuracy and in turn, have much different computational requirements that limit the system size and the length of the time evolution. The study of silicon SPER involves thousands of atoms and a time scale up to 100ns, which means that empirical potentials are the only feasible choice.

Empirical potentials usually consist of simple analytic functionals of atom positions. These functionals are often ad hoc with little physical insight. Thus empirical potentials usually have limited transferability and are only valid in applications related to what their parameters are fitted for. To better understand and model SPER, it is critical that a suitable potential is used.

There are many interatomic potentials that have been developed for silicon. Six of them are widely used, including the original Stillinger-Weber parameterization (SW) [11], the re-parameterized SW (SW115) [12], the third Tersoff parameterization (T3) [13], the environment dependent interatomic potential (EDIP) [14], the Lenosky modified embedded atom method (MEAM) [15], and the bond order potential (BOP) [16]. Krzeminski *et al.* [17] did a comprehensive study comparing the first five potentials on their ability to model SPER in <001> direction and Gillespie *et al.* [18] did a similar study on BOP. These work demonstrated that SW, EDIP, MEAM and BOP transformed the original amorphous Si into liquid Si instantly and became liquid-phase epitaxial regrowth (LPER) instead of SPER. Among the other two potentials, T3 gave an activation energy of 2.99eV, much closer to the value of 2.70eV obtained by measurement [4] than the SW115's 1.87eV. Also the absolute growth rate of T3 is on the same order of magnitude as the measurement, while that of SW115 is two orders of magnitude faster. Therefore, the T3 is chosen in this study.

SPER MD SIMULATION SETUP

The system used for SPER MD simulation consists of four regions as shown in Figure 1. The bottom 2 monolayers are designated as substrate region. The atoms in this region are held fixed in the perfect crystalline positions. They serve as a bulk substrate to stabilize the system above it.

The 10 monolayers above the substrate region make up the heat bath region. The temperature of this region is regulated by a thermostat to the desired value that serves as the substrate temperature.

Above the heat bath region is the crystal region. It consists of 4 to 8 monolayers depending on the melting-quenching process described below. This region serves as the buffer zone to reduce the artificial effect of the thermostat on the kinetics of the a/c interface, especially during the crystallization of the first several monolayers.

On the top of the crystal region is the regrowth region. Initially, it is mostly made up of amorphous Si and contains the a/c interface. The time evolution of this interface is used to determine the growth rate. The number of atoms in this region is equivalent to the atom number in 24 to 28 monolayers of crystalline Si.

Periodic boundary conditions are applied in x and y directions, whereas there is a free surface in z direction with a reflective wall far above the top of the regrowth region in case any atom sublimes.

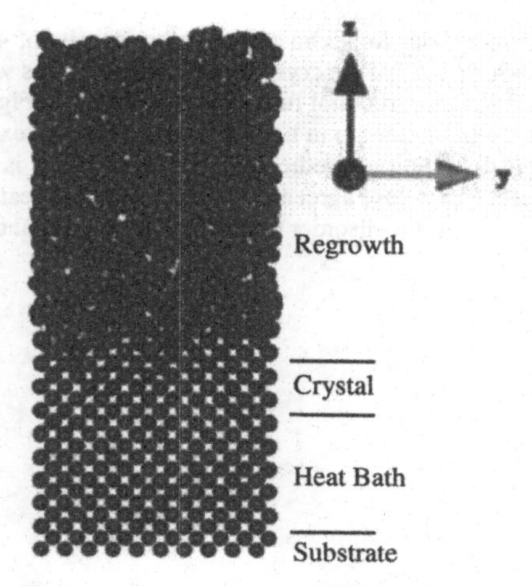

Figure 1. System setup for SPER MD.

GENERATION OF AMORPHOUS SI

The amorphous Si is generated by quenching liquid Si to solid state. First, atoms in the regrowth region are randomly displaced from their perfect crystal position with a maximum displacement of 0.5Å. The potential energy generated by this displacement converts to kinetic energy large enough to melt the crystalline Si. Then, the regrowth region is cooled down by keeping the heat bath region at a low temperature to obtain the amorphous Si.

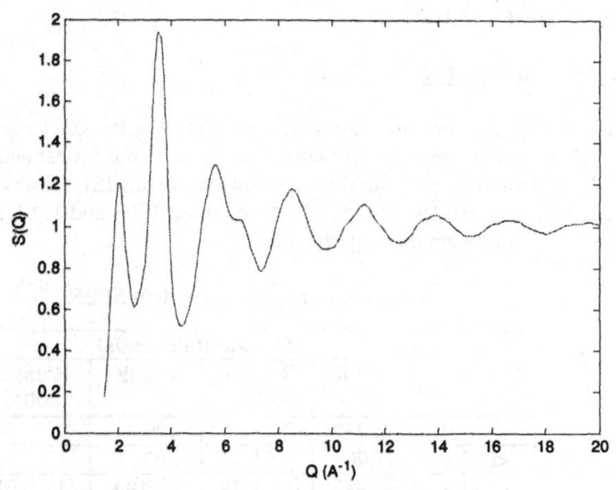

Figure 2. Structural factor S(Q) of amorphous Si.

The resulting amorphous Si has an atom number density of 4.86e22 cm^{-3}, a 2.61% reduction from the 4.99e22 cm^{-3} of the crystalline Si which agrees with the 1.76% reduction from the experiment [19]. The structural factor S(Q) is shown in Figure 2, the peak positions and magnitudes of the structural factor are in line with those from the experiments [20, 21].

The bond angle distribution together with the Gaussian fit is plotted in Figure 3. It has a mean bond angle of 108.5°, in close agreement with the 108.6° mean angle in deposited amorphous Si [20]. However, the disorder width is 14.3°, larger than the 11.0° from experiment [20].

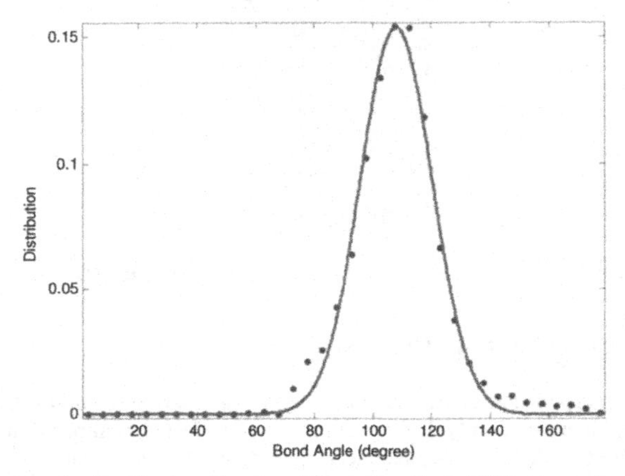

Figure 3. Bond angle distribution (dots) and Gaussian fit (solid) in amorphous Si.

RESULTS AND DISCUSSION

Orientation Dependent SPER

Three different growth orientations were studied: <001>, <011> and <111>. The results are shown in Table 1. The growth rate in <001> direction is the fastest and the <111> direction is the slowest. This is in agreement with the experimental results [5]. However, the ratios of growth rates between <001> and <011>, <111>, which are about 1.7:1 and 1.9:1, respectively, are far less than the 3:1 and 20:1 from the experiments [5].

Table 1. Orientation dependent SPER

Direction	Growth Rate (cm/s)			
	1800K	Ratio to <001>	1900K	Ratio to <001>
<001>	62.272		179.41	
<011>	37.908	1 : 1.64	102.18	1 : 1.76
<111>	34.987	1 : 1.78	92.194	1 : 1.95

From Table 1, it also can be seen that the growth rate ratios are not the same at different temperature. This may suggest an orientation-dependent activation energy for SPER. More simulations are needed to verify this.

Uniaxial Stress Dependent SPER

The uniaxial stress dependent SPER was done in the <001> direction at 1700K. The stress was applied in the x direction while keeping the y and z directions stress free.

The MD results of the uniaxial stress dependent SPER are shown in Figure 4. There were not defects found in the crystallized regions for all stress values. For tensile stress, the growth rate can be regarded as constant, which agrees with the experiment done by Rudawski [6] but not the experiment done by Aziz [7]. When compressive stress is applied, the growth rate decreases, which agrees with both experiments. Also, this part of the data can be fitted with either the Arrhenius-type (used by Aziz) or the Rudawski type function. However, the rate reduction is only about 13% at -1GPa, much smaller than the 50% by Rudawski [6] and 30% by Aziz [7]. One reason for this may be that in both experiments, the stress was applied through bending the samples [7, 22]. Such stress is not uniform throughout the sample. In addition, bending also creates shear stress that may change the growth rate. All these effects were not included in the MD simulation.

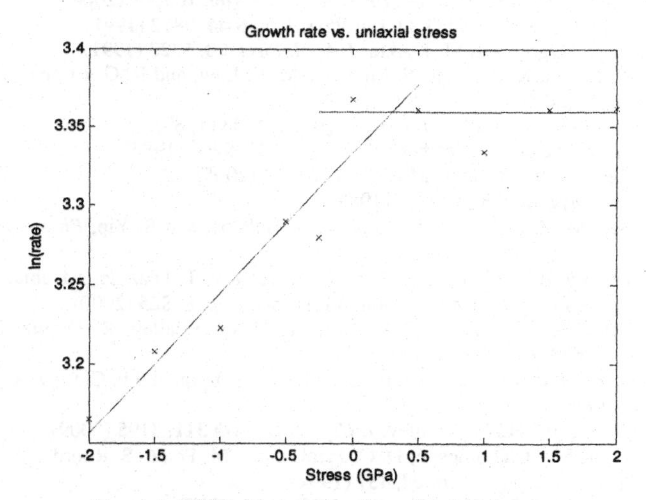

Figure 4. SPER rate as a function of uniaxial stress.

CONCLUSIONS

The SPER process has been modeled using MD simulation. Based on the studies of temperature dependent SPER rate in <001> direction available in the literature, the T3 empirical interatomic potential was chosen. The initial amorphous Si and a/c interface were created using the method of melting-quenching. The properties of the amorphous Si were analyzed and found to be in good agreement with experimental data. The SPER rate was studied as a function orientation and uniaxial. The regrowth was fastest in <001> direction and slowest in <111> direction. The ratios of <001> to <011> and <111> were 1.7:1 and 1.9:1 respectively. The regrowth rate remained nearly constant for tensile uniaxial stress and decreased as the compressive uniaxial stress increased, with a 13% rate reduction at -1GPa observed. Compared to the experimental data, these results indicate that MD with Tersoff (T3) potential can qualitatively model SPER. However, for a more quantitatively accurate model, a better interatomic potential are needed.

REFERENCES

1. H. Cerva and K. H. Kusters, *J. Appl. Phys.* **66**, 4723 (1989).
2. E. Simoen, M. B. Gonzalez, B. Vissouvanadin, M. K. Chowdhury, P. Verheyen, A. Hikavyy, H. Bender, R. Loo, C. Claeys, V. Machkaoutsan, P. Tomasini, S. Thomas, J. P. Lu, J. W. Weijtmans, and R. Wise, *IEEE Trans. Electron Devices* **55**, 925 (2008).
3. J. P. Liu, J. Li, A. See, M. S. Zhou, and L. C. Hsia, *Appl. Phys. Lett.* **90**, 261915 (2007).
4. J. A. Roth and G. L. Olson, *Appl. Phys. Lett.* **57**, 1340 (1990).
5. L. Csepregi, E. F. Kennedy and J. W. Mayer, *J. Appl. Phys.* **49**, 3906 (1978).
6. N. G. Rudawski and K. S. Jones, *Phys. Rev. Lett.* **100**, 165501 (2008).
7. M. J. Aziz, P. C. Sabin and G. Q. Lu, *Phys. Rev. B* **44**, 9812 (1991).
8. G. Q. Lu, E. Nygren and M. J. Aziz, *J. Appl. Phys.* **70**, 5323 (1991).
9. N. G. Rudawski, K. S. Jones, S. Morarka, M. E. Law, and R. G. Elliman, *J. Appl. Phys.* **105**, 081101 (2009).
10. H. Cerva and K. H. Kusters, *J. Appl. Phys.* **66**, 4723 (1989).
11. F. H. Stillinger and T. A. Weber, *Phys. Rev. B* **31**, 5262 (1985).
12. E. J. Albenze and P. Clancy, *Mol. Simul.* **31**, 11 (2005).
13. J. Tersoff, *Phys. Rev. B* **38**, 9902 (1988).
14. J. F. Justo, M. Z. Bazant, E. Kaxiras, V. V. Bulatov, and S. Yip, *Phys. Rev. B* **58**, 2539 (1998).
15. T. J. Lenosky, B. Sadigh, E. Alonso, V. V. Bulatov, T. Diaz de la Rubia, J. Kim, A. F. Voter, and J. D. Kress, *Model. Simul. Mater. Sci. Eng.* **8**, 825 (2000).
16. B.A. Gillespie, X.W. Zhou, D.A. Murdick, H.N.G.Wadley, R. Drautz, D.G. Pettifor, *Phys. Rev. B* **75**, 155207 (2007).
17. C. Krzeminski, Q. Brulin, V. Cuny, E. Lecat, E. Lampin, and F. Cleri, *J. Appl. Phys.* **101**, 123506 (2007).
18. B.A.Gillespie and H.N.G.Wadley, *J. Crystal Growth* **311**, 3195 (2009).
19. J. S. Custer, M. O. Thompson, D. C. Jacobson, J. M. Poate, S. Roorda, W. C. Sinke and S. Spaepen, *Appl. Phys. Lett.* **64**, 437 (1994).
20. J. Fortner and J. S. Lannin, *Phys. Rev. B* **39**, 5527 (1989).
21. K. Laaziri, S. Kycia, S. Roorda, M. Chicoine, J. L. Robertson, J. Wang and S. C. Moss, *Phys. Rev. Lett.* **82**, 3460 (1999).
22. N. G. Rudawski and K. S. Jones, *Appl. Phys. Lett.* **91**, 172103 (2007).

Mater. Res. Soc. Symp. Proc. Vol. 1245 © 2010 Materials Research Society 1245-A16-18

Self-catalyzed Tritium Incorporation in Amorphous and Crystalline Silicon

Baojun Liu[1], Nazir P. Kherani[2], Kevin P. Chen[1], Tome Kosteski[2], Keith Leong[2] and Stefan Zukotynski[2]
[1]Department of Electrical and Computer Engineering, University of Pittsburgh, Pittsburgh, PA 15261, U.S.A.
[2]Department of Electrical and Computer Engineering, University of Toronto, Toronto, ON M5S 3G4, Canada

ABSTRACT

Tritiated amorphous and crystalline silicon is prepared by exposing silicon samples to tritium gas (T_2) at various pressures and temperatures. Total tritium content and tritium concentration depth profiles in the tritiated samples are obtained using thermal effusion and Secondary Ion Mass Spectroscopy (SIMS) measurements. The results indicate that tritium incorporation is a function of the material microstructure rather than the tritium exposure condition. The highest tritium concentration attained in the amorphous silicon is about 20 at.% on average with a penetration depth of about 50 nm. In contrast, the tritium occluded in the c-Si is about 4 at.% with a penetration depth of about 10 nm. The tritium concentration observed in a-Si:H and c-Si is higher than reported results from post-hydrogenation experiments. The beta irradiation appears to catalyze the tritiation process and enhance the tritium dissolution in silicon material.

INTRODUCTION

Tritiated hydrogenated amorphous silicon (a-Si:H:T) is a candidate nuclear fuel for chip-scale radioisotope micropower sources (RIMS) [1]. The preparation of a-Si:H:T was previously attained by introducing tritium gas in a silane glow discharge DC saddle field plasma enhanced chemical vapor deposition (PECVD) process [2], and more recently using a simple and versatile method where both hydrogenated amorphous silicon (a-Si:H) and crystalline silicon (c-Si) were tritiated through direct exposure to tritium gas (T_2) [3]. In this paper, we report the tritium effusion and SIMS (Secondary Ion Mass Spectroscopy) study of tritiated silicon through the post-tritiation process of both amorphous and crystalline silicon, and further clarify the tritiation mechanism.

EXPERIMENT

Three difference amorphous silicon films were deposited on c-Si substrates, at various substrate temperatures, using the dc saddle-field PECVD technique. The substrate temperature, thickness of the film and hydrogen concentration of the a-Si:H films are listed in Table I. The a-Si:H samples, as well as the c-Si samples, were exposed to tritium gas at pressures up to 120 bar and temperatures up to 250 °C. Thermal effusion measurements were conducted in order to determine the total tritium content, and hence the atomic concentration, and its bonding characteristics [4]. SIMS (Secondary Ion Mass Spectroscopy) was carried out to determine the tritium depth profiles.

Table I. Substrate temperature, thickness and hydrogen concentration in the a-Si:H films.

Sample Name	aSiH1	aSiH2	aSiH3
Substrate Temperature (°C)	115	200	315
Thickness of Film (μm)	0.6	0.3	2.0
Hydrogen Concentration (at.%)	35	20	17

RESULTS

The a-Si:H samples deposited under different conditions were exposed to tritium gas for 84 hours at a temperature of 250 °C and a pressure of 120 bar. Thermal effusion measurements were then carried out to determine the tritium concentration and bonding characteristics. Similar profiles are observed for all three samples prepared at different deposition conditions. Figure 1 shows a typical tritium effusion profile and tritium evolution rate as a function of temperature for sample aSiH2.

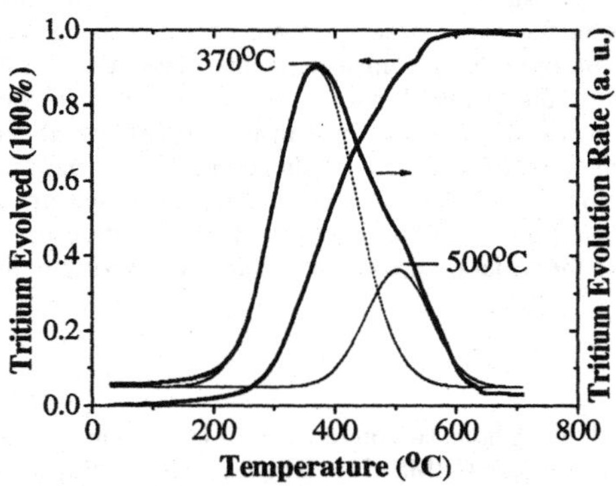

Figure 1 Cumulative tritium effusion and rate of tritium effusion as a function of temperature for sample aSiH2 exposed to tritium gas at 120 bar and 250 °C for 84 hours. The dashed lines show the effusion peaks obtained by Gaussian deconvolution; the corresponding peak temperatures are indicated.

The profile shows that significant tritium effusion commences above the loading temperature of 250 °C, and that the majority of the tritium has evolved out of the sample before it reaches 700 °C. Gaussian deconvolution was employed to analyze the tritium evolution rate profile. For all three tritiated a-Si:H samples, the low-temperature (LT) peak is located around 370 °C and the high temperature (HT) peak is centered around 500 °C, which agrees with previous effusion studies on tritiated hydrogenated amorphous silicon (a-Si:H:T) prepared by the PECVD process [1, 4-5]. The LT peaks suggest the presence of higher order silicon tritides and tritium clusters, and the HT peak could be attributed predominantly to the Si-T monotritide bonds [3, 5]. By comparing the fractional areas under both HT and LT peaks, it is estimated that around 70% of tritium in a-Si:H films exists in forms of higher order silicon tritides and tritium clusters. Despite identical tritium loading conditions and similar effusion profiles, tritium concentration in the three samples are significantly different. The effusion measurement yielded tritium concentrations of 1.2, 3.3 and 0.6 mCi/cm^2 for samples aSiH1, aSiH2 and aSiH3, respectively. The a-Si:H film deposited at 200 °C yields the highest tritium concentration after the tritium exposure.

To study the influence of temperature and pressure on tritium incorporation, film samples aSiH2 were exposed to T_2 gas at two other conditions: temperature of 100 °C at a pressure of 70 bar, and temperature of 70 °C at a pressure of 2 bar. The exposure duration was maintained at 84 hours, the same as the exposure described above at 250 °C. The former condition yielded a tritium concentration of 3.1 mCi/cm^2 while the latter that of 3.3 mCi/cm^2. Despite drastically different exposure conditions, there is no significant difference in the total tritium absorbed in the film under different tritium pressures and temperatures. The results suggest that the tritium incorporation in a-Si:H is predominately determined by the material microstructure, which is determined by the deposition condition.

Considering a host which hasan extremely tight atomic structure and is free of hydrogen, c-Si was also studied using the post-tritiation process. Tritiation was performed at 250 °C and 100 bar for 132 hours. The effusion profile, shown in Figure 2, reveals a much lower tritium incorporation concentration of 0.1mCi/cm^2. Gaussian deconvolution of the effusion rate profile shows two effusion peaks at 370 °C and 480 °C. In this case, the larger effusion peak at the higher temperature indicates that tritium is predominantly bonded in the mono-tritide form.

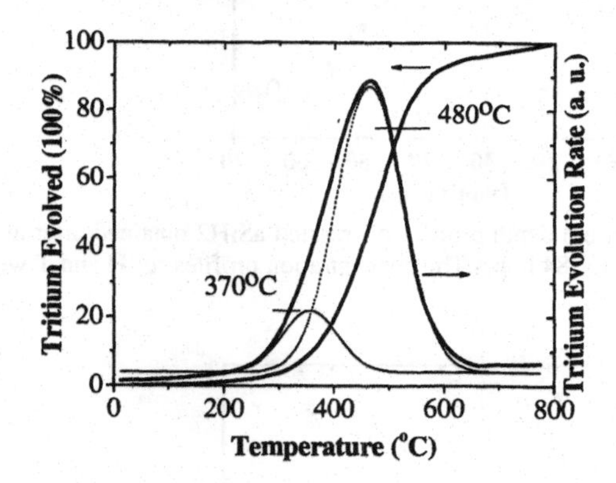

Figure 2 Cumulative tritium effusion and rate of tritium effusion as a function of temperature for crystalline silicon sample exposed to tritium gas at 100 bar and 250 °C for 132 hours. The dashed lines show the effusion peaks obtained by Gaussian deconvolution; the corresponding peak temperatures are indicated.

To study tritium diffusion profiles, SIMS measurements were performed on both tritiated a-Si:H and c-Si. Concentration profiles of sample aSiH2 for hydrogen (H), tritium (T), and the total concentration of hydrogen and tritium (H+T) are shown in Figure 3. The *a*-Si:H sample was exposed to T_2 gas at 250 °C, 120 bar for 84 hours. As shown in the figure, an extremely tritium-rich layer forms near the surface, the tritium concentration reaches 7×10^{22} cm^{-3} at a depth of 3 nm from the surface, which is likely for the most part ditritides and tritium clusters. The concentration drops rapidly moving inwards, reaching 3×10^{22} cm^{-3} at a depth of about 6 nm. The tritium concentration then falls off at a slower rate into the sample. The H loss at the surface in relation to the deeper bulk is about two-thirds. The loss of H due to out-diffusion is over-compensated by the incorporation of T and a net increase in the total hydrogen isotope (H+T) concentration of 2×10^{22} cm^{-3} is observed at a depth of 6 nm from the surface in relation to the bulk H concentration. Figure 4 shows tritium profiles in tritiated c-Si exposed to tritium gas at 250 °C and 100 bar for 132 hours. A tritium concentration of 8 at.%, with a number density of 4×10^{21} cm^{-3}, is observed near the surface; also, the tritium concentration drops off rapidly with depth. The concentration drops to $1/10^{th}$ of the surface concentration at a depth of ~5 nm. Tritium dissolution is confined with the superficial surface (~10nm) of the c-Si. The tight regular covalent atomic lattice of crystalline silicon limits tritium infiltration.

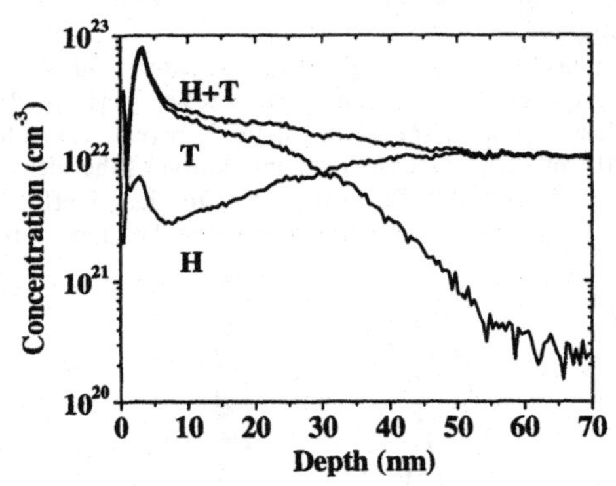

Figure 3. Hydrogen and tritium depth profiles of tritiated aSiH2 obtained after an exposure to T_2 gas at 250 °C and 120 bar for 84 hours.The concentration profiles for H and T were measured simultaneously.

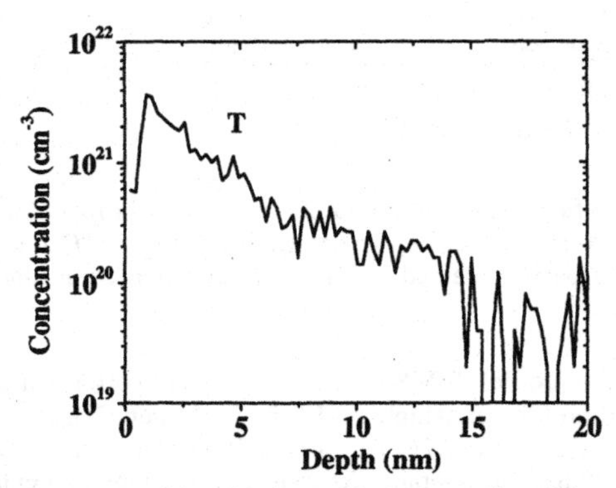

Figure 4. Tritium concentration depth profile in crystalline silicon obtained after an exposure to T_2 gas at 250 °C and 100 bar for 132 hours.

The depth profiles reveal that the tritium permeation depths in amorphous and crystalline silicon are 50 nm and 10 nm, respectively. Using the tritium content obtained from the effusion measurements, the average tritium concentrations are calculated to be 20 at. % and 4 at.%, respectively.

DISCUSSION

Tritium was incorporated in amorphous and crystalline silicon by exposure to the molecular form of tritium. The level of hydrogen/tritium concentration attained here is higher than that reported in post-hydrogenation processes. The previously reported post-hydrogenation experiments were performed using atomic or plasma hydrogen since silicon is chemically inactive to molecular hydrogen. In our experiment the distinguishing condition, tritium beta self-catalysis appears to play an important role in the tritiation of the samples.[6] The combination of tritium decay and beta-induced ionizations results in the formation of reactive species of tritium (tritium atoms, radicals, and ions) that readily adsorb on silicon. The electron bombardment of the silicon surface and sub-surface is expected to render it chemically active thereby promoting surface adsorption and sub-surface diffusion of tritium, thus leading to tritium occlusion in the silicon matrix.

In the tritium exposure system, there is approximately a 1 cm thick T_2 layer above the silicon sample. Beta particles emitted from tritium decay can travel up to 35 mm in tritium gas at one atmosphere. The average range of the beta particles expectedly decreases with increasing tritium pressure. For pressurized tritium gas, the electron flux will saturate at about $\Phi = 5\times10^{11}$ cm^{-2} s^{-1} [7-8]. The electron bombardment excites the silicon surface and makes it chemically active to the ionized or dissociated tritium. Thus, tritium can be absorbed at the sample surface. The beta radiation is also likely to promote tritium permeation in the silicon material. Beta particles impinging and traversing the material lose energy in collisions with the lattice, leading to the production of extremely large density of electron-hole pairs (EHPs). In case the of a-Si:H, an EHP is created for every 4.3 eV.[9] Electrons of 5.7 keV can produce about 1300 pairs. Energy released from electron-hole pair recombination can break Si-H bonds to produce mobile H and DBs (dangling bonds) [10]. As the betas penetrate the structure, the loss in beta energy occurs rapidly with depth and accordingly the effective depth for electrons with energy of ~5.7 keV is around 50 nm in a-Si:H[11]. Within this depth, additional DBs will be created and thus promote hydrogen (tritium) incorporation.

CONCLUSIONS

Tritium effusion and SIMS techniques are employed to study tritium evolution from various a-Si:H films and crystalline silicon tritiated over a range of tritium pressures and temperatures. The results indicate that tritium incorporation is determined mainly by the material microstructure which depends on the deposition condition. Gaussian deconvolution of tritium effusion spectra yield two peaks, one low temperature (LT) peak which is attributed to tritium clusters and weakly bonded tritium, and another high temperature peak (HT) which is attributed to the monotritides. Due to a much tighter atomic lattice, c-Si absorbed far less tritium than a-Si:H. The average concentration near the surface is 20 at.% and 4 at.% in a-Si:H and c-Si, respectively. These values are higher than the reported results from post-hydrogenation and post-deuteration experiments. The solubility is enhanced by electron bombardment due to beta-decay of tritium. The high energy electrons serve to activate the material surface, promotes the formation of tritium radicals, creates dangling bonds within the lattice, and thus enhance solubility of tritium in silicon.

ACKNOWLEDGMENTS

This work was supported by NSF grant #0826289 and NSERC discovery grants. The authors wish to thank Dr. Armando B. Antoniazzi and Mr. Clive Morton for their assistance in tritium loading. The help in preparation of the a-Si:H samples from Dr. D. Yeghikyan is also gratefully acknowledged.

REFERENCES

1. T. Kosteski, *et al.*, "Tritiated amorphous silicon betavoltaic devices," *IEE Proc.: Circuits Devices Syst.*, vol. 150, pp. 274-281, Aug 2003.
2. N. P. Kherani, *et al.*, "Tritiated Amorphous-Silicon for Micropower Applications," *Fusion Technology*, vol. 28, pp. 1609-1614, Oct 1995.
3. B. Liu, *et al.*, "Tritiation of amorphous and crystalline silicon using T_2 gas," *Appl. Phys. Lett.*, vol. 89, p. 044104, Jul 24 2006.
4. N.P.Kherani, *et al.*, "Hydrogen Effusion From Trititated Amorphous Silicon," *Journal of Applied Physics*, 2008.
5. T. Kosteski, "Tritiated Amorphous Silicon Films and Devices," Ph.D, Department of Electrical and Computer Engineering, University of Toronto, Toronto, 2001.
6. N. P. Khernai and W. T. Shmayda, "Tritium-Materials Interacitons," *Nuclear Science and Technology, Safety in Tritium Handling Technology, Euro Course Series,* pp. 85-105, 1993.
7. K. E. Bower, *et al.*, *Polymer, Phosphors, and Voltaics for Radioisotope Microbatteries*. Boca Raton, FL: CRC Press, 2002.
8. B. Liu, *et al.*, "Power-scaling performance of a three-dimensional tritium betavoltaic diode," *Applied Physics Letters*, vol. 95, p. 233112, 2009.
9. J. Dubeau, *et al.*, "Radiation ionization energy in alpha-Si:H," *Physical Review B*, vol. 53, pp. 10740-10750, Apr 15 1996.
10. A. Yelon, *et al.*, "Electron beam creation of metastable defects in hydrogenated amorphous silicon: hydrogen collision model," *Journal of Non-Crystalline Solids*, vol. 266, pp. 437-443, May 2000.
11. S. Najar, *et al.*, "Electronic Transport Analysis by Electron-Beam-Induced Current at Variable Energy of Thin-Film Amorphous-Semiconductors," *Journal of Applied Physics*, vol. 69, pp. 3975-3985, Apr 1 1991.

Poster Session: Characterization

Mater. Res. Soc. Symp. Proc. Vol. 1245 © 2010 Materials Research Society 1245-A17-02

Accuracy of the Fatigue Lifetime of Polysilicon Predicted from its Strength Distribution

Vu Le Huy[1], Joao Gaspar[2], Oliver Paul[2], Shoji Kamiya[1]
[1]Materials Nagoya Institute of Technology, Nagoya, Aichi, Japan.
[2]Department of Microsystems Engineering (IMTEK), University of Freiburg, Germany

ABSTRACT

This paper discusses the accuracy of the distribution of the fatigue lifetime of polysilicon thin films predicted from their strength distribution. On the basis of the authors' previous studies, where the fatigue process determining the lifetime was formulated using the well-known fatigue crack extension Paris' law, prediction error ranges for polysilicon specimens with different levels of strength are determined. The errors of the predicted fatigue lifetime in the logarithmic scale, defined as $\Delta\log N = |\log_{10}N_{exp}-\log_{10}N_{pred}|$ where N_{exp} and N_{pred} were the experimental and predicted number of cycle, were found to be less than 1 in the range of the cumulative fracture probability F between 0.1 and 0.9. Therefore, based on the measured Paris' law parameters of polysilicon, the fatigue lifetimes of different polysilicon thin film structures can be predicted from their strength distributions with errors of roughly 10% in the logarithmic scale, which was average of percentages of $\Delta\log N$ to $\log_{10}N$ of experimental data.

INTRODUCTION

The fatigue lifetime of micro electro-mechanical systems (MEMS) structures is one of the most serious concerns for their long-term reliability in many applications. Silicon is the most common structural material in MEMS. It is typically brittle and was recently found to be susceptible to fatigue [1,2]. Tsuchiya et al. [3] discovered that fatigue lifetime under different stress levels could be statistically explained on the basis of Paris' law, which is well suited to describe fatigue crack extension. On the other hand, Izumi et al. [4] performed static strength tests on monocrystalline silicon (c-Si) specimens shaped by etching under different conditions. They reported a clear correlation between the state of damage on etched surface and the strength. Recently, fatigue behavior of polycrystalline silicon (polysilicon, poly-Si) thin films was found to be formulated as a fatigue crack extension process starting from the initial damage and being controlled by Paris' law [5]. Therefore, fatigue lifetime of polysilicon as well as ceramic materials could be predicted under cyclic loading condition from the static parameters [6].

Paris' law has already been used with two unknown parameters to predict the fatigue behavior of polysilicon by using the results of fatigue experiment with constant amplitude [5,7]. The fatigue behavior was then predicted from static strength distributions under monotonically increasing stress. Based on these studies, where the fatigue process determining the lifetime was formulated by using Paris' law, prediction error ranges for polysilicon specimens with different levels of strength are now examined. The accuracy of the prediction is discussed on the basis of the results of tensile fatigue tests performed on two groups of samples fabricated using different conditions leading to different etching damages.

SPECIMENS AND EXPERIMENT

The tensile test structure design previously developed by the authors was used for the experiments consisting both static tensile tests and fatigue tests [8]. The test structure is shown in figure 1. It consists of a c-Si stationary outer frame and inner frame connected by parallel springs. They are separated by 400 μm-wide gaps with width and bridged by four poly-Si specimens. The specimens are loaded when the inner frame is pushed by external actuation.

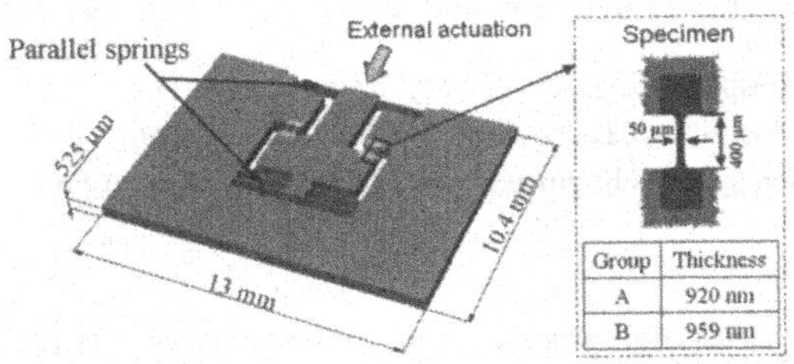

Figure 1. Microtensile test structure used to extract static and cyclic fracture parameters of poly-Si specimens.

Two groups of polysilicon thin film specimens were characterized, named Groups A and B [5,7]. Polysilicon film was deposited on c-Si wafer by low-pressure chemical vapor deposition at a temperature of 625°C, then annealed at 1050°C for 1 hour. The film was patterned into the specimens by inductively coupled plasma (ICP) etching after the annealing and the wafer was subsequently through-etched. All the specimens were designed with 50 μm-wide and 400 μm-long gauges, as shown in figure 1. The average thickness values of the samples from Groups A and B were 920 nm and 959 nm, respectively. The two groups of specimens were fabricated at the different times, therefore the two etching conditions were not the same even by using the same etching method. Since this reason leading to they had different strength levels.

The evaluation of the static strength distribution was first performed using monotonically increasing load under lab-air conditions. Data were ranked from the weakest to the strongest sample and the cumulative fracture probability F was evaluated using $F = i/(I+1)$, where i is the ranked number of a specimen, and I is the total number of tested specimens. The experimental results are plotted in figure 2 along with the curves obtained from fits using the Weibull distribution [9].

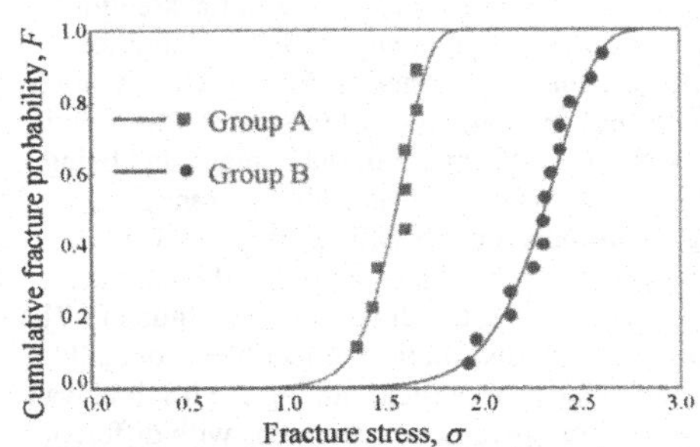

Figure 2. Static strength test results and related Weibull distribution fits.

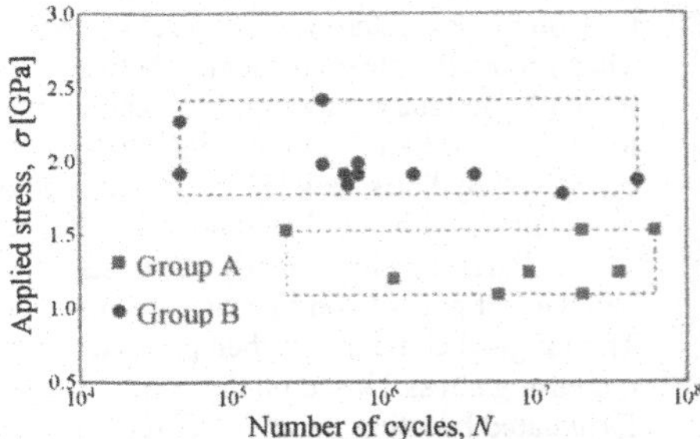

Figure 3. Raw data distribution of the fatigue lifetime.

The fatigue experiments were realized at a temperature of 22°C and relative humidity of 80%. All the specimens were cyclically loaded between a common minimum of zero and maximum stress, which was kept at different constants for the individual ones. Therefore the stress ratio R, ratio of minimum stress to maximum stress, was therefore zero for all the specimens. The specimens from Groups A and B were loaded with different frequencies of 100 and 250 Hz, respectively. The effect of frequency on fatigue behavior was reported to be negligible [10]. Results on the applied stress as a function of number of cycles until specimen failure of the two groups are shown in figure 3.

STATISTICAL ANALYSIS

In Weibull statistics, the cumulative fracture probability F of the specimens is defined as

$$F = 1 - \exp[-(\sigma/\sigma_0)^m], \tag{1}$$

where σ is the fracture stress, σ_0 is the scale parameter related to the average strength and m is the Weibull modulus [9]. The optimum values of σ_0 and m, which were obtained by fitting equation 1 to the actual experimental data, from each group of specimens tested under static conditions, as shown in figure 2, are listed in Table 1 together with their standard deviation.

Table 1. The optimum values of m and σ_0.

Parameters	Group A	Group B
σ_0 [GPa]	1.60 ± 0.033	2.38 ± 0.058
m	12.1 ± 3.90	11.6 ± 1.76

When a specimen breaks, the stress intensity factor at the tip of the largest crack is expected to be equal to the toughness K_{Ic}. The material strength can therefore be formulated in terms of an equivalent crack length a as

$$\sigma = K_{Ic} / \left(\beta\sqrt{\pi a}\right), \tag{2}$$

where β denotes a dimensionless correction factor reflecting the geometry of both the cracks and the structures.

Furthermore, the extension rate of the equivalent crack in ceramic materials under cyclic loading, named the crack growth rate da/dN, is formulated by Paris' law as

$$da/dN = C\left(\Delta K / K_{Ic}\right)^n, \tag{3}$$

where C and n are material parameters [6,11]. ΔK is the amplitude of the stress intensity factor defined as $\Delta K = \beta\sigma(\pi a)^{1/2}$, where a is the crack length at the cycle N. In this study, K_{Ic} and β were 1.1 $MPam^{1/2}$ and 1.12, respectively [5]. By integrating equation 3 with respect to the crack length from the initial crack length a_0 to the critical length $a_c = (K_{Ic}/\beta\sigma\pi^{1/2})^2$, where the specimen broke, i.e., the number of cycles from 0 to N, the initial crack length is obtained as

$$a_0 = \left(\frac{K_{Ic}}{\beta\sigma\sqrt{\pi}}\right)^2 \left[1 + \frac{C(n-2)}{2}\left(\frac{\beta\sigma\sqrt{\pi}}{K_{Ic}}\right)^2 N\right]^{2/(2-n)}. \tag{4}$$

By combining equations 1 and 2, and then substituting equation 4 into it, the cumulative fracture probability F is obtained as

$$F = 1 - \exp\left\{-\left(\frac{\sigma}{\sigma_0}\right)^m \left[1 + \frac{C(n-2)}{2}\left(\frac{\beta\sigma\sqrt{\pi}}{K_{Ic}}\right)^2 N\right]^{m/(n-2)}\right\}. \tag{5}$$

When the number of cycles N is 0, equation 5 becomes identical to equation 1, showing the static strength distribution.

By rearranging equation 5, the number of cycle N until failure can then formulated as

$$N = \frac{2}{C(n-2)}\left(\frac{K_{Ic}}{\beta\sigma\sqrt{\pi}}\right)^2\left\{\left(\frac{\sigma_0}{\sigma}\right)^{n-2}\left[-\ln(1-F)\right]^{(n-2)/m} - 1\right\}. \tag{6}$$

Equation 6 predicts the fatigue lifetime when the parameters C and n, the static probability of failure F and applied stress σ, and σ_0 and m obtained from static tests are known.

RESULTS AND DISCUSSION

The optimization of n and C was performed so as minimize the square sum of errors between the values of cumulative fracture probability F determined from experimental data and those obtained by equation 5 with the corresponding applied stress σ and number of cycles N. These are listed in Table 2, "common" means that equation 5 was fitted to both groups at the same time, assuming that they share the same values of C and n. The other values, simply denominated as individual optimal values, were obtained by fitting equation 5 to the experimental data of the individual groups.

Table 2. Optimum values of parameters C and n in Paris' law.

Parameters	Group A	Group B	Common
C [m/cycle]	1.43×10^{-13}	1.86×10^{-13}	7.96×10^{-13}
n	26.9	18.0	30.9

The cumulative fracture probabilities F for cyclic fatigue were calculated for the two specimen groups by equation 5 with the obtained values of C and n in Table 2 and m and σ_0 in Table 1. For example, figure 4.a shows the result of calculation with the common optimum values of C and n together with the experimental data for the case of Group B. The calculated fatigue behavior reproduces the experimental data reasonably well, even though using the common optimum values. The same holds for the samples from Group A.

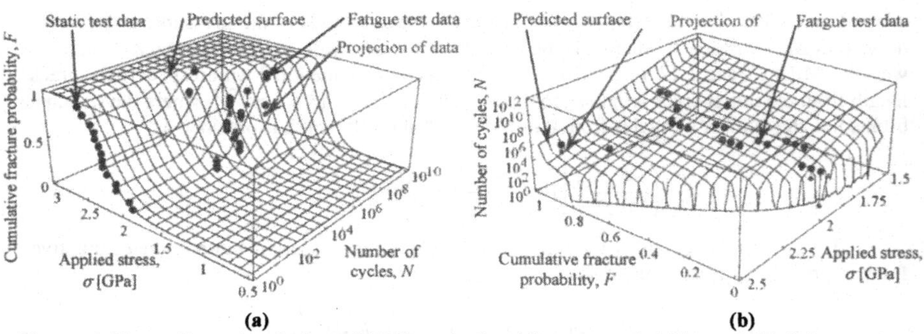

Figure 4. Three-dimensional plots of **(a)** the calculated fracture probability and **(b)** fatigue lifetime prediction of specimens from Group B, using common values of C and n.

Figure 4.b shows the calculated fatigue lifetime of Group B by using equation 6 using the common optimum of parameters C and n. The errors between the calculated fatigue lifetime and the experimental data, defined as $\Delta\log N = |\log_{10}N_{exp}-\log_{10}N_{pred}|$ where N_{exp} and N_{pred} were the experimental and predicted number of cycle, are plotted in figure 5 for the two groups using the individual and common optimum values of C and n. For the case of specimens from Group A, the errors were found to be smaller than 1 in the range of the cumulative fracture probability F between 0.1 and 0.7. In contrast, for the case of Group B, the errors were found to be smaller than 1 between 0.1 and 0.9, probably because of the large number of data as the number of data is 10 and 22 for the case of Groups A and B, respectively. The average relative errors of the fatigue lifetime in the logarithmic scale defined as

$$\frac{1}{I}\times\sum_{i=1}^{I}\frac{\Delta\log N_i\times 100}{\log_{10}N_i} \tag{7}$$

between the data and prediction of two groups were evaluated as in Table 3. Therefore the fatigue lifetime of polysilicon thin films was able to be predicted from their strength within the error range of roughly 10% with the cumulative fracture probability F lies in the range from 0.1 to 0.9, as shown in Table 3, not only for the two cases examined in this study but also in general. As a natural consequence, the prediction becomes more exactly when the number of reference data is larger.

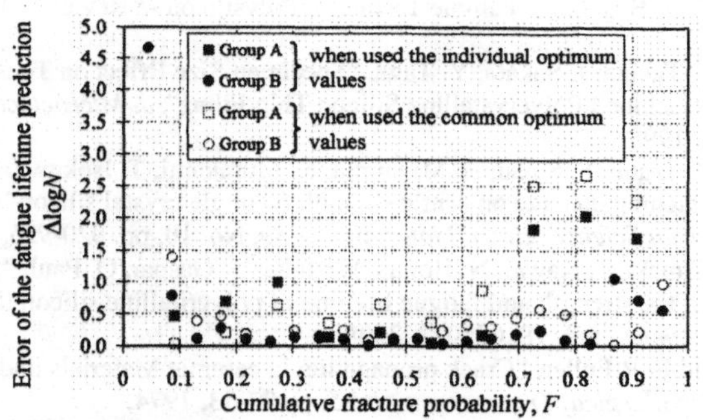

Figure 5. Error in fatigue lifetime $\Delta\log N$ of the experimental results compared to data estimated from the static.

Table 3. Average relative errors defined by equation 7 in the range of the cumulative fracture probability F from 0.1 to 0.9.

Relative error values when	Group A	Group B
using individual optimum values	10.0%	2.7%
using common optimum values	12.8%	4.2%

Figure 5 also shows that the errors in the fatigue lifetime prediction are not so different between the two cases of individual and common optimum values of C and n. Table 3 shows that the average relative errors difference between the two cases is about 2%. This means that the fatigue lifetime of polysilicon thin films have a unique combination of the values of C and n, i.e., the common optimum values of *C* and *n* are expected to be the parameters inherent to the material and independent of strength level determined by etching conditions.

CONCLUSION

Both individual and common optimum *C* and *n* values in Paris' fatigue crack extension law for the two groups of specimens were applied to predict the fatigue lifetime distribution of each individual group. The errors between the calculated fatigue lifetime and experimental data were found to be less than 1 in logarithmic scale in the range of the cumulative fracture probability *F* between 0.1 and 0.9. The average relative errors of the fatigue lifetime between the data and prediction showed that the fatigue lifetime of polysilicon thin films can be predicted from their strength within the error range of roughly 10% by using common optimum values of *C* and *n* those are expected to be material inherent constants. It is concluded that the fatigue behavior of polysilicon can be well estimated with sufficient accuracy in a time-saving manner from knowledge about its static strength.

REFERENCES

1. C. L. Muhlstein, S. B. Brown, R. O. Ritchie, "High-Cycle Fatigue of single-crystal silicon thin films", *J. Microelectromech. Syst.*, vol. 10, pp. 593-600, 2001.
2. W. N. Sharpe Jr, J. Bagdahn, "Fatigue Testing of Polysilicon -a Review", *Mech. Mater.*, vol. 36, 3-11, 2004.
3. T. Tsuchiya, O. Tabata, J. Sakata, Y. Taga, "Specimen Size Effect on Tensile Strength of Surface Micromachined Polycrystalline Silicon Thin Films", *J. Microelectromech. Syst.*, vol. 7, pp. 106-113, 1998.
4. Y. Kubodera, S. Izumi, S. Sakai, H. Miyajima, K. Murakami, T. Isokawa, "Influence of distribution of etching damage on brittle strength of single crystal silicon" (in Japanese), *The Society of Material Science Japan, Material*, vol. 59, No. 10, pp. 920-925, 2007.
5. S. Kamiya, S. Amaki, T. Kawai, N. Honda, P. Ruther, J. Gaspar, O. Paul, "Seamless interpretation of the strength and fatigue lifetime of polycritalline silicon thin films", *J. Micromech. Micoeng.*, vol. 18, 095023, 2008.
6. A. G. Evans and E. R Fuller, "Crack propagation in ceramic Materials under cyclic loading conditions", *Metallurgical Transactions*, vol.5, pp.27-33, 1974.
7. T. Kawai, S. Amaki, J. Gaspar, P. Ruther, O. Paul, S. Kamiya, "Prediction of Fatigue Lifetime Based on Static Strength and Crack Extension Law", in *Tech. Digest MEMS 2008 Conference*, pp. 431-434, Tucson, January 13-17, 2008.
8. S. Kamiya, J. Kuypers, A. Trautmann, P. Ruther, O. Paul, "Process Temperature Dependent Mechanical Properties of Polysilicon Measured Using a Novel Tensile Test Structures", *J. Microelectromech. Syst.*, vol. 16, pp. 202-212, 2007.
9. W. Weibull, "A statistical distribution function of wide applicability" *J. Appl. Mech.* 18 293-7, 1951.
10. J. Bagdahn, W. N. Sharpe Jr, "Fatigue of polycrystalline silicon under long-term cyclic loading", *Sensors Actuators*, vol.A 103, pp. 9–15, 2003.
11. R.W. Davidge, *Mechanical Behavior of Ceramics*, Camdridge University Press, 1979.

Mater. Res. Soc. Symp. Proc. Vol. 1245 © 2010 Materials Research Society 1245-A17-04

Sub-gap photoconductivity in Germanium-Silicon films deposited by low frequency plasma.

Andrey Kosarev and Francisco Avila
Electronics Department, Institute National for Astrophysics, Optics and Electronics, Puebla, 72840, Mexico.

ABSTRACT

(Ge_xSi_{1-x}:H) films are of much interest for many device applications because of narrow band gap and compatibility with films deposited by plasma. However, electronic properties of Ge_xSi_{1-x}:H films for high Ge content $x > 0.5$ have been studied less than those of Si films. In this work, we present a study of sub-gap photoconductivity (σ_{pc}) in Ge_xSi_{1-x}:H films for $x = 1$ and $x = 0.97$ deposited by low frequency plasma enhanced chemical vapor deposition (LF PECVD) with both various H-dilution (R_H) during growth (non-doped films) and boron (B) incorporation in the films. Spectra of sub-gap photoconductivity $\sigma_{pc}(h\nu)$ were measured in the photon energy range of $h\nu = 0.6$ to 1.8 eV. $\sigma_{pc}(h\nu)$ spectra were normalized to constant intensity. For $h\nu < E_g$ two regions in $\sigma_{pc}(h\nu)$ can be distinguished: "A", where σ_{pc} is related to transitions between tail and extended states, and "B", where photoconductivity is due to defect states. $\sigma_{pc}(h\nu)$ in "A" region showed exponential behavior that could be described by some characteristic energy E_U^{PC} similar to Urbach energy E_U in spectral dependence of optical absorption. $E_U^{PC} > E_U$ was observed in all the films studied. This together with higher relative values (i.e. normalized by the maximum value at $h\nu = E_g$) for photoconductivity comparing with those for α means that mobility-lifetime product ($\mu\tau$) depends on photon energy $\mu\tau = f(h\nu)$ that was determined from $\alpha(h\nu)$and $\sigma_{pc}(h\nu)$. $\mu\tau(h\nu)$ increases by factor of 20 to 40 depending on the sample with reducing $h\nu$ from 1.1 to 0.7 eV. In some samples, this dependence was monotonous, while in others demonstrated maxima related to both interference and density of states. Effects of both R_H and boron incorporation have been found and are discussed.

INTRODUCTION

Hydrogenated germanium film (Ge:H) deposited by plasma enhanced chemical vapor deposition (PE CVD) are of much interest because of possible applications in devices such as long wavelength part of tandem solar cells, photo-detectors suitable for optical communication, thermo-voltaic devices, IR micro-bolometers etc. Some devices with Ge:H films have been reported [1, 2], where Ge:H was used as the intrinsic layer in p-i-n structures, however, doped n- and p-layers were fabricated of silicon. Dalal et al. [3] have reported the important role of ion bombardment during growth from plasma for high quality Ge:H films. In this respect low frequency (LF) PECVD with its inherent higher ion bombardment than that in conventional RF plasma is very attractive. In our previous studies [4,5] we have demonstrated a LF PECVD fabrication of Ge:H and Ge_xSi_{1-x}:H films with both low tail and deep localized states. It has been also demonstrated that Ge_xSi_{1-x}:H films with x>0.95 have shown superior electronic properties in comparison with those in Ge:H films having practically the same optical gap [4-6].

This work is devoted to the study of sub-gap photoconductivity in Ge:H and Ge_xSi_{1-x}:H films (x= 0.97) deposited with different hydrogen dilution, and boron doped Ge:H films with different boron concentration.

EXPERIMENT

The samples were deposited by LF PECVD in the system from "Applied Materials" (mod. AMP 3300). The substrate temperature was $T_S = 300°C$ and the frequency was f = 110 kHz. The samples were deposited on glass substrates (Corning 1737). The samples of $Ge_{0.97}Si_{0.03}$:H and Ge:H were deposited in the range of hydrogen dilution from $R_H = 20$ to 80. The samples of Ge:H (B) were deposited in the wide range of boron $C_B= [B/Ge]_{solid} = 0$ to $1.4x10^{-3}$. The thickness of the samples of Ge_xSi_{1-x}:H, Ge:H, and Ge:H (B) was in the range of d = 0.38–0.85μm. Titanium (Ti) electrodes in the forms of stripes with inter-electrode distance L= 1.5mm were deposited by e-gun evaporation on the glass substrate prior to the film deposition. Before measuring spectral dependence of photoconductivity, it was examined that all the films obey the Ohm law and exhibit an almost linear dependence on intensity ($\sigma_{pc} = I_0^{\gamma}; \gamma \approx 0.9$). During measurements, the films were placed in a micro-probe system (LTMP-2) at the room temperature and atmospheric pressure. The monochromatic light was provided by the monochromator (Jobin Yvon , Triax-320) with wavelength varied in the range of λ = 700nm-1600nm with steps of $\Delta\lambda$ = 20nm. The light beam was modulated at the frequency of $F=4Hz$ with the optical chopper. All the samples during the measurements were polarized with the electrometer Keithley 6517A, so that the DC current was 1μA. The spectral dependence of photoconductivity was normalized to the constant intensity of $I_0 = (1.64 \pm 0.01)*10^{-3} W/cm^2$. Urbach energy E_U^{PC} was obtained in the region, where the photoconductivity exhibits an exponential dependence $\sigma_{pc}= \sigma_0 \exp(h\nu/E_U^{PC})$. Considering no back surface reflection, and $\eta \approx 1$, mobility-lifetime product $\mu\tau$ can be obtained from $\mu\tau \approx \sigma_{pc}/[(I_0/h\nu)*[1-R]*[1-\exp(-\alpha d)]$, and for low absorption when $1-\exp(\alpha d) \approx \alpha d$, from $\mu\tau \approx (\sigma_{pc}*h\nu)/\alpha$ Optical gap of the films was $E_g \approx 1.1eV$ [4]. Maximum observed value of photoconductivity reduced to light intensity $I_0 = 100mW/cm^2$ was $\sigma_{pc} = 3.4x10^{-4}$ $(\Omega*cm)^{-1}$ for Ge films, and $\sigma_{pc} = 8.4x10^{-4}$ $(\Omega*cm)^{-1}$ for $Ge_{0.97}Si_{0.03}$:H film. The latter is slightly higher than the value (also reduced to $I_0= 100mW/cm^2$) $\sigma_{pc} = 7.8x10^{-4}$ $(\Omega*cm)^{-1}$ reported in ref. [7] for Ge films deposited in the glow discharge. $\alpha(h\nu)$ was reported in refs.[4, 6] for the same films.

RESULTS AND DISCUSSION

PC spectra for sub-gap photon energies $\sigma_{pc}(h\nu)$ (measured at the same light intensity) are shown in figure 1 for $Ge_{0.97}$ $Si_{0.03}$:H films deposited with different H-dilution. In order to characterize spectra we used slope E_U^{PC} of the curves near E_g (determined similar to Urbach energy E_U for optical absorption spectra $\alpha(h\nu)$) and defect ratio $R_{def.} = [\sigma_{PC}(h\nu = E_g)]/[\sigma_{PC}(h\nu = 0.8 eV)]$, where PC at $h\nu$ = 0.8eV corresponds to deep defects. As we can see in figure 1 H-dilution effects significantly on PC sub-gap spectra. Figure 2a shows behaviour of both E_U and E_U^{PC} as a function of H-dilution. Both curves show similar trends of reducing with H-dilution, while E_U^{PC} demonstrate higher values within the range of R_H studied. If the slope of $\alpha(h\nu)$ characterized by E_U is determined by mainly density of states in mobility gap, the slope of PC spectra depends also on $\mu\tau$ mobility-lifetime product $\sigma_{pc}(h\nu) \sim \alpha(h\nu)\, \mu\tau$. Therefore dependence of $\mu\tau=f(h\nu)$ can be extracted from comparison of $\sigma_{pc}(h\nu)$ and $\alpha(h\nu)$.

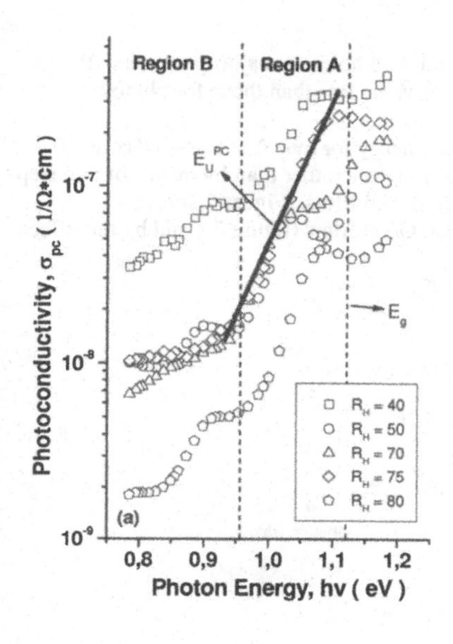

Figure1. Sub-gap PC spectra for $Ge_{0.97}Si_{0.03}$:H films. Regions A and B related to band tail, and defect absorption, respectively.

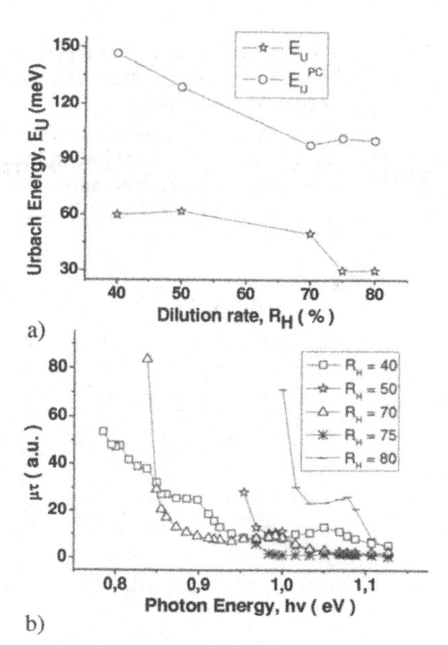

Figure 2. E_U and E_U^{PC} versus R_H (a), and $\mu\tau = f(h\nu)$ for films of $Ge_{0.97}Si_{0.03}$:H (b) with different R_H.

The results of this are demonstrated in figure 2b for the samples deposited with different H-dilution. All the samples studied show general trend of increasing $\mu\tau$ with reducing photon energy ($h\nu < E_g$) for illumination. Figure 3 shows sub-gap PC spectra for Ge:H films deposited with different H-dilution. In these samples a strong effect of H-dilution was also observed. However, both $E_U(R_H)$ and $E_U^{PC}(R_H)$ showed the trends different from those observed for $Ge_{0.97}Si_{0.03}$:H films, as it is shown in figure 4a. $\mu\tau(h\nu)$ in Ge:H films decreases with $h\nu$ only for $R_H = 20\text{-}60$ and shows no change with $h\nu$ for $R_H = 70\text{-}80$ (see figure 4b). Dependence of $\mu\tau(h\nu)$ shown in figures 2 b and 4 b suggests the photo- generated electron for $h\nu < E_g$ occupies a localized state below the conductance band (CB) edge $E_{e\text{-}ph} = E_{CB} - (E_g - h\nu)$. In order to recombine it should be excited thermally to the conductance band with time $\tau_{ex\text{-}th} \sim exp(E_{e\text{-}ph}/kT)$. Because of $\tau_{ex\text{-}th}$ time is longer than recombination time from the E_{CB} it will control the entire recombination process and results in dependence of $\mu\tau$ on photon energy $h\nu$ as it was observed. Figure 5 shows PC spectra $\sigma_{pc}(h\nu)$ for Ge:H films doped with different C_B. It can be seen that increasing boron concentration C_B results in significant changes of the spectra. Both $E_U(R_H)$ and $E_U^{PC}(R_H)$ show similar general trend of growth with C_B (see figure 6a). Only weak changes of $\mu\tau$ with $h\nu$ were observed in Ge:H films doped with boron (see figure 6b). It should be noted that Ge:H films are typically slightly n-type i.e. Fermi level lies in upper half of the optical gap. B-doping with $C_B \leq 3x10^{-4}$ results in compensation and shifting Fermi level to the middle of the optical gap.

Further increase of $C_B > 3x10^{-4}$ results in p-type material with holes as majority carriers. Photo-generated holes show $(\mu\tau)_h$ values by about one order of value less than those for photo-generated electrons (see figures 2 and 4).

This together with weak dependence on photon energy for $h\nu<E_g$ shown in figure 6 b suggests that holes recombine via tunneling in localized states rather than by means of two step process with thermal excitation described above for un-doped films. Figure 7 shows comparatively effect of H-dilution in $Ge_{0.97}Si_{0.03}$:H and Ge:H films (figure 7 a and b) and effect

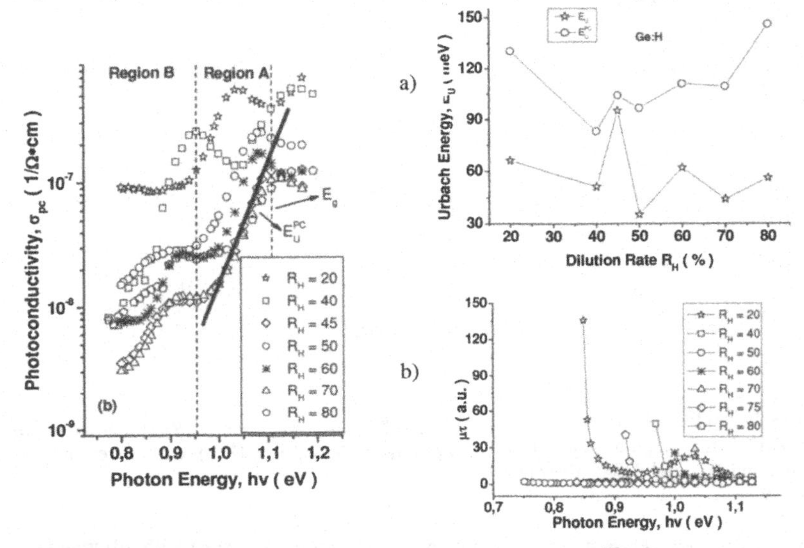

Figure 3. Sub gap PC spectra for Ge:H films deposited with different R_H.

Figure 4. E_U and E_U^{PC} versus R_H (a), and $\mu\tau= f(h\nu)$ for Ge:H films deposited with different R_H (b).

of boron incorporation (figure 7 c). If compare $Ge_{0.97}Si_{0.03}$:H and Ge:H films we can see different behavior of these materials with H-dilution. For $Ge_{0.97}Si_{0.03}$:H reducing monotonously E_U^{PC}, R_{def} and σ_{pc} was observed with H-dilution ratio. Reducing both E_U^{PC}, and R_{def} suggests decreasing density of states related to both conductance band tail states and deep defects. For Ge:H films, E_U^{PC} and R_{def} reduce, when R_H increases from 20 to 50, and then increase with R_H further change from 50 to 80 that is different from the behavior observed in $Ge_{0.97}Si_{0.03}$:H films. Small amount of boron ($C_B< 3x10^{-4}$) has no practically effect on E_U^{PC} and results in increase of R_{def} from ~40 to ~60 a.u. (it means reducing deep defects) and decrease of photoconductivity by about two orders of value. At $C_B\approx 3x10^{-4}$ the films become compensated when Fermi level is in

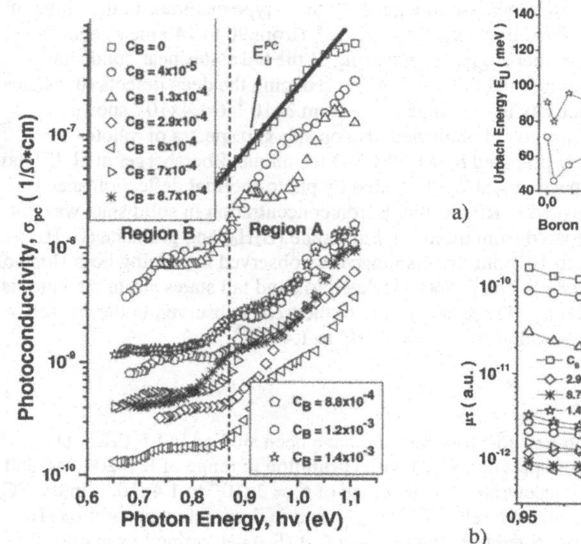

Figure 5. Sub gap PC spectra for Ge:H (B) films with different C_B.

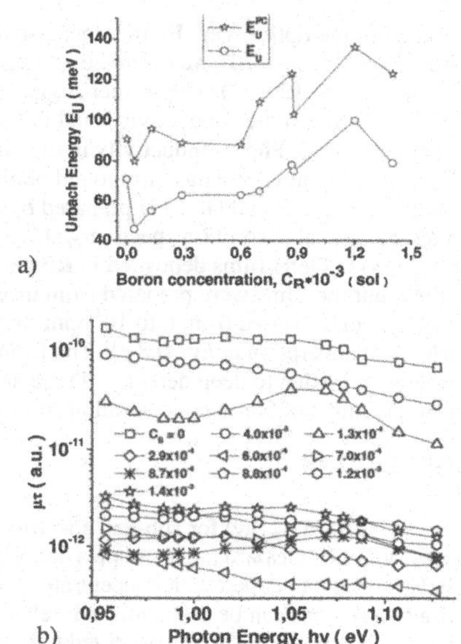

Figure 6. E_U and E_U^{PC} versus H-dilution (a), and $\mu\tau$ $(h\nu)$ for films of Ge:H (B) with different C_B (b).

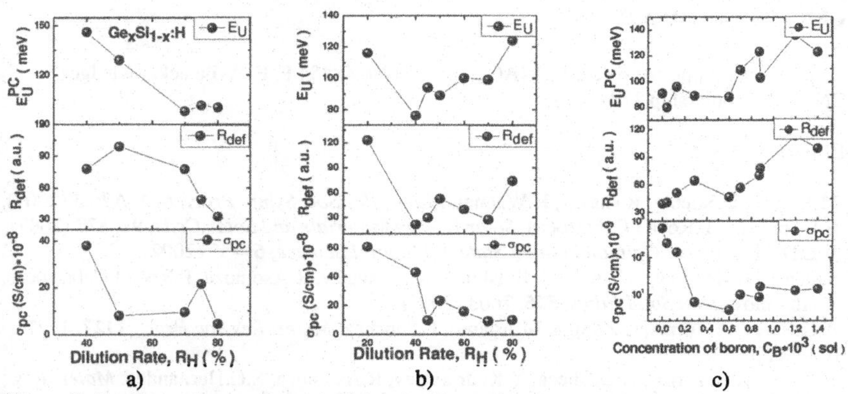

Figure 7. E_U, R_{def} and σ_{PC} at $h\nu$=1.1eV (σ_{PC}(1.1)) as a function of R_H in the films of $Ge_{0.97}Si_{0.03}$:H (a) and Ge:H (b), and as a function of boron concentration in Ge:H film (c).

the middle of the optical gap. Further increase of C_B results in p –type material. In the range of C_B from $3x10^{-4}$ to $1.4x10^{-3}$ we observed increasing both E_U^{PC} (from 90 to 140 meV) and R_{def} (from about 60 to 120a.u.). The former suggests increasing of the tail states near conductance band (because of p-type conductivity) and the latter means reducing the deep defects (the states occupied by holes). Photoconductivity in the range of C_B from $3x10^{-4}$ to $1.4x10^{-3}$ shows practically no change. We have not found published data on sub-gap spectra of photo-conductivity in Ge:H and Ge-Si:H prepared by LF PECVD technique. Ebersberger et al. [7] have reported on optical absorption spectra $\alpha_{PDS}(h\nu)$ measured by photo-thermal deflection spectroscopy (PDS) in Ge:H films deposited in RF plasma. Boron concentration in solid state was not specified, but the films were prepared from mixture of diborane (B_2H_6) and germane (GeH_4) with B_2H_6/GeH_4 varied from 0 to 10^4 ppm. In this range they observed increasing both slope of $\alpha_{PDS}(h\nu)$ and absorption at $h\nu \approx 0.8$ eV The former is related to band tail states while the latter is considered to be due to deep defects. These are similar to the trends observed in the studied films in the range of boron concentration from $C_B= 3x10^{-4}$ to $1.4x10^{-3}$.

CONCLUSIONS

PC spectra $\sigma_{pc}(h\nu)$ for sub-gap absorption $h\nu<E_g$ have been studied in LF PECVD samples of Ge:H, $Ge_{0.97}Si_{0.03}$:H samples deposited with H-dilution in range of R_H=20 to 80 and in Ge:H(B) films B-doped with concentration in the range of $C_B= 3x10^{-4}$ to $1.4x10^{-3}$. In the PC spectra two regions can be distinguished: related to tail (A) and to deep defect absorption (B). Characteristic energy E_U^{PC} was observed to be higher than that (E_U) determined from optical absorption $\alpha(h\nu)$. This is due to contribution of mobility-lifetime product ($\mu\tau$) in $\sigma_{pc}(h\nu)$ spectra. The significant effect of H-dilution on both tail deep defect states and the dependence of $\mu\tau$ on photon energy were observed in two groups of the samples: Ge:H and $Ge_{0.97}Si_{0.03}$:H. This effect revealed itself in different ways in these groups of the samples. Ge:H(B) samples demonstrated effect of B incorporation on $\sigma_{pc}(h\nu)$, but only weak effect on $\mu\tau(h\nu)$.

ACKNOWLEDGMENTS

This work was supported by the CONACyT project No.48454 F. F. Avila acknowledges CONACyT scholarship 212471.

REFERENCES

1. M.Krause, H.Stiebig, R.Carius, H.Wagner. *Mater. Res.Soc. Symp. Proc.*, **664**, A26.5 (2001).
2. E.V.Johnson, P.Roca I Cabarrocas. *Solar Energy Materials and Solar Cells*, **91**, 877 (2007).
3. V.L.Dalal. *Current Opinion in Solid State & Material Science*, **6**, 455 (2002).
4. Sanchez, A.Kosarev, A.Torres, A.Ilinskii, Y.Kudriavtsev, R.Asomoza, P.Roca I Cabarrocas, A. Abramov. *Thin Solid Films*, **515**, 7603 (2007).
5. A.Kosarev, A.Torres, C.Zuniga, M.Adamo, L.Sanchez. *Mater. Res.Soc.Proc.*, **1127**, 1127-T04-03 (2009).
6. A.Kosarev, A.Torres, A.D.Checa, Y.Kudriavtsev, R.Asomoza, S.G.Hernandez. *Mater. Res. Soc. Symp.Proc.*, **1066**, 1066-A05-04 (2008).
7. B.Ebersberger, W.Kruelhler. *Appl.Phys. Lett.*, **65**(13), 1683 (1994).

Mater. Res. Soc. Symp. Proc. Vol. 1245 © 2010 Materials Research Society 1245-A17-10

Thermoelectric properties of doped and undoped mixed phase hydrogenated amorphous/nanocrystalline silicon thin films

Y. Adjallah, C. Blackwell, and J. Kakalios
School of Physics and Astronomy, University of Minnesota,
University of Minnesota, Minneapolis, MN, 55455

ABSTRACT

The Seebeck coefficient and dark conductivity for undoped, and n-type doped thin film hydrogenated amorphous silicon (a-Si:H), and mixed-phase films with silicon nanocrystalline inclusions (a/nc-Si:H) are reported. For both undoped a-Si:H and undoped a/nc-Si:H films, the dark conductivity is enhanced by the addition of silicon nanocrystals. The thermopower of the undoped a/nc-Si:H has a lower Seebeck coefficient, and similar temperature dependence, to that observed for undoped a-Si:H. In contrast, the addition of nanoparticles in doped a/nc-Si:H thin films leads to a negative Seebeck coefficient (consistent with n-type doping) with a positive temperature dependence, that is, the Seebeck coefficient becomes larger at higher temperatures. The temperature dependence of the thermopower of the doped a/nc-Si:H is similar to that observed in unhydrogenated a-Si grown by sputtering or following high-temperature annealing of a-Si:H, suggesting that charge transport may occur via hopping in these materials.

INTRODUCTION

Thermoelectric (TE) devices enable the direct conversion of thermal energy into electrical power using only solid-state materials with no moving parts. Temperature measurements, power generation and electronic refrigeration [1] in semiconductors may be performed through the Seebeck effect and the Peltier effect. The Seebeck coefficient S, also called the thermoelectric power or thermopower, is related to an intrinsic property of the material, and represents the voltage generated across two points on a material by a temperature gradient. The Seebeck coefficient is very small for metals, on the order of a few $\mu V/K$ and for semiconductors is on the order of few mV/K.

We have investigated the thermoelectric properties of undoped and doped hydrogenated amorphous silicon (a-Si:H) films in which are embedded silicon nanocrystallite inclusions (a/nc-Si:H) [2,3]. These materials are synthesized in a dual chamber co-deposition system where the silicon nanocrystals are generated in one plasma reactor and then injected into a second plasma chamber in which the surrounding a-Si:H matrix is deposited. Hydrogenated amorphous silicon is an attractive material for photovoltaic applications due to the ease of large area thin film deposition on a wide variety of substrates, which would also be advantageous for certain thermoelectric applications. In this report, we describe the dependence of the dark conductivity and thermopower on the nanocrystallite volume fraction in undoped and phosphorus doped n-type a-Si:H and a/nc-Si:H films.

MATERIAL PREPARATION

The mixed phase amorphous/nanocrystalline hydrogenated silicon thin films (a/nc-Si:H) are synthesized in a dual chamber co-deposition system described previously in detail [2,3]. Briefly, as sketched in Fig. 1, silicon nanocrystallites with a diameter of 5 nm are synthesized in a dual-chamber Plasma Enhanced Chemical Vapor Deposition (PECVD) system, where one chamber is operated at high silane gas pressure and high RF power for particle production while the second chamber is operated at low silane gas pressure and low RF power. An inert carrier gas of argon entrains the nanoparticles in the first chamber, and they are injected into the second plasma chamber, in which the a-Si:H matrix is deposited. For the undoped a/nc-Si:H studied here, the silicon nanocrystallite density is varied by changing the substrate location relative to the particle injection tube.

To avoid contamination in the undoped plasma reactor, the doped films were deposited in a different dual chamber co-deposition reactor. In this system the concentration of embedded silicon nanocrystals in the doped a-Si:H is varied by changing the silane flow rate. Doping is achieved by dynamically mixing the reactive gases silane (SiH_4) and phosphine (PH_3). In this paper we report measurements for a doping concentration of $[PH_3]/[SiH_4] = 6.2\times10^{-4}$. The gases are also diluted with an inert carrier gas of argon, such that a total gas flow rate of 90 sccm is maintained through the PECVD chamber. The particle chamber was held at 1.7 Torr, while the film chamber pressure was 700 mTorr where the substrates resided on the heated (T = 525K), grounded electrode. The RF power (13.56 MHz) applied to the ring electrodes in the particle chamber is 70 Watts, while the RF power density applied to the non-heated electrode in the second, PECVD chamber is 4Watts The a-Si:H films are deposited onto Corning 7059 glass substrates. The deposition time was 60 minutes for each sample with the resulting film thicknesses ranging from 0.2 to 1.5 microns.

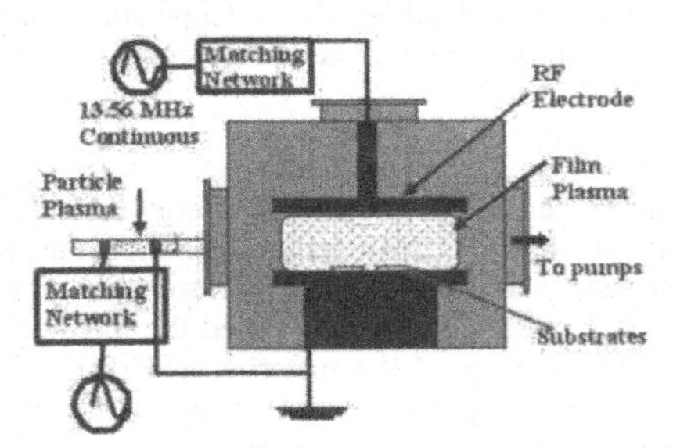

Figure 1: Sketch of the dual chamber co-deposition system. Dopant gases and silane pass through the particle deposition chamber (left) and then flow into the second chamber where the amorphous silicon film is deposited on substrates located on the bottom-heated electrode.

The presence and volume concentration of silicon nanocrystals in the mixed phase thin films is confirmed by measurements of the Raman spectrum. The peak shift of the sharp TO crystalline silicon Raman peak is consistent with the presence of crystallites of diameter 5 nm. The crystalline fraction X_c is the percent, by volume, of the mixed phase film that is crystalline silicon, and can be defined as the ratio of the area under the crystalline silicon peak A_{nc} over the sum of the area under the crystalline silicon peak and the area of the amorphous silicon peak A_{am}, that is, $X_c = (A_{nc})/[A_{nc} + \lambda A_{am}]$ (where λ represents the variation in the Raman backscattering cross-section for the amorphous and nanocrystalline phases) determined by using Gaussian fits of the Raman spectra as reported elsewhere [2,3]. Data are reported for an undoped and n-type doped film with no nanocrystalline inclusions ($X_c = 0$) and for undoped mixed phase a/nc-Si:H films for which the crystalline volume fraction as determined by Raman spectroscopy is $X_c = 0.01$ and $X_c = 0.18$ respectively. The n-type doped a/nc-Si:H films synthesized in the deposition system shown in Fig. 1 have crystal fractions of $X_c = 0.15$ and $X_c = 0.21$. Co-planar chromium electrodes (0.2 cm wide, 1 cm long, 50 nm thick, 4 mm gap) for electronic measurements were deposited using a shadow mask via e-beam evaporation. These electrodes exhibited a linear current-voltage characteristic for the voltages investigated here.

MEASUREMENT TECHNIQUES

The Seebeck coefficient measurements are performed in an experimental system described previously [4,5]. The sample rests across two separate copper blocks; inside of each, a 50W cartridge heater is embedded. The copper blocks are 4 mm apart, which corresponds to the separation of the co-planar metal electrodes, evaporated onto the top film surface. A type T (copper/constantan) thermocouple is attached at one end of each electrode, where each thermocouple is connected to one input of a temperature controller .The temperature of each block is then controlled independently via a dual-channel temperature controller. A temperature gradient is maintained across the film, inducing a thermoelectric voltage between the two electrodes deposited on the sample [1,6].

To measure the Seebeck coefficient $S = \Delta V/\Delta T$ at $T_{ave} = (T_1 + T_2)/2$, we choose the temperature of the blocks T_1 and T_2 such that their average is T_{ave}. By creating a temperature gradient $\Delta T = T_2 - T_1$, a thermoelectric voltage is induced between the electrodes. That voltage is measured by an electrometer, recorded and the temperature gradient is then adjusted, always maintaining the same average set temperature T_{ave}. For each average temperature, we generate thermal gradients of $\Delta T = \pm 6°C, \pm 3°C$ and 0 and record the respective induced voltage. The Seebeck coefficient is defined as the slope of a linear plot of the induced voltage against the temperature gradient. This procedure eliminates the contribution of any small temperature-dependant voltage offsets to the signal [7]. All the measurements are performed following annealing of the film at 450K under vacuum for two hours and cooled slowly to 350K at 1K/min, in order to remove the effects of any prior light exposure [8,9], or surface adsorbates. The electrical conductance and Seebeck coefficient are measured upon warming in five-degree steps from 350K to 450K. We are able to measure the thermoelectrical properties of high impedance thin films down to $10^{-8}\,\Omega^{-1}cm^{-1}$.

RESULTS

The temperature dependence of the electrical conductivity and thermopower of a pure undoped a-Si:H film and undoped mixed phase a/nc-Si:H films, are shown in figure 2. The dark conductivity increases with the addition of nanocrystalline inclusions, accompanied by a corresponding reduction in the activation energy from 0.91 for the pure a-Si:H ($X_c = 0$) to 0.89 for the $X_c = 0.01$ film, to 0.72 eV in the $X_c = 0.18$ a/nc-Si:H film. The observed negative Seebeck coefficient confirms that the majority carrier is negatively charged electrons in these films. There is a decrease in the Seebeck coefficient with nanocrytalline inclusions. The thermopower activation energy is 0.5 eV for the $X_c = 0$ film, and 0.43 – 0.45 for the a/nc-Si:H films. The lowest thermopower is observed for the film containing a small concentration of nanoparticles. The large difference between the activation energy obtained from isothermal dark conductivity measurements and the temperature dependence of the thermopower has been previously observed in undoped a-Si:H[10].

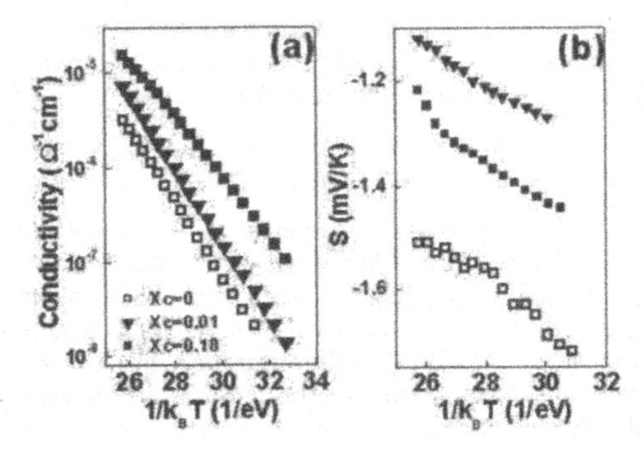

Figure 2: Arrhenius plot of the dark conductivity (Fig.2a) and of the temperature dependence of the thermopower (Fig. 2b) for undoped a-Si:H and mixed-phase a/nc-Si:H films, deposited in a dual chamber co-deposition system for films containing Xc=0.01 and 0.18 crystalline fraction, as determined by Raman spectroscopy measurements.

The dark conductivity and thermopower for doped mixed phase amorphous/nanocrystalline materials (a/nc-Si:H) are shown in Figure 3. An enhancement of the electrical conductivity and Seebeck coefficient from the nanocrystalline silicon inclusions is also observed. The doped a/nc-Si:H exhibit a modest increase in the dark conductivity at room temperature for a nanocrystalline content of $X_c = 0.15$, compared to the $X_c = 0$ film, but a dramatic rise of nearly two orders of magnitude for the $X_c = 0.21$ a/nc-Si:H film. The doped film with $X_c = 0$ has an activation energy of 0.31 eV, which decreases to 0.13 eV for the $X_c = 0.21$ film. The thermopower of the doped a/nc-Si:H films remains negative and is reduced in magnitude compared to the n-type $X_c = 0$ a-Si:H. However, the slope of the thermopower becomes positive with the addition of

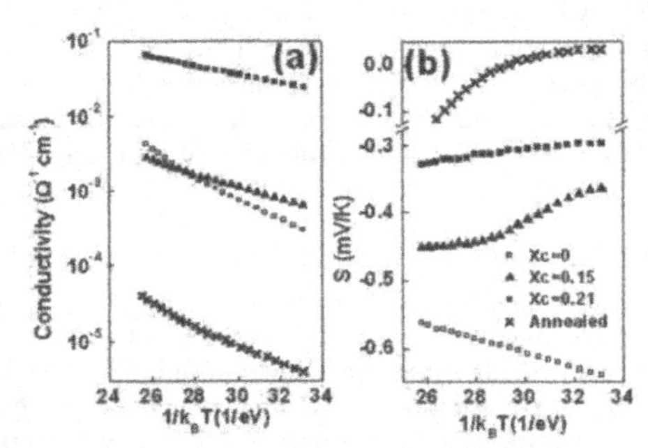

Figure 3: Arrhenius plot of the dark conductivity (Fig. 3a) and of the temperature dependence of the thermopower (Fig. 3b) for n-type phosphorus doped a-Si:H films as a function of nanoparticles inclusion concentration. Also shown are data (x data points) for an a-Si:H film annealed at 925K, for which all of the bonded hydrogen has evolved .

nanocrystallites, that is, the Seebeck coefficient is lower at room temperature than at 450 K, in contrast to the temperature dependence observed in the n-type a-Si:H $X_c = 0$ film. A similar behavior, with a negative magnitude and a positive temperature dependence has been observed in sputtered a-Si films grown without hydrogen, and for a-Si:H films annealed for extended periods at temperatures at 925K, for which all of the bonded hydrogen has evolved[11,12]. In these latter two systems charge transport occurs by hopping through a high density of localized defect states in the mobility gap.

DISCUSSION

As shown in Figure 2, and reported previously [2], the addition of silicon nanocrystals in undoped a/nc-Si:H increases the dark conductivity of the surrounding a-Si:H matrix. We have previously proposed that this enhancement arises from the thermal generation of electrons in the nanocrystallites that are able to hop into the surrounding amorphous silicon. Due to band offsets, the barrier for electron injection from the nanocrystallite is much lower than for holes [2]. The new result reported here is that the thermopower activation energy also decreases with the addition of nanocrystallites.

The observed difference between the thermopower and conductivity activation energy for nominally homogenous undoped a-Si:H is much larger than for n-type doped a-Si:H; we have confirmed this result on several undoped films made in differing deposition runs. The conventional interpretation of the activation energy difference, though by no means universally accepted, is that the difference results from the influence of long-range disorder on the electronic transport properties. Alternative explanations for this difference include polaron hopping [11,13], or electronic hopping transport through band tail states [14]. The activation energy difference between conductivity and thermopower measurements is essentially unchanged for the $X_c = 0$ and $X_c = 0.01$ a/nc-Si:H films (0.41 eV and 0.46 eV, respectively), consistent with the changes in the conductance resulting from the addition of free electrons from the nanoparticles, which are trapped by midgap defects in the a-Si:H matrix, shifting the dark conductivity activation energy. In contrast, the activation energy difference is 0.27 eV for the a/nc-Si:H film with $X_c = 0.18$. This result is surprising, as one would expect the addition of a density of approximately 3×10^{18} nanocrystals/cm^3 to result in a significant increase in the long range disorder at the mobility edge, which has traditionally been ascribed as the source of this activation energy difference. Experiments are in progress to systematically investigate the activation energy difference as a function of nanocrystalline concentration.

The conductivity of the doped mixed-phase a/nc-Si:H is higher than for the $X_c = 0$ doped a-Si:H, suggesting that a similar mechanism of electron donation to the surrounding a-Si:H matrix occurs in the doped mixed phase films. The very low conductivity activation energy of 0.13 eV for the $X_c = 0.21$ a/nc-SiH film indicates that the Fermi energy resides in a high density of bandtail states just below the conduction band mobility edge in this film. Calculations suggest that the conduction band edge for the Si nanocrystal is approximately 0.1 eV below that for the a-Si:H [15,16], roughly at the location of the Fermi energy in the $X_c = 0.21$ film. The average spacing between nanocrsytallites for the $X_c = 0.21$ film is roughly 7 nm, on the same order as the localization length of bandtail states at this energy in the mobility gap. The temperature dependence of the thermopower for the $X_c = 0.15$ and 0.21 films is similar to that observed in unhydrogenated a-Si, where transport occurs via hopping through a high density of dangling bond defects. We thus speculate that the transport in the doped mixed phase a/nc-Si:H is via hopping motion, possibly through the bandtail states as well as the nanocrystalline inclusions, though more work on the possible transport mechanism is needed.

SUMMARY

The dark conductivity and thermoelectric power (Seebeck effect) of undoped and n-type doped mixed phase a/nc-Si:H thin films are reported. The materials are synthesized in a dual chamber co-deposition system, where the silicon nanocrystallites, of diameter 5 nm, are produced in a plasma chamber and, entrained by a flowing inert carrier gas, are injected into a second PECVD chamber in which the surrounding a-Si:H matrix is deposited. As reported previously, the addition of silicon nanocrsytallites enhances the dark conductivity of the a/nc-Si:H, independent of the doping level, consistent with the addition of electrons from the nanocrystallites to the surrounding a-Si:H. The temperature dependence of the thermopower for the n-type doped a/nc-Si:H suggests that electronic transport occurs via hopping through bandtail states in these mixed phase thin films, as opposed to thermal excitation to a mobility edge.

ACKNOWLEDGEMENTS

We gratefully thank C. Anderson for deposition of the undoped mixed-phase amorphous/nanocrystalline silicon thin films and L. Wienkes for growth of the undoped a-Si:H film and helpful comments. This work was partially supported by NSF grants NER-DMI-0403887, DMR-0705675, the NINN Characterization Facility, the Xcel Energy grant under RDF contract #RD3-25, and the University Of Minnesota Center Of Nanostructure Applications.

REFERENCES

1. D.K.C. MacDonald, *Thermoelectricity: An introduction to the principles*,Wiley,New York,1962.
2. Y. Adjallah, C. Anderson, U. Kortshagen, J. Kakalios, J. Appl. Phys. **107** 043704 (2010).
3. C. Blackwell, X. Pi, U. Kortshagen, J. Kakalios, Mat. Res. Soc. symp. **1066** 155-160 (2008).
4. Harold Dyalsingh. Thermoelectric Effects in Amorphous Silicon. PhD dissertation. (1996).
5. H. Dyalsingh, J. Kakalios Physical Review B. **54** 7630-7633 (1996).
6. H. Overhof, W. Beyer. Phil. Mag. B **47** 377-92 (1983).
7. R. Fletcher, V.M. Pudalov, A.D.B. Radcliffe, C. Possanzini, Semicond. Sci. Technol. **16** 386-393 (2001).
8. D.L. Staebler, C.R. Wronski, Appl. Phys. Lett. **31** 292 (1977).
9. D.L. Staebler, C.R. Wronski, J. Appl. Phys. **51** 3262 (1980).
10. R. Meaudre, M. Meaudre, R. Butté, S. Vignoli,Thin Solid Films **366** 207-210 (2000).
11. D. Emin, C. H. Seager, R. K. Quinn, Phys. Rev. Lett. **28** 813-816 (1972).
12. C. H. Seager, D. Emin, R.K. Quinn, J. Non Cryst. Solids. **8-10** 341-346 (1972).
13. R.A. Street, Hydrogenated Amorphous Silicon, Cambridge University Press, Cambridge England, 1991.
14. D. Monroe, Band-edge conduction in Amorphous Semiconductors , (1987).
15. A. H. Mahan and M. Vanecek, *International Meeting on Stability of Amorphous Silicon Materials and Solar Cells,* **234**, 195 (1991).
16. D. Kwon, J. D. Cohen, B. P. Nelson, and E. Iwaniczko, Mat. Res. Soc., **377**, 301 (1995).

Poster Session: Novel Devices

Mater. Res. Soc. Symp. Proc. Vol. 1245 © 2010 Materials Research Society 1245-A18-01

Optimization of the *a*-SiC *p*-layer in *a*-Si:H-based *n-i-p* photodiodes

Y. Vygranenko[1], A. Sazonov[2], G. Heiler[3], T. Tredwell[3], M. Vieira[1,4], A. Nathan[5]
[1]Electronics, Telecommunications and Computer Engineering, ISEL, Lisbon, 1950-062, Portugal
[2]Electrical and Computer Engineering, University of Waterloo, Waterloo, N2L 3G1, Canada
[3]Carestream Health Inc., Rochester, NY, 14652-3487, USA
[4]CTS-UNINOVA, Quinta da Torre, 2829-516, Caparica, Portugal
[5]London Centre for Nanotechnology, UCL, London, WC1H 0AH, United Kingdom

ABSTRACT

Our work is aimed at enhancing the external quantum efficiency (EQE) of *n-i-p* photodiodes by reducing the absorption losses in the *p*-layer and the recombination losses in the *p-i* interface. We have applied boron-doped and undoped hydrogenated amorphous silicon carbon alloy (*a*-SiC:H) grown in hydrogen-diluted, silane-methane plasma to both the *p*-layer and undoped buffer layer, thus tailoring the *p-i* interface. The current-voltage, capacitance-voltage, and spectral-response characteristics of fabricated photodiodes are correlated with the doping level, optical band gap, and deposition conditions for *a*-SiC:H layers. The optimized device exhibits a leakage current of about 110 pA/cm^2 at the reverse bias of 5 V, and a peak value of 89% EQE at a wavelength of 530 nm. At shorter wavelengths, the EQE decreases down to 56% at a 400 nm wavelength. Calculations of transmission/reflection losses at the front of the photodiode show that observed short-wavelength sensitivity enhancement can be attributed to improved separation of electron-hole pairs in the *p*-layer depletion region.

INTRODUCTION

Hydrogenated amorphous silicon (*a*-Si:H) *p-i-n* photodiodes are commonly used as pixel sensors in digital radiographic flat-panel imaging detectors [1]. Photodiode performance is one of the factors limiting the signal-to-noise ratio and image quality. In particular, a high sensor sensitivity in the visible spectral range is required to provide an efficient optical coupling with conventional phosphors such as CsI:Tl or Gd_2O_2S:Tb [2]. One of the approaches to minimize the absorption losses in the *p*-layer is to use an *a*-$Si_{1-x}C_x$:H alloy having a wider band gap than *a*-Si:H. This approach has been widely used for solar cells by engineering the heterojunction *a*-SiC:H/*a*-Si:H *p-i* interface in order to accommodate the band offset, to optimize the electric field profile in this region, to passivate the interface defects, and to reduce recombination [3–5]. One of the findings is that the implementation of a thin buffer layer of intrinsic *a*-SiC:H results in a higher open circuit voltage and fill factor [6]. The similar interface design has been applied for *a*-Si:H-based *n-i-*δ_i*-p* photodiodes. In this device, a thin (~4 nm) undoped *a*-SiC:H buffer (δ_i) significantly reduces the reverse dark current and recombination losses at the *p-i* interface [7]. A 220°C deposition process has been developed for *n-i-*δ_i*-p* photodiode arrays [8].

In this paper, we report on blue-enhanced *n-i-*δ_i*-p* photodiodes deposited at a substrate temperature of 150°C. The deposition conditions of boron-doped and undoped *a*-SiC:H layers were optimized analyzing their impact on the device performance.

EXPERIMENT

A series of boron-doped *a*-SiC:H films were prepared to study their electrical and optical properties. Then, using the same recipes for *p*-layers, a series *a*-Si:H-based *n-i-*δ_i*-p* photodiodes were fabricated. The films and devices were deposited at 150°C onto Corning 1737 glass substrates using a multichamber, 13.56 MHz PECVD system, manufactured by MVSystems Inc.

The trimethylboron ($B(CH_3)_3$) (TMB) and phosphine (PH_3), diluted in hydrogen to a concentration of 1%, were used as the doping gases. Boron-doped *a*-SiC:H films for electrical and optical characterization were deposited onto Corning 1737 glass substrates using an SiH_4 + CH_4 + H_2 + TMB gas mixture at a pressure of 600 mTorr and an rf power of 4 W. The gas flow ratios and depositions rates are shown in Table 1.

Table 1. Deposition conditions of the boron-doped *a*-SiC films.

Sample N°	[CH_4] / [SiH_4]	[H_2] / [SiH_4]	[TMB] / [SiH_4]	Dep. Rate, nm
#1	1.0	4	0.01	0.124
#2	1.0	4	0.02	0.129
#3	1.2	4	0.01	0.120
#4	1.4	4	0.01	0.116

The *n-i-*δ_i*-p* photodiodes were fabricated by the following deposition sequence. First, a 100 nm thick Mo film was sputtered on the glass substrate, followed by the deposition of the *n-i-*δ_i*-p* stack. Finally, a 65 nm thick ZnO:Al film was sputtered and patterned to form the top electrodes with an area ranging from 1 × 1 to 5 × 5 mm^2.

In order to avoid cross-contamination, the doped and undoped layers of the *n-i-*δ_i*-p* stack were deposited in different chambers of the cluster tool system without breaking the vacuum. A 25 nm thick *n*-layer was prepared using a 1:4:0.01 mixture of SiH_4 / H_2 / PH_3. The deposition pressure and rf power were 500 mTorr and 2 W, respectively. Then, a 500 nm of undoped *a*-Si:H layer was deposited in hydrogen-diluted silane plasma at [H_2] / [SiH_4] = 3, with a deposition pressure of 400 mTorr, and an rf power of 2 W. In order to avoid the band offsets in the *i-p* region, the δ_i-layer was deposited by gradually increasing the flow rate of CH_4 from zero to a higher value until reaching the band gap up to that in the *p*-layer. The [H_2] / [SiH_4] ratio, process pressure, and rf power were set the same as that for the *p*-type *a*-SiC:H film series.

Samples for conductivity measurement were prepared by sputtering coplanar Al electrodes through a shadow mask. A Dektak 8 surface profiler was used for film thickness measurements. Dark conductivity of the films and current-voltage characteristics of the photodiodes were measured at room temperature using a Keithley 4200-SCS semiconductor characterization system. The capacitance-voltage (C-V) characteristics of selected devices were measured at 1 kHz frequency using an Agilent 4284A LCR-meter.

The absorption coefficient α, for α above 10^4 cm^{-1}, was obtained by transmission/reflection spectroscopy using a UV-visible 2501 PC Shimadzu spectrophotometer. The spectral response measurements were performed with a PC-controlled setup based on an Oriel 77 200 grating monochromator, a Stanford Research System SR540 light chopper, and an SR530 DSP lock-in amplifier. The system was calibrated in the spectral range of 300–1100 nm using a Newport 818-UV detector.

RESULTS AND DISCUSSION

Figure 1 shows the typical quasi-static current-voltage characteristics of the *p*-δ_p-*i*-*n* photodiodes. The forward and reverse bias sweeps were performed starting at zero bias. In order to minimize the transient current induced by the trapped charge in the *i*-layer, the sweep delay was set to 20 s, and the bias voltage was varied at 25 mV increments.

All photodiodes show an exponential dependence of the forward current over seven orders of magnitude in the biasing range from 0.1 to 0.6 V. At higher biases, the TCO series resistance and the space-charge limited current effect are factors defining the current-voltage dependence. The saturation current density (J_0) and diode ideality factor (n) values, determined through a fitting procedure, are shown in Table 2. The table also includes a reverse dark current density at −2 V along with the conductivity (σ) and optical band gap (E_{04}) values of *a*-SiC:H *p*-layers. The diode ideality factor is usually used as a figure of merit for diodes. In the case of *a*-Si:H-based diodes, this parameter has been interpreted considering the drift/diffusion and recombination-conduction mechanisms [9]. Indicating the improvement in *p*-type *a*-SiC:H, the achieved n and J_0 values are lower than the best data, $n = 1.4$ and $J_0 = 0.3$, reported for 200°C *n*-*i*-δ_i-*p* diodes with *a*-SiC:H *p*-layers deposited without hydrogen dilution [7].

Table 2. Key parameters of the *n*-*i*-δ_i-*p* photodiodes along with the conductivity and optical bandgap values of the *p*-layer.

Sample N°	σ, µS/cm	E_{04}, eV	n	J_0, fA/cm^2	J(−2 V), pA/cm^2
#1	1.30	2.12	1.31	150	107
#2	2.10	2.11	1.18	126	708
#3	0.11	2.18	1.17	260	79
#4	0.05	2.21	1.33	220	47

Two bias regions are observed in the reverse current-voltage characteristics: the reverse dark current increases, albeit slowly, with increasing bias up to a threshold value of 3–6 V; then, the current increases nearly exponentially at higher biases. The magnitude of leakage current in the low-bias region strongly depends on the doping level in the *p*-layer. In comparison to Sample #1,

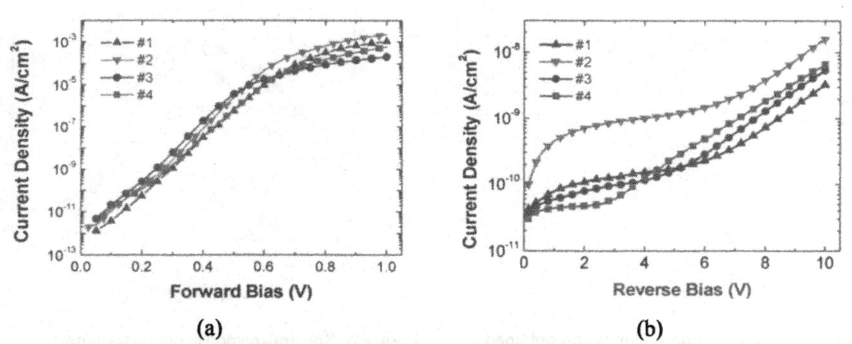

Figure 1. *J*-*V* characteristics of the *n*-*i*-δ_i-*p* photodiodes: (a) forward, and (b) reverse curves.

Sample #2, with a *p*-layer deposited at a doubled doping gas flow shows a reverse dark current density that is a factor of seven higher. Furthermore, at the same [TMB] / [SiH_4] ratio (see Table 1, Samples #1, #3, and #4), the increase in the band gap of *a*-SiC:H reduces the leakage current in the low-bias region but causes a negative voltage shift of the exponential component of *J-V* curve. The values of leakage current at −2 V correlates with conductivity data for doped *a*-SiC:H (see Table 2), i.e., the leakage current decrease is mainly due to a reduction of the surface current component. The exponential leakage current component is likely related to a field-enhance generation of free charge carriers in the *i-p* region [10].

For more detailed study of the *i-p* interface, the capacitance-voltage characteristics were measured. Under moderate reverse bias the *i*-layer is fully depleted; therefore, a small decrease in the diode capacitance with increasing applied voltage is due to expansion of the depleted region into the doped contacts. In the devices under study, the conductivity of phosphine-doped *a*-Si:H, $\sigma \approx 10^{-3}$ S/cm, is much higher than that of boron-doped *a*-SiC:H. In this case, the increase in the depletion width ΔW_P within *p*-layer is

$$\Delta W_P = \varepsilon_0 \varepsilon\, A \left(\frac{1}{C(V)} - \frac{1}{C_0} \right), \tag{1}$$

where ε is the static dielectric constant of the semiconductor material, ε_0 the permittivity of free space, A is the area of the junction. Figure 2 shows a $\Delta W_P - V$ plot for Samples #1 and #2 having different doping levels in *p*-layers. The dependences are close to linear in some biasing range above 1 V, when the *i*-layer is fully depleted. Here, the curve slop depends on the doping level. An acceptor density N_A can be estimated using expression

$$N_A = \frac{C(V)}{e \cdot A} \left(\frac{\delta W_P}{\delta V} \right)^{-1}, \tag{2}$$

where e is the electron charge. The obtained N_A values are 6.8×10^{17} and 8.2×10^{17} cm^{-3} for Samples #1 and #2, respectively.

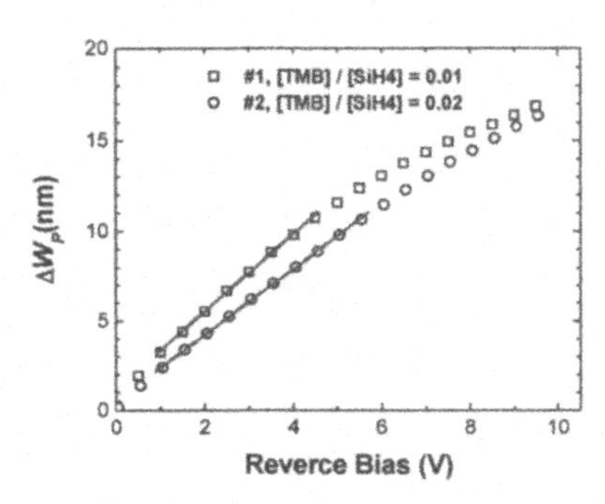

Figure 2. Variation of depletion width obtained from C-V measurements.

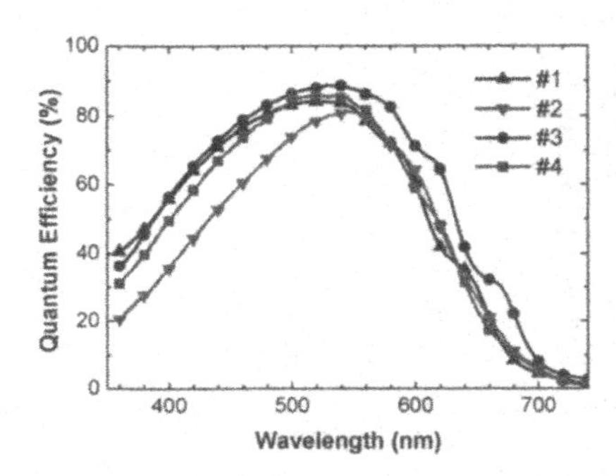

Figure 3. Spectral-response characteristics measured at a reverse bias of 5 V.

Figure 3 shows the EQE spectra of the photodiodes at a reverse bias of 5 V. Peak values of the curves vary from 81 to 89% in the narrow spectral interval of 530–550 nm. The spectral response decays in the red region due to decreasing absorption in the *i*-layer. Besides the light absorption in the *p*-layer, the recombination losses at the *i-p* interface can be considered as a limiting factor for decreasing short-wavelength response [10,11]. Samples #1 and #3 exhibit a sensitivity enhancement in the blue range indicating a good quality of the i-δ_i-p interface. In contrast, Sample #2 shows a sensitivity deterioration in the short wavelength range presumably because the increased boron content leads to a higher number of boron-induced defects at the vicinity of the *i-p* interface, thus enhancing the recombination losses.

Figure 4 shows the EQE as a function of the reverse bias voltage measured at a 400 nm wavelength. The degree of signal enhancement with increasing reverse bias for a short wavelength response reflects the relative degree of recombination at the front of the photodiode [12]. Sample #3 shows the lowest sensitivity variation and the highest EQE in the moderate bias range.

To estimate the optical losses, the transmittance/reflectance of the *p*-layer/ZnO:Al-layer stack on the infinitely thick *a*-Si:H substrate were calculated. Figure 5 shows calculated transmittance and reflectance spectra along with the EQE spectra of Sample #3 at zero bias and 5 V reverse bias. The reflectance reaches a minimum of ~0.1% at 493 nm, indicating a perfect match for refractive indexes of ZnO:Al and *a*-SiC films. The calculated transmittance is consistent with measured EQE in the spectral range from 400 to 500 nm, when the light absorption is mainly in the *i*- and *p*-layers. At longer wavelengths, the reflection and absorption losses related with a bottom-metal electrode became dominant. The transmittance and zero bias EQE became the same (about 50%) at a wavelength of 400 nm. At this wavelength, the EQE increases from 50% at zero bias to 56% at a reverse bias of 5 V. The EQE exceeds the calculated transmittance at longer wavelengths up to 465 nm. The observed enhancement in EQE under reverse bias conditions cannot be explained only by an increase in the charge collection within the *i*-layer. Apparently, not all electron-hole pairs generated in the depletion region within the *p*-layer recombine, thus giving a noticeable contribution in the photocurrent. This conclusion is consistent with measured C-V characteristics, which indicate about 10 nm depletion-region expansion within the *p*-layer under 5 V reverse bias.

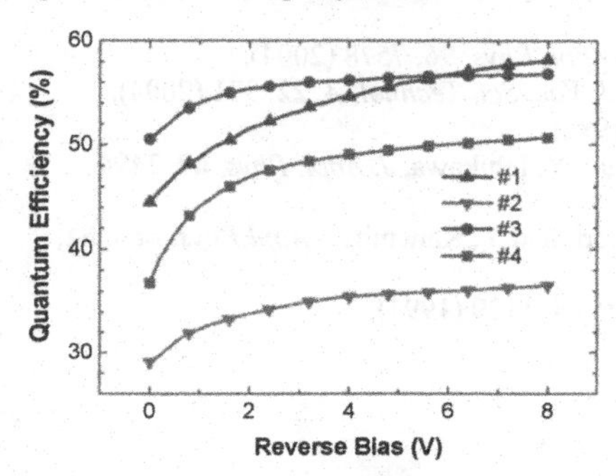

Figure 4. EQE of the photodiodes as a function of reverse bias at a 400 nm wavelengths.

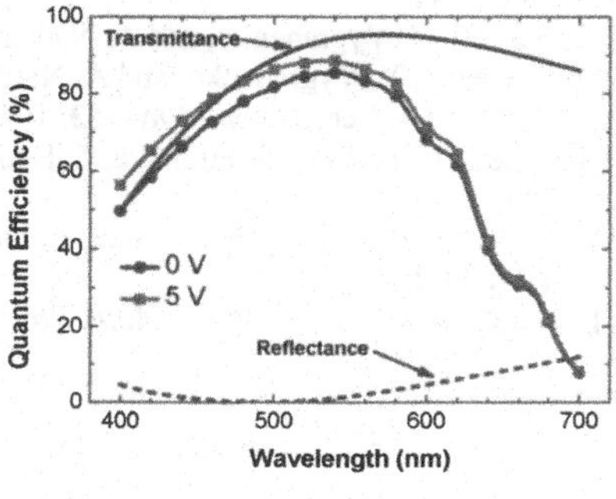

Figure 5. EQE spectra along with calculated transmittance/reflectance of the *a*-SiC:H/ZnO:Al interface.

CONCLUSIONS

Applying wide-gap (E_{04} = 2.1–2.2 eV) *a*-SiC:H grown in hydrogen-diluted, silane-methane plasma as a p^+ and buffer layer material, a series of *a*-Si:H-based *n-i-*δ_i*-p* photodiodes has been fabricated and characterized. By varying the methane and doping gas flows, the deposition condition for *a*-SiC:H layers were optimized targeting both low-leakage current and high short-wavelength sensitivity. The optimized device (Sample #3) exhibits a leakage current of about 110 pA/cm^2 at the reverse bias of 5 V, and 56% EQE at a wavelength of 400 nm. Device modelling shows that the separation of electron-hole pairs in the *p*-layer depletion region leads to the observed short wavelength sensitivity enhancement under reverse bias conditions.

ACKNOWLEDGMENTS

The authors are grateful to the Portuguese Foundation of Science and Technology through fellowship BPD20264/2004 for financial support of this research, and to the Giga-to-Nanoelectronics Centre at the University of Waterloo for providing the necessary equipment and technical help to carry out this work.

REFERENCES

[1] R. A. Street, Ed., *Technology and Applications of Amorphous Silicon* (Berlin: Springer-Verlag, 2000).
[2] J. Beutel, H. L. Kundel, and R. Van Metter, Eds., *Handbook of Medical Imaging*, (Washington, DC.: SPIE Press, 2000).
[3] H. Stiebig, F. Siebke, W. Beyer, C. Beneking, B. Rech, and H. Wagner, *Sol. Energy Mater. Sol. Cells* **48**, 351 (1997).
[4] J. M. Pearce, R. J. Koval, A. S. Ferlauto, R. W. and Collins, *Appl. Phys. Lett.* **77**, 3093 (2000).
[5] B. Vet and M. Zeman, *Thin Solid Films* **516**, 6873 (2008).
[6] G. Munyeme, M. Zeman, R. E. I. Schropp, and W. F. van der Weg, *Phys. Stat. Sol.* (c) **1**, 2298 (2004).
[7] P. Servati, Y. Vygranenko, and A. Nathan, *J. Appl. Phys.* **96**, 7578 (2004).
[8] J. H. Chang, Y. Vygranenko, and A. Nathan, *J. Vac. Sci. Technol. A*, **22**, 971 (2004).
[9] I. Chen and S. Lee, *J. Appl. Phys.* **53**, 1045 (1982).
[10] H. Sakai, T. Yoshida, S. Fujikake, T. Hama, and Y. Ichikawa, *J. Appl. Phys.* **67**, 3494 (1990).
[11] Y. Nasuno, M. Kondo, A. Matsuda, H. Fukuhori, and Y. Kanemitsu, *Appl.Phys. Lett.* **81**, 3155 (2002).
[12] N. Kramer and C. van Berkel, *Appl. Phys. Lett.* **64**, 1129 (1994).

Mater. Res. Soc. Symp. Proc. Vol. 1245 © 2010 Materials Research Society 1245-A18-04

Silicon Germanium Oxide ($Si_xGe_yO_{1-x-y}$) Infrared Sensitive Material for Uncooled Detectors

R. Anvari[1], Q. Cheng[1], M. L. Hai[1], T. Bui[1], A. J. Syllaios[2], S. Ajmera[2] and M. Almasri[1]
[1]Department of Electrical and Computer Engineering, University of Missouri, Columbia, MO, almasrim@missouri.edu
[2]L-3 Communications Electro-Optical Systems, Dallas, TX, AJ.Syllaios@l-3com.com

ABSTRACT

This paper presents the formation and the characterization of silicon germanium oxide ($Si_xGe_yO_{1-x-y}$) infrared sensitive material for uncooled microbolometers. RF magnetron sputtering was used to simultaneously deposit Si and Ge thin films in an Ar/O_2 environment at room temperature. The effects of varying Si and O composition on the thin film's electrical properties which include temperature coefficient of resistance (TCR) and resistivity were investigated. The highest achieved TCR and the corresponding resistivity at room temperature were -5.41 %/K and 3.16×10^3 Ω cm using $Si_{0.039}Ge_{0.875}O_{0.086}$ for films deposited at room temperature.

INTRODUCTION

Infrared imaging cameras have a broad range of commercial and military applications. These cameras are developed by companies such as L-3 Communications [1], Raytheon [2], DRS [3] and BAE Systems [4], and are mainly based on vanadium oxide (VOx) or amorphous silicon (a-Si) technology. Several other materials have been used in uncooled infrared detection, including yttrium barium copper oxide (YBaCuO), silicon germanium (SiGe), silicon germanium oxide (Si-Ge-O), metals, and poly silicon (Poly: Si). A detailed TCR summary of common materials used in uncooled infrared imaging technology is shown in Table I.

Table I. Temperature coefficient of resistance (TCR) of common uncooled infrared microbolometer materials.

Infrared Materials	TCR (%/K)	References
VOx	2 - 2.4	[5, 6]
a:Si	2 - 3.2	[1, 7-9]
YBaCuO	2.88 - 3.5	[10, 11]
Si-Ge	2 – 3	[12, 13]
Si-Ge-O	See table II	
Metals	0.2	[14]

Other research groups have studied $Si_xGe_yO_{1-x-y}$ compound as an infrared sensitive material. Clement *et al.* sputtered Ge-Si compound target (15% atomic Si) in argon/oxygen environment [15]. A similar deposition method was performed by Rana *et al.* [16] but instead of using Si-Ge target, they fixed a piece of silicon (cut from a silicon wafer) to a Ge target and deposited using one power source. A third group has deposited $Si_xGe_yO_{1-x-y}$ using co-sputtering system in Ar/O_2 environment [17, 18]. The three groups have achieved relatively high TCR. However, the corresponding resistivities were too high to be compatible with microbolometer readout electronics.

Table II. Recent results of TCR and resistivity of Si-Ge-O.

TCR (%/K)	Resistivity (Ω.cm)	References
- (2.27 - 8.69)	4.22×10^2 - 3.47×10^9	[15]
-5	10^4	[19]
-5	3.8×10^4	[17, 18]
- 6.43	3.34×10^2	[20, 21] Our group

This paper utilizes amorphous $Si_xGe_yO_{1-x-y}$ as the infrared sensing layer because of its excellent infrared radiation absorption, and mechanical and electrical properties at room temperature. Si and Ge based compounds are standard materials in silicon integrated circuit technology providing a wide range of established knowledge for fabrication of the microbolometer arrays (FPA). $Si_xGe_yO_{1-x-y}$ is compatible with CMOS technology due to the low deposition temperature and the use of conventional dry-etch processing. Therefore, the focal plane array can be easily integrated with the readout electronics.

The performance of microbolometer can be enhanced by achieving high infrared absorption at specific spectral wavelength windows if the IR sensing material has high TCR with relatively low resistivity and low 1/*f*-noise. TCR shows how rapidly the resistance of a material responds to a change in temperature. It is given by:

$$TCR = \frac{1}{R}\frac{dR}{dT} = \frac{1}{R}\frac{\Delta R}{\Delta T} = -\frac{E_a}{kT^2} \quad (1)$$

$$R(T) = R_0 \exp\left(\frac{E_a}{kT}\right) \quad (2)$$

Where E_a is the activation energy, k is the Boltzmann's constant, $R(T)$ is the resistance at temperature T,and R_0 is the initial resistance.

This paper investigates TCR, and electrical resistivity of $Si_xGe_yO_{1-x-y}$ thin films. The influence of changing silicon and oxygen contents on TCR and resistivity are discussed. $Si_xGe_yO_{1-x-y}$ was grown by RF magnetron sputtering Si and Ge simultaneously from two targets in an oxygen and argon environment. The depositions were performed at room temperature, and at low pressure (4 mTorr). Film composition was varied by adjusting RF power applied to the targets and by varying the oxygen flow of the gas mixture in the deposition chamber. The atomic compositions of Si, Ge and O in the deposited thin film were determined and analyzed using energy dispersive X-ray spectroscopy (EDX). The TCR and resistivity were determined using four point probe measurements.

FABRICATION AND EXPERIMENTAL DETAILS

Thin films of amorphous $Si_xGe_yO_{1-x-y}$ were deposited on a p-type silicon substrate, an oxidized silicon substrate (200 nm), and on a glass slide, using a radio frequency (RF) magnetron sputtering system. The targets were 3" undoped 99.999% pure silicon and n-type germanium. Prior to deposition, the substrates were cleaned with pirhana solution. The sputtering chamber was evacuated to base pressure between 0.5-3 μTorr. The deposition pressure was fixed at 4 mTorr. The RF powers of Si and Ge targets were varied in order to change the elemental concentrations.

The thin films that were deposited directly on silicon substrates were used to determine the element's concentrations using EDX. A 4 KeV acceleration voltage with 120 sec scanning period was applied to create scanning element spectrum. A typical EDX spectrum of $Si_xGe_yO_{1-x-y}$ is shown in Figure 1. The thickness of the thin film was measured using KLA profiler. The substrate ($Si/SiO_2/Si_xGe_yO_{1-x-y}$) was then diced to 1cm × 0.5cm rectangular shape and mounted in a ceramic flat pack package by conductive epoxy. The package was heated at 55° for 15 min to cure the epoxy.

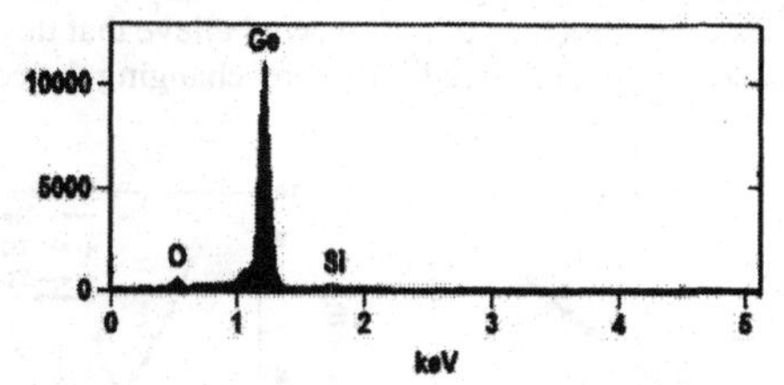

Figure 1. EDS for $Si_{0.039}Ge_{0.875}O_{0.086}$

The resistance versus temperature (R-T) characteristics, and the corresponding TCR were performed using four-point probe method as follow: the sample was mounted firmly on the base stage inside a closed-cycle cryostat (APD Cryogenics' DISPLEX DE 202), a thermometer and a temperature controller were used to control the temperature inside the cryostat to heat the sample, and to measure its temperature, respectively, a Keithley 220 programmable current source was used to force a constant current ranging from 50 nA to 200 nA into the device, a Keithley 2182 nano-voltmeter was used to measure the voltage across the two inner probes. The current magnitude was verified with Keithley 6485 pico-ammeter. The temperature was varied from 20°C to 60°C with 2°C intervals.

The relation between the measured resistance and the calculated resistivity of $Si_xGe_yO_{1-x-y}$ can be expressed by the following two equations:

$$R = \int_s^{2s} \frac{\rho}{2\pi t}\frac{dx}{x} = \frac{\rho}{2\pi t}\ln 2 = \frac{V}{I} \tag{3}$$

$$\rho = \frac{\pi t}{\ln 2}\left(\frac{V}{I}\right) \tag{4}$$

where R is the resistance of the film, ρ is the sheet resistivity, t is the thickness of the thin film, V and I are the measured voltage and current respectively.

RESULTS

The activation energy of the deposited $Si_xGe_yO_{1-x-y}$ thin films (300 nm) was calculated from the slope of Arrhenius plot, resistance versus 1/kT, with a value of 0.4216 eV as shown in Figure 2.

The solid line in the figure is the measured data and the gray shaded line is the corresponding linear fitting of that curve. TCR was determined in two ways. First, it was measured experimentally using four point probe and is called "measured", which was achieved via the left side of Eq. 1. Second, it was deduced from the right side of the same equation and is called "calculated". The measured and calculated TCR are shown in Figure 3. The resistivity in Figure 3 was obtained by applying measured data to Eq. 4. The highest achieved TCR along with the corresponding resistivity were -5.41 %/K and 3.16×10^3 Ω cm respectively using $Si_{0.039}Ge_{0.875}O_{0.086}$ for films deposited at room temperature (Figure 3). This resistivity value along with the corresponding TCR value indicates significant improvement from previously published results from other groups [15, 17-19]. The E_a was used to study the effects of silicon and oxygen on the electrical properties of $Si_xGe_yO_{1-x-y}$ thin film. We believe that this is not the optimum value for TCR. These values can be improved further by changing the composition ratio of Si, Ge, and O.

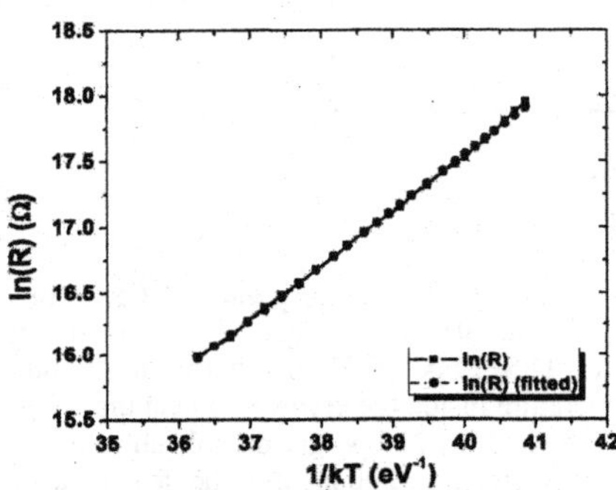

Figure 2. Variation of $ln(R)$ with $1/kT$ for $Si_{0.039}Ge_{0.875}O_{0.086}$ deposited at room temperature

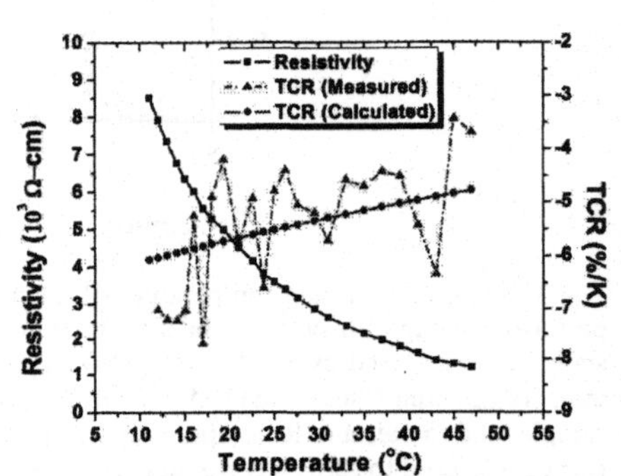

Figure 3. TCR (measured, and calculated from E_a) and Resistivity versus Temperature for $Si_{0.039}Ge_{0.875}O_{0.086}$ deposited at room temperature.

The variations of TCR and activation energy with changes in oxygen concentration for a fixed Si concentration of 10% are shown in Figure 4. As the oxygen concentration increases both TCR and activation energy increases, which is related to the increase in the band gap of the compound. This agrees with results from other group [16, 18]. The effect of changing the Si concentration in $Si_xGe_yO_{1-x-y}$ thin film was investigated in this work. Figure 5 shows the variations of TCR while Si concentration was varied at two different levels of O concentrations. It can be observed that for a fixed oxygen concentration (7%, and 9%), TCR initially increases as the Si concentration increases but then starts to decrease after a certain concentration between 9% and 10% range.

CONCLUSION

Thin films of $Si_xGe_yO_{1-x-y}$ were grown using RF magnetron sputtering from Si and Ge targets in Ar/O_2 environment. EDX was used to measure the samples' compositions. TCR, resistivity and activation energy at room temperature were measured. Based on the measurements, the highest TCR value achieved and the corresponding resistivity were -5.41 %/K and 3.16×10^{3} Ω cm using $Si_{0.039}Ge_{0.875}O_{0.086}$. As the O concentration in the film increases, both TCR and E_a increase. Moreover, as the Si concentration increases, TCR increases initially and then deceases. In the future, noise properties of this material will be investigated.

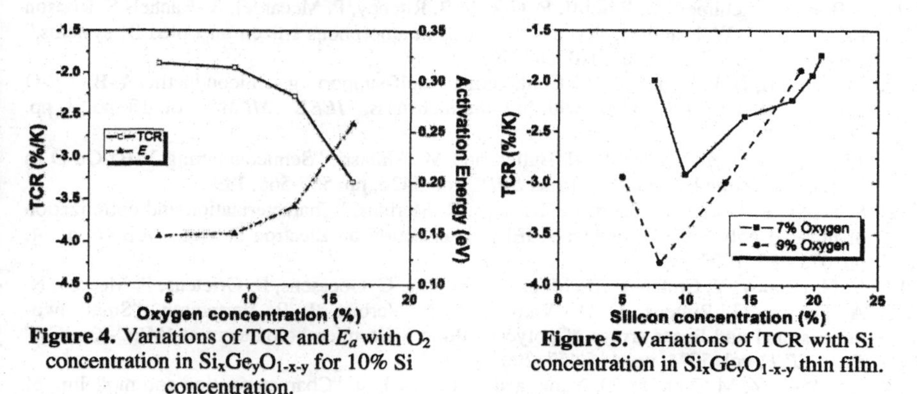

Figure 4. Variations of TCR and E_a with O_2 concentration in $Si_xGe_yO_{1-x-y}$ for 10% Si concentration.

Figure 5. Variations of TCR with Si concentration in $Si_xGe_yO_{1-x-y}$ thin film.

ACKNOWLEDGMENTS

This material is based upon work funded by the Army Research Office. The authors would like to thank Dr. William Clark of the Army Research Office for the guidance he provided in this project.

REFERENCES

1. T. Schimert, C. Hanson, J. Brady, T. Fagan, M. Taylor, W. McCardel, R. Gooch, M. Gohlke, and A. J. Syllaios, "Advances in small-pixel, large-format -Si bolometer arrays," *Proc. SPIE*, vol. 7298, 72980T, 2009.
2. D. Murphy, M. Ray, J. Wyles, et al., "640 × 512 17 μm Microbolometer FPA and sensor development," *Proc. SPIE*, vol. 6542, 65421Z, 2007.
3. Chuan Li, George Skidmore, Christopher Howard, Elwood Clarke, and C. J. Han, "Advancement in 17-micron pixel pitch uncooled focal plane arrays," *Proc. SPIE*, Vol. 7298, 72980S, 2009.
4. R. Blackwell, D. Lacroix, T. Bach, J. Ishii, S. Hyland, T. Dumas, S. Carpenter, S. Chan, B. Sujlana, "17 μm microbolometer FPA technology at BAE Systems," *Proc. SPIE*, vol. 7298, 72980P, 2009.
5. D. Murphy, M. Ray, J. Wyles, et al., "640 × 512 17 μm Microbolometer FPA and sensor development," *Proc. SPIE*, vol. 6542, 65421Z, 2007.
6. W. A. Radford, R. Wyles, J. Wyles, J. B. Varesi, M. Ray, D. F. Murphy, A. Kennedy, A. Finch, E. A. Moody, F. Cheung, R. Coda, S. T. Baur, "Microbolometer uncooled infrared camera with 20-mK NETD," *Proc. SPIE*, vol. 3436, 636, 636645, 1998.

7. M. Russ, J. Bauer, H. Vogt, "The geometric design of microbolometer elements for uncooled focal plane arrays," *Proc. SPIE*, vol. 6542, 654223, 2007.
8. C. Vedel, J. L. Martin, J. L. Ouvrier-Buffet, J. L. Tissot, M. Vilain, J. J. Yon, "Amorphous-silicon-based uncooled microbolometer IRFPA," *Proc. SPIE*, vol. 3698, 276, 1999.
9. J. Brady, T. Schimert, D. Ratcliff, R. Gooch, B. Ritchey, P. Mccardel, K. Rachels S. Ropson M. Wand, M. Weinstein, J. Wynn, "Advances in amorphous silicon uncooled IR systems," *Proc. of SPIE*, vol. 3698, pp. 161-167, 99.
10. M. Almasri, D. P. Butler, and Z. Çelik-Butler, "Self-supporting semiconducting Y-Ba-Cu-O uncooled IR microbolometers with low-thermal mass," *IEEE/ JMEMS*, vol. 10, no. 3, pp. 469-476, 2001.
11. J.E. Gary, Z. Çelik-Butler, D. P. Butler, and M. Almasri, "Semiconducting Y-Ba-Cu-O as infrared detecting bolometers," *Proc. SPIE*, vol. 3436, pp. 555-566, 1998.
12. S. Sedky, P. Fiorini, K. Baert, L. Hermans, R. Mertens, "Characterization and optimization of infrared poly SiGe bolometers," IEEE Transaction on Electron Devices, vol. 46, no. 4, pp. 675-682, 1999.
13. V. N. Leonov, Y. Creten, P. De Moor, B. Du Bois, C. Goessens, B. Grietens, P. Merken, N. A. Perova, G. Ruttens, C. A. Van Hoof, A. Verbist, J. P. Vermeiren, "Small two-dimensional and linear arrays of polycrystalline SiGe microbolometers at IMEC-XenICs," *Proc. SPIE*, vol. 5074, pp. 446-457, 2003.
14. J. S. Shie, Y. M Chen, M. O. Yang, and B. C. S. Chou, "Characterization and modeling of metal-film microbolometer," *IEEE/ JMEMS*, vol 5, No. 4 December 1996.
15. M. Clement, E. Iborra, J. Sangrador, I. Barberan, "Amorphous $Ge_xSi_{1-x}O_y$ sputtered thin films for integrated sensor applications," J. Vac. Sci. Technol., vol. 19, no. 1, pp. 294-298, 2001.
16. M. M. Rana, D. P. Butler, "Radio frequency sputtered $Si_{1-x}Ge_x$ and $Si_{1-x}Ge_xO_y$ thin films for uncooled infrared detectors," *Thin Solid Films*, vol. 514, pp. 355-360, 2006.
17. A. H. Z. Ahmed, R. N. Tait, Tania B. Oogarah, H. C. Liu, Mike W. Denhoff, G. I. Sproule, and M. J. Graham, "A Surface micromachined amorphous $Ge_xSi_{1-x}O_y$ bolometer for thermal imaging applications," *Proc. SPIE*, vol. 5578, pp. 298-308, 2004.
18. A. H. Ahmed, and R. N. Tait, "Characterization of amorphous $Ge_xSi_{1-x}O_y$ for micromachined uncooled bolometer applications," J. appl. Phys., vol. 94, no. 8, pp. 5326-5332, 2003.
19. T. A. Enukova, N. L. Ivanova, Y. V. Kulikov, V. G. Marlyarov, I. A. Khrebtov, "Amorphous silicon and germanium films for uncooled microbolometers," Technical Physics Letter, vol. 23, pp. 504-506, 1997
20. Q. Cheng and M. Almasri, "Silicon germanium oxide ($Si_xGe_{1-x}O_y$) infrared material for uncooled infrared detection," *Proc. SPIE*, vol. 7298, 72980K , 2009.
21. Q. Cheng and M. Almasri, "Characterization of radio frequency sputtered $Si_xGe_{1-x}O_y$ thin films for uncooled micro-bolometer," *Proc. SPIE*, vol. 6940, 694011, 2008.

Mater. Res. Soc. Symp. Proc. Vol. 1245 © 2010 Materials Research Society 1245-A18-05

Demultiplexer/photodetector Integrated system based on a-SiC:H multilayered structures

P. Louro[1,2], M. Vieira[1,2,3], M. A. Vieira[1,2], J. Costa[1,2], M. Fernandes[1,2], M. Barata[1,2]
[1] Electronics Telecommunications and Computer Dept, ISEL, Lisbon, Portugal.
[2] CTS-UNINOVA, Lisbon, Portugal.
[3] DEE-FCT-UNL, Quinta da Torre, Monte da Caparica, 2829-516, Caparica, Portugal

ABSTRACT

In this paper we present results on the use of multilayered a-SiC:H heterostructures as an integrated device for simultaneous wavelength-division demultiplexing and measurement of optical signals. These devices are useful in optical communications applications that use the wavelength division multiplexing technique to encode multiple signals into the same transmission medium. The device is composed of two stacked p-i-n photodiodes, both optimized for the selective collection of photo generated carriers. The generated photocurrent signal using different input optical channels was analyzed at reverse and forward bias and under steady state illumination. A demux algorithm based on the voltage controlled sensitivity of the device was proposed and tested. An electrical model of the WDM device is presented and supported by the solution of the respective circuit equations. Other possible applications of the device in optical communication systems are also proposed.

INTRODUCTION

Wavelength division multiplexing (WDM) devices are used when different optical signals are encoded in the same optical transmission path, in order to enhance the transmission capacity and the application flexibility of optical communication and sensor systems. The use of WDM technologies not only provides high speed optical communication links, but also offers advantages such as higher data rates, format transparency, and self-routing. For these reasons WDM devices are crucial in fiber-optics sensing systems and in optical communication systems. Various types of available wavelength-division multiplexers and demultiplexers include prisms, interference filters, and diffraction gratings. Currently modern optical networks use Arrayed Waveguide Grating (AWG) as optical wavelength (de)multiplexers [1] that use multiple waveguides to carry the optical signals. In this paper we report the use of a monolithic WDM device based on an a-Si:H/a-SiC:H multilayered semiconductor heterostructure that combines the demultiplexing operation with the simultaneous photodetection of the signal. The device makes use of the fact that the optical absorption of the different wavelengths can be tuned by means of electrical bias changes or optical bias variations. This capability was obtained using adequate design of the multiple layers thickness, absorption coefficient and dark conductivities [2, 3]. The device described herein operates from 400 to 700 nm which makes it suitable for operation at visible wavelengths in optical communication applications.

DEVICE CONFIGURATION

The device is a multilayered heterostructure based on a-Si:H and a-SiC:H produced by PE-CVD at 13.56 MHz radio frequency. The configuration of the device includes two stacked p-i-n structures between two electrical and transparent contacts: p(a-SiC:H)-i(a-SiC:H)-n(a-SiC:H)-p(a-

SiC:H)-i(a-Si:H)-n(a-Si:H). The front intrinsic layer based on a-SiC:H is 200 nm thick, while the back one of a-Si:H is five times thicker (1000 nm). The thickness and optical gap of the intrinsic layers were adjusted to ensure in the front p-i-n structure high short wavelength absorption and high transparency to longer wavelengths, as well as high absorption of long wavelengths in the back structure. As a result, both front and back structures act as optical filters confining, respectively, the short and the long optical carriers, while the intermediate wavelengths are absorbed across both [4, 5]. The device was operated within the visible range using as optical signals, to simulate the transmission optical channel, the modulated light (external regulation of frequency and intensity) supplied by a red (R), a green (G) and a blue (B) LED with wavelengths of 470 nm, 524 nm and 626 nm, respectively.

INFLUENCE OF THE OPTICAL BIAS

Figure 1 shows the time dependent photocurrent signal measured under reverse (-8V, symbols) and forward (+1V, dotted lines) bias using different input optical signals without (no bias) and with (λ_L) red, green and blue steady state additional optical bias. Both optical signals and steady state bias were directed onto the device by the side of the a-SiC:H thin structure. The optical signals were obtained by wave square modulation of the LED driving current and the optical power intensity of the red, green and blue channels adjusted to 51, 90, 150 $\mu W/cm^2$, respectively. The steady state light was brought in LEDS driven at a constant current value (R: 290 $\mu W/cm^2$, G: 150 $\mu W/cm^2$, B: 390 $\mu W/cm^2$).

Results show that the blue steady state optical bias amplifies the signals carried out by the red (Fig. 1a) and the green channels (Fig. 1b) and reduces the signal of the blue channel (Fig. 1c). Red steady state optical bias has an opposite behavior, reinforcing the blue channel and decreasing the blue and the green channels. The green optical bias mainly affects the green channel, as the output signal is reduced while the signals of the red and blue channels show negligible changes.

The behavior of the device under steady state optical bias can be explained attending to the dependence of the internal electric field distribution. When an optical bias is applied it mainly enhances the field distribution within the less photo excited sub-cell: the back under blue irradiation and the front under red steady bias. Therefore, the reinforcement of the electric field under blue irradiation and negative bias increases the collection of the carriers generated by the red channel and decrease

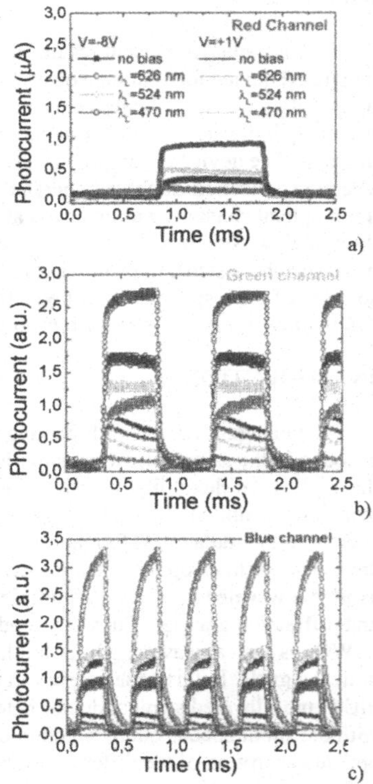

Figure 1 Red (a), green (b) and blue (b) channels under reverse and forward voltages without and with (λ_L) red, green and blue steady state bias.

the blue one. Under red optical bias, an opposite behavior is observed. The green bias absorption is balanced in both front and back cells and the collection of the carriers generated by the green channel strongly reduced.

The effect of the applied voltage on each optical channel depends also on the wavelength of the channel. In the blue channel it is observed a strong increase of the signal under reverse bias, while the red channel remains constant. The green channel is also voltage sensitive, although not as much as the blue channel.

BIAS SENSITIVE WAVELENGTH DIVISION MULTIPLEXING

A chromatic time dependent wavelength combination of red (λ_R=624 nm), green (λ_G=526 nm) and blue (λ_B=470 nm) input channels with different bit sequences, was used to generate a multiplexed signal on the device. The output photocurrent was measured under reverse and forward bias. In Figure 2 it is displayed the input channels (symbols and the transient multiplexed signals (lines) under reverse (-8 V) and forward (+1V) applied voltages. The bit rate used to transmit the optical signals along each channel was 4000bps. The reference level was assumed to be the signal when all the input channels were OFF (dark level). As expected from Figure 1 the red signal remains constant while the blue and the green decrease as the voltage changes from reverse to forward. The lower decrease in the green channel when compared with the blue one is probably due to its red-like behavior under forward bias.

The multiplexed signal depends on the applied voltage and on the ON-OFF state of each channel. Under reverse bias, there are eight separate levels while under positive bias they were reduced to one half. The highest level appears when all the channels are ON and the lowest if they are OFF. Furthermore, the levels ascribed to the mixture of three (R&G&B) or two input channels (R&B, R&G, G&B) are higher than the ones due to the presence of only one (R, G, B). It is interesting to notice that the sum of the R, G and B input channels is lower than the multiplexed signal showing capacitive effects due to the time-varying input channels. Under forward bias, the blue component of the combined spectra falls into the dark level, tuning the red/blue input channels.

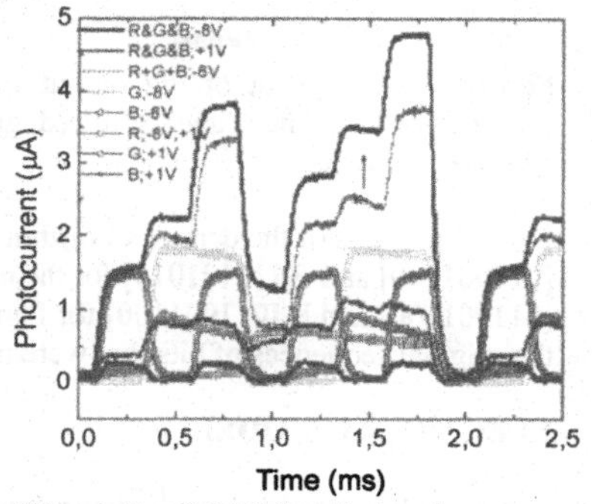

Figure 2 Multiplexed signal (symbols) and experimental (solid lines) results under positive and negative dc bias. The input channels (symbols) are displayed at-8V and +1V.

As expected from Figure 1 as the reverse bias increases the multiplexed signal exhibits a sharp increase if the blue component is present. By comparing the signals under forward and reverse bias and using a simple algorithm that takes into account the different sub-level behaviors under reverse and forward bias it is possible to split the red from the green component and to decode their RGB transmitted information.

RECOVERY OF THE INPUT SIGNALS

To recover the transmitted information at 4000bps per channel, i.e., to demultiplex the combined output signal, the multiplexed electric signal was divided into time slots. A demux algorithm was implemented in Matlab that receives as input the measured photocurrent and derives the sequence of bits that originated it. The algorithm makes use of the variation of the photocurrent instead of its absolute intensity to minimise errors caused by signal attenuation. A single linkage clustering method is applied to find automatically eight different clusters based on the measured current levels in both forward and reverse bias. This calibration procedure is performed for a short calibration sequence. Each cluster is naturally bound to correspond to one of the known eight possible combinations of red, green and blue bits. Following this procedure the sequence of transmitted bits can be recovered in real time by sampling the photocurrent at the selected bit rate and finding for each sample the cluster with closest current levels. In Figure 3 we present an example of the output obtained for two different sequences at 4000 bps.

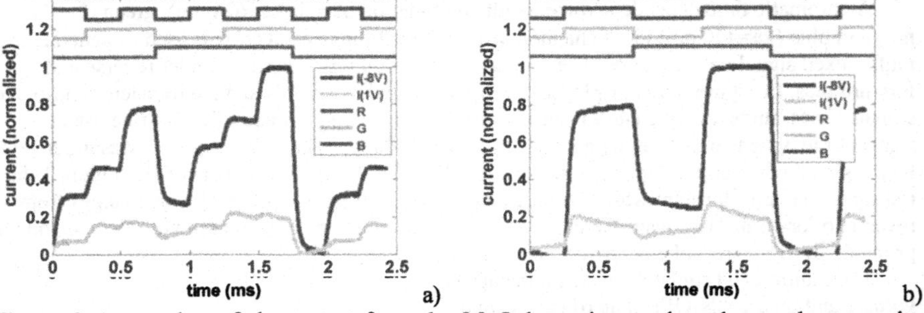

Figure 3 A snapshot of the output from the MatLab routine used to demux the transmitted sequence of bits. The sequence of red, green and blue bits (shown at the top) was derived from the measured currents.

Output data of the demux algorithm show that the derived sequences are R [00011100], G [011001100] and B [10101010] for the multiplexed signal of Figure 3a) and R [00011100], G [011001100] and B [011001100] for Figure 3b). Both were found to be in exact agreement with the original sequences of bits that were transmitted.

ELECTRICAL MODEL

The silicon-carbon pinpin device can be considered as a monolithic double pin photodiode structure with two red and blue optical connections for light triggering (Figure 4a). Based on the experimental results and device configuration an electrical model was developed [6]. Operation is explained in terms of the compound connected phototransistor equivalent model displayed in Figure 4 b. The current, i(t), under positive (open symbols) and negative (solid symbols) *dc* bias is displayed in Figure 4c. The input transient current sources used to simulate the photons absorbed in the front (blue, I_1), back (red, I_2), or across both (green, I_3 and I_4) photodiodes are also displayed (dash lines). R_1 and R_2 model the dynamic resistance of the internal and back junctions, respectively. The capacitive effects due to the transient nature of the input signals are simulated through C_1 and C_2 capacitors. We have used as input parameters the experimental values of Figure 2. To validate the model the experimental multiplexed signals at -8V and +1V are also shown (lines). Good agreement between experimental and simulated data was observed. The expected levels, under reversed bias, and their reduction under forward bias are clearly seen (Figure 3).

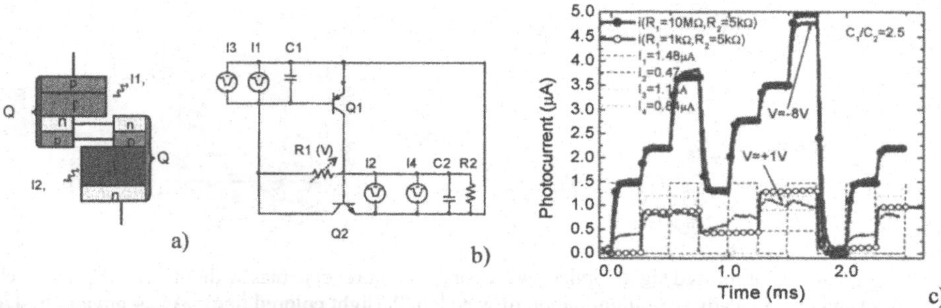

Figure 4 a) Compound connected phototransistor equivalent model. b) *ac* equivalent circuit. c) Multiplexed simulated (symbols) and experimental (solid lines) results under positive and negative dc bias. The current sources used as input channels (dash lines) are displayed.

If not triggered ON by light the device is nonconducting (low levels), when turned ON it conducts through different paths depending on the applied voltage (negative or positive) and trigger connection (Q_1 Q_2 or both).

Under negative bias (low R_1) the base emitter junction of both transistors are inversely polarized and conceived as phototransistors, taking, so, advantage of the amplifier action of neighboring collector junctions, which are polarized directly. This results in a charging current gain proportional to the ratio between both collector currents (C_1/C_2). The device behaves like a transmission system able to store and transport all the minority carriers generated by the current pulses, through the capacitors C_1 and C_2.Under positive bias (high R_1) the device remains in its non conducting state unless a light pulse is applied to the base of Q_2. This pulse causes Q_2 to conduct because the reversed biased n-p internal junction behaves like a capacitor inducing a charging current across both collector junctions.

OPTICAL TUNING OF THE INPUT CHANNELS

Besides demultiplexation purposes, other applications can be predicted using the device operation mechanism. One possible application is the optical tuning of the green and blue channels using adequate steady state optical bias. In Figure 5 it is displayed the multiplexed signal obtained at reverse bias using the same optical signal combinations of Figure 2. The output signal was acquired without steady state optical bias and with green (Fig. 5a) and blue (Fig. 5b) optical bias.

Results show that the presence of steady state illumination changes the output signal. Under green optical bias the multiplexed signal decreases when the green channel is ON and remains unchanged when it is OFF (Fig. 5a). Thus, the difference between the signals without and with green bias indicates the presence and absence of the green channel. A similar behavior is observed under blue steady state illumination (Fig. 5b), and thus the difference signal represents the successive ON-OFF states of the blue channel. It is important to remark that when the three channels are ON the signal with and without blue optical bias are identical, which results in a mismatch for the recognition of the blue channel.

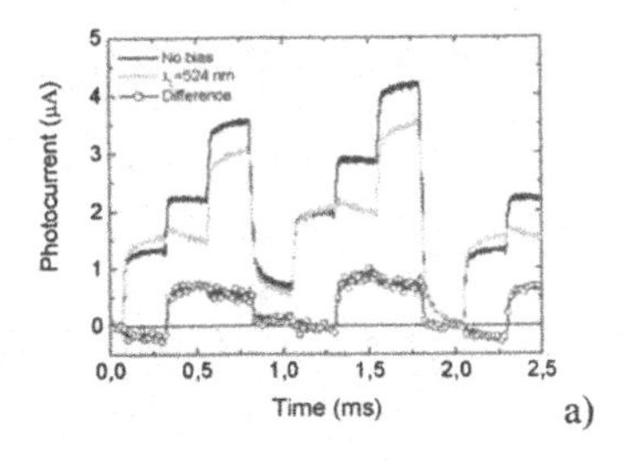

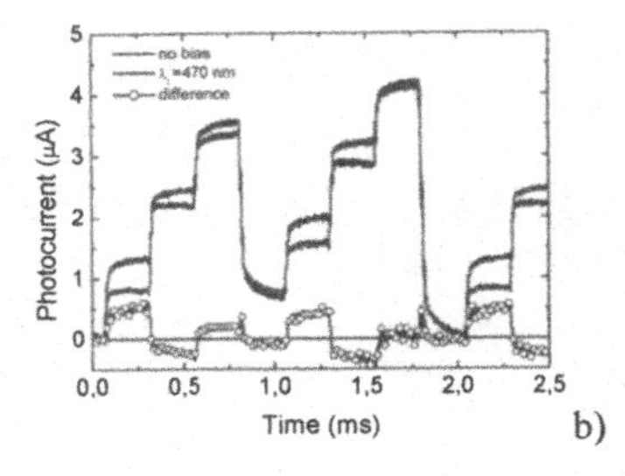

Figure 5 Multiplexed signal under reverse obtained at reverse bias without optical bias (black line) and under steady state illumination of wavelength (light colored line): a) 524 nm and b) 470 nm. The difference between both signals is also plotted (dark colored line).

This mechanism can be used for the detection of the presence of the input green and blue channels, which integrated in the optical communication systems can work as an alarm identifying any anomalous transmission of these channels.

CONCLUSIONS

A multilayered device based on a-SiC:H/a-Si:H was used for demultiplexing optical signals operating in the visible range at 4000 bps. The effect of the electrical applied voltage and of the use of steady illumination was analyzed. A recovery algorithm to demultiplex the optical signals was proposed and tested. An electrical model based on a two-input cascode circuit was developed to support the device operation. Other applications were proposed.

ACKNOWLEDGEMENTS

This work was supported by Fundação Calouste Gulbenkian and by POCTI/FIS/70843/2006.

REFERENCES

1 MichaBel as, Fiber Optics Handbook, Fiber, Devices and Systems for Optical Communication, Chap, 13, Mc Graw-Hill, Inc. 2002.
2 P. Louro, M. Vieira, Yu. Vygranenko, A. Fantoni, M. Fernandes, G. Lavareda, N. Carvalho Mat. Res. Soc. Symp. Proc., 989 (2007) A12.04.
3 M. Vieira, M. Fernandes, P. Louro, A. Fantoni, Y. Vygranenko, G. Lavareda, C. Nunes de Carvalho, Mat. Res. Soc. Symp. Proc., Vol. 862 (2005) A13.4.
4. P. Louro, M. Vieira, M.A. Vieira, M. Fernandes, A. Fantoni, C. Francisco, M. Barata, Physica E: Low-dimensional Systems and Nanostructures, 41 (2009) 1082-1085.
5 P. Louro, M. Vieira, M. Fernandes, J. Costa, M. A. Vieira, J. Caeiro, N. Neves, M. Barata, , Phys. Status Solidi C 7, No. 3–4, 1188– 1191 (2010).
6 M. A. Vieira, M. Vieira, M. Fernandes, A. Fantoni, P. Louro, M. Barata, Amorphous and Polycrystalline Thin-Film Silicon Science and Technology 2009, MRS Proceedings Vo. 1153, A08-0.

Mater. Res. Soc. Symp. Proc. Vol. 1245 © 2010 Materials Research Society 1245-A18-06

Light-triggered silicon-carbon pi'npin devices for optical communications: Theoretical and electrical approaches

M. A. Vieira[1,2], M. Vieira[1,2,3], J. Costa[1,2], P. Louro[1,2], M. Fernandes[1,2], A. Fantoni[1,2]
[1]Electronics Telecommunications and Computer Dept, ISEL, 1959-007, Lisbon, Portugal.
[2]CTS-UNINOVA, Quinta da Torre, 2829-516, Caparica, Portugal.
[3]DEE-FCT-UNL, Quinta da Torre, 2829-516, Caparica, Portugal.

ABSTRACT

In this paper a light-activated multiplexer/demultiplexer silicon-carbon device is analysed. An electrical model for the device operation is presented and used to compare output signals with experimental data. An algorithm that takes into accounts the voltage and the optical bias controlled sensitivities is developed. The device is a double pi'n/pin a-SiC:H heterostructure with two optical gate connections for light triggering in different spectral regions. Multiple monochromatic pulsed communication channels were transmitted together, each one with a specific bit sequence. The combined optical signal was analyzed by reading out, under different applied voltages and optical bias, the generated photocurrent across the device. Experimental and simulated results show that the output multiplexed signal has a strong nonlinear dependence on the light absorption profile, i.e. on the incident light wavelength, bit rate and intensity under unbalanced light generation of carriers. By switching between positive and negative voltages the input channels can be recovered or removed from the output signal.

INTRODUCTION

The current need for communication demands the transmission of huge amounts of information. To increase the capacity of transmission and allow bidirectional communication over one strand fiber, wavelength-division multipexing (WDM) is used [1]. In the WDM technique multiple optical signals are transmitted on a single optical fiber using different wavelengths (colors) of the light source to encode different signals [2]. A WDM system uses a multiplexer at the transmitter to join the signals together and a demultiplexer at the receiver to split them apart. High optical nonlinearity makes semiconductor amplifiers attractive for all optical signals. There has been much research on semiconductor optical amplifiers as elements for optical signal processing, wavelength conversion, clock recovery, signal demultiplexing and pattern recognition [3], where a particular band, or spread, or frequencies need to be filtered from a wider range of mixed signals. This paper reports results on the use of a double pi'n/pin a-SiC:H WDM heterostructure as an active band-pass filter transfer function depending on the wavelength of the trigger light and device bias.. The dynamic response can range from a positive feedback (regeneration) under positive bias, to two different behaviors under negative bias. Under negative bias the device acts either as an active multiple-feedback filter with internal gain or in a mode that preserves the amplitude of the signal, depending on the trigger lights. An electrical model gives insight on the physics of the device.

EXPERIMENT

The tunable optical devices were produced by PECVD and optimized for a proper fine tuning of a specific wavelength. The active device consists of a p-i'(a-SiC:H)-n / p-i(a-Si:H)-n heterostructure with low conductivity doped layers ($<10^{-7}\Omega^{-1}cm^{-1}$). The thicknesses and optical gap of the thin i'- (200nm; 2.1 eV) and thick i- (1000nm; 1.8eV) layers are optimized for light absorption in the blue and red ranges, respectively [4]. Transparent contacts have been deposited on front and back layers to allow the light to enter and leave from both sides (see insert in Figure 1). To test the sensitivity of the device under different applied voltages and optical bias three modulated monochromatic lights channels: red (R: 626 nm; 51μW/cm^2), green (G: 524 nm; 73μW/cm^2) and blue (B: 470nm; 115μW/cm^2) and their polychromatic combinations (multiplexed signal) illuminated separately the device and the generated photocurrent was measured under positive and negative voltages (+3V<V<-10V), with and without steady state green optical bias (G: 524 nm; 73μW/cm^2). The light modulation frequency of each channel was chosen to be multiple of the others to ensure a synchronous relation of ON-OFF states along each cycle. The optical powers were adjusted to give different output signal magnitudes at -8V.

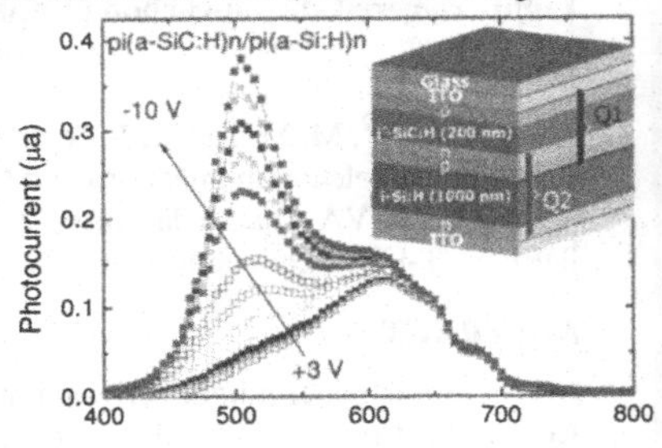

Figure 1 p-i'-n-p-i-n spectral photocurrent under different applied voltages. Insert: device configuration.

RESULTS

Voltage controlled wavelength discrimination

Figure 1 displays the spectral photocurrent under different applied voltages. Results show that for long wavelengths (> 600 nm) the spectral response is independent of the applied bias whereas for short wavelengths the photocurrent increases with negative voltages (reverse bias).

Figure 2 displays the single and multiplexed signals under negative (-8V) and positive (+1V) electrical bias. As expected from Figure 1, the input red signal remains constant while the blue and green ones decrease as the voltage changes from positive to negative. The output multiplexed signal, obtained with the combination of the three optical sources, depends on both the applied voltage and on the ON-OFF state of each input optical channel.

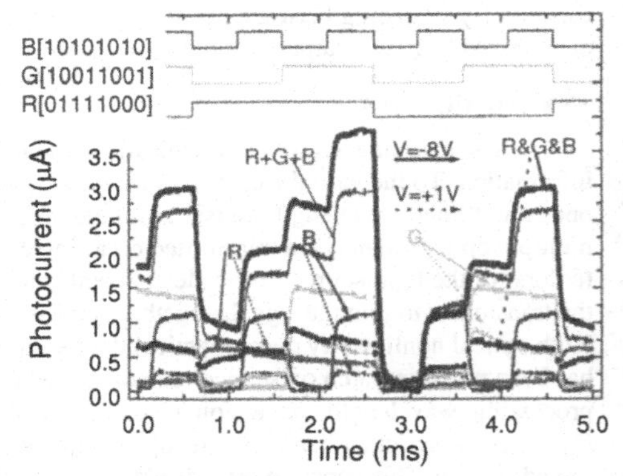

Figure 2 Single (R, G and B) and combined (R&G&B) signals under -8V (solid arrows) and +1V (dotted arrows) applied voltage.

Under negative bias, the multiplexed signal presents eight separate levels. The highest level appears when all the channels are ON and the lowest if they are all OFF. Furthermore, the levels ascribed to the mixture of three or two input channels are higher than the ones due to the presence of only one (R, G, B). Optical nonlinearity was detected; the sum of the input channels (R+B+G) is lower than the correspondent multiplexed signals (R&G&B). This optical amplification, mainly on the ON-ON states, suggests capacitive charging currents due to the time-varying nature of the incident lights. Under positive bias the levels were reduced to one half since the blue component of the combined spectra falls into the dark level, the red remains constant and the green component decreases.

To recover the transmitted information (8 bit per wavelength channel) the multiplexed signal, during a complete cycle, was divided into eight time slots, each corresponding to one bit where the independent optical signals can be ON (1) or OFF (0). Under positive bias, the device has no sensitivity to the blue channel (Figure 1-2) and because of it the red and green sequences can be identified. The highest level corresponds to both channels ON (R=1, G=1), and the lowest to the OFF-OFF stage (R=0; G=0). The two levels in-between are related with the presence of only one channel ON, the red (R=1, G=0) or the green (R=0, G=1). To distinguish between these two situations and to decode the blue channel, the correspondent sub-levels, under reverse bias, have to be analyzed. The highest increase at -8V corresponds to the blue channel ON (B=1), the lowest to the ON stage of the red channel (R=1) and the intermediate one to the ON stage of the green (G=1). Using this simple key algorithm the independent red, green and blue bit sequences were decoded as: R[01111000], G[10011001] and B[10101010], as shown on the top of Figure 2, which are in agreement with the original sequence of the independent channels.

Optical bias controlled wavelength discrimination

In Figure 3 the input and the multiplexed channels, with and without green bias, are displayed at -8V. The sequence of bits is shown at the top of the figure to guide the eyes. Results show that the presence of the optical bias reduces significantly the amplitude of green channel while a slight increase is observed for the other two. The sum of the input channels (R+G+B; symbol) shows that when the green channel is ON no amplification occurs. This suggests that the green channel can be tuned by making the difference between the multiplexed signal with and without green irradiation (Δ, symbols).This nonlinearity is due to the transient asymmetrical light penetration of the input channels into the device and to its optical filters properties. When an external optical bias is applied it influences the field distribution within the less photo excited sub-cell. Under green light irradiation, since the green photons are absorbed across front and back photodiodes, the electric field decreases on both sub-cells. So, some of the carriers generated by the green channel, also in both sub-cells, recombine and the collection decreases. When the red or blue channels are ON, the generation occurs only in one sub-cell. The electrical field, in the presence of the red and blue channels, lowers, respectively, in the back and front photodiodes, while the correspondent front and back photodiodes reacts by assuming a reverse bias configuration

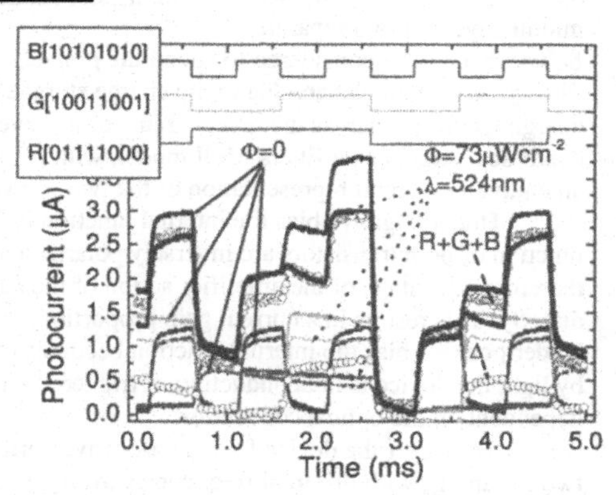

Figure 3 Single and combined signals @-8V; without (solid arrows) and with (dotted arrows) green optical bias.

compensating the effect of the green optical bias [5]. This self bias effect explains the slightly increase on the red and blue collection under green optical bias. This nonlinearity provides the possibility for selectively removing and adding a particular wavelength and can be used to boost signal power after multiplexing or before demultiplexing which usually introduce optical loss into the system.

THEORETICAL MODEL AND VALIDATION

The silicon-carbon pi'npin device was considered as a monolithic double pin photodiode structure with two optical connections, the intrinsic layers, for light triggering. Operation is explained in terms of the compound connected phototransistor equivalent model displayed in Figure 4a. The two-transistor model (Q_1-Q_2) is obtained by bisecting the two middle layers in two separate halves that can be considered to constitute *pinp* (Q_1) and *npin* (Q_2) phototransistors separately. The dynamical effects are due to the charge storage and leakage currents in the transistors. So, charging currents across the reversed junctions have to be considered. Supported in the complete dynamical large signal Ebers-Moll model, with resistances, R_1 and R_2, and capacities, C_1 and C_2, an equivalent circuit representation of the two-transistor model is displayed in Figure 4b [6].

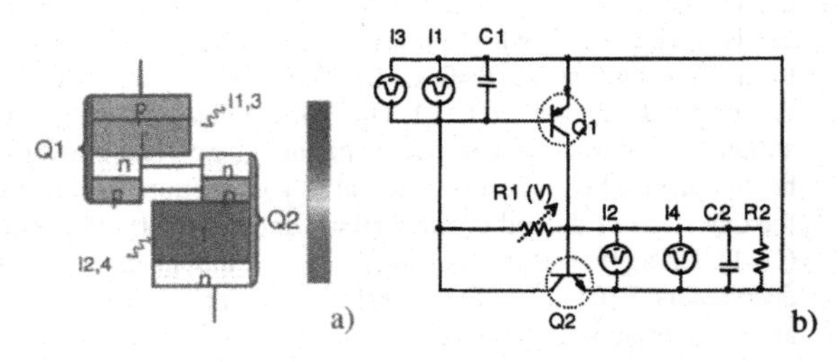

Figure 4 Schematics diagram of the device: a) two-transistor model; b) *ac* circuit representation.

Under negative bias the internal junction is forward bias (low R_1) and the base-emitter junction of both transistors are inversely polarized and conceived as phototransistors, taking therefore advantage of the amplifier action of the adjacent collector junction which is polarized directly. This results in a current gain proportional to the ratio between both collector currents. Under positive bias the internal junction becomes reverse-biased (high R_1). So, if not triggered ON by light the device is nonconducting, if triggered ON it goes to its ON state with an output current dependent on the optical connection.

To trigger the device four square-wave current sources with different intensities are used; two , I_1 and I_2, with different frequencies to simulate the input blue and red channels and other two, I_3 and I_4, with the same frequency to simulate the green channel due to its asymmetrical absorption across both front and back phototransistors. When a pulse of light triggers either Q_1 or Q_2 during a small time interval ∂t, the voltage across the capacitor changes form v_{off} to v_{on}. The charge that flows across them is not $(i_{on}-i_{off})\,\partial t$. More charge has to flow. At higher voltage it needs to store higher charge and that charge also has to be supplied. This charge appears as a current component. So, a leaky capacitor is considered and should be represented as a parallel combination of an ideal capacitor C and a resistance R (R//C).

Under negative bias, when only I_1 is used to trigger the device, the voltage drops across the emitter base of Q_1. The device acts as an active multiple-feedback filter (with two feedback paths R_2 and C_2), without changing the amplitude of the signal due to the attenuation of the signal trough R_1. If I_2 triggers Q_2, the device acts as an active multiple-feedback filter with internal gain.

Under positive bias (high R_1) the device remains in its non conducting state unless a light pulse (I_2 or I_2+I_4) is applied to the base of Q_2. This pulse causes Q_2 to conduct because the reversed biased n-p internal junction behaves like a capacitor inducing a charging current (I_2+I_4) across both collector junctions. The collector of the conducting transistor pulls low, moving the Q_1 base toward its collector voltage, which causes Q_1 to conduct. The collector of the conducting Q_1 pulls high, moving the Q_2 base in the direction of its collector. This positive feedback (regeneration) reinforces the Q_2 already conducting state and a current I_2+I_4 will flow on the external circuit. If not triggered ON (all the input channels OFF), either under positive or negative bias, the device is nonconducting (low level).

The multiplexed signal was simulated by applying the Kirchhoff's laws for the simplified *ac* equivalent circuit and the four order Runge-Kutta method to solve the corresponding state equations. MATLAB was used as a programming environment and the input parameters chosen in compliance with the experimental results (Figure 2, Figure 3). The simulated transient currents (symbols) under negative and positive *dc* bias are displayed in Figure 5a. In Figure 5b the currents under negative bias, with ($\Phi \neq 0$) and without ($\Phi=0$), green bias, are compared. To simulate the green background, the intensities of the current sources were multiplied by the on/off ratio between the input channels with and without optical bias (Figure 3). The same bit sequence of Figure 2 was used. To validate the model the experimental multiplexed signals are also shown (solid lines).

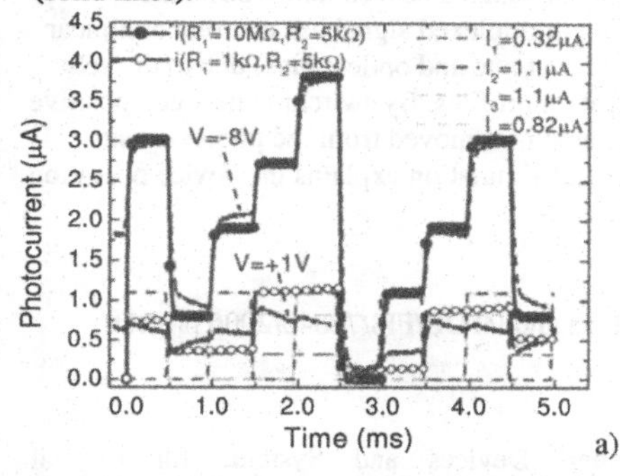

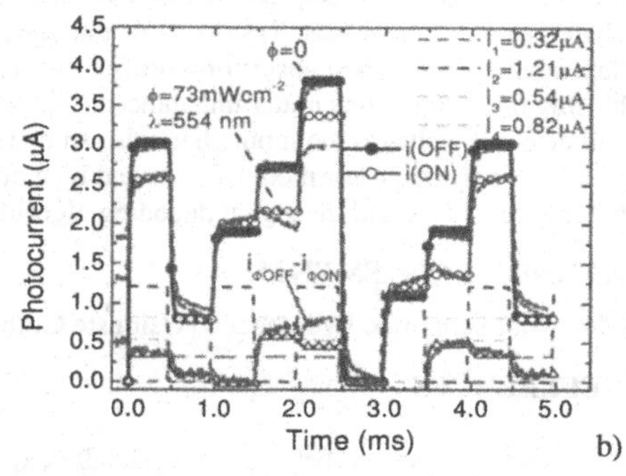

Figure 5 Simulated multiplexed (symbols), current sources (dash lines) and experimental (solid lines) signals under: a) positive (R_1=10MΩ; +1V) and negative (R_1=1KΩ; -8V) *dc* bias; b) negative *dc* bias (R_1=1kΩ; -8V) with and without green irradiation. C_1/C2=2.5

Good agreement between experimental (figures 2 and 3) and simulated data was observed. The eight expected levels, under reversed bias, and their reduction under forward bias or green irradiation, are clearly seen. The device behaves like a transmission system able to store and transport the minority carriers generated by the current pulses, through the capacitors C_1 and C_2.

Under negative bias (low R_1) and $\Phi=0$, the expected eight levels are detected, each one corresponding to the presence of three, two, one or no color channel ON. The expected optical amplification (Figure 2) is observed due to the effect of the active multiple-feedback filter when the back diode is light triggered. Green steady state irradiation ($\Phi \neq 0$) and negative bias moves asymmetrically Q_1 and Q_2 bases toward their emitter voltages (self-forward effect), resulting in lower values of I_3 and I_4 when compared to no optical bias (≅75%). Opposite behaviour occurs with I_1 and I_2 that slightly increase due to the self reverse effect of the less absorbing diode. The effect of the green bias is negligible if I_1 or I_2 are ON since Q_1 and Q_2 are triggered separately and the discharge of the capacitors independent. During the green pulse (I_3 and I_4 ON) only residual charges are transferred through R_1. So, only the charges generated in the base of Q_2 (I_4) are collected as can be confirmed by the good agreement between simulated and experimental curves of the multiplexed signals with and without optical bias (Figure 5b).

Results show that the two-transistor model explains the difference between the conduction mechanisms, under positive and negative bias, helping to understand the signal decoding algorithm. Under positive bias the red and green channels are immediately decoded. Under negative bias, different charging currents have to be considered. The balance between them depends on the presence of three, two or one channel ON (Figure 2-3). So, by comparing the different signal sublevels under positive and negative bias the input channels are recovered.

CONCLUSIONS

A double pi'n/pin a-SiC:H heterostructure with two optical gate connections for light triggering in different spectral regions was presented.

Multiple monochromatic pulsed communication channels were transmitted together, each one with a specific bit sequence and the combined optical signal analyzed under different electrical and optical bias. Results show that the output multiplexed signal has a strong nonlinear dependence on the light absorption profile (wavelength, bit rate and optical bias) due to the self biasing of the junctions under unbalanced light generation profiles. By switching between positive and negative voltages the input channels can be recovered or removed from the photocurrent.

A two-transistor model, supported by a numerical simulation explains the device operation helping to understand the signal decoding algorithm.

ACKNOWLEDGEMENTS

This work supported by Fundação Calouste Gulbenkian and PTDC/FIS/70843/2006 project.

REFERENCES

1. Michael Bas, Fiber Optics Handbook, Fiber, Devices and Systems for Optical Communication, Chap, 13, Mc Graw-Hill, Inc. 2002.
2. Mark G. Kuzyk, Polimer Fiber Optics, Materials Physics and Applications, Taylor and Francis Group, LLC; 2007.
3. M. J. Connelly, Semiconductor Optical Amplifiers. Boston, MA: Springer-Verlag, 2002. ISBN 978-0-7923-7657-6.
4. M. Vieira, A. Fantoni, M. Fernandes, P. Louro, G. Lavareda, C. N. Carvalho, Journal of Nanoscience and Nanotechnology, Vol. 9, , Number 7, July 2009 , pp. 4022-4027(6).
5. M. Vieira, A. Fantoni, P. Louro, M. Fernandes, R. Schwarz, G. Lavareda, C.N. Carvalho, Vacuum, Volume 82, Issue 12, 8 August 2008, Pages 1512-1516.
6. M A Vieira, M. Vieira, M. Fernandes, A. Fantoni, P. Louro, M. Barata, Amorphous and Polycrystalline Thin-Film Silicon Science and Technology — 2009, Mater. Res. Symp. Proc. Vol. 1153, pp 73-178.

Poster Session: Thin Film Transistors

Mater. Res. Soc. Symp. Proc. Vol. 1245 © 2010 Materials Research Society 1245-A19-02

Threshold voltage shift variation of a-Si:H TFTs with anneal time

A. Indluru[1], S. M. Venugopal[2], D. R. Allee[2,3], and T.L. Alford[1, 2,3]
[1] School of Mechanical, Aerospace, Chemical and Materials Engineering, Arizona State University, Tempe, Arizona 85287
[2] Flexible Display Center, Arizona State University, Tempe, Arizona 85287
[3] School of Electrical, Computer, and Energy Engineering, Tempe, Arizona 85287

ABSTRACT

Hydrogenated amorphous silicon (a-Si:H) thin-film transistors (TFTs) are widely used in many areas and the most important application is in active matrix liquid crystal display. However, the instability of the a-Si:H TFTs constrains their usability. These TFTs have been annealed at higher temperatures in hope of improving their electrical performance. But, higher anneal temperatures become a constraint when the TFTs are grown on polymer-based flexible substrates. This study investigates the effect of anneal time on the performance of the a-Si:H TFTs on PEN. Thin-film transistors are annealed at different anneal times (4 h, 24 h, and 48 h) and were stressed under different bias conditions. Sub-threshold slope and the off-current improved with anneal time. Off-current was reduced by two orders of magnitude for 48 hours annealed TFT and sub-threshold slope became steeper with longer annealing. At positive gate-bias stress (20 V), threshold voltage shift (ΔV_t) values are positive and exhibit a power-law time dependence. High temperature measurements indicate that longer annealed TFTs show improved performance and stability compared to unannealed TFTs. This improvement is due to reduction of interface trap density and good a-Si:H/insulator interface quality with anneal time.

INTRODUCTION

Hydrogenated amorphous silicon (a-Si:H) thin-film transistors (TFTs) is the dominant switching element in active matrix liquid crystal displays (AMLCD) and also used in pixel circuits of organic light emitting displays (OLED). [1-4] There is a considerable industrial interest in exploring the performance of the low temperature fabricated TFTs on polymer substrates for flexible applications. However, the low temperature fabricated TFTs produce a substantial amount of defects in a-Si:H layer and the insulator compared to TFTs fabricated at standard temperatures on glass substrate. [5-6]

Fabricating low-leakage insulators with low defect density a-Si:H layer at low temperatures is a great challenge. The most important drawback of a-Si:H TFTs is the threshold-voltage shift (ΔV_t) as a result of prolonged application of gate-bias stress. The trapped charge in the insulator and the defect creation in the a-Si:H layer are two main reasons for threshold voltage shift. It is important to minimize these defects in the TFTs for better performance. Annealing at elevated temperatures and high temperature deposition can be an option to improve the performance of the TFTs. [5-9] However, high anneal temperatures become a constraint when the TFTs are grown on polymer-based flexible substrates, as they are sensitive to high temperatures. In this study, low temperature fabricated (180 °C) a-Si:H TFTs on flexible substrate have been annealed for different times (4, 24, and 48 hours) to investigate their performance and stability with anneal time.

EXPERIMENTAL DETAILS

The a-Si:H TFTs employed in this study have a bottom gate inverted staggered structure. Figure 1 shows the cross sectional schematic of the TFT. Bottom-gate inverted staggered structures were fabricated on flexible substrate like stainless steel and heat stabilized PEN (Polyethylene Naphthalate) at 180 °C. First, the gate-metal molybdenum was deposited and patterned, followed by the deposition of silicon nitride (SiN_x), the a-Si:H active layer and n^+ amorphous silicon/ aluminum bilayer as source/drain contacts by plasma enhanced chemical vapor deposition. A nitride passivation layer is deposited before the drain/source contacts are etched. Finally after fabrication, the TFTs are annealed at 180 °C in nitrogen atmosphere for 3 hours. The width and length of all the TFTs in this study is 96 μm and 9 μm, respectively. To study the effect of anneal, the TFTs are annealed in 5 % H_2/Ar (reducing atmosphere) at 150 °C for 4, 24, and 48 hours. The bias-stress measurements are performed for up to 10^4 seconds and interrupted at regular intervals to measure the transfer characteristics. The source and the drain electrodes were grounded during the stress experiments. The TFTs were stressed with both positive (20 V) and negative (-20 V) gate bias voltage. Transfer characteristics were measured at source-drain voltage of 10 V and gate voltages from -20 V to 20 V. The threshold voltage is extracted from the intercept of the extrapolation of the linear region of the square root of I_{DS} vs V_{GS} with the horizontal axis. The sub-threshold slope is derived from the inverse of the maximum slope of the logarithmic scale I_{DS} vs V_{GS}. Subsequently, the stress induced shifts in these TFT electrical parameters as a function of stress time and gate-bias stress are determined.

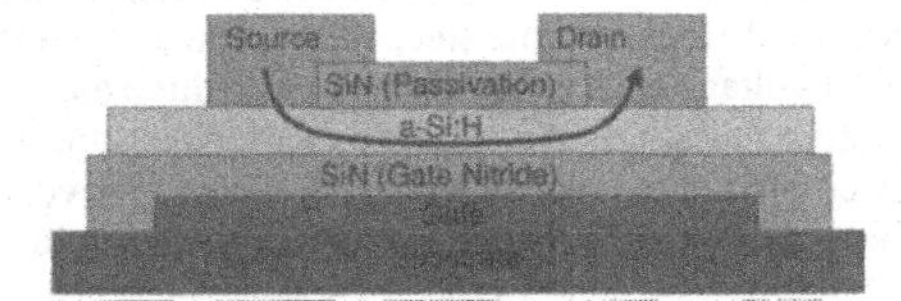

Fig. 1 Schematic showing the cross sectional of the a-Si:H TFT.

RESULTS AND DISCUSSION

Table 1 lists the variation of off-current and sub-threshold slope (*S*) of TFTs annealed for different times. The performance of the TFTs are often limited by off-current and sub-threshold slope. The combination of these two parameters typically determines the on/off ratio of the devices. In AMLCDs, the TFTs are desired to have high on-off ratio and low leakage current for high quality display. [11] We find that the sub-threshold slope and off-current improves considerably for 48 h annealed TFTs when compared to unannealed TFTs. The sub-threshold slope of the unannealed TFT is 1.5 V/decade and for 48 hour annealed TFT it improves to 0.6 V/decade. In addition, the off-current values reduces by two orders of magnitude with longer anneal times. The improvement of these parameters after annealing is substantial to device performance. The off-current is mainly dictated by both deep localized states in the a-Si:H layer and interface states at the a-Si:H/SiN_x interface. Figure 2 shows the band diagram of a-Si:H TFT under different bias conditions. At zero or very low positive gate voltage, the Fermi-level is in the middle of the band gap close to its intrinsic level; hence, most of the induced carriers go into either the deep localized states in the a-Si:H layer and/or into interface states. Only a very small fraction of electrons close to the front a-Si:H/insulator interface participate in the conduction. As the positive bias on the gate increases, the density of electrons increases and this leads to an exponential growth of current and subsequent transition to the above-threshold regime of operation. Therefore increased anneal times result in a decrease in either the number of deep states in the a-Si:H layer or the interface states at the a-Si:H/SiN_x interface and leads to a reduction in the off-current with a steeper sub-threshold slope.

Table 1. Summary of sub-threshold slope and off-current of a-Si:H TFTs at different anneal times.

Anneal time (hours)	Sub-threshold slope (V/decade)	Off-current (A)
Unannealed	1.53	2.1×10^{-8}
4 h anneal	1.4	8.2×10^{-9}
24 h anneal	1.2	6.2×10^{-9}
48 h anneal	0.6	8.3×10^{-10}

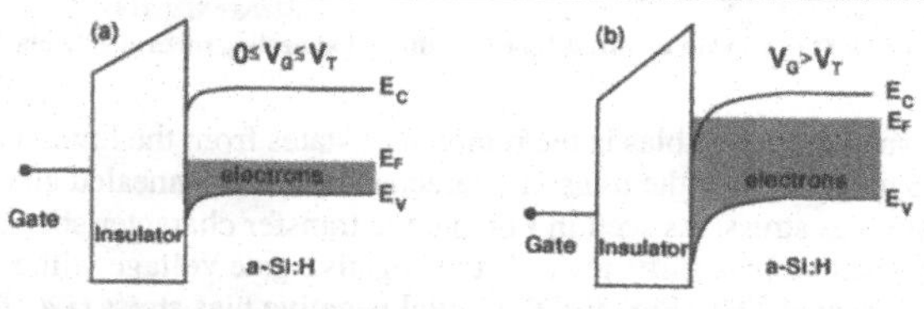

Figure 2. a-Si:H TFT band-diagram under different gate-bias voltages: (a) for gate voltage less than the threshold voltage b) for gate voltage higher than the threshold voltage.

Figure 3 illustrates the variation in ΔV_t with the bias stress time for different anneal times and gate-bias stress (20 V and -20 V). For the positive gate-bias stress (Fig. 3a), the threshold voltages are all positively shifted and increases with bias stress times. The positive shift in the threshold voltage is due to creation of states (more silicon dangling bonds) in the lower part of the band gap. Hence, the threshold voltage increases with longer bias-stress time and follows a power-law dependence with respect to stress time [12], $\mathbf{\Delta V_t \sim t^{\beta}}$ where β is a constant for a given gate bias stress. The ΔV_t data fits well to this equation for different β values corresponding to different anneal times. The straight line fits to unannealed, 4 h annealed, 24 h annealed and 48 h annealed TFTs have β values of 0.6, 0.54, 0.54 and 0.4, respectively. These values are in good agreement with the literature. [12-14] The reduction in the β value from 0.6 for unannealed to 0.4 for 48 h annealed TFT indicates improvement in the properties of the a-Si:H channel and/or the a-Si:H/insulator interface. The straight line extrapolation of ΔV_t for 48 h annealed TFTs show a better life time when compared to unannealed TFTs. The life time of 48 h annealed TFTs improves by 3 factors when compared to unannealed TFTs for a ΔV_t of 10 V. For the negative gate-bias stress (Fig. 3b), the ΔV_t is smaller when compared to the positive gate-bias stress. The threshold voltages are negatively shifted for unannealed and 4 hours annealed TFTs. However for longer annealed TFTs (24 and 48 hours), there is a positive ΔV_t and a turnaround. Tai *et al.* [14] have observed a similar turnaround phenomenon of ΔV_t at higher negative gate-bias stress. This can also be associated with a combination of both charge trapping and state creations.

Figures 4a and b show transfer characteristics of the a-Si:H TFTs for a drain voltage of 10 V and positive gate-bias of 20 V for the unannealed and 48 hours annealed cases. The threshold voltage increase with bias-stress time is depicted in the transfer characteristic as a nearly parallel shift towards the positive gate-voltage direction. Unlike positive bias, the

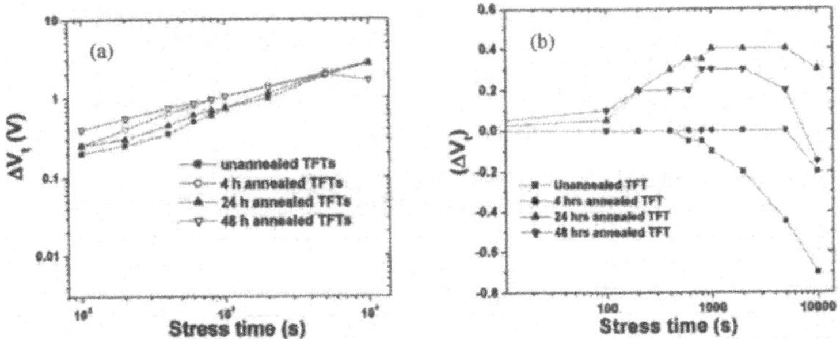

Fig. 3 Threshold voltage shift (ΔV_t) with the stress time for different annealing times at gate bias stress of (a) 20 V, and (b) -20 V.

dominant mechanism at negative bias is the removal of states from the lower part of the band gap. [15] Figures 5 a and b show the transfer characteristics of unannealed and 48 hours annealed TFTs for -20 V gate bias stress. As seen in Fig. 5a, the transfer characteristics of the unannealed a-Si:H TFT under negative bias shifts towards the negative gate voltage with stress times. However for 48 h annealed TFT (Fig. 5b), the initial negative bias stress (*i.e.*, for less than 600 s stress time) results in a shift of transfer characteristics in the positive gate voltage. Also for longer annealed TFTs (24 and 48 h) unlike unannealed and 4 h annealed TFTs, a positive ΔV_t (Fig. 5b) is observed. This indicates state creations and is similar to what is seen during positive gate-bias stress. However upon further increase in the gate-bias

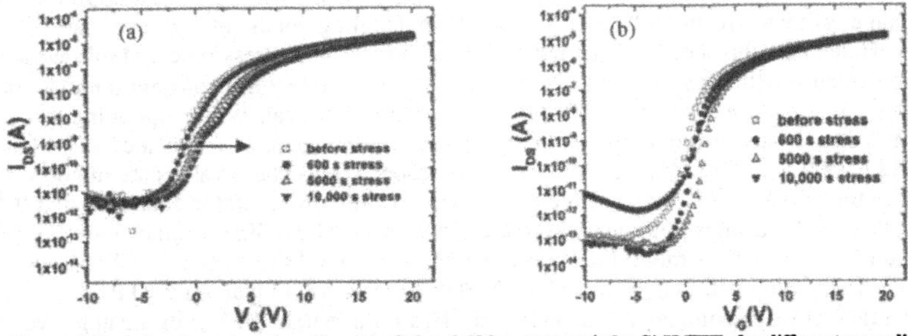

Figure 4 Transfer characteristics of unannealed and 48 hours annealed a-Si:H TFTs for different annealing times at gate bias stress of 20V; (a) unannealed, and (b) 48 hours annealed.

stress, the transfer curves shift towards the negative gate voltage. A similar shift (not shown) is also observed in the a-Si:H TFTs annealed for 24 h. This is reflected in the ΔV_t vs stress time (Fig. 3 b) as a decrease in ΔV_t at longer stress times for 24 and 48 hours annealed TFTs (negative bias-stress). However for longer annealed TFTs the ΔV_t is small compared to unannealed TFTs. This shows that longer anneals can be an effective tool to improve the performance of the low temperature fabricated TFTs on polymer substrates.

To further investigate the effect of anneal time on the performance and stability of the a-Si:H TFTs, high temperature measurements have been performed at every 25 °C from room temperature to 125 °C. Figure 6 shows the variations of normalized sub-threshold slope (S/S_o) as a function of temperature for different anneal times; where, S_o is the slope at room temperature.

As shown in the Figure 6, there is negligible increase in the sub-threshold slope value with temperature for TFTs annealed for extended times (24 and 48 hour annealed) when compared to the unannealed and 4 h annealed TFTs. In the latter case, the sub-threshold slope increases by a factor of 2.4 at 100 °C when compared to room temperature. However for 24 hour and 48 hour

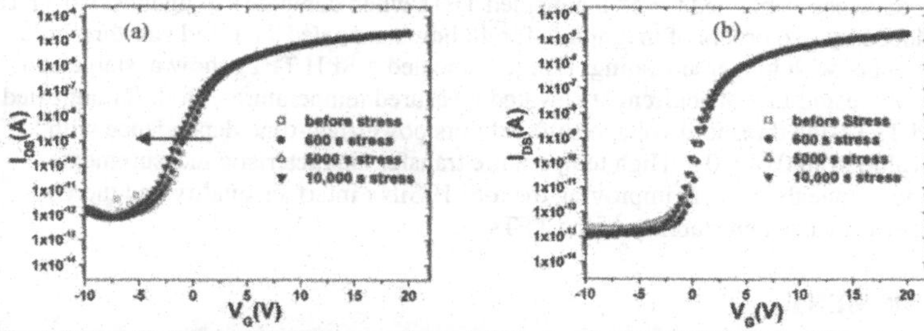

Figure 5 Transfer characteristics of unannealed and 48 hours annealed a-Si:H TFTs for different annealing times at gate bias stress of -20 V; (a) unannealed and (b) 48 hours annealed.

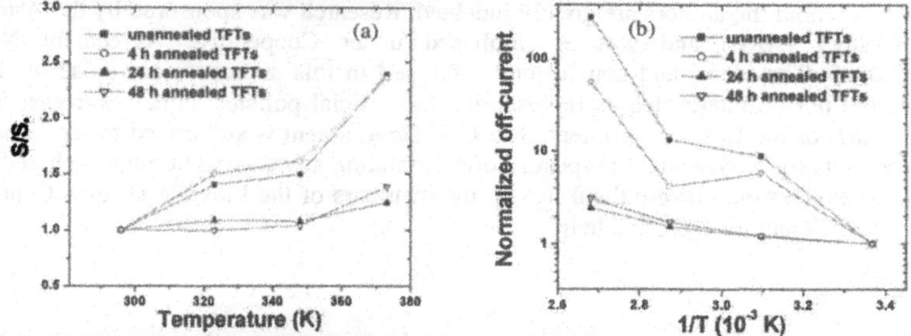

Figure 6 (a) Variations of normalized sub-threshold slope (S/S_o) with temperature for different anneal times; (b) Arrhenius plot of off-current with temperatures for all the TFTs.

annealed TFTs, the sub-threshold slope increases only by afactor of 1.4 at 100 °C when compared to room temperature. Sub-threshold slope is largely decided by the quality of a-Si:H/SiN_x interface and localized state density near the interface. [7] The longer annealed TFTs display a stable sub-threshold slope value with temperature indicating a negligible degradation of the properties at the a-Si:H/SiN_x interface.

The TFTs fabricated at low temperatures have a less stable gate insulator due to a substantial amount of trapped charge inside the insulator. [5] The trapped charges are themionically excited at high temperatures which results in increase in the off-current. Figure 6 (b) shows the Arrhenius plot of off-current with measured temperature for all the TFTs. This temperature dependence of off-current indicates that the leakage mechanism is dominated by Poole-Frenkel emission. [12] The longer anneal times show much lower leakage current when compared to unannealed TFT at even elevated measurement temperatures. This indicates that longer anneal times promote the reduction of the trapped charge inside the insulator for low-temperature fabricated TFTs. Therefore, stable sub-threshold slope and low leakage currents for longer anneals of these TFTs show an improved high temperature stability and reduction in the trapped charges inside the insulator. Longer anneals can be an important tool to improve the performance and stability of the low-temperature fabricated TFTs on plastic substrates.

CONCLUSION

In conclusion, the performance and stability of the low temperature fabricated a-Si:H TFT has been improved by longer anneals. Considerable improvements in sub-threshold slope and off-current have been obtained for 48 h annealed TFTs when compared to unannealed TFTs. Off-current reduced by two orders of magnitude for 48 hours annealed TFT and sub-threshold slope become steeper with longer annealing. Longer annealed a-Si:H TFTs show a stable sub-threshold slope value and low off-current at elevated measured temperatures. Both for annealed and unannealed TFTs, the threshold voltage shift exhibits power-law-time dependence with reduction in β value from 0.6 to 0.4. High temperature transfer characteristic measurements indicate that longer anneals helps in improving the a-Si:H/SiNx interface quality and thereby improving the performance and stability of the TFTs.

ACKNOWLEDGMENT

This work was partially supported by National Science Foundation (L. Hess, Grant No. DMR-0902277) to whom the authors are greatly indebted. Research was sponsored by the Army Research Laboratory (ARL) and was accomplished under Cooperative Agreement No. W911NG-04-2-0005. The views and conclusions contained in this document are those of the authors and should not be interpreted as representing the official policies, either expressed or implied, of the ARL or the U.S. Government. The U.S. Government is authorized to reproduce and distribute reprints for Government purposes notwithstanding any copyright notation hereon. We would like to express our sincere thanks to all the members of the Flexible Display Center for providing the TFTs and the technical help.

REFERENCES

1. J. T. Rahn, F. Lemmi, J. P. Lu, P. Mei, R. B. Apte, R. A. Street, R. Lujan, R. L. Weisfield, and J. A. Heanue, *IEEE.Tran.Nuc. Sci.* **55**, 457 (1999).
2. N. Ibaraki, *Mater. Res. Soc. Symp. Proc.* **336**, 749 (1994).
3. Y. He, R. Hattori, J. Kanicki, *IEEE Electron Device Lett.* **21**, 590 (2000).
4. P. Servati, S. Prakash, A. Nathan, *J. Vac. Sci. Technol. A* **20**, 1374 (2002).
5. J. Z. Chen, and I.-C. Cheng, *J. Appl. Phys.* **104**, (2008).
6. K. Long, A. Z .Kattamis, I.-C. Cheng, H. Gleskova, S. Wagner, and J. C. Sturm, *IEEE Electron Device Lett.* **47**, 387 (1996).
7. S. Nishizaki, K. Ohdaira, and H. Matsumura, *Jap. J. Appl. Phys.* **47**, 8700 (2008).
8. H. N. Chern, C. L. Lee, and T. F. Lei, *IEEE. Electron Dev. Lett.* **15**, 181 (1994).
9. K. H. Cherenack, A. Z. Kattamis, B. Hekmatshoar, J. C. Sturm, and S. Wagner, *IEEE Electron Device Lett.* **28**, 1004 (2007).
10. S. K. Kim, Y. J. Choi, K. S. Cho, and J. Jang, *J. Appl. Phys.*, **84**, 4006 (1998).
11. S. Musa, *Sci. Am.* **277**, 87 (1997).
12. M. J. Powell, C. Van Berkel, and J. R. Hughes, *Appl. Phys. Lett.*, **54**, 1323 (1989).
13. K. S. Karim, A. Nathan, M. Hack, and W. I. Milne, *IEEE Electron Device Lett.*, **25**, 188 (2004).
14. Y. H. Tai, J. W. Tsai, and H. C. Cheng, and F. C. Su, *Appl. Phys. Lett.*, **67**, 76 (1995).
15. M. J. Powell, C. van Berkel, A. R. Franklin, S. C. Deane, and W. I. Milne, *Phys. Rev. B* **45**, 4160 (1992).

Mater. Res. Soc. Symp. Proc. Vol. 1245 © 2010 Materials Research Society 1245-A19-06

A Full p-Type Poly-Si TFT Shift Register for Active Matrix Displays

Myoung-Hoon Jung[1], Hoon-Ju Chung[2], Young-Ju Park[1], and Ohyun Kim[1]
[1]Department of Electronic and Electrical Engineering,
Pohang University of Science and Technology, Pohang, Gyeongbuk 790-784, Republic of Korea
[2]School of Electronic Engineering, Kumoh National Institute of Technology,
Gumi, Gyeongbuk 730-701, Republic of Korea

ABSTRACT

We propose a new shift register using p-type poly-Si thin-film transistors (TFTs) for active matrix displays. It utilizes only p-type TFTs to simplify the fabrication process, and provides time-shifted output signals with a voltage swing from VSS to VDD without loss of signal-level. In the proposed shift register, output is structurally separated from carry and therefore is very insensitive to output signal distortion caused by output load capacitance. We also propose a new method of controlling light emission using this shift register; this method can be used in high image quality active-matrix organic light emitting diode displays. We verified the proposed shift register by simulation and measurement.

INTRODUCTION

Low temperature poly-silicon (LTPS) technology has potential application as backplanes for active-matrix organic light emitting diode (AMOLED) displays and active-matrix liquid crystal displays. The high mobility of the LTPS thin-film transistor (TFT) enables on-panel integration of peripheral driving circuits. This allows fabrication of high-resolution displays that are slim and compact [1, 2]. The pMOS process is simpler than that for CMOS, and pMOS TFTs are more reliable than nMOS TFTs because n-type TFTs suffer more from hot carrier degradation. Therefore, pMOS technology is more attractive than CMOS technology for integrated circuits on panels because pMOS logic has low cost and high production yield [3]. However, pMOS circuit design is more difficult than conventional CMOS circuit design [4].

Many studies have evaluated integrated gate drivers using only p-type poly-Si TFTs [1], [5, 6]. However, these shift registers are optimized to shift single pulses for a display scan driver, so they are functionally different from the general shift register. We propose a general-purpose shift register that utilizes only p-type TFTs, and an application of this shift register to control light emissions to achieve low motion-blur AMOLED displays.

PROPOSED SHIFT REGISTER

The proposed shift register (Fig. 1) consists of two scan drivers and an output driver, and performs the data shift operation; SET START and RESET START signals trigger set and reset scan drivers respectively, and also control the pulse width of the outputs. These set and reset scan drivers are compatible with conventional scan drivers that use p-type TFTs. We assumed that the set and reset scan drivers are conventional 4-phase scan drivers, and that the shift operation begins when START and CLK4 go low simultaneously [1]. These drivers can shift only a single pulse with fixed pulse width.

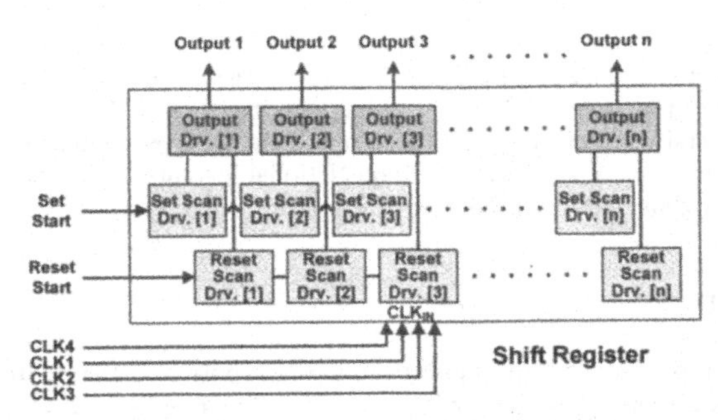

Figure 1. Schematic diagram of the proposed p-type poly-Si TFT shift register realized by combining two scan drivers (Set Scan Drv., Reset Scan Drv.) and an output driver (Output Drv.)

The Set Scan driver shifts a SET pulse to pull down the output voltage to the low level; the reset scan driver shifts a reset pulse to pull up the output voltage to the high level. The output driver is a memory device: if it is triggered by the SET pulse, it maintains a low output level until the next reset pulse arrives; if it is triggered by the RESET pulse, it maintains a high output level until next SET pulse arrives. Therefore, the pulse width of output waveforms is determined by the duration between the SET START pulse and the RESSET START pulse, and the outputs of the proposed shift registers are shifted by the clock signals due to the shift of the SET and RESET signals by the clocks. Because a datum can be loaded every CLK4 period, the data rate of the proposed shift register is the same as the frequency of the Set and Reset Scan drivers. Therefore, the pulse width of the shift register's output can be shortened by using a set and reset scan driver that includes a conventional 2-phase scan driver operated by a clock having a shorter period [6].

The output driver provides a final output that depends on the SET and RESET signals from the scan drivers (Fig. 2). When the SET signal goes low, T4 is turned on and this causes pull-up TFT T2 to turn off, and the low set signal through diode-connected T5 causes pull-down TFT T1 to turn on. Thus, the output is pulled down to VSS. Although the low level of the SET signal is equal to VSS, output can be pulled down to VSS by bootstrapping in T1. In contrast, when the RESET signal goes low, T2 turns on and T1 is turned off. As a result, the output is pulled up to VDD. The SET signal provides the gate voltage of T1, which is stored on capacitor C1; the RESET signal provides the gate voltage of T2, which is stored on capacitor C2. C1 and C2 maintain the output level after the SET or RESET signal ceases. Because the output of the proposed shift register is not used as a carry for the next stage, it is very insensitive to output signal distortion that can be caused by output load capacitance [2].

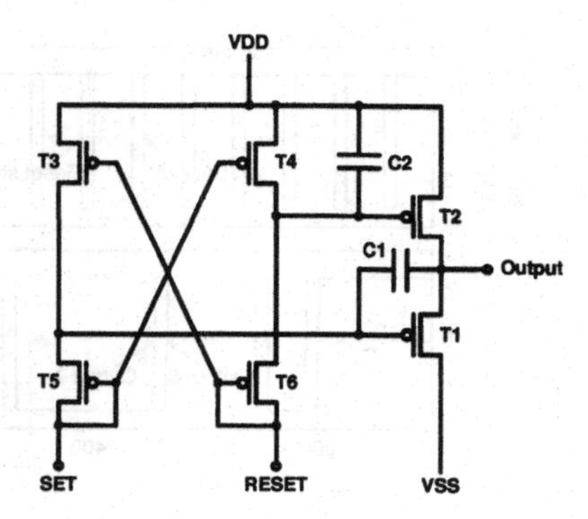

Figure 2. The output driver circuit for a final output stage of the proposed shift register. The set and reset signals are both control signals, and are generated from the set and reset scan drivers.

RESULTS AND DISCUSSION

SPICE simulation was performed to verify the proposed circuit. A conventional 4-phase shift register was used for the SET and RESET scan drivers, and the scan drivers and the output driver both used VDD = 12 V, and VSS = -3 V. The clock swung from 12 V to -3 V and the output load capacitance was 20 pF. The pulse width of output waveforms was well-controlled by the product of the number of SET start pulses and the period of the clocks, and the outputs of the proposed shift registers were shifted by the clock signals (Fig. 3a).

The proposed output driver was fabricated and measured to verify its operation. P-type poly-Si TFTs were fabricated using advanced solid phase crystallization [7]. The threshold voltage was -2.4 V and the mobility was 16.3 $cm^2/V\cdot s$. The sizes of the TFTs were 240/7 μm/μm for T1 and T2, 7/7+7 μm/μm for T3 and T4, and 20/7+7 μm/μm for T5 and T6. Capacitor C1 was 0.5 pF and capacitor C2 was 0.6 pF. To suppress the leakage current, double gate structure was employed in TFTs T3 ~ T6. When the output is low, leakage current flows through T3 and T5 and when the output is high, leakage current flows through T4 and T6. The SET and RESET signals swing from 12 V to -3 V with a period of 100 μs and a pulse width of 10 μs. Output was loaded by a 20 pF capacitor. The output state was successfully switched by the SET and RESET signals, and the bootstrapping caused the output to swing from VDD (12 V) to VSS (-3 V) (Fig. 3b). Measured rise time was 3.2 μs and measured fall time was 5.0 μs.

Because the desired output waveform can be easily generated, it can also be used as a scan driver for a pixel circuit requiring special scan signals, such as wide overlapping scans [8].

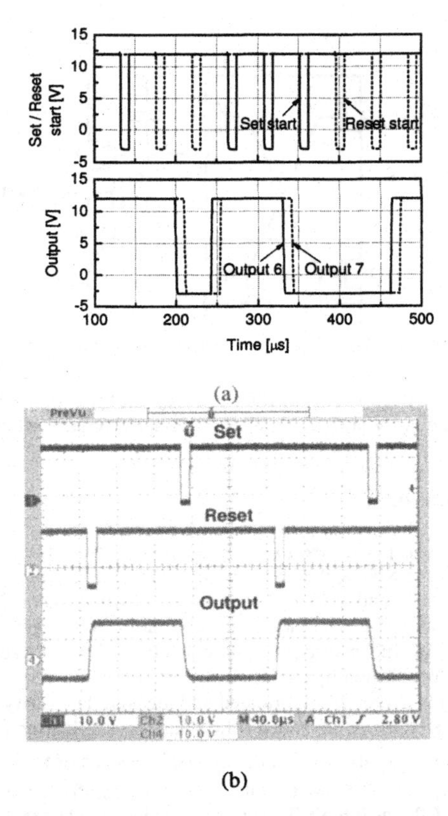

Figure 3. (a) Simulation results of the proposed shift register: SET START and RESET START, 6th, and 7th output waveforms. The RESET start and 7th output waveforms are represented by dashed lines. (b) Measured result of the fabricated output driver in Fig. 2.

Another application of the proposed shift register is a multi-purpose light emitting control driver for AMOLED displays (Fig. 4). Many conventional voltage and current programmed pixel circuits require a light-emitting control TFT to block current flow from the driving TFT to the OLED during the programming period [9, 10]. The conventional light-emitting control driver only has the function of blocking current path between the driving TFT and the OLED during the programming period, but by using the proposed shift register as a light-emitting control driver, the current path can be formed after the programming period for emitting, and blocked during a frame period to allow insertion of black data between successive frames. Therefore, application of the proposed shift register as a light-emitting control driver does not require additional TFTs and control lines, or high speed data and scan drivers to reduce motion blur and improve picture quality [11].

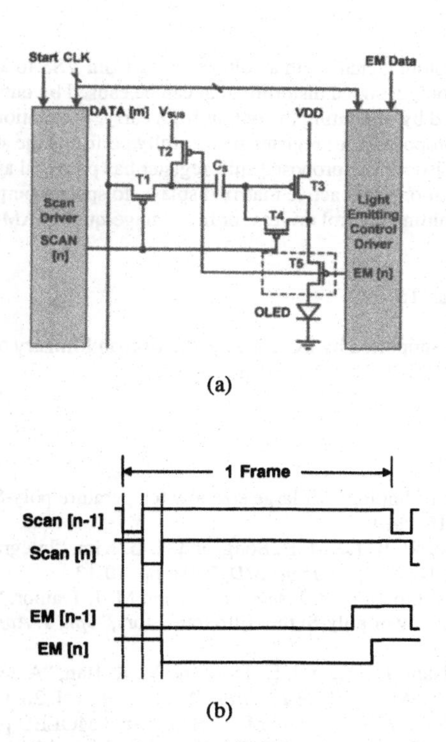

Figure 4. (a) Application of the proposed shift register as a light emitting control driver of AMOLED to reduce motion blurring. For simplicity, only one pixel circuit is shown. Light emitting control TFT is shown in dotted box. (b) Light emitting control signal waveforms during a frame time. The OLED can be turned on and off by light emitting control signal.

In a scan driver, output voltage retention capability is very important. The output voltage should not change during each frame period of 16.7 ms. To measure the holding time of the proposed shift register, an active matrix display with full HD resolution was assumed. In this case, the pulse period is 64 μs and the durations of the signals are both 15 μs. Assuming that the holding time is the time required for the output voltage to change by 1 % of its amplitude, the proposed shift register shows a sufficient holding time of 280 ms.

CONCLUSIONS

We have proposed a new shift register for active matrix displays. This register uses only p-type TFTs to simplify the fabrication process, and is composed of two conventional p-type TFT scan drivers and a new output driver. The proposed shift register uses bootstrapping to provide time-shifted output signals with a voltage swing from VSS to VDD. It is also designed to be very insensitive to output signal distortion that can be caused by output load capacitance; this insensitivity is achieved by separating the output from carry. Simulation and measurement results show that the proposed shift register successfully performs the shift operation and has stable output characteristics. The proposed shift register has potential applications including logic circuits for system-on-glass active matrix displays, to special-purpose scan drivers and multi-purpose light emitting control drivers for high image-quality AMOLED displays

ACKNOWLEDGMENTS

This work was supported by LG Display, the Korean Ministry of Education (BK21 Program), and the POSTECH Core Research Program.

REFERENCES

1. Y. M. Ha, "P-type technology for large size low temperature poly-Si TFT-LCDs," in *Proc. SID*, 2000, pp. 1116-1119.
2. J. Jeon, K. S. Choo, W. K. Lee, J. H. Song, and H. G. Kim, "Integrated a-Si gate driver circuit for TFT-LCD panel," in *Proc. SID*, 2004, pp. 10-13.
3. J. R. Ayres, S. D. Brotherton, D. J. McCulloch, and M. J. Trainor, "Analysis of drain field and hot carrier stability of poly-Si thin film transistors," *Jpn. J. Appl. Phys.*, vol.37, pp. 1801-1808, 1998.
4. S. H. Jung, W. J. Nam, J. H. Lee, J. H. Jeon, and M. K. Han, "A new low-power pMOS poly-Si inverter for AMDs," *IEEE Electron Device Lett.*, vol. 26, no. 1, pp. 23-25, Jan. 2005.
5. S. H. Jung, H. S. Shin, J. H. Lee, and M. K. Han, "An AMOLED pixel for the V_T compensation of TFT and a p-type LTPS shift register by employing 1 phase clock signal," in *Proc. SID*, 2005, pp. 300-303.
6. W. J. Nam, H. J. Lee, H. S. Shin, S. G. Park, and M. K. Han, "Low-voltage driven p-type polycrystalline silicon thin-film transistor integrated gate driver circuits for low-cost chip-on-glass panel," *Jpn. J. Appl. Phys.*, vol.45, pp. 4389-4391, 2006.
7. H. S. Choi, J. S. Choi, S. K. Hong, B. K. Kim, and Y. M. Ha, "LTPS technology for improving the performance of AMOLEDs," in *Proc. IMID*, 2007, pp. 1781-1784.
8. S. Ono, K. Miwa, Y. Maekawa, and T. Tsujimura, "V_T compensation circuit for AM OLED displays composed of two TFTs and one capacitor," *IEEE Trans. Electron Devices*, vol. 54, no. 3, pp. 462-467, Mar. 2007.
9. S. M. Choi, O. K. Kwon, and H. K. Chung, "An improved voltage programmed pixel structure for large size and high resolution AM-OLED displays," in *Proc. SID*, 2004, pp. 260-263.
10. Y. Hong, J. Kanicki, and R. Hattori, "Novel poly-Si TFT pixel electrode circuits and current programmed active-matrix driving methods for AM-OLELs," in *Proc. SID*, 2002, pp. 618-621.
11. J. J. L. Hoppenbrouwers, F. P. M. Budzelaar, C. N. Cordes, W. H. M. van Beek, F. J. Vossen, A. A. M. Hoevenaars, R. G. H. Boom, N. C. van der Vaart, D. A. Fish, and M. J. Childs, "Polymer OLED television image quality," in *Proc. IDW*, 2004, pp. 1257-1260.

Mater. Res. Soc. Symp. Proc. Vol. 1245 © 2010 Materials Research Society 1245-A19-07

High mobility a-Si:H TFT fabricated by Hot Wire Chemical Vapor Deposition

Chun-Yuan Hsueh, Chieh-Hung Yang, and Si-Chen Lee, *Fellow, IEEE*
Department of Electrical Engineering & Graduate Institute of Electronics Engineering
National Taiwan University, Taipei, Taiwan, Republic of China

ABSTRACT

The hydrogenated amorphous silicon (a-Si:H) thin film transistors (TFTs) having a very high field-effect mobility of 1.76 cm^2/V-s and a low threshold voltage of 2.43 V have been fabricated successfully using the hot wire chemical vapor deposition (HWCVD).

INTRODUCTION

Hydrogenated amorphous silicon (a-Si:H) thin-film transistors (TFTs) fabricated by plasma enhanced chamical vapor deposition (PECVD) have been widely used as a switching element in liquid crystal display (LCD)[1,2] and active matrix organic LED (AMOLED) application[3-5]. To enable the flat-panel display industry to advance to next-generation displays while using existing a-Si:H TFT backplane process technology, we propose a convenient technology to fabricate the inverted-staggered back-channel etched (BCE) type of a-Si:H TFTs by hot wire chamical vapor deposition (HWCVD)[6]. Moreover, the performance of electrical properties of the proposed TFT is much improved.

EXPERIMENT

The cross section of the bottom gate a-Si TFT structure is shown in figure 1.The detailed process flow are as follows: First, 75 nm thick Cr is evaporated and patterned as a bottom gate. Next, 300 nm thick SiN_x insulator, 40 nm thick undoped a-Si:H, and 20 nm thick n^+ a-Si:H layer are deposited sequentially by HWCVD. For SiN_x insulator deposition, the ratio of reaction gas is SiH_4:NH_3:H_2 = 0.7:4.3:5 sccm, and the tungsten filament is kept at 1800°C. For undoped a-Si:H deposition, the gas flow rate of H_2 and SiH_4 are 2 and 3 sccm, respectively, at a process pressure of 0.05 torr. The tungsten filament is kept at 1400 °C. The filament-substrate distance is 6 cm. Substrate temperature is around 300°C. For n^+ a-Si:H deposition, the gas flow rate is 1(H_2), 3(SiH_4) and 1(1.5% PH_3 in H_2) with [PH_3] to [PH_3+ H_2 + SiH_4] gas ratio of 3000 ppm. The crystallinity of the Si film is checked by Raman scattering, and it is totally amorphous.The device is patterned and mesa etched by RIE to define the active region. After opening the bottom gate contact window, 200 nm thick Al is evaporated and patterned to define the source, gate and drain electrodes. Finally, after the patterned n^+ a-Si:H layer was etched by RIE to define the channel region, the TFT is completed.

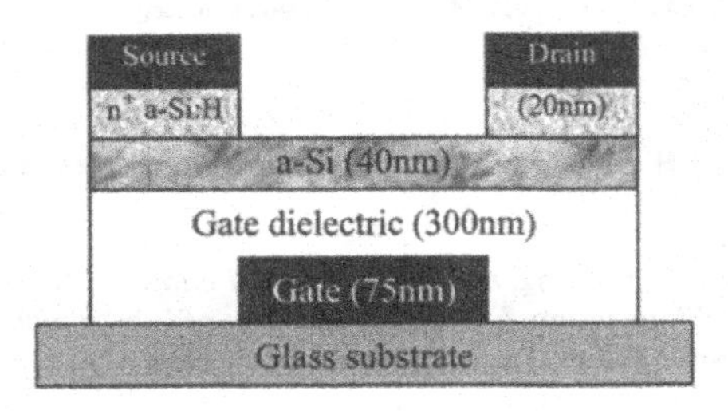

Figure 1. Cross section of the bottom gate TFT structure.

DISCUSSION

Figure 2 shows the drain current (I_{DS}) versus drain voltage (V_{DS}) curve under different gate voltage (V_{GS}) from 0 to 15 V. The dashed and solid lines indicate the performance of a-Si:H TFTs fabricated with conventional PECVD as well as our proposed process, respectively. The TFT has the channel width of 200μm and length of 40μm.The driving currents of the TFT fabricated by HWCVD are apparently higher than that of the conventional TFT under the same gate voltage(V_{GS}). Moreover, in Fig. 2, the proposed TFT has lower contact resistance.

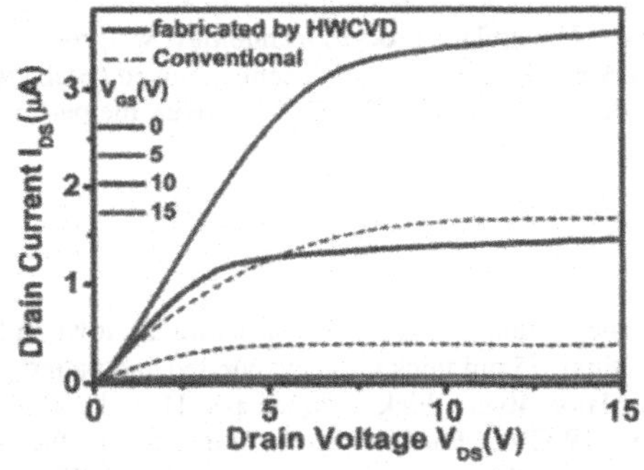

Figure 2. The drain current (I_{DS}) versus drain voltage (V_{DS}) curve under different gate voltage (V_{GS}) from 0 to 15 V. The dashed and solid lines indicate the performance of a-Si:H TFTs fabricated with conventional PECVD as well as our proposed process, respectively.

Fig. 3 shows the drain current (I_{DS}) as a function of gate biases V_{GS} at V_{DS} = 1 and 10 V. The dashed and solid lines indicate the performance of a-Si:H TFTs fabricated with conventional PECVD and our proposed process, respectively. From Fig 3, the ON/OFF current ratio of our proposed TFT is more than 10^6 at V_{DS} = 1V, and the subthreshold slope of our proposed TFT (0.11 V/decade) is steeper than that of the conventional TFT. However, the OFF current of our proposed TFT is large (~10^{-9}A) at V_{DS} = 10V. This is not due to the a-Si:H film. We suppose that it is due to the low SiN_x bandgap, making charge tunneling and/or trapping more likely.

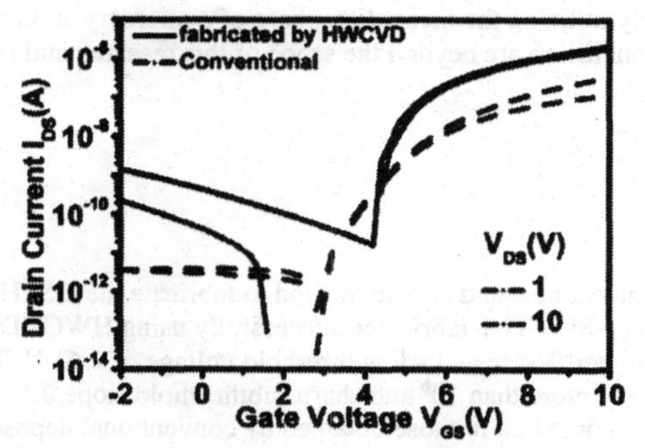

Figure 3. The drain current (I_{DS}) as a function of gate biases V_{GS} at V_{DS} = 1 and 10 V. The dashed and solid lines indicate the performance of a-Si:H TFTs fabricated with conventional PECVD and our proposed process, respectively.

Fig 4 shows the relation between square root of I_{DS} and V_{GS} of the TFT at the $V_{GS} = V_{DS}$ condition. By using equations 1, the current formula of the MOSFET,

$$I_{DS} = \frac{1}{2}\mu_{FE}\frac{W}{L}C_i(V_{GS} - V_T)^2 \text{ (in saturation mode)} \quad (1)$$

the field effect mobility (μ_{FE}) and the threshold voltage (V_{th}) can be extracted from the slope and intercept with voltage axis in Fig. 4, respectively. The field effect mobility (μ_{FE}) of our proposed TFT is about 1.76 cm^2/V-sec, and threshold voltage (V_{th}) is about 2.43 V. The field effect mobility of conventional TFT fabricated by PECVD is only about 0.86 cm^2/V-sec.

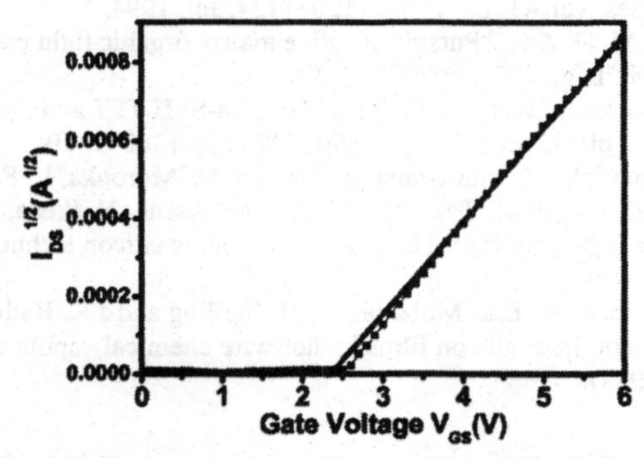

Figure 4. The relation between square root of I_{DS} and V_{GS} of the TFT at the $V_{GS} = V_{DS}$ condition.

Therefore, the proposed TFT exhibits superior current driving capability. This result demonstrates that HWCVD a-Si:H can replace PECVD a-Si:H in TFTs with similar device structure and a better field effect mobility. We suppose that the good performance of the a-Si:H film is caused by effectively utilizing the incubation phase of a microcrystalline Si deposition. However, more detailed conclusion are beyond the scope of this research and require a more thorough study.

CONCLUSIONS

We have demonstrated a new and simple method to fabricate the a-Si:H TFT with high electrical performance. The a-Si:H TFT fabricated successfully using HWCVD has a very high field-effect mobility of 1.76 cm^2/V-sec and a low threshold voltage of 2.43 V. The TFT has a high ON/OFF-current ratio of more than 10^6 and sharp subthreshold slope 0.11 V/decade. High performance of the devices, compared to those obtained by conventional deposition methods, makes HWCVD an attractive method for industrial application.

ACKNOWLEDGMENTS

This work was supported by the National Science Council of the Republic of China under Contract No. NSC 97-2221-E-002-227-MY3.

REFERENCES

1. H.Yamamoto, H. Matsumaru, K. Shirahashi, M. Nakatani, A. Sasano, N.Konishi, K. Tsutsui, and T. Tsukada, "A new a-Si TFT with Al_2O_3/SiN double-layered gate insulator for 10.4-inch diagonal multicolor display," in IEDM Tech. Dig., 1990, pp. 851–854.
2. G. Kawachi, E. Kimura, Y. Wakui, N. Konishi, H. Yamamoto, Y. Matsukawa, and A. Sasano, "A novel technology for a-Si TFT-LCDs with buried ITO electrode structure," IEEE Trans. Electron Devices, vol. 41, no. 7, pp. 1120–1124, Jul. 1994.
3. R. M. A. Awson and M. G. Ane, "Pursuit of active matrix organic light emitting diode displays," in SID Tech. Dig., 2001, pp. 372–375.
4. J. A. Nichols, T. N. Jackson, M. H. Lu, and M. Hack, "a-Si:H TFT active-matrix phosphorescent OLED pixel," in SID Tech. Dig., 2002, pp.1368–1370.
5. T. Tsujimura, Y. Kobayashi, K. Murayama, A. Tanaka, M. Morooka, E. Fukumoto, H. Fujimoto, J. Sekine, K. Kanoh, K. Takeda, K. Miwa, M. Asano, N. Ikeda, S. Kohara, and S. Ono, "A 20-inch OLED display driven by super-amorphous-silicon technology," in SID Tech. Dig., 2003, pp. 6–9.
6. R.E.I. Schropp, K.F. Feenstra, E.C. Molenbroek, H. Meiling and J.K. Rath, "Device-quality polycrystalline and amorphous silicon films by hot-wire chemical vapour deposition", Philos. Mag. B 76, p. 309 (1997).

Crystalline Si Film

Mater. Res. Soc. Symp. Proc. Vol. 1245 © 2010 Materials Research Society **1245-A20-01**

Crystallization kinetics in high-rate electron beam evaporated poly-Si thin film solar cells on ZnO:Al

T. Sontheimer, C. Becker, C. Klimm, S. Gall, B. Rech

Helmholtz Zentrum Berlin fuer Materialien und Energie, Silicon Photovoltaics,
Kekuléstr. 5 12489 Berlin, Germany.

ABSTRACT

The microstructure and crystallization kinetics of electron beam evaporated Si on ZnO:Al coated glass for polycrystalline solar cells was studied by electron backscatter diffraction and optical microscopy at various deposition temperatures. A time dependent analysis of the dynamics of the crystallization allowed for the individual determination of growth and nucleation processes. The nucleation process of Si on ZnO:Al was found to be influenced by a variation of the deposition temperature of the amorphous Si in a critical temperature regime of 200 °C to 300 °C. The nucleation rate decreased significantly with decreasing deposition temperature, while the activation energy for nucleation increased from 2.9 eV at a deposition temperature of 300 °C to 5.1 eV at 200 °C, resulting in poly-Si which comprised grains with features sizes of several μm.

INTRODUCTION

A major challenge for the low-cost thin film Si solar cell production is the fast and effective deposition of Si films with excellent electronic quality which are embedded in a smart contacting scheme and efficient light trapping structure. The preparation of Si thin film solar cells with excellent electronic quality has been investigated by numerous poly-crystalline Si (poly-Si) approaches. One poly-Si concept is based on the crystallization of deposited amorphous Si (a-Si) by thermal annealing in a solid phase crystallisation (SPC) process. With plasma enhanced chemical vapor deposition (PECVD) used for the deposition of a-Si, the concept was already investigated and a conversion efficiency of 10.4 percent has been achieved [1-3]. However, the low deposition rate of 0.5 - 2 nm/s curtails the effectiveness of the poly-Si concept's production process. A poly-Si solar cell device prepared by high-rate electron beam (e-beam) evaporation with deposition rates of 1 μm/min or even higher overcomes this obstacle. Recently, an efficiency of 6.7 percent has been achieved for e-beam evaporated poly-Si solar cells on planar glass, illustrating the technology's potential for the fabrication of high quality Si [4, 5]. The additional implementation of ZnO:Al as a front-contact layer into the superstrate concept allows for the incorporation of light trapping structures and a realization of a smart contacting scheme. Moreover, the charge carrier mobility of Si-capped ZnO:Al is significantly improved by thermal annealing, making it the ideal transparent conductive oxide to be implemented in a poly-Si solar cell [6]. The compatibility of the front contact layer ZnO:Al with poly-Si solar cells in a superstrate configuration has been demonstrated by a systematic analysis of the photovoltaic performance at various deposition parameters [7]. The amorphous-to-crystalline phase transformation represents a very crucial process in the fabrication of the poly-Si solar cell since its kinetics governs the grain size of the poly-Si films and hence influences the quality of the complete solar cell device.

This paper presents a systematic analysis of the crystallization kinetics and the microstructure of Si on ZnO:Al-coated glass and addresses the influence of the deposition temperature of the e-beam evaporated a-Si on the crystallization behavior. This crystallization process is subsequently compared to the phase transition of Si on SiN-coated glass, a substrate commercially used for the production of poly-Si solar cells. The crystallization process was analyzed by applying a combination of two ex-situ characterization techniques. First, employing electron backscatter diffraction (EBSD) revealed essential information about the grain size and grain orientation of the fully crystallized samples. Second, the analysis of the phase transition in multiple consecutive annealing steps and at various annealing temperatures allowed for the individual determination of the three crucial parameters that govern the crystallization process, namely the nucleation rate, time-lag and the crystal growth velocity. Ex-situ optical microscopy operating in transmission mode was used to detect crystals by distinguishing the crystalline material from the amorphous material based on their dissimilar optical absorption.

EXPERIMENTAL

Sample preparation

ZnO:Al films (thickness 800 nm) were deposited on SiN-coated Borofloat 33 glass at a substrate temperature of 300 °C by non-reactive (RF) sputtering from a ceramic target containing 1% wt Al_2O_3 [8]. Subsequently, 1 μm thick 2 x 10^{16} cm^{-3} boron doped a-Si films were deposited by e-beam evaporation at a deposition rate of 300 nm/min at three different substrate temperatures ranging from 200 °C to 300 °C. For comparison, an a-Si film was prepared on SiN-coated glass at 300 °C, using identical deposition parameters. Raman spectroscopy confirmed that the structure of all deposited films was amorphous.

Crystallization kinetics

For a systematic analysis of the dynamics of the crystallization process, the samples were alternately isothermally annealed in the tube furnace and analyzed by ex-situ optical microscopy. The annealing temperatures at which the analysis was conducted ranged from 560 °C to 610 °C. The influence of multiple consecutive anneals on the crystallization process was analyzed by comparing the crystalline fraction of a sample which underwent multiple anneals to a sample which was tempered for the same time in one single annealing step. Since the samples have shown to comprise equivalent crystalline fractions, the annealing procedure was carried out in multiple anneals. The microscopic images that served as a basis for this investigation contained 200 to 1000 crystals at a crystalline fraction of 5 to 10 percent, representing a crystallization stage at which the crystals have not yet impinged each other.
The characteristic crystal growth velocity $v=\Delta r/\Delta t$ for a specific temperature was determined by measuring the change in radius Δr_i of a crystal at two crystallization steps t and $(t+\Delta t)$. The analysis of the nucleation process was carried out in conformity with classical nucleation theory [9] and was based on one microscopic image containing up to 1000 individual crystals. The moment of nucleation of each crystal was calculated by employing the time-independent crystal growth velocity v. The largest crystal naturally determines the starting point of the crystallization and is therefore an indication for the time-lag τ. This method enabled us to calculate the number

of crystals N as a function of time t. The derivative of N at a specific time t_i normalized by the still amorphous fraction $(1-\chi(t_i))$ represented the nucleation rate I at t_i, with I reaching a steady state I_{ss} after passing a transient regime.

In a previous study the crystal growth velocities v for Si on ZnO:Al and Si on SiN were found to be identical with respective activation barriers E_v=(2.5±0.2) eV and E_v =(2.6±0.2) eV, while the nucleation rate differed significantly [10]. At a deposition temperature of 300 °C, the steady state nucleation rate of Si on ZnO:Al was found to be two orders of magnitude higher than on SiN and the time-lag is significantly shorter, resulting in a faster crystallization process. In addition, the temperature dependence differed drastically in the investigated temperature regime of 560 °C to 610 °C, with activation energies E_{Iss} of Si on ZnO:Al and Si on SiN being (2.9±0.4) eV (I_0= 10^{12} μm^{-2} sec^{-1}) and (5.0±0.6) eV (I_0= 10^{23} μm^{-2} sec^{-1}), respectively (Figure 1) [10]. I_0 denotes the pre-factor of the fitting function of the Arrhenius plot.

Figure 1(a) and 1(b) illustrate the steady state nucleation rates I_{ss} and time-lags τ as a function of annealing temperature for Si on ZnO:Al and SiN deposited at T_{dep} = 300 °C and display the temperature dependent I_{ss} and τ for Si on ZnO:Al samples for various deposition temperatures. The nucleation process in Si on ZnO:Al showed a unique dependence on the deposition temperature. We obtained a significantly smaller I_{ss} by reducing the deposition temperature to 250 °C for Si on ZnO:Al, which was accompanied by an increase of E_{Iss} to (3.6±0.2) eV (I_0= 10^{16} μm^{-2} sec^{-1}). In addition, the time-lag increased, resulting in an increase of the pre-factor from τ_0=10^{-21} sec for Si deposited at 300 °C to τ_0=10^{-18} sec for Si deposited at 250 °C at similar activation energies of E_τ=(3.8±0.2) eV and (3.6±0.1) eV. By reducing the temperature to 200 °C, the nucleation rate declined even further. With a very low I_{ss} and an increased E_{Iss} of (5.1±0.7) eV (I_0= 10^{23} μm^{-2} sec^{-1}), the material exhibited a nucleation behavior similar to Si on SiN. The crystallization process showed an even larger time-lag, leading to a τ_0 = 10^{-16} sec at an activation energy of E_τ= (3.4±0.6) eV. For deposition temperatures of 150 °C and below the nucleation rate was found to be identical to the nucleation rate at T_{dep}=200 °C. These findings demonstrate that a variation of the deposition temperature between 200 to 300 °C has a crucial impact on the crystallization behavior and the final microstructure of the poly-Si.

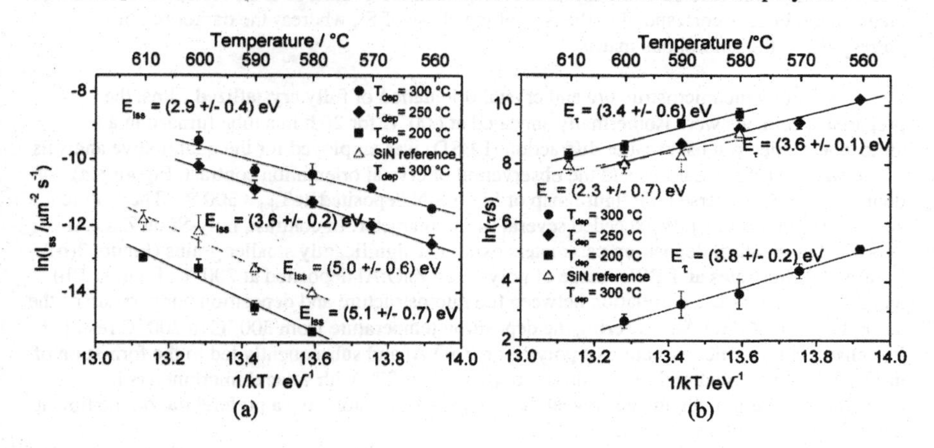

(a) (b)

Figure 1.The temperature dependence of the steady state nucleation rate I_{ss} (a) and the time-lag τ (b) are depicted for Si on ZnO:Al prepared at deposition temperatures 200 °C, 250 °C and 300 °C. For comparison, I_{ss} and τ of Si on SiN are shown.

Microstructure and crystal grain orientation

Cross-sectional transmission electron microscopy (TEM) was performed on Si on SiN and ZnO:Al, which were deposited at 300 °C and subsequently annealed at 600 °C for 3.5 h and 45 min, respectively. TEM on partially crystalline films showed that the Si crystals on SiN could grow up to several μm before they impinged each other, while the Si crystals on ZnO:Al were considerably smaller [10]. Figures 2(a) and 2(b) display TEM images of partially crystalline Si on SiN and ZnO:Al, respectively. Since Si crystals on ZnO:Al did not nucleate preferentially at the Si/ZnO:Al interface, the low E_{Iss} cannot be ascribed only to heterogeneous nucleation at the Si/ZnO:Al interface.

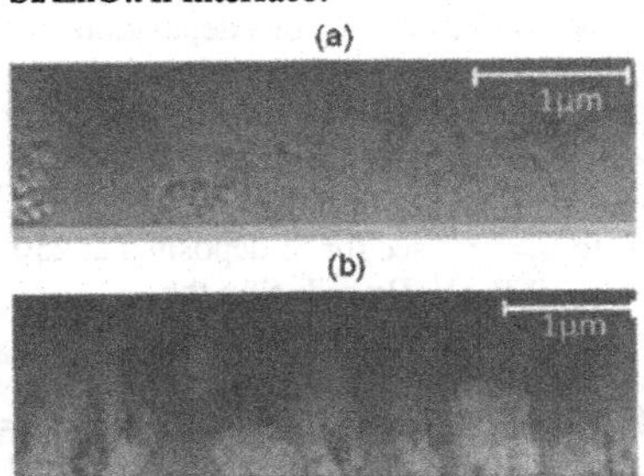

Figure 2. Cross-sectional TEM micrographs of partially crystallized Si on SiN (a) and ZnO:Al (b). The amorphous films were deposited at 300 °C. The SiN/Si interface on (a) and ZnO:Al/Si interface on (b) are located at the top of the image and the Si surface is on the bottom. The bright areas on the images correspond to the crystalline phase of Si, whereas the darker regions correspond to the amorphous matrix.

To study the microstructure and crystal orientation of fully crystallized films, the prepared specimens were isothermally annealed at 600 °C for 20 h in a tube furnace in a N_2 atmosphere. Electron backscatter diffraction (EBSD) was employed for the quantitative analysis of the size of surface features and the observation of crystal orientation contrast. Figure 3(a) depicts an EBSD inverse pole figure map of Si on SiN deposited at T_{dep}= 300 °C. The feature size of the grains evidently exceeded several μm in diameter. In contrast, poly-Si on ZnO:Al deposited at identical deposition parameters exhibited significantly smaller grains (Figure 3(b)). Figure 3(c) illustrates an EBSD image of poly-Si on ZnO:Al deposited at 200 °C. Figures 3(b) and 3(c) clearly show a correlation between the microstructure and deposition temperature of the amorphous Si on ZnO:Al. Lowering the deposition temperature from 300 °C to 200 °C resulted in a change of the microstructure of poly-Si on ZnO:Al and subsequently led to the formation of grains which exhibited grain sizes similar to poly-Si on SiN with several micrometers in diameter. In addition, the images reveal that the crystals do not have a preferential orientation at either deposition temperature or substrate.

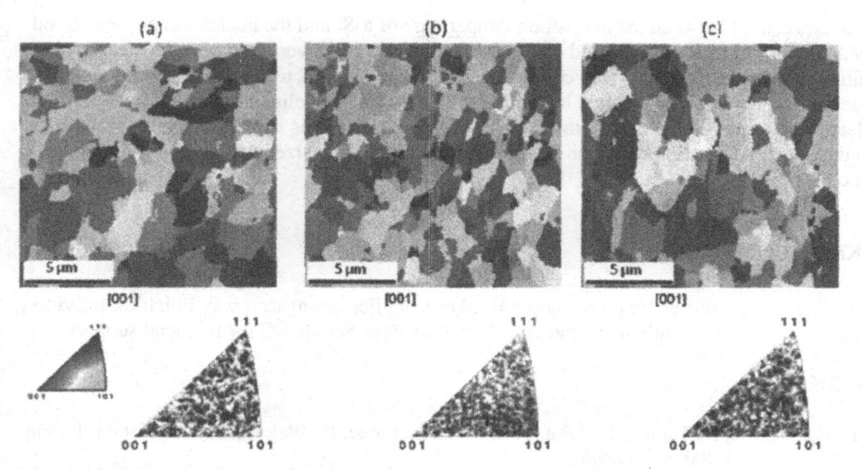

Figure 3. EBSD inverse pole figure maps of Si on SiN (a) and Si on ZnO:Al deposited at 300 °C (b) and 200 °C (c) are shown.

DISCUSSION

At a constant crystal growth velocity, the nucleation rate dominates the crystallization process and controls the microstructure of the fully crystallized Si film. The very high nucleation rate of Si on ZnO:Al at T_{dep}=300 °C is directly linked to a microstructure with small grains, while a lower nucleation rate translates into the formation of large grained Si (Figure 3). Koester reported an activation energy for the steady state nucleation rate of E_{Iss}= 4.9 eV [11]. This value is in perfect accordance to the determined value for Si on SiN [10]. The work to form a critical nucleus $W_c = E_{Iss} - E_v = (2.4\pm0.6)$ eV resembles a value which has been reported in previous studies on amorphous Si [9]. In contrast, W_c for Si on ZnO:Al is well below 1 eV and differs drastically from W_c of Si on SiN. However, a gradual decrease of the temperature from 300 °C and 200 °C not only resulted in a peculiar decrease of I_{ss} and the formation of large grains, but also led to a remarkable increase of E_{Iss} and W_c. W_c incrementally increased from below 1 eV at T_{dep}= 300 °C to the characteristic W_c of 2.4 eV at T_{dep}=200 °C. The work to form a critical nucleus W_c is directly influenced by the fundamental mechanism that controls cluster formation in amorphous Si. The characteristic dependence of the nucleation mechanism in the amorphous Si on ZnO:Al on the deposition temperature might be ascribed to a transition of the structural properties of amorphous Si on ZnO:Al at gradually lowered deposition temperatures.

CONCLUSION

In conclusion, we presented a detailed analysis of the influence of the deposition temperature on the crystallization behavior and microstructure of Si on ZnO:Al. We revealed a unique correlation between the deposition temperature of a-Si and the nucleation process. Si on ZnO:Al deposited at 300 °C showed a high nucleation rate and a very low activation energy E_{Iss}, resulting in a microstructure which comprises grain sizes of 1 to 2 µm. A gradual reduction of the deposition temperature resulted in an increase of E_{Iss} and a decline of the steady state nucleation rate and finally led to a microstructure with large grains which exceed several µm. This unique crystallization behavior allows us to tailor the grain size in poly-Si thin film solar cells on ZnO:Al.

ACKNOWLEDGMENTS

We express our gratitude to Jürgen Hüpkes from Forschungszentrum Jülich for providing ZnO:Al films. Tobias Sontheimer gratefully acknowledges Schott AG for financial support.

REFERENCES

1. T. Matsuyama, N. Terada, T. Baba, T. Sawada, S. Tsuge, K. Wakisaka and S. Tsuda, J. Non-Cryst. Sol. 198-200, 940 (1996).
2. M.A. Green, P.A. Basore, N. Chang, D. Clugston, R. Egan, R. Evans, D. Hogg, S. Jarnason, M. Keevers, P. Lasswell, J. O`Sullivan, A. Turner, U. Schubert, S.R. Wenham and T. Young, Solar Energy 77, 857 (2004).
3. M.J. Keevers, T.L. Young, U. Schubert, M.A. Green in: Proc. of the 22nd European Photovoltaics Solar Energy Conference, Milan, Italy (2007), p.1783.
4. T. Sontheimer, P. Dogan, C. Becker, S. Gall, B. Rech, U. Schubert, T. Young, S. Partlin, M. Keevers, R. J. Egan in: Proc. of the 24th European Photovoltaics Solar Energy Conference, Hamburg, (2009), p.2478.
5. R. Egan, M. Keevers, U. Schubert, T. Young, R. Evans, S. Partlin, M. Wolf, J. Schneider, D. Hogg, B. Eggleston, M. Green, F. Falk, A. Gawlik, G. Andrä, M. Werner, C. Hagendorf, P. Dogan, T. Sontheimer, S. Gall in: Proc. of 24th European Photovoltaic Solar Energy Conference, Hamburg, (2009) 2280 -2285
6. F. Ruske, M. Roczen, K. Lee, M. Wimmer, S. Gall, J. Hüpkes, D. Hrunski, B. Rech, J. Appl. Phys. **107**, 013708 (2010)
7. C. Becker, F. Ruske, T. Sontheimer, B. Gorka, U. Bloeck, S. Gall, B. Rech, J. Appl. Phys. 106, 084506 (2009).
8. M. Berginski J. Hüpkes, M. Schulte, G. Schöpe, H. Stiebig, B. Rech, M. Wuttig, J. Appl. Phys 101, 074903 (2007).
9. C. Spinella, S. Lombardo and F. Priolo, J. Appl. Phys. 84, 10 (1998).
10. T. Sontheimer, C. Becker, U. Bloeck, S. Gall, B. Rech, Appl. Phys. Lett. 95, 101902 (2009).
11. U. Köster, Phys. Stat. Sol. (a), 48, 313 (1978).

Mater. Res. Soc. Symp. Proc. Vol. 1245 © 2010 Materials Research Society 1245-A20-03

Al-mediated Solid-Phase Epitaxy of Silicon-On-Insulator

Agata Šakić, Yann Civale, Lis K. Nanver, Cleber Biasotto, and Vladimir Jovanović
Laboratory of Electronic Components, Technology and Materials, Delft Institute of Microsystems and Nanoelectronics – DIMES, Delft University of Technology, Feldmannweg 17, 2628CT Delft, The Netherlands. Tel: +31 (0)15 27 82185, Fax: +31 (0)15 26 22 163, E-mail: a.sakic@tudelft.nl

ABSTRACT

Silicon-on-insulator (SOI) regions have been grown on lithographically predetermined positions by Al-mediated Solid-Phase Epitaxy (SPE) of amorphous silicon (α-Si). A controllable Si lateral overgrowth is induced from windows formed in silicon dioxide (SiO_2) to the crystalline Si substrate. The resulting hundred-of-nanometer large areas of high-quality monocrystalline SOI are formed at the temperatures that can be as low as 400 °C. The as-obtained SOI regions were found to take on the same crystal orientation as the (100)-Si substrate and have the ability to merge seamlessly over the oxide.

INTRODUCTION

Recently an Al/α-Si SPE process was demonstrated, where an Al-mediation of the α-Si on a monocrystalline Si substrate was shown to provide high-quality Al-doped crystalline Si islands on predetermined positions. With this process, near-ideal, ultra-shallow, ultra-abrupt p^+n-junctions down to sub-100-nm dimensions were formed in contact windows to an n-substrate [1, 2]. In the present paper this Al-mediated SPE process is used to obtain silicon-on-insulator (SOI) regions around the contact windows. Such SOI processes are particularly interesting for ultra-high-scale integration systems and three-dimensional integrated circuits (3D-IC) [3]. Moreover, large investments are at present being made in bonded SOI wafers for the purpose of serving fully-depleted CMOS SOI production. Compared to bulk-Si devices, the SOI version is usually more latch-up resistant, has reduced parasitic capacitances, facilitates the fabrication of shallow junctions and has better radiation hardness.

Among many studied SOI technologies, Lateral Solid-Phase Epitaxy (L-SPE) of α-Si silicon films deposited on SiO_2 and Si surfaces has been promising mainly due to its low process temperatures. It has also been found that the annealing under ultra-high pressure up to 2 GPa increases further the growth and nucleation rate. However, reported cases suffer from unexpected oxide films formed on the surface of the grown silicon layers [4], or are limited to long annealing times up to 120h at about 600 °C [5]. In this paper, we characterize and evaluate for the first time the Al-mediated L-SPE for anneal cycles as short as 30 min at 400 °C, and for a range of different geometries and SiO_2/Al/α-Si stack thicknesses.

EXPERIMENTAL PROCEDURES

To investigate the possibility of controlling the epitaxial Si lateral overgrowth on SiO_2, the samples were subjected to variety of process modifications. A schematic of the basis process flow is shown in Fig. 1. Contact windows were etched through 30 nm, 70 nm, and 100 nm of oxide grown on a (100)-Si substrate. Thermally-grown SiO_2 is chosen because it can be formed with a very low number of defects that may function as parasitic nucleation centers for polycrystalline Si island growth on the oxide. Only when such parasitic growth is avoided, can a good control of growth at the desired nucleation site be obtained. The process is studied for different contact window size and Al/α-Si layer stacks. The defined contact window openings are 700 nm with pitches of either 700 nm or 1 μm.

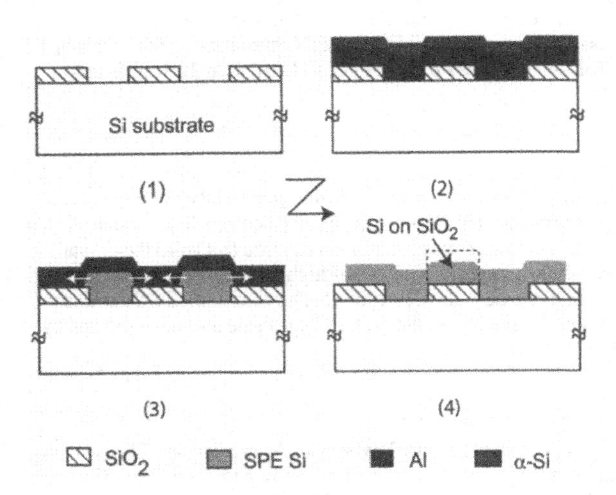

Fig. 1. Schematic cross-section of SPE Silicon-On-Insulator (SOI) process flow.

After patterning and plasma etching of the windows, a 4 min dip-etching in HF 0.55% is performed before depositing 200 nm Al (containing 1% Si) and 20 nm, 50 nm or 100 nm of α-Si. Both layers were deposited by physical vapour deposition (PVD) at room temperature using an Ar flow of 100 sccm, without breaking the vacuum between the two depositions in order to prevent the formation of aluminium oxide (Al_2O_3) interface. Finally, the crystallization was performed in a nitrogen (N_2) atmosphere at 400 °C and 500 °C for 30 min and the Al transport layer was removed using conventional metal etchants (Fig. 1). As-obtained surfaces were analyzed by the use of scanning electron microscopy (SEM), high-resolution transmission-electron-microscopy (HRTEM), selective area diffraction (SAD), and electron back-scattering diffraction (ESBD) analyses.

RESULTS

The growth mechanism governing the selective epitaxial deposition of SPE-Si on Si has been described by a semi-empirical theory in [6]. In that work the SPE growth was localized and limited to filling contact window surfaces by patterning the *α*-Si/Al layer stack around the windows and optimizing growth parameters such as the total amount of *α*-Si, the Al thickness, and the anneal temperature. Then, the Si deposits preferably within contact windows rather than on the surrounding SiO_2 [1]. However, a limited lateral overgrowth of SPE-Si on the surrounding

SiO_2 was often observed. In case of a non-patterned α-Si/Al layer stack, it is clear from the results that the Si atoms which have diffused to the oxide surface are kinetically capable of traveling along this interface. On thermally grown SiO_2 there are practically no nucleation centers and the Si atoms will with high probability reach the c-Si surface before precipitating to form crystals. Thus the oxide remains free of precipitates.

As the SPE-Si first nucleates in the contact windows, the step coverage of the Al over the SiO_2 window-height is a limiting aspect of the lateral overgrowth. The analysis of fabricated test structures of different SiO_2 thicknesses - 30 nm, 70 nm, and 100 nm - substantiate that the Si SPE lateral growth over SiO_2 decreases when the SiO_2 thickness increases, for a given Al/α-Si stack thickness. Although the surface of the thermally grown SiO_2 is smooth and without defects that can act as nucleation centers for the Si atoms, the formation of the windows in the SiO_2 demands an aggressive chemical etching in lithographically defined positions. For thick SiO_2 films, the sidewalls of the openings etched to the Si substrate are higher; therefore, the area exposed to the chemical etching is larger. Consequently, the parasitic nucleation on SiO_2 sidewalls is enhanced which leads to the formation of large polycrystalline Si islands on SiO_2 initiated in the region close to the contact windows edges as seen in Figs. 2b and 2c. Moreover, for much thicker dielectric layers, the issues related to the poor conformality of Al thin films deposited by PVD at the edges of the contact window also may significantly affect the SPE-Si growth sequence. In this work, the most favorable overgrowth conditions were achieved for a 30 nm thick oxide as seen in Fig. 2a.

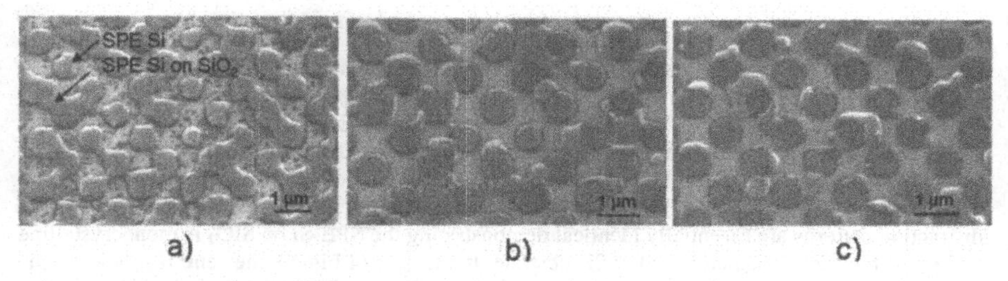

Fig. 2. SEM micrographs of SPE-Si performed on 700-nm-wide contact windows with a pitch of 700 nm, for 200 nm Al, 100 nm α-Si, and (a) 30 nm, (b) 70 nm, and (c) 100 nm of SiO_2 layer.

Also the thickness of the α-Si layer has a large effect on the resulting overgrowth as seen in Fig. 3. It is evident that 20 nm of Si partially fills the inside of the contact openings and practically no overgrowth is observed (Fig. 3a) whereas 50 nm deposited Si provides enough to fully overlay the contact openings (Fig. 3b). Finally, for a 100 nm α-Si deposition, the SPE-Si overgrows the SiO_2 and merges, following a diagonal pattern corresponding to the (100)-preferential direction of growth (Fig. 3c). A large sensitivity of the overgrowth process to the pitch between windows of a fixed size is also observed. In Fig. 4 a pitch of 1 µm is compared to 700 nm. In both cases the windows are filled but a very unevenly distributed overgrowth is seen for the 1 µm pitch while for 700 nm the overgrowth is fairly limited to the diagonals between the windows. Increasing the anneal temperature from 400 °C to 500 °C also has a detrimental effect on the uniformity of the overgrown SPE-Si.

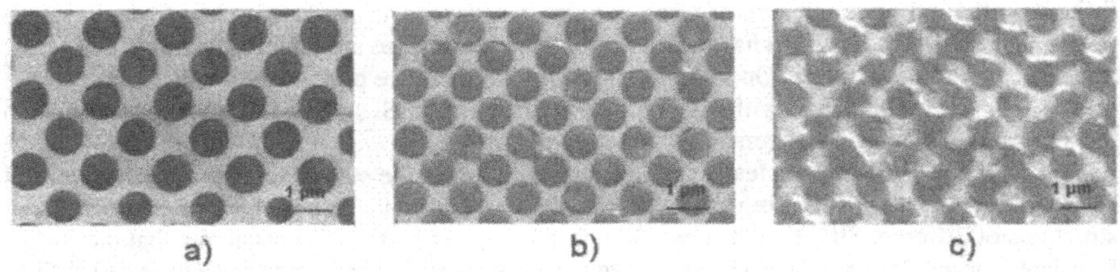

Fig. 3. SEM micrograph of SPE-Si performed on 700-nm-wide contact windows with a pitch of 700 nm, for 200 nm Al, 33 nm SiO_2, and (a) 20 nm, (b) 50 nm, and (c) 100 nm of α-Si layer.

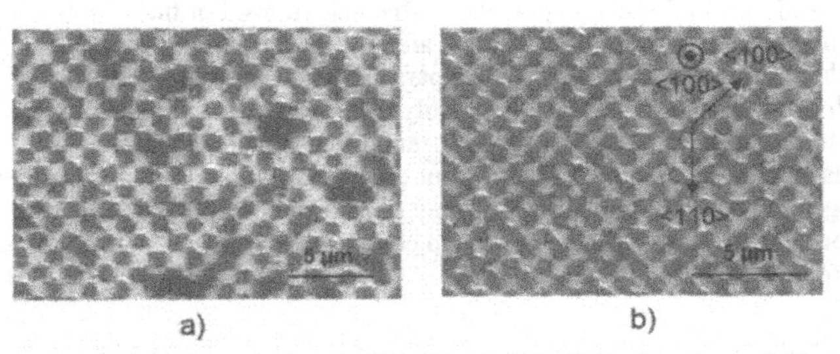

Fig. 4. SEM micrographs of SPE-Si with SiO_2/Al/α-Si values of 30 nm/200 nm/100 nm, contact openings of 700 nm, and a pitch of (a) 1 μm and (b) 700 nm, annealed for 30 min at 400 °C.

The crystallinity of the SPE-Si on SiO_2 was verified by EBSD analysis, the results of which show that both the SPE-Si located within the contact windows and on the SiO_2 is (100)-oriented (Fig. 5). Additionally, transmission electron microscopy (TEM) including selective-area diffraction (SAD) with a spot size as small as 200 nm, was performed (Fig. 6). The diffraction patterns are essentially identical demonstrating the SPE-Si on SiO_2 is monocrystalline and keeps the same orientation of the Si substrate. In the case of Fig. 6b the central spot is much more intense due to the diffraction from the amorphous-Si layer (PECVD) deposited above SPE-Si as a scratch protection layer.

Fig. 5. SEM micrograph of the SPE-Si regions on a SiO_2 surface with windows to the Si substrate (with 75° tilt). The inset is the result of EBSD analysis, indicating the (100)-orientation of the SPE-Si deposited both within the contact windows and on surrounding SiO_2 (red) and the absence of grain boundaries where the SPE islands merge.

Microtwin-like defects were sometimes observed on HR-TEM images, as seen in Fig. 7. They are probably caused by phenomenon of stress release during the TEM sample preparation or atomic scale roughness – less than a few lattice planes (0.5 nm) - at the SiO_2/Si interface that is proven to cause {111} growth planes [7].

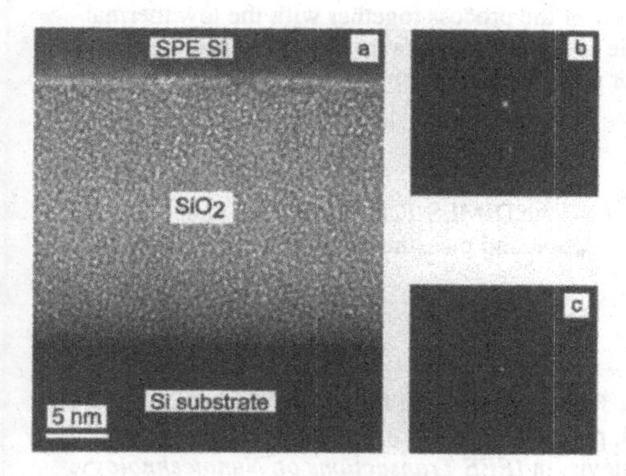

Fig. 6. (a) High-resolution TEM image of the SPE-Si/ SiO_2/Si substrate interfaces. Selective area diffraction (SAD) pattern obtained (b) on SPE-Si island on SiO_2 (c) on the Si substrate.

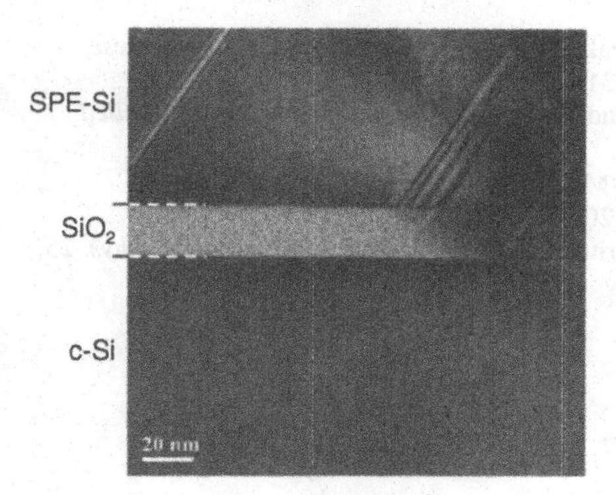

Fig. 7. High-resolution TEM image of the SPE-Si/ SiO_2/Si substrate interfaces showing microtwins on L-SPE region.

CONCLUSIONS

A controllable Si lateral overgrowth on silicon dioxide (SiO_2) has been demonstrated with an Al-mediated SPE process. Hundreds-of-nanometer large areas of high-quality monocrystalline silicon-on-insulator (SOI) were obtained on thermally-grown SiO_2 using a maximum process temperature of 400 °C. The SPE-Si first nucleates in the contact windows and, after completely filling them, the lateral overgrowth begins. The overgrowing Si proceeds preferentially in the (100) direction and keeps the lattice structure of the underlying c-Si over micron large distances. As a result, adjacent overgrowth regions can join together in a manner free of grain boundaries or dislocations, which has been verified by high-resolution transmission-electron-microscopy (HRTEM), selective area diffraction (SAD), and electron back-scattering diffraction (ESBD) analyses. The controllability of the process together with the low thermal budget involved make this a promising module to locally incorporate SOI regions either in the front- or back-end Si processing, as well as for thin-film device applications.

ACKNOWLEDGMENTS

The authors would like to thank the staff of the DIMES-ICP cleanrooms and measurement room for their support in the fabrication and measurement of the experimental material.

REFERENCES

1. Y. Civale, L. K. Nanver, P. Hadley, E. J. G. Goudena and H. Schellevis, in *IEEE Electron Device Letters*, vol. 27, no. 5, pp. 341-343 (2006).
2. Y. Civale, L. K. Nanver, and H. Schellevis, in *IEEE Transactions on Nanotechnology*, vol. 6, no. 2, pp. 196-200 (2007).
3. V. W. C. Chan, P. C. H. Chan, and M. Chan, in *IEEE Trans. Electron Devices* 48, 1394 (2001).
4. H. Ishiwara, H. Wakabayashi, K. Miyazaki, K. Fukao, and A.Sawaoka, in *Japanese Journal of Applied Physics 32,* pp. 308-311 (1993)
5. M. Miyao, M. Moniwa, K. Kusukawa and W. Sinke, in *Japanese Journal of Applied Physics* 64, pp.3018-3023 (1988)
6. Y. Civale, G. Vastola, L. K. Nanver, R. Mary-Joy, and J.-R. Kim, in J*ournal of Electronic Materials*, vol. 38, pp. 2052-2062 (2009)
7. H. Kawarada, T. Ueno, Y. Kunii, S. Horiuchi and I. Ohdomari *Japan. J. Appl. Phys.* 25, pp. L814-L817 (1986)

Mater. Res. Soc. Symp. Proc. Vol. 1245 © 2010 Materials Research Society 1245-A20-06

Biaxial Texturing of Inorganic Photovoltaic Thin Films Using Low Energy Ion Beam Irradiation During Growth

James R. Groves [1,2], Garrett J. Hayes [1], Joel B. Li [4], Raymond F. DePaula [2], Robert H. Hammond [3], Alberto Salleo [1] and Bruce M. Clemens [1]
[1]Department of Materials Science and Engineering, Stanford University
[2]Superconductivity Technology Center, Los Alamos National Laboratory
[3]Geballe Laboratory for Advanced Materials
[4]Department of Electrical Engineering, Stanford University

ABSTRACT

We describe our efforts to control the grain boundary alignment in polycrystalline thin films of silicon by using a biaxially textured template layer of CaF_2 for photovoltaic device applications. We have chosen CaF_2 as a candidate material due to its close lattice match with silicon and its suitability as an ion beam assisted deposition (IBAD) material. We show that the CaF_2 aligns biaxially at a thickness of ~10 nm and, with the addition of an epitaxial CaF_2 layer, has an in-plane texture of ~15°. Deposition of a subsequent layer of Si aligns on the template layer with an in-plane texture of 10.8°. The additional improvement of in-plane texture is similar to the behavior observed in more fully characterized IBAD materials systems. A germanium buffer layer is used to assist in the epitaxial deposition of Si on CaF_2 template layers and single crystal substrates. These experiments confirm that an IBAD template can be used to biaxially orient polycrystalline Si.

INTRODUCTION

Solar cell efficiency is a strong function of minority carrier lifetime, since photo-generated carriers that recombine before reaching the p-n junction do not contribute to photocurrent. Grain boundaries in polycrystalline silicon films provide electron traps that act as recombination centers that reduce minority carrier lifetimes [1]. This recombination is a function of the grain boundary structure. In particular, the high dislocation density of high angle grain boundaries result in higher recombination rate than low angle grain boundaries. Dimitriadis et al. showed that the effective carrier lifetime increases as the dislocation density decreases [2], and in an elegant experiment using electron beam induced current contrast ratios in polycrystalline silicon films, Seifert et al. showed that recombination is a strong function of grain boundary defect density [3].

Grain boundaries can be described as having both out-of-plane and in-plane misorientation known as tilt and twist, respectively. Both types of misorientation result in defect densities that lead to recombination. The degree of tilt and twist in a thin film grain boundary population reflects the crystallographic texture of the film. Biaxial texture, which has a preferred crystallographic direction for both out-of-plane and in-plane directions, can decrease both twist and tilt misorientation between grains. One way to develop biaxial texture is application of an ion beam during the initial stages of nucleation of a thin film. This ion beam assisted deposition (IBAD) process uses a low energy (< 1keV), inert (Ar^+) ion beam to develop in-plane texture in a growing thin film during concurrent physical vapor deposition of the desired source material. The ion beam is aligned along a particular crystallographic direction at an oblique angle relative to the desired out-of-plane growth direction. The ion beam sputters away unfavorably oriented crystallites and allows favorably oriented crystallites to survive and grow. If the correct channeling angle is selected then bi-axial texture can be developed.

The IBAD process has been used to form MgO template layers for seeding crystallographic texture in the high temperature superconductor $YBa_2Cu_3O_{7-\delta}$ (YBCO), as its superconducting properties are dependent upon the amount of in-plane alignment. Typically, IBAD MgO can be deposited with an in-plane texture of 5-6° phi-scan FWHM and an out-of-plane texture of 1-2° omega scan FWHM [4]. This texture develops in about 10 nm and the subsequent deposition of YBCO has a phi-scan FWHM of about 1°, which is very near single crystal quality.

Choi et al. used an IBAD MgO template layer, optimized for high-temperature superconductor coated conductors, as a template layer for the deposition of polycrystalline silicon [5]. Silicon films deposited on this template layer have reduced grain boundary misorientation and increased carrier mobility [6].

Here, we use the IBAD process to develop a template layer for the subsequent deposition of polycrystalline silicon for photovoltaic applications. We chose CaF_2 as our starting template material because it fulfills some of the empirically accepted criteria for a good IBAD candidate material [7]. CaF_2 is a cubic material with well-defined channeling planes, is highly ionic in bond character, and CaF_2 is a good lattice match with Si with lattice parameters of 0.5451 nm and 0.5431 nm, respectively. In this report, we describe the development of IBAD CaF_2 as a template layer for the subsequent deposition of heteroepitaxial polycrystalline silicon with low angle grain boundaries associated with biaxial crystallographic texture.

EXPERIMENT

Four types of substrates were used in these experiments: fused silica; silicon (100) coated with 800 nm of thermally grown SiO_2; single crystal yttria-stabilized zirconia (YSZ) (111) or (100); or CaF_2 (111) or (100) single crystals. All substrates used in these experiments were nominally 1 x 1 cm in size.

Depositions for these experiments were performed in a PVD high vacuum system with a typically base pressure of 7.0×10^{-6} Pa (5.0×10^{-8} torr) at room temperature. A four-pocket 7 cc Temescal SuperSource provided the deposit vapor flux. A two-grid collimated Kaufman ion source at an incidence angle of either 35.3°, 45° or 54.7° (corresponding to particular crystallographic directions in the CaF_2 crystal) relative to the substrate normal provided an Ar ion flux to the substrate. The ion current density was monitored with a separate Faraday cup. The Faraday cup was biased at -20 V to eliminate contributions from electrons to the ion current reading.

The CaF_2 layer was deposited with concurrent Ar ion and CaF_2 fluxes. The ion energy range for these studies was varied between 200 and 900 eV with a current density of ~80 μA/cm^2. The electron beam evaporator provided the CaF_2 vapor flux at 0.06 nm/s to 0.11 nm/s. The flow rate of Ar gas into the system was kept constant at 10 sccm, which corresponded to a chamber pressure of ~5.0×10^{-3} Pa. Subsequent Ge and Si films were deposited in-situ using e- beam evaporation at 570°C and 0.05 nm/s. In some cases, the Ge and Si layers were sputter deposited at temperature between 500°C and 800°C. The film growth was monitored in-situ using reflection high-energy electron diffraction (RHEED). The RHEED beam is aligned along an axis 90° relative to the ion beam. All patterns were taken at an electron beam energy of 28 keV.

The crystal structure of the films was determined ex-situ using X-ray diffraction (XRD). Samples were characterized for both in-plane and out-of-plane texture. Symmetric theta-2theta scans were used to verify that the desired phases were present in the deposited films.

DISCUSSION

To determine if texture could be developed in CaF_2 by IBAD processing we deposited CaF_2 IBAD films onto fused silica with a deposition rate of 0.06 nm/s and ion beam energy of 500 eV. The beam current density was ~80 μA/cm^2. The CaF_2 oriented with a (111)-type texture out-of-plane in ~10 nm of deposited film as shown in the RHEED image captured at the end of the IBAD run in the upper left of Figure 1. A subsequent 30 nm homoepitaxial layer of CaF_2 was deposited at 400°C and its in-plane texture was measured to be ~15° FWHM for the (220) in-plane peaks as shown in Figure 1. The RHEED diffraction spots increased in intensity and sharpened as the epitaxial CaF_2 layer was added indicating that texture improved from the initial IBAD layer. This behavior is similar to the improvement in texture observed for homoepitaxially deposited layers of MgO on IBAD [8].

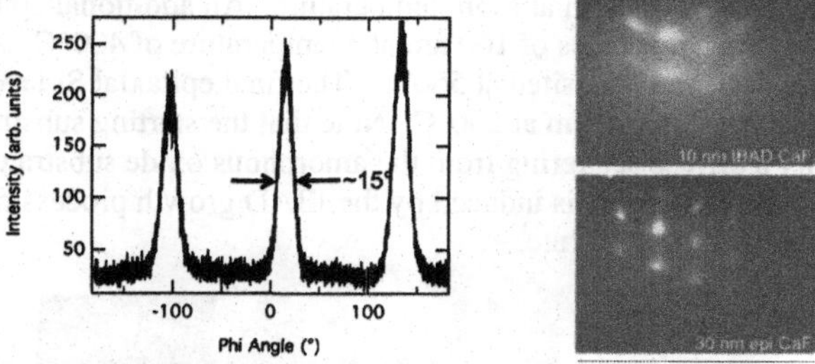

Figure 1. RHEED images and (220) phi scan for an IBAD CaF_2 film with a 30 nm homoepitaxial layer.

In a parallel effort to demonstrate proof-of-principle, we used single crystal substrates of (111) and (100) CaF_2 to determine if CaF_2 is a suitable seed layer for epitaxial growth of silicon. Preliminary results, however, were inconclusive. Little separation exists between CaF_2 and Si X-ray peaks and the CaF_2 single crystal substrate peaks were so intense that the Si peaks could not be easily distinguished using our standard laboratory X-ray diffraction methods. In order to reduce this diffraction interference, we used yttria-stabilized zirconia (YSZ) single crystal substrates capped with thin (30 nm) layer of CaF_2. The YSZ peaks are sufficiently removed from the Si peaks and the small X-ray diffraction signal from the thin CaF_2 layers will not swamp the signal from the thin Si films. The CaF_2 aligned well on YSZ (111) and (100) single crystal substrates, but silicon did not grow epitaxial on these capped single crystal seeds [9]. Ge, however, did grow epitaxially on the CaF_2/YSZ substrates, and provided an excellent seed for subsequent growth of Si [9]. Deposition of the Ge at 700°C produced an epitaxial layer with good (<1° FWHM) in-plane alignment as shown in Figure 2. The subsequent deposition of Si on this Ge-buffered substrate resulted in an epitaxial film as indicated by the spot pattern for the Si in the upper RHEED image in Figure 2. This high degree of orientation achieved indicates that growth of Si on textured CaF_2 proceeds with little-to-no degradation in crystallographic orientation.

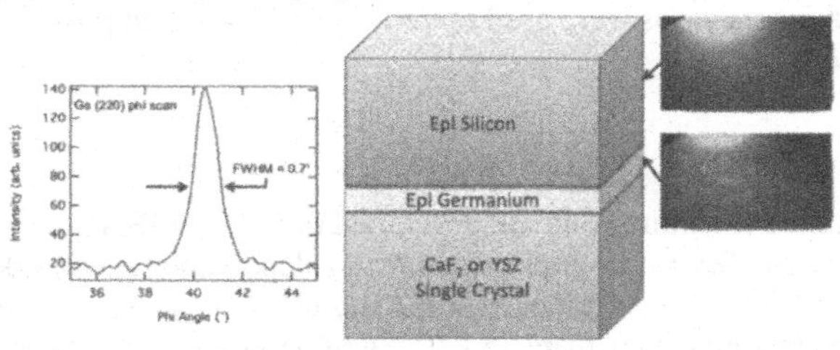

Figure 2. An illustration of the architecture used for the proof of principle experiments used to determine whether Si can be deposited on CaF_2. The XRD phi-scan on the left shows the epitaxial nature of the Ge layer deposited on CaF_2.

The final experiment was to assemble the IBAD CaF_2 film with the Ge buffer layer and silicon thin film as shown in the illustration in Figure 3. RHEED images taken at the conclusion of each deposition step are included in Figure 3. The IBAD CaF_2 was deposited to a thickness of 10 nm at room temperature. An additional epitaxial layer of CaF_2 was deposited to a thickness of 100 nm at a temperature of 400°C. A 50 nm thick epitaxial Ge layer was then deposited at 560°C. The final epitaxial Si layer was deposited to a thickness of 150 nm at 560°C. Note that the starting substrate RHEED image shows only a diffuse scattering from the amorphous oxide substrate. Thus the biaxial texture in the film layers is induced by the IBAD growth process rather than any influence from the starting substrate.

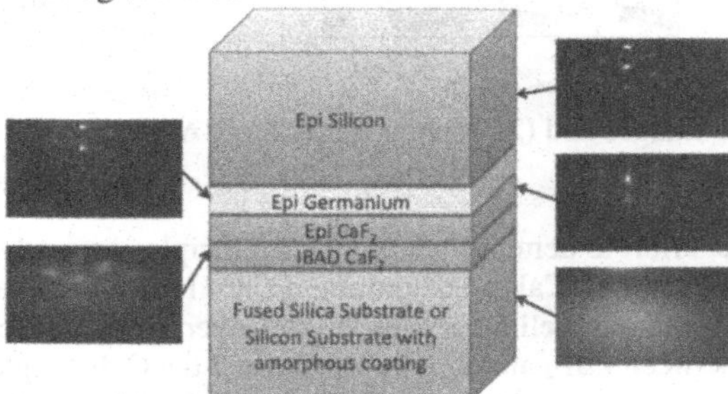

Figure 3. An illustration of the architecture used for the IBAD CaF_2-templated polycrystalline Si structure. The accompanying RHEED images were taken at the conclusion of each layer's growth.

The XRD analysis of the film structure described in Figure 3 showed the presence of good in-plane alignment throughout the film structure. We used the (220) peaks to characterize the in-plane texture for both the Si and CaF_2 films. As shown in Figure 4, the CaF_2 (220) peak had a full-width-at-half-maximum (FWHM) of 16.3° (with a 50 nm homoepitaxial CaF_2 layer). Deposition of the Si layer improves the texture by ~6° in-plane to a value of 10.8°, which indicates that the texture improves with each subsequent layer. This is similar to the improvement observed with homoepitaxial growth on IBAD seed layers [10]. This suggests that the texture is improved by grain growth competition and the overgrowth of misoriented grains, similar to that which is commonly observed in thin film microstrucutral evolution [11]. Further evidence of this effect is indicated in the sharpening of the RHEED image spots shown in Figure 3 for each additional layer.

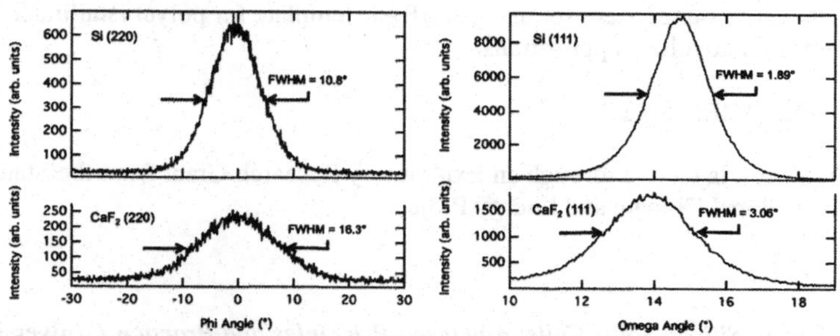

Figure 4. The XRD spectra of a film with the architecture illustrated in Figure 4. On the left are phi-scans of the CaF_2 (lower left) and Si (upper left) (220) peaks with FWHM of 16.3° and 10.8°, respectively. On the right-hand side are the omega (out-of-plane) scans for the (111) peaks of the two layers.

The films exhibit excellent out-of-plane alignment and no additional phases are detected as indicated by the theta-two theta XRD scan of Figure 5. Only the (111) reflections are observed for the Ge and Si films. Only two additional peaks are contributed to the scan by the substrate. The Si (111) most likely includes a contribution from the CaF_2 IBAD and homoepitaxial layers, but this is indistinguishable from the Si peak without the use of high-resolution optics and is the subject of a future investigation.

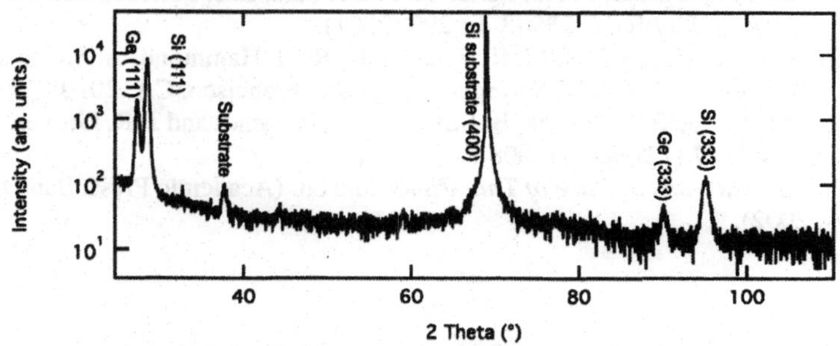

Figure 5. The theta-2 theta XRD scan for a Si (100) wafer with native oxide/IBAD CaF_2/epi CaF_2/30 nm Ge/150 nm Si.

CONCLUSIONS

We have shown that CaF_2 is a suitable material for the IBAD process and useful as a template layer for subsequent silicon deposition. Even without full process optimization, CaF_2 films can be deposited on fused silica substrates with an in-plane texture of ~15° FWHM. Silicon can be deposited heteroepitaxially at 570°C on a Ge buffered CaF_2 template with an in-plane texture of 10.8°. Additionally, an improvement in the in-plane texture of ~6° is observed as subsequent layers are deposited on the CaF_2 IBAD template. Further optimization of the IBAD processing conditions for CaF_2 is expected to improve the in-plane texture. Combining process optimization with the appropriate buffer layer and Si deposition conditions are expected to yield a Si in-plane texture comparable to that achieved by Choi et al. on the IBAD MgO template [5]. These experiments verify our concept for using an IBAD template for polycrystalline Si deposition for photovoltaic applications.

ACKNOWLEDGEMENTS

This work is funded through an Exploratory Research Grant from the Stanford University's Global Climate and Energy Project.

REFERENCES

1. M. A. Green, *Silicon Solar Cells: Advanced Principles and Practice*. (University of New South Wales, Sydney, NSW, Australia, 1995).
2. C. A. Dimitriadis, Solid State Commun. **56** (11), 925 (1985).
3. W. Seifert, G. Morgenstern and M. Kittler, Semicond. Sci. Technol. **8** (9), 1687 (1993).
4. P. N. Arendt and S. R. Foltyn, MRS Bull. **29** (8), 543 (2004).
5. W. Choi, V. Matias, J. K. Lee and A. T. Findikoglu, Appl. Phys. Lett. **87** (15), 262111 (2005).
6. A. T. Findikoglu, W. Choi, V. Matias, T. G. Holesinger, Q. X. Jia and D. E. Peterson, Adv. Mater. **17** (12), 1527 (2005).
7. H. Matzke, Radiat Eff. Defects Solids **64** (1-4), 3 (1982).
8. J. R. Groves, P. C. Yashar, P. N. Arendt, R. F. DePaula, E. J. Peterson and M. R. Fitzsimmons, Physica C **355** (3-4), 293 (2001).
9. J. R. Groves, G. J. Hayes, J. B. Li, R. F. DePaula, R. H. Hammond, A. Salleo and B. M. Clemens, in *2010 MRS Spring Meeting* (San Francisco, CA, 2010).
10. S. Gsell, M. Schreck, R. Brescia, B. Stritzker, P. N. Arendt and J. R. Groves, Jpn. J. Appl. Phys. **47** (12), 8925 (2008).
11. M. Ohring, *Materials Science of Thin Films*, 2nd ed. (Academic Press, San Diego, CA, 2002).

Solar Cell: Fundamental

Mater. Res. Soc. Symp. Proc. Vol. 1245 © 2010 Materials Research Society 1245-A21-01

High Efficiency Hydrogenated Nanocrystalline Silicon Solar Cells Deposited at High Rates

Guozhen Yue, Laura Sivec, Baojie Yan, Jeffrey Yang, and Subhendu Guha
United Solar Ovonic LLC, 1100 West Maple Road, Troy, MI 48084, U.S.A.

ABSTRACT

We report recent progress on hydrogenated nanocrystalline silicon (nc-Si:H) solar cells prepared at different deposition rates. The nc-Si:H intrinsic layer was deposited, using a modified very high frequency (MVHF) glow discharge technique, on Ag/ZnO back reflectors (BRs). The nc-Si:H material quality, especially the evolution of the nanocrystallites, was optimized using hydrogen dilution profiling. First, an initial active-area efficiency of 10.2% was achieved in a nc-Si:H single-junction cell deposited at ~5 Å/s. Using the improved nc-Si:H cell, we obtained 14.5% initial and 13.5% stable active-area efficiencies in an a-Si:H/nc-Si:H/nc-Si:H triple-junction structure. Second, we achieved a stabilized total-area efficiency of 12.5% using the same triple-junction structure but with nc-Si:H deposited at ~10 Å/s; the efficiency was measured at the National Renewable Energy Laboratory (NREL). Third, we developed a recipe using a shorter deposition time and obtained initial 13.0% and stable 12.7% active-area efficiencies for the same triple-junction design.

INTRODUCTION

nc-Si:H is a good candidate for high efficiency multi-junction solar cells because of its superior long wavelength response and improved stability as compared to hydrogenated amorphous silicon (a-Si:H) and hydrogenated amorphous silicon germanium (a-SiGe:H) alloys. However, its indirect bandgap inherently results in lower absorption coefficients; thus a thicker intrinsic layer is required to achieve high photocurrents. It is known that thick intrinsic layers will weaken the built-in electrical field, reduce carrier collection, and consequently decrease the fill factor (FF) of solar cells. A good light trapping technique using a textured BR can enhance the short-circuit current density (J_{sc}) without the need of increasing the intrinsic layer thickness; thus playing a critical role in improving the efficiency of nc-Si:H cells. However, nc-Si:H materials deposited on textured surfaces usually have a high defect density due to crystallite collisions [1,2] and result in a poor cell performance. An optimized BR surface morphology is needed to minimize this effect. In previous work [3,4], we systematically studied the effect of Ag and ZnO texture and thicknesses on cell performance and found that textured Ag with thin ZnO was the most desirable BR structure for nc-Si:H. Using such an optimized BR, we have further optimized the deposition parameters for nc-Si:H solar cells and significantly improved efficiencies for single-junction and a-Si:H/nc-Si:H/nc-Si:H triple-junction. In this paper, we report our recent progress on nc-Si:H based cell efficiencies made at different rates. In addition, a comparison study of the stability against prolonged light-soaking for nc-Si:H single- and triple-junction solar cells made at different rates is presented.

EXPERIMENTAL

A multi-chamber system with three RF chambers and one MVHF chamber is used to deposit nc-Si:H single-junction and multi-junction solar cells. Solar cells with an *nip* structure were deposited on Ag/ZnO BR coated stainless steel substrates. The a-Si:H and nc-Si:H intrinsic layers and buffer layers were prepared using the MVHF glow discharge. The doped layers were deposited using RF glow discharge. The solar cells were completed with indium-tin-oxide dots having an active-area of 0.25-0.26 cm^2 on the top *p* layer. Current density versus voltage J-V characteristics were measured under an AM1.5 solar simulator at 25 °C. Quantum efficiency (QE) was measured in the wavelength range from 300 to 1200 nm. The integrals of the QE data with AM1.5 spectrum were used to calculate J_{sc}. Cells were light-soaked under 100 mW/cm^2 white light at 50 °C for more than 1000 hours to evaluate stability.

RESULTS AND DISCUSSION

BR optimization

We first review the results of Ag/ZnO BR optimization. Details of the procedure have been published elsewhere [4]. We found that the optimized BR for growing nc-Si:H cells includes a bi-layer of textured Ag and thin ZnO. Compared to a conventional BR with a thick ZnO layer, the optimized BR has a lower optical absorption loss as shown by the total reflection measurements in Fig.1 (a). In addition, the conventional BR has a more textured surface than the optimized BR. Solar cells deposited on such BRs have poor intrinsic material quality, which reduces the FF and open-circuit voltage (V_{oc}) [3]. Also, solar cells with a defective intrinsic layer suffer from poor carrier collection. Figure 1 (b) shows QE curves of cells deposited with identical recipe on conventional and optimized BRs. To check the carrier collection loss, we conducted QE measurements with a 2 V reverse bias for each sample. The difference between the QEs at short circuit and -2 V conditions reflects the carrier collection loss. The results show that the cell on the optimized BR has very little current loss (<1.0%), whereas the cell on the conventional BR has a >4% loss. As a result, the conventional BR gave a lower J_{sc} than the optimized BR although the former is more textured, as evidenced by less distinct interference fringes in the QE. Our previous study [3] showed that the crystalline volume fraction for nc-Si:H cells does not strongly depend on the texture of substrates. However, textured surface gives rise to poorer film quality due to crystallite collisions, leading to poorer carrier collection in nc-Si:H solar cells.

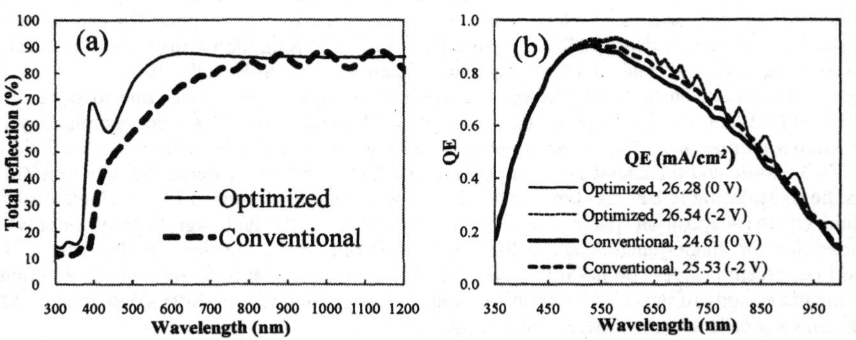

Figure 1 (a) Total reflections of two BRs; (b) QE measured at short cirucit and -2 bias conditions of nc-Si:H cells deposited on two BRs.

nc-Si:H based cells made at ~5 Å/s

Targeting high efficiencies, we first made nc-Si:H solar cells at a relatively low rate of ~5 Å/s using the optimized BRs. We employed an improved seed layer and hydrogen dilution profiling technique [5] to ensure an optimum structure along the growth direction. Using these techniques, the solar cell performance improved significantly. J_{sc}=29.11 mA/cm^2 has been achieved with V_{oc}=0.516 V, FF=0.647 and efficiency=9.72%. Using the same BR, we further optimized the growth recipe to obtain the highest efficiency. An initial active-area efficiency of 10.21% (V_{oc}=0.540 V, FF=0.680, and J_{sc}=27.80 mA/cm^2) was obtained as shown in Fig. 2. These two solar cells were sent to NREL for efficiency confirmation. The results are summarized in Table I. It should be pointed out that the J_{sc} values at United Solar are based on the QE measurements. NREL measures the total-area efficiency only; the active-area current density is obtained by first subtracting the grid coverage from the total area and then calculating its active-area current. It is readily seen that the active-area efficiency values of the two laboratories agree well within measurement error. The first high J_{sc} solar cell was light soaked under 100 mW/cm^2 white light at 50 °C for 1000 hours. The result is also listed in Table I. One can see that the cell displayed very little degradation. The FF and V_{oc} values do not show any light-induced change. The J_{sc} is reduced by < 1% from 29.11 to 28.84 mA/cm^2, occurring at short wavelengths according to the QE curves. Correspondingly, the efficiency degraded from 9.7% to 9.6%.

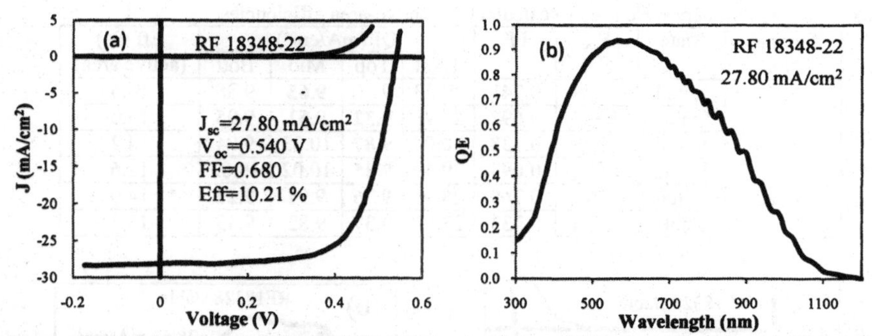

Figure 2. (a) J-V characteristics and (b) QE curve of a nc-Si:H solar cell with an initial active-area efficiency of 10.2%.

Table I. High efficiency nc-Si:H single-junction solar cells. Note that NREL measures total area only. Conversion from total-area to active-area is described in the text.

Sample No.	State	Measurement	V_{oc} (V)	FF	J_{sc} (mA/cm^2) Total	Active	Eff(%) Total	Active
18345	Initial	United Solar	0.516	0.647	27.07	29.11	9.04	9.72
		NREL	0.5169	0.6484	26.786	28.80	8.98	9.65
	Stable	United Solar	0.516	0.647	26.82	28.84	8.96	9.63
18348	Initial	United Solar	0.540	0.680	25.85	27.80	9.50	10.21
		NREL	0.5369	0.6742	25.821	27.76	9.35	10.05

Having developed high J_{sc} and high efficiency recipes, we proceeded to fabricate a-Si:H/nc-Si:H/nc-Si:H triple-junction cells on the optimized BRs, where the nc-Si:H bottom cell was made using the high J_{sc} recipe, and the nc-Si:H middle cell was made using the high efficiency recipe. The initial and the stabilized results after 1000 hours of light soaking are shown in Table II. Sample 18321 shows the best initial active-area efficiency of 14.5%, and sample 18323 shows the highest total-current density of >30 mA/cm^2. Regarding stability results, the light-induced degradations are <1% in V_{oc} and 4-7% in FF, which we believe originate mainly in the a-Si:H top cells. The top cell current degraded by ~4%, while the bottom cell current did not show any degradation. This is consistent with our previous results, where we demonstrated that red light with a photon energy smaller than the a-Si:H bandgap did not cause degradation in nc-Si:H cells [6]. We noticed that the middle cell current degraded by 1-2%. This is probably because the middle cell also sees some blue light. As shown in Table I, the blue light will induce a small amount of J_{sc} degradation in our high quality nc-Si:H cells. Overall, depending on the current mismatch, the efficiency degrades by 4-11%. The best stabilized active-area efficiency is 13.5%, which is from sample 18328 having a strong bottom cell limited current. This efficiency is higher than our previously reported efficiency of 13.2% from the same triple-junction structure [7]. The stable J-V characteristics and the QE curves of the best device are plotted in Figure 3.

Table II. Initial and stable J-V characteristics of a-Si:H/nc-Si:H/nc-Si:H triple-junction solar cells. The bold numbers were used to calculate the active-area efficiencies.

Sample No.	State	V_{oc} (V)	FF	QE(mA/cm^2) Total	Top	Mid	Bott	Eff (%) (active area)
18321	Initial	2.024	0.741	29.23	9.70	**9.65**	9.88	14.5
	Stable	2.012	0.692	28.78	**9.32**	9.51	9.95	13.0
18323	Initial	1.943	0.738	30.04	**9.87**	10.23	9.94	14.2
	Stable	1.924	0.693	29.49	**9.45**	10.02	10.02	12.6
18328	Initial	1.971	0.768	28.95	9.76	9.94	**9.25**	14.0
	Stable	1.964	0.733	28.62	**9.37**	9.82	9.43	13.5

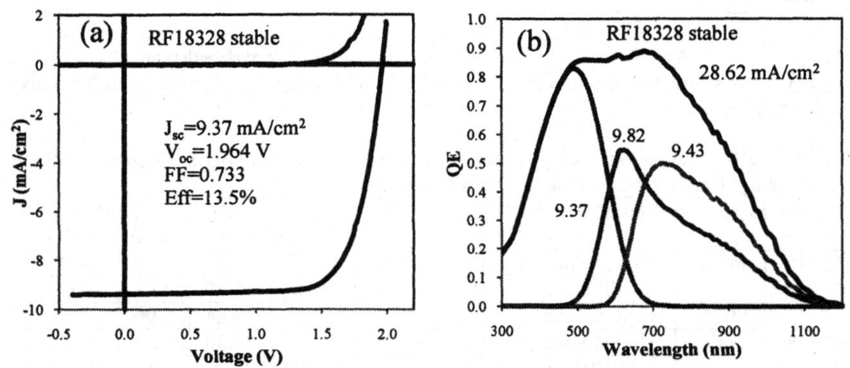

Figure 3. (a) Stable J-V characteristics and (b) QE curves of an a-Si:H/nc-Si:H/nc-Si:H triple-junction solar cell with the stabilized active-area efficiency of 13.5%.

nc-Si:H based cells made at high rate of ~10 Å/s

As mentioned before, nc-Si:H has a low optical absorption coefficient and hence a thick intrinsic layer is required to obtain sufficient current in solar cells. Therefore, the deposition rate has to be increased significantly in order to be viable for commercial manufacturability. Moreover, the feasibility of manufacturing nc-Si:H depends also on the deposition time. For a throughput similar to the a-SiGe:H product, the deposition time of nc-Si:H intrinsic layers should be limited to 20 minutes for the bottom cell. For this purpose, two nc-Si:H cell recipes with different intrinsic layer deposition times of 35 and 20 minutes, respectively, have been developed at a high rate of ~10 Å/s. Table III lists typical J-V characteristics of nc-Si:H single junction cells with different deposition times. Initial active-area efficiencies of ~9.0% with the 35 minute recipe and 8.7% with the 20 minute recipe were obtained. The a-Si:H/nc-Si:H/nc-Si:H triple-junction cells were fabricated using these recipes, where the deposition times in the top, middle, and bottom intrinsic layers are 5, 20, and 35 minutes for the 35 minute recipe, and 4, 15, and 20 minutes for the 20 minute recipe. Light-soaking was carried out on these cells. A stabilized total-area efficiency of 12.5% was achieved using the 35 minute recipe. The efficiency was measured at NREL. It should be pointed out that this is the highest efficiency NREL has measured for a-Si:H based thin film solar cells [8]. Figure 4 shows its J-V characteristics which NREL provided.

Table III. J-V characteristics of nc-Si:H cells made with different deposition times.

Run No.	V_{oc} (V)	FF	QE (mA/cm^2)	Efficiency (%)	Time (min)
18449	0.517	0.654	26.44	8.94	35
18712	0.559	0.665	24.25	9.01	35
18448	0.529	0.662	24.82	8.69	20
18444	0.553	0.655	24.05	8.71	20

The initial and stable J-V characteristics of a-Si:H/nc-Si:H/nc-Si:H triple-junction solar cells made with the 20 minute recipe are listed in Table IV. The best initial active-area efficiency is ~13%. Stability results after 1000 hours of light soaking show that the efficiency degradation is in the range of 2-5%, depending on the current mismatch, which is smaller than the degradation of 4-11% for the low rate triple-junction cells shown in Table II. Since the degradation of a-Si:H/nc-Si:H/nc-Si:H triple-junction cells occurs only in the top cells, the larger degradation can be attributed to a thicker top cell in the low rate cells than in the high rate cells. As shown in Table IV, the stabilized active-area efficiency of 12.7% was achieved at a high rate with a 20 minute deposition time for the bottom cell intrinsic layers.

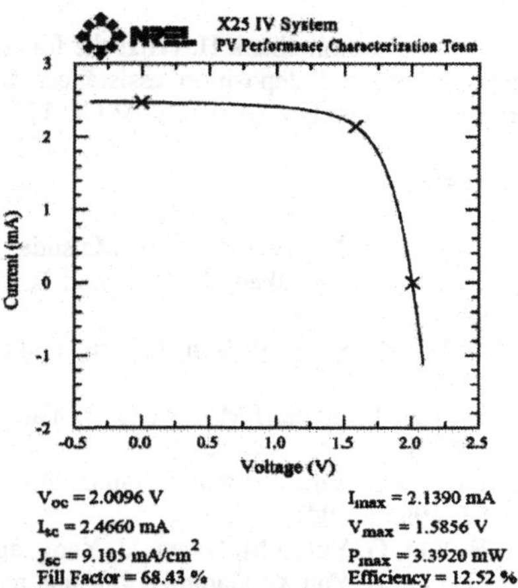

Figure 4. J-V characteristics of an a-Si:H/nc-Si:H/nc-Si:H triple-junction cell with a stable total-area efficiency of 12.5%.

Table IV. Initial and stable characteristics of a-Si:H/nc-Si:H/nc-Si:H triple-junction solar cells made at high rates. The bold numbers were used to calculate the efficiency.

Run No.	State	V_{oc} (V)	FF	Q(mA/cm^2)				Efficiency (%) (active-area)
				Total	Top	Mid	Bott	
18596	Initial	2.032	0.734	26.01	**8.66**	**8.64**	8.71	12.89
	Stable	2.030	0.715	25.68	**8.44**	8.55	8.69	12.25
18601	Initial	2.015	0.767	26.24	9.66	**8.29**	8.29	12.81
	Stable	2.008	0.759	25.85	9.20	8.36	**8.31**	12.63
18603	Initial	1.983	0.780	26.36	9.48	8.64	**8.24**	12.75
	Stable	1.972	0.770	25.65	9.00	8.44	**8.21**	12.47
18617	Initial	2.015	0.738	26.54	8.98	**8.70**	8.86	12.94
	Stable	2.018	0.724	26.21	8.70	**8.70**	8.81	12.71

SUMMARY

We have made nc-Si:H single junction solar cells on optimized Ag/ZnO BRs at different deposition rates and used an improved nc-Si:H cell as the bottom and middle cells in triple-junction structures. Initial and stable active-area efficiencies of 14.5% and 13.5% have been achieved in a-Si:H/nc-Si:H/nc-Si:H triple-junction cells made at a relatively low rate of ~5 Å/s. At a deposition rate of ~10 Å/s with a 35 minute bottom cell intrinsic layer, we have achieved 12.5% stabilized total-area efficiency as measured at NREL. This is the highest efficiency NREL has measured for a-Si:H based thin film solar cells. In addition, with the deposition time of the bottom intrinsic layer limited to 20 minutes, an initial active-area efficiency of 13.0% and stable active-area efficiency of 12.7% have been achieved.

ACKNOWLEDGMENT

The authors thank H. Fritzsche for fruitful discussions, and J.M Owens and T. Palmer for measurement and deposition assistance. This work was supported by U.S. DOE under SAI Program contract No. DE-FC36-07 GO 17053.

REFERENCES

1. Y. Nasuno, M. Kondo, and A. Matsuda, Proc. of the 28th IEEE PVSC, 142 (2000).
2. H. Li, R.H. Franken, J. Rath, and R.E.I. Schropp, Sol. Energy Mater. Sol. Cells **93**, 338 (2009).
3. G. Yue, L. Sivec, B. Yan, J. Yang, and S. Guha, Mater. Res. Soc. Symp. Proc. **1153**, A10-05 (2009)
4. G. Yue, L. Sivec, J.M. Owens, B. Yan, J. Yang, and S. Guha, Appl. Phys. Lett. **95**, 263501 (2009).
5. B. Yan, G. Yue, J. Yang, S. Guha, D.L. Williamson, D. Han, and C. Jiang, Appl. Phys. Lett. **85**, 1955 (2004).
6. B. Yan, G. Yue, J.M. Owens, J. Yang, and S. Guha, Appl. Phys. Lett. **85**, 1925 (2004).
7. G. Yue, B. Yan, G. Ganguly, J. Yang, and S. Guha, Appl. Phys. Lett. **88**, 263507 (2006).
8. M.A. Green, K. Emery, Y. Hishikawa, and W. Warta, Prog. Photovolt: Res. Appl. **17**, 320 (2009).

N-type Microcrystalline Silicon Oxide (μc-SiO_x:H) Window Layers with Combined Anti-reflection Effects for n-i-p Thin Film Silicon Solar Cells

V. Smirnov, W. Böttler, A. Lambertz, O. Astakhov, R. Carius and F. Finger
IEF-5 Photovoltaik, Forschungszentrum Jülich, D-52425, Jülich, Germany

ABSTRACT

N-type hydrogenated microcrystalline silicon oxide (μc-SiO_x:H) films were used as window layers in n-side illuminated microcrystalline silicon n-i-p solar cells. Optical, electrical and structural properties of μc-SiO_x:H films were investigated by Photothermal Deflection Spectroscopy, conductivity and Raman scattering measurements. μc-SiO_x:H layers were prepared over a range of carbon dioxide (CO_2) flow and film thickness, and the effects on the solar cell performance were investigated. By optimising the μc-SiO_x:H window layer properties, an improved short-circuit current density of 23.4 mA/cm^2 is achieved, leading to an efficiency of 8.0% for 1 μm thick absorber layer and Ag back contact. The correlation between cell performance and reduced reflection in μc-SiO_x:H n-i-p solar cells is discussed. The results are compared to the performance of solar cells prepared with alternative optimised window layers.

INTRODUCTION

As a wide optical gap material, silicon oxide (SiO_x:H) has been the subject of research as a material for photovoltaic applications [1-9]. Electrical, optical and structural properties of both amorphous (a-SiO_x:H) and microcrystalline (μc-SiO_x:H) silicon oxide films have been studied. The properties of silicon oxide thin films are influenced by the doping and the oxygen content [2-4, 7]. Optical band gap (E_{04}), refractive index (n), and conductivity can be modified over a wide range by varying input gas ratios during material growth. Several reports concerned a preparation of silicon oxide thin films as p-type window layers [1-3]. An optical gap (E_{04}) of 1.95-2.06 eV for p-type silicon oxide films was reported [2, 3]. Recently, n-type μc-SiO_x:H has attracted attention as an intermediate reflector in tandem solar cells [6-9]. In that case, the SiO_x layer was incorporated between amorphous silicon (a-Si:H) and microcrystalline silicon (μc-Si:H) sub-cells. This approach allows the current of the a-Si sub-cell to be increased, allowing the absorber layer to be kept reasonably thin to minimize the impact of light-induced degradation.

While optimal performance a-Si solar cells need to be illuminated through the p-side, solar cells utilising μc-Si:H absorber layers may be illuminated from either p or n side, since the hole drift mobility in μc-Si films is much higher then in a-Si films [10, 11]. It was recently shown [12], for example, that n-type microcrystalline silicon carbide (μc-SiC), having a wide optical gap of around 2.8eV and grown by Hot Wire Chemical Vapour Deposition (HWCVD), can be successfully used as a window layer in single junction microcrystalline silicon n-i-p solar cells.

The possibility to tune optical (such as E_{04} and n) and electrical properties in n-type μc-SiO_x:H offers potential for its application as a window layer in n-side illuminated n-i-p solar cells. In this work we present the results of investigation of n-type μc-SiO_x:H thin films and n-i-p solar cells prepared with μc-SiO_x:H window layers.

EXPERIMENTAL DETAILS

n-type μc-SiO_x:H layers were deposited by RF (13.56 MHz) plasma enhanced chemical vapour deposition (PECVD) technique at 185 °C substrate temperature and a power of 50 W, using a mixture of phosphine (PH_3), silane (SiH_4), hydrogen (H_2) and carbon dioxide (CO_2) gases. Additional details on the deposition conditions can be found elsewhere [7, 13]. The layers were deposited at varied CO_2 flows (up to 2.5 sccm) and thickness, and subsequently used as window layers in microcrystalline silicon solar cells. Silicon thin films were deposited by PECVD technique using either RF or VHF (94.7 MHz) excitation for different intrinsic and extrinsic layers. p-type and intrinsic μc-Si layers were prepared by VHF-PECVD; the deposition conditions of p-layer remained unaltered during this study. Intrinsic μc-Si:H layers (nominally 1 μm thick) were prepared from a mixture of SiH_4 and H_2 gases, at the silane concentration SC = [SiH_4] / ([SiH_4 + [H_2]) ratio of 4.5%.

Solar cells were prepared on textured glass/ZnO [14] substrates in the sequence: glass/ZnO/μc-SiO_x(n)/μc-Si(i)/μc-Si(p). Additionally, n-side illuminated n-i-p solar cells, employing μc-Si:H n-layer and p-side illuminated p-i-n cells were prepared to enable a comparison. The area of the individual devices was defined by the 1 cm^2 Ag back contacts. Solar cells were characterised by current-voltage (*J-V*) measurements under AM 1.5 illumination, and also under modified AM 1.5 illumination with red cut-on filter OG590 ($\lambda > 590$ nm) and blue band filter OG7 (λ centred at around 480 nm). The total optical reflectance of the cells was measured on a Perkin-Elmer photospectrometer, type lambda 950, within a spectral range from 320 to 2000 nm.

The window layers concerned in this work (μc-SiO_x:H, μc-Si:H (n) and μc-Si:H (p)) were also prepared on Corning Eagle 2000 glass substrates. Optical and electrical properties of these layers were investigated by Photothermal Deflection Spectroscopy (PDS) and conductivity measurements, respectively. Conductivity measurements were performed on films equipped with coplanar electrodes 5 mm in length separated by a 0.5 mm gap. The structural properties of the layers were probed using Raman spectroscopy, with excitation at 647 nm supplied by an Ar-ion laser. The ratio of integrated intensities attributed to crystalline and amorphous regions, $I_{CRS} = I_c / (I_c + I_a)$ was used as semi-quantitative value of the crystalline volume fraction [15].

RESULTS AND DISCUSSION

Fig. 1 summarizes the variations in dark conductivity (σ_{dark}), crystallinity (I_{CRS}) and optical gap E_{04} of μc-SiO_x:H films, prepared at different CO_2 flows. It is seen that E_{04} increases with an increase in CO_2 flow, while σ_{dark} and I_{CRS} decrease. For the μc-SiO_x:H film, prepared at CO_2 flow of 2 sccm, optical gap E_{04} is around 2.65 eV, but σ_{dark} already drops significantly down to around 10^{-11} S/cm and no crystalline content can be detected. At CO_2 flow of 1 sccm, E_{04} = 2.3 eV, σ_{dark} is around $5x10^{-4}$ S/cm and I_{CRS} is around 35%. The CO_2 flows of 1 and 1.5 sccm (indicated by dashed lines in Fig. 1) were subsequently selected for the preparation of μc-SiO_x:H window layers, incorporated into solar cells. To enable a comparison, Fig. 1 also includes the data for n- and p- type μc-Si films and a μc-SiC layer, taken from ref. [12].
The performance of solar cells prepared with μc-SiO_x, p-type and n-type μc-Si window layers is presented in Table 1. μc-SiO_x:H window layers were deposited for 12 minutes and μc-Si:H absorber layers were grown under identical deposition conditions. One can see that the cells A

and B, prepared with μc-SiO_x window layers show enhanced short circuit current densities (J_{sc}) and improved conversion efficiencies. The results of total reflection measurements for these cells are displayed in Fig. 2. It can be seen that for the solar cells employing μc-SiO_x window layers, the reflection in long wavelength region of the spectrum is significantly reduced. A minimum total reflectance of only 6% at 570 nm wavelength was achieved with μc-SiO_x:H window layers prepared at CO_2 flow of 1 sccm. In this case, the refractive index n of μc-SiO_x:H material is around 2.5, and thus, anti-reflection conditions [16] are fulfilled when μc-SiO_x:H window layer is introduced between ZnO ($n_{ZnO} \sim 1.9$) and μc-Si:H absorber ($n_{\mu c\text{-}Si} \sim 3.5$) layers.

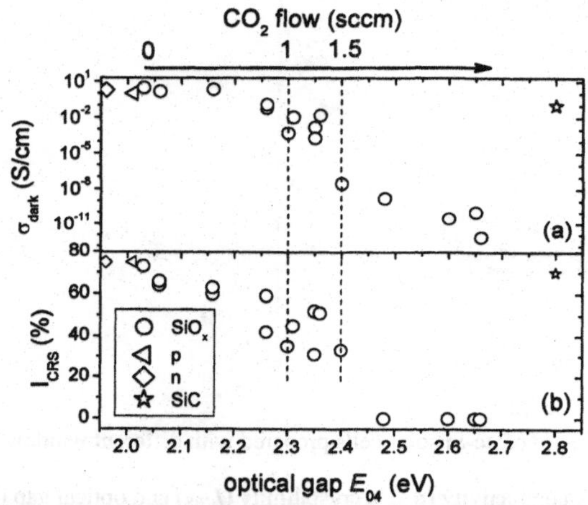

Figure 1. Dark conductivity (σ_{dark}) and crystallinity (I_{CRS})of μc-SiO_x:H films with varied optical gap (E_{04}). The direction of an increase in CO_2 flow is indicated by the arrow. Data of p- and n-type μc-Si and μc-SiC are shown for comparison.

Table 1. *J-V* characteristics, η , FF, J_{sc} and V_{oc} of μc-Si solar cells prepared with different window layers.

Cell	window layer	η (%)	FF (%)	V_{oc} (mV)	J_{sc} (mA/cm^2)
A	SiO_x (CO_2=1)	7.2	65.4	510	21.5
B	SiO_x (CO_2=1.5)	7.2	65.9	509	21.3
C	μc-Si n	6.2	67.2	512	18.1
D	μc-Si p	7.0	69.8	515	19.4

Subsequently, the effects of μc-SiO_x:H layer thickness (growth time t) on cell performance were investigated in more detail. The effects on short circuit current density (J_{sc}), fill factor (FF), open circuit voltage (V_{oc}) and conversion efficiency (η) are illustrated in Fig. 3. The CO_2 flow of 1 sccm was maintained during the growth of μc-SiO_x:H layers. Solar cells were also measured under the illuminations of a modified AM1.5 spectrum with OG590 or BG7 filters, to evaluate filtered short circuit current densities, red J_{sc} and blue J_{sc}, respectively. It is seen in Fig. 3e that the blue response of the cells (J_{sc} blue) tends to rise with reduction of μc-SiO_x:H layer thickness, while the red- filtered J_{sc} remains at high level above 12 mA/cm^2. The figure also includes the J_{sc} values of the cells prepared with alternative window layers (cells 1, 2 and 3). While the highest J_{sc} values are obtained for the cells prepared with 3 mins μc-SiO_x:H layers, the performance of these cells is poor due to dramatic drop in FF values, as evident in Fig. 3a and Fig. 3b. The best cell in this series, fabricated with μc-SiO_x:H window layer, is prepared at t = 4 mins.

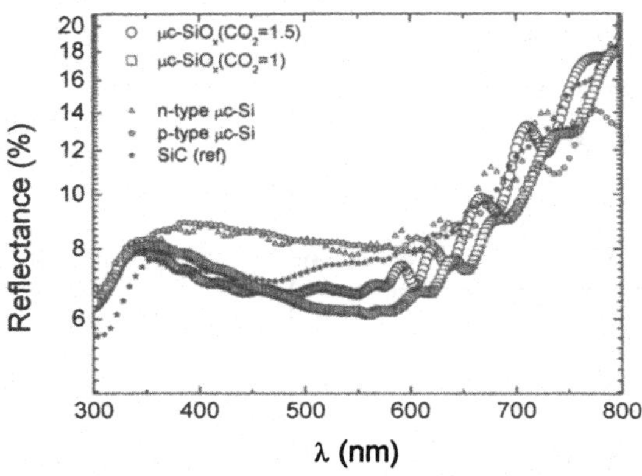

Figure 2. Total reflectance spectra of μc-Si solar cells prepared with different window layers.

The correlation of dark conductivity (σ_{dark}), crystallinity (I_{CRS}) and optical gap (E_{04}) with CO_2 flow is evident in Fig. 1: the trend indicates that when CO_2 flow increases, I_{CRS} decreases while E_{04} increases, resulting in a reduction of σ_{dark}. Such a trend has been previously observed for silicon oxide films and explained in terms of the effects of oxygen incorporation into silicon network [4, 5]. Additional information on the properties of μc-SiO_x:H can be found in ref. [17]. Considering the application of μc-SiO_x:H as window layers in solar cells, there is the trade off between electrical (σ_{dark}), structural (I_{CRS}) and optical (E_{04}) properties. While the optical band gap of SiO_x:H layer is highest for the CO_2 flow of 2.5 sccm, σ_{dark} and I_{CRS} are rather low to employ this material as a doped and a nucleation layer for subsequent intrinsic μc-Si layer growth [14] in solar cells. It is evident from Table 1 and Fig. 3 that μc-SiO_x:H films, prepared with CO_2 flow of 1 and 1.5 sccm, can provide sufficient transparency, conductivity and crystallinity to function well as window, doped, and also nucleation layer for subsequent μc-Si absorber layers growth. A reduction of μc-SiO_x:H layer thickness (see Fig. 3) results in slight (within 10%) improvement in red current (red J_{sc}), while a substantial (around 30%) increase in blue response of the cell is observed, that would be consistent with reduced optical absorption in thinner films. Using μc-SiO_x:H as a window layer, high J_{sc} values as compared to the solar cells utilising alternative window layers, namely n- and p-type μc-Si and also μc-SiC can be achieved, as evident in Fig. 3. Particularly, the observed high red response (red J_{sc} above 12.5 mA/cm^2) is consistent with the high transparency due to wide optical gap (see Fig. 1) and low total reflection (see Fig. 2) for the wavelengths above 450 nm. This is much above the current values obtained for cells with μc-Si window layers (cells 2 and 3). We note slightly higher blue J_{sc} values obtained for the cell prepared with μc-SiC window layer, that may be related to reduced total reflection (see Fig. 2) in short wavelength range and/or wider optical bang gap in SiC.

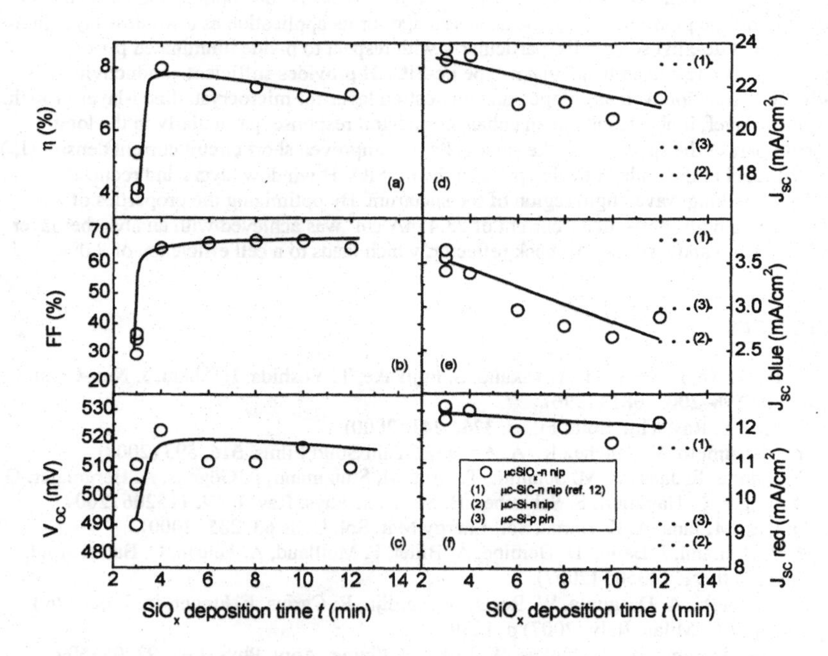

Figure 3. η, FF, V_{oc} and J_{sc} vs. the deposition time (*t*) of μc-SiO_x n-layer. The values are determined from the *J–V* characteristics measured under: (a), (b), (c) and (d) AM1.5 illumination without filters and filtered AM1.5 with (e) BG7 (blue) and (f) OG590 (red) filters. J_{sc} values of cells 1,2 and 3 are shown by dotted lines for comparison. The lines are to guide the eye.

The FF values drop dramatically (see Fig. 3b) nearly twice, down to around 30-35% when μc-SiO_x:H layer is deposited for less then 4 mins, which is also accompanied by a slight reduction in V_{oc}. In this case, the n-layer is possibly too thin to provide an appropriate built-in field in the device. Optimal μc-SiO_x:H layer in this series of cells allows 8% conversion efficiency to be achieved (FF = 65.1%, V_{oc} = 523 mV, J_{sc} = 23.4 mA/cm^2) for 1 μm thick absorber layer and Ag back contact. Further improvement in device performance may be possible by using highly reflective ZnO/Ag back contact, which is currently under investigation.

CONCLUSIONS

n-type μc-SiO_x:H layers have been used as window layers in n side illuminated microcrystalline silicon n-i-p solar cells. The possibility to easily tune optical and electrical properties in n-type μc-SiO_x:H suggests an advantage for its application as a window layer in n-side illuminated μc-Si:H solar cells, particularly with respect to p-side illuminated p-i-n configuration. The results indicate that n-type μc-SiO_x:H provides sufficient conductivity and crystallinity to function well as a doped and nucleation layer for microcrystalline i-layer growth. As a window layer, it also results in an enhanced spectral response, particularly in the long wavelength part of the spectrum of the solar cells. An improved short circuit current density (J_{sc}) can be attributed to the wide optical gap E_{04} in the μc-SiO_x:H window layers and reduced reflection in the long wavelength region of the spectrum. By optimizing the properties of μc-SiO_x:H layers, a high short-circuit current of 23.4 mA/cm^2 was achieved with an absorber layer thickness of 1 μm and a simple Ag back reflector, which leads to a cell efficiency of 8.0%.

REFERENCES

1. Y. Ichikawa, K. Tabuchi, A. Takano, S. Fujikake, T. Yoshida, H. Sakai, J. Non-Cryst. Solids **198–200**, 1081 (1996).
2. T. Jana, S. Ray, Thin Solid Films, **376**, 241 (2000).
3. Y. Matsumoto, V. Sanchez R., A. Avila G., Thin Solid Films, **516**, 593 (2008).
4. A. Janotta, R. Janssen, M. Schmidt, T. Graf, M. Stutzmann, L. Gögens, A. Bergmaier, G. Dollinger, C. Hammerl, S. Schreiber, B. Stritzker, Phys. Rev. **B 69**, 115206 (2004).
5. D. Das, M. Jana, A. K. Barua, Sol. Energy Mat. Sol. Cells **63**, 285 (2000).
6. P. Buehlmann, J. Bailat, D. Domine, A. Billet, F. Meillaud, A. Feltrin, C. Ballif, Appl. Phys. Lett. **91**, 143505 (2007).
7. A. Lambertz, A. Dasgupta, W. Reetz, A. Gordijn, R. Carius, F. Finger, in: Proc. 22nd EUPVSEC, Milan, Italy (2007) p. 1839.
8. C. Das, A. Lambertz, J. Hüpkes, W. Reetz, F. Finger, Appl. Phys. Lett. **92**, 053509 (2008).
9. D. Domine, P. Buehlmann, J. Bailat, A. Billet, A. Feltrin, C. Ballif, Phys. Stat. Sol. (RRL) **2 (4)**, 163 (2008).
10. T. Dylla, F. Finger, E. A. Schiff, Appl. Phys. Lett. **87**, 32103 (2005).
11. T. Dylla, S. Reynolds, R. Carius, F. Finger, J. Non-Cryst. Solids **352**, 1093 (2006).
12. Y. Huang, A. Dasgupta, A. Gordign, F. Finger, R. Carius, Appl. Phys. Lett. **90**, 203502 (2007).
13. V. Smirnov, W. Böttler, A. Lambertz, H. Wang, R. Carius, and F. Finger, Phys. Stat. Sol. (C) **7**, 1053 (2010).
14. O. Kluth, G. Schoepe, J. Huepkes, C. Agashe, J. Mueller, and B. Rech, Thin Solid Films **442**, 80 (2003).
15. L. Houben, M. Luysberg, P. Hapke, R. Carius, F. Finger and H. Wagner, Philos. Mag. **A 77**, 1447 (1998).
16. D. Bouhafs, A. Moussi, A. Chikouche, J. Ruiz, Sol. Energy Mat. Sol. Cells **52**, 79 (1998).
17. L. Xiao, O. Astakhov, R. Carius, A. Lambertz, and F. Finger, Phys. Stat. Sol. (C) **7**, 941 (2010).

Mater. Res. Soc. Symp. Proc. Vol. 1245 © 2010 Materials Research Society 1245-A21-03

Research Progresses on High Efficiency Amorphous and Microcrystalline Silicon-Based Thin Film Solar Cells

Ying Zhao, Xiaodan Zhang, Guofu Hou, Huizhi Ren, Hong Ge, Xinliang Chen, Xinhua Geng

The Institute of Photoelectronic Thin Film Devices and Technology, Nankai University, 300071, Tianjin, China

ABSTRACT

This paper reviews our research progresses of hydrogenated amorphous silicon (a-Si:H) and microcrystalline (μc-Si:H) based thin film solar cells. It coves the three areas of high efficiency, low cost process, and large-area proto-type multi-chamber system design and solar module deposition. With an innovative VHF power profiling technique, we have effectively controlled the crystalline evolution and made uniform μc-Si:H materials along the growth direction, which was used as the intrinsic layers of *pin* solar cells. We attained a 9.36% efficiency with a μc-Si:H single-junction cell structure. We have successfully resolved the cross-contamination issue in a single-chamber system and demonstrated the feasibility of using single-chamber process for manufacturing. We designed and built a large-area multi-chamber VHF system, which is used for depositing a-Si:H/μc-Si:H micromorph tandem modules on 0.79-m^2 glass substrates. Preliminary module efficiency has exceeded 8%.

INTRODUCTION

Amorphous silicon (a-Si:H) and microcrystalline silicon (μc-Si:H) tandem (micromorph) solar cells with an a-Si:H top cell and a μc-Si:H bottom cell technology has been gradually transferred from research laboratories to manufactures [1,2]. The a-Si:H/μc-Si:H micromorph structure has the advantage of higher efficiency and better stability than the cells with pure a-Si:H and amorphous silicon germanium alloy (a-SiGe:H). However, On the way towards real mass production, several technical issues need to be resolved. Among the hurdles, high efficiency, large-area uniformity, and reduction of manufacturing cost are the three major areas for further studied. Our group has focused on these three areas. In this paper, we review our progress made in the last a few years.

EXPERIMENT

All μc-Si:H films and μc-Si:H single-junction solar cells with a small area of 0.25 cm^2 were deposited in a cluster-tool system with a background pressure of 5×10^{-6}Pa. The intrinsic μc-Si:H layers were deposited with plasma enhanced chemical vapor deposition (PECVD) with a very high frequency (60 MHz) excitation. The solar cells were deposited on SnO_2 or Al doped ZnO coated glass substrates with a *pin* structure, where the μc-Si:H *p*-layer and a-Si:H *n*-layer were deposited with VHF (60MHz) and RF (13.56MHz), respectively. The a-Si:H top cells were deposited on the μc-Si:H single-junction bottom cells in a seven-chamber in-line PECVD system with the excitation frequency of 13.56 Hz. Chemically etched Al doped ZnO layers were also used to enhance the light trapping. Intrinsic μc-Si:H films were deposited on Eagle-2000 glass substrates for material characterization. Raman spectroscopy was used to measure the crystallinity. The crystalline volume fraction (X_C) was estimated by the decomposition of the Raman spectra and grain size from X-ray diffraction (XRD) spectra.

RESULTS AND DISCUSSION

High Rate Deposition of High Efficiency Thin Film Silicon Solar Cell

Crystalline evolution has been widely reported in the literature. In our study, we also checked the X_C as a function film thickness. A series of intrinsic μc-Si:H films with various thicknesses were deposited at 12Å/s on Eagle-2000 glass substrates under the same condition. Fig. 1 shows the Xc and grain size of the μc-Si:H films as a function of film thickness. Both Xc and grain size increase sharply with the increase of thickness when the films are thinner than 1 μm, and then they saturate when the films are thicker more than 1 μm. Except the initial growth, the Xc evolution along the growth direction in high-rate deposited μc-Si:H films is not as serious as originally suspected. Actually, the X_C increases from 52.3% to 61.5% with the thickness increase from 1 μm to 3 μm. The same trend is also observed in grain sizes. The (220) grain size increase sharply from 12 nm to 21 nm in the first 1 μm deposition, and then it saturates with further deposition.

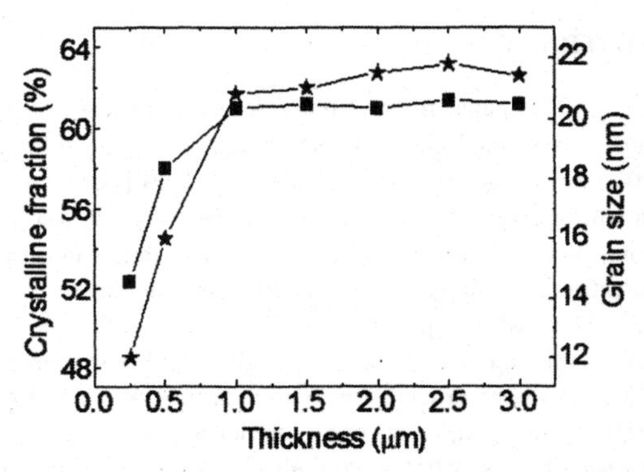

Figure 1. Xc(square) and (220) grain size(star) as a function of film thickness. The lines are guides to the eyes.

The same μc-Si:H films as shown in Fig. 1 were used as the absorber layers in single-junction *pin* solar cells on SnO_2/ZnO coated glass substrates. Al dots were deposited directly on the *n* layer as back contact and defined the solar cell area. All solar cells were prepared under the same deposition conditions except the *i*-layer thickness that was varied by the deposition time. Figure 2 shows the current density versus voltage (J-V) characteristic parameters as a function of *i*-layer thickness. It is observed that both *FF* and V_{oc} decreases with the increase of *i*-layer thickness, while J_{sc} increases first and then saturates. These results are in agreement with previous reports [3,4].

The decrease of *FF* suggests that carrier extraction efficiency decreases in the thick devices, probably due to the reduced built-in electric field in the *i*-layer. The higher J_{sc} results from more light absorption and the saturation could be attribute to a balance between the increased optical absorption and reduced built-in electric field. One reason for the V_{oc} decreases with *i*-layer thickness is the reduced built-in electric field in a thick *i*-layer, the other reason could be an increased defect density in the thicker intrinsic layer caused by large grain sizes, which have been believed to have a poor grain boundary passivation. In order to prove the second reason, the dark J-V characteristics of these solar cells were measured and analyzed to get the information about the carrier transport in the solar cells. The diode ideality factor *n* and saturation current density J_0, which are calculated from the fitting of exponential part of the dark *J-V* curves to the diode equation of $J_{dark}=J_0\,[\exp(eV/nkT)-1]$, are plotted in Fig. 3. With the increase of *i*-layer thickness from 1 μm to 3 μm, the diode factor n monotonously increases from 1.34 to 1.73, while the J_0 increases from 41 mA/cm^2 to 262 mA/cm^2. The higher n values than 1.5 in a *pin* diode suggest that the carrier transport is dominated by the bulk recombination. The high J_0 values indicate an increased defect density, which could be the main reason of the decrease of V_{oc} with the *i*-layer thickness. Besides the two reasons mentioned above, we suppose another factor for the lower V_{oc} and FF in the thicker cells. In order to get a high deposition rate, a high power density is normally used. However, the high-energy ion

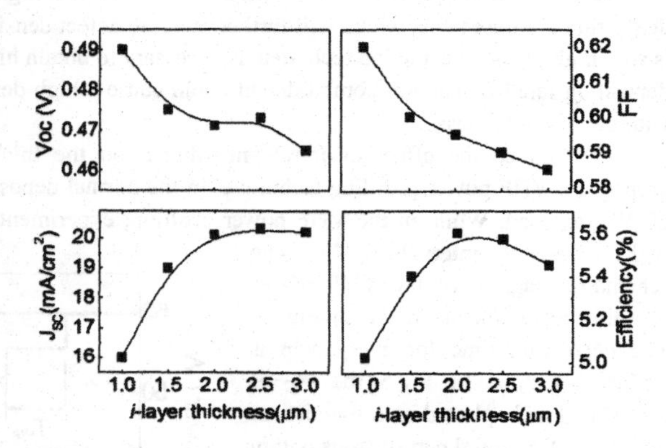

Figure 2. J-V characteristic parameters of μc-Si:H single-junction solar cells as a function of i-layer thickness. The lines are guides to the eyes

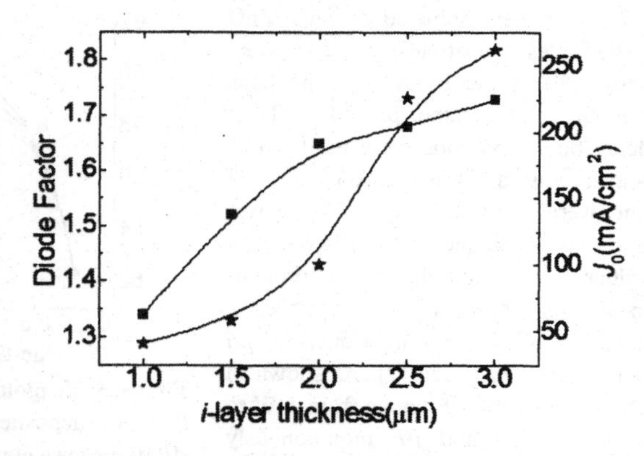

Figure 3. Diode ideality factor n(square) and saturation current density J_0(star) plotted as a function of i-layer thickness. The lines are guides to the eyes.

bombardment on the growth surface associated with the high power density during *i*-layer deposition is foreseeable, which will further increase defect density in the bulk *i*-layer [5,6]. This is why high pressure depletion technique is necessary to obtain high quality μc-Si:H films at high deposition rate.[6]. The ion bombardment could cause a high defect density as the X_C increases with the film thickness.

To reduce the effect of ion bombardment on the thick μc-Si:H intrinsic layers, we proposed a VHF power profiling technique. In the normal deposition, a constant VHF power of 60 W was used. While in the VHF power profiling experiment, it was decreased step by step with a power intervals (ΔP). The schematic diagram of the VHF power profiling technique is shown in Fig. 4. The deposition time for every step is defined as T_{step}. In Fig.5 the X_C of μc-Si:H films is plotted as a function of thickness. The initial part of films within 500 nm were deposited with the same VHF power, the other parts of films were deposited with different ΔP. The results demonstrate that the microstructure evolution along the growth direction can be effectively controlled by varying the steps, ΔP and T_{step}.

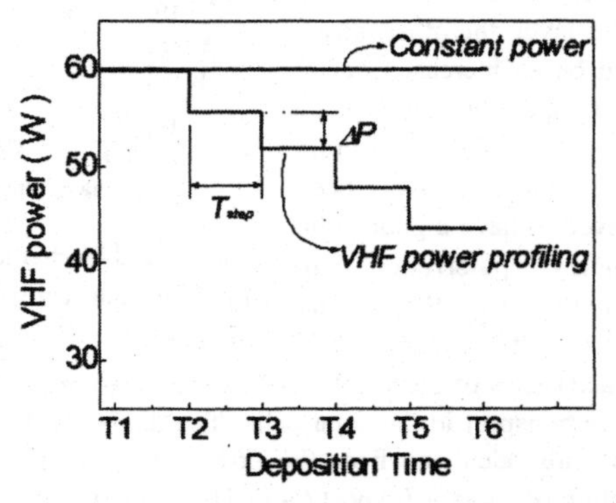

Figure 4. Schematic diagram of the VHF power profiling technique.

A series of single-junction μc-Si:H solar cells were prepared on SnO_2/ZnO coated glass substrates. Al dots were deposited on the n layer as the back contact. The same *p* and *n* layer deposition conditions were used, while *i*-layers with a similar thickness of ~2 μm were deposited by VHF power profiling technique with different ΔP values. In order to reduce the amorphous incubation layer at *p/i* interface, a lower-rate-deposited high-quality *p/i* buffer layer was used [7]. As shown in Fig. 6, when the ΔP varies from 1 W to 4 W, the V_{oc} and *FF* monotonously increase and the J_{sc} also shows a slight increase up to 3 W. From Fig. 5, one can see that when the VHF power profiling

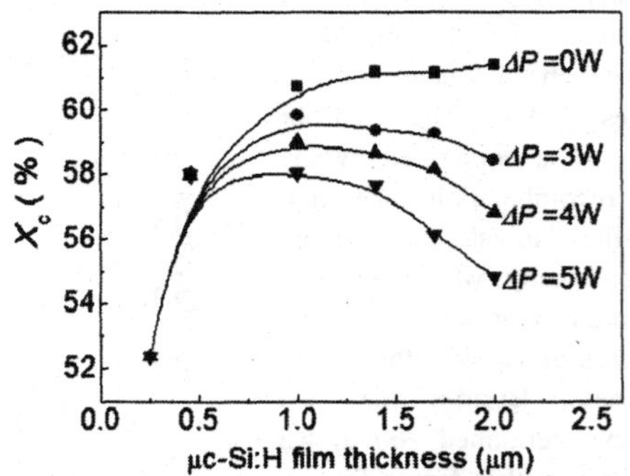

Figure 5. Xc plotted as a function of μc-Si:H film thickness deposited under different ΔP values. ΔP=0 means a constant power. The lines are guides to the eyes.

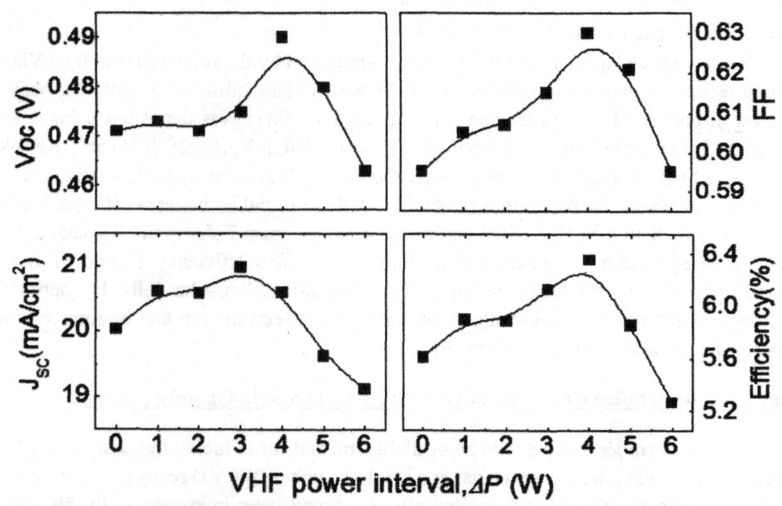

Figure 6.J-V characteristic parameters as a function of ΔP values. ΔP=0 means a constant power. The lines are guides to the eyes.

technique is applied, the X_C in the part of the film beyond 1 μm decreases with the film grows thicker. A decreased average X_C for the whole *i*-layer certainly results in an increase of V_{oc}. Because of the decrease of VHF power, the high-energy ion bombardment on the growing surface is reduced and defect density in the bulk *i*-layer is believed to be lower. Thus the transport property of the *i*-layer is improved, which results in the increases of V_{oc} and *FF*. When ΔP is higher than 4W, V_{oc}, *FF* and J_{sc} begin to decrease. This is probably caused by the too lower X_C in the latter part of the *i*-layer, which deteriorates the charge carrier transport, the same behavior as an amorphous incubation layer does at *p/i* interface [7,8]. The series studies show that a significant improvement of the solar cell

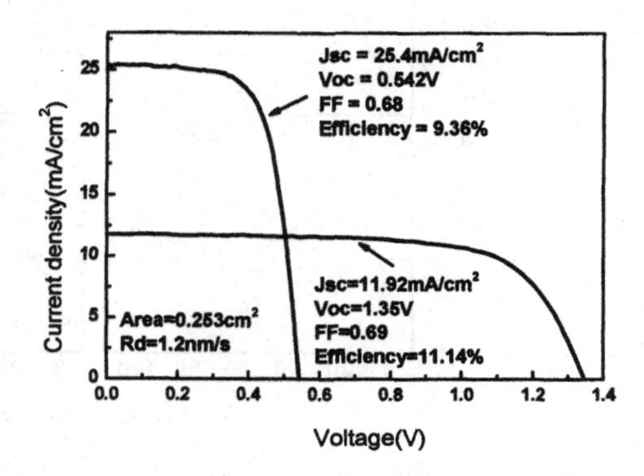

Figure 7 J-V characteristics of a μc-Si:H single-junction and an a-Si:H/μc-Si:H tandem solar cells with μc-Si:H i-layer deposited with a primary optimized VHF power profile

performance has been achieved with a ΔP of 4W.

We have transferred the μc-Si:H solar cell prepared by the primary optimized VHF power profiling technique onto chemically etched ZnO coated glass substrate. A ZnO/Ag back contact was used to enhance the light trapping, where the ZnO layer was deposited using a MOCVD method. An initial active-area efficiency of 9.36% (V_{oc}=542mV, J_{sc}=25.4mA/cm^2, *FF*=68%) has been obtained with a single-junction μc-Si:H *pin* solar cell at an average deposition rate over 10 Å/s. Using this recipe for depositing the bottom cell, an initial active-area efficiency of 11.14% was achieved with an a-Si:H/μc-Si:H tandem solar cell. Figure 7 shows the J-V characteristics of the best single-junction and tandem solar cells. The 9.36% efficiency is one of the highest reported efficiency in the literature for μc-Si:H single-junction solar cells. By optimizing the current match and *n/p* tunnel recombination junction between the top and bottom cell, the solar cell performance will be further improved.

Thin Film Silicon Solar Cells Deposited with PECVD Single-Chamber System

The single-chamber technique is an available method for reducing the cost of thin film solar cell product in large scale. However, using a single-chamber PECVD reactor to fabricate *pin* type silicon solar cells, the *i*-layer is deposited after p^+ doped layer in sequence. In this case, boron contamination subsequently affects subsequent *i*-layer deposition [9,10]. We propose an intrinsic interface layer with a relatively high Xc and a strong (220) orientation, deposited at a low rate at

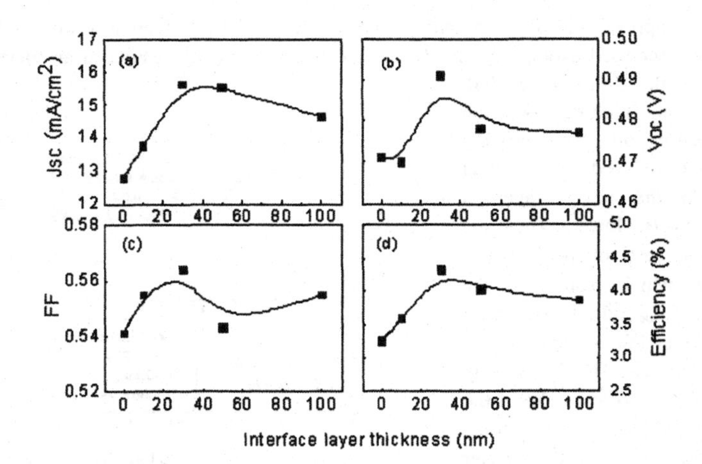

Figure 8. Performance parameters of μc-Si:H solar cells under different high crystalline interface layer thicknesses.

the *p/i* interface, to check if it can reduce the boron contamination and improve the performance of μc-Si:H solar cells, where the μc-Si:H intrinsic layer is deposited under the condition close to the amorphous/microcrystalline transition.

Figure 8 shows the J-V characteristic parameters of μc-Si:H solar cells with different high crystalline interface layer thicknesses, but with the same deposition parameters in all other layers. The total thickness for all solar cells is almost the same. As expected, the J_{sc} shows a maximum value at a certain thickness, which could be due to the reduction of boron contamination at the *p/i* interface. The optimized interface layer thickness is around 30-40 nm. The *Voc* and *FF*, as shown in Figs. 8 (b) and (c), also show maximum values at 30 nm. Finally, the conversion efficiency (in Fig.8 (d)) demonstrates the highest value at the same optimized interface layer. Although the real reasons for the improvement is not very clear at this moment, we believe that it results from the reduced the boron contamination, which mainly arises from atom hydrogen etching away boron-related materials. Of course, initial high crystalline interface layer also covers the surface of electrode and inhibits the release of boron-related materials on the substrate holder and other some places.

Quantum efficiency (QE) was measured for all the samples shown in Fig. 8. The results show that the increase of J_{sc} by the high crystalline interface layer was mainly from the spectral response enhancement in the short wavelength region from 400 nm to 600 nm. The short wavelength QE is usually affected by the *p/i* interface layer. Using Raman measurement, we found that increasing the high crystalline interface layer thickness leads to an increase of crystalline phase in the first part of the absorber layer. As a consequence, the extraction of charge carriers in this part becomes more efficient because of the decrease of boron contamination effect

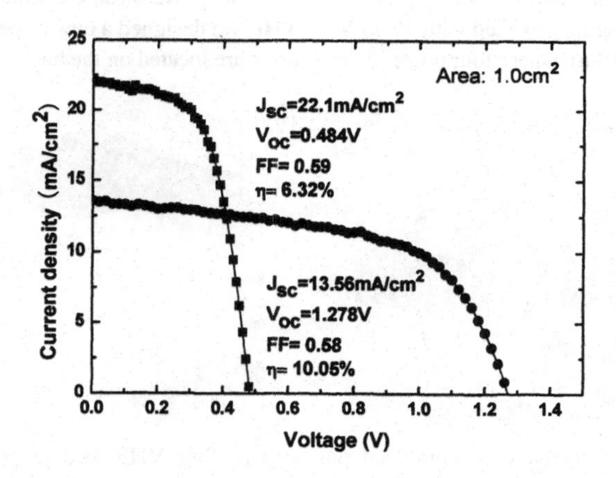

Figure 9. J-V characteristics of a μc-Si:H single-junction solar cell and an a-Si:H/μc-Si:H micromorph tandem solar cell made in the single chamber system with a suitable high crystalline interface layer.

and improved crystallinity. In general, for illumination from the *p* layer side of *p-i-n* type solar cells, as used here, the generation rates of short wavelength light are high in the initial part of the intrinsic layer, the corresponding spectral response is strongly increased by the optimized interface layer. Therefore, the *p/i* interface layer is one of the critical areas for efficiency improvement with a single-chamber system.

In order to increase the conversion efficiency of μc-Si:H solar cells further, high quality ZnO coated glass substrates from Jülich IPV were used (the resistance of etched ZnO is a little large because of the long etching time). Combined the improved deposition parameters, especially the *p/i* interface layer, we have improved the efficiency of μc-Si:H single-junction and a-Si:H/μc-Si:H double-junction solar cells. Figure 9 shows the J-V characteristics of a μc-Si:H single-junction solar cell and an a-Si:H/μc-Si:H micromorph tandem solar cell, where the single-junction cell has a 6.3% efficiency and the a-Si:H/μc-Si:H micromorph tandem cell has an efficiency of 10.05%. Although the efficiencies are still not as high as those achieved with multi-chamber systems, they are still very respectable. The series resistance of solar cells is still high, which partly comes from the chemically etched ZnO. We believe that the efficiency will be further improved when a good light trapping and low resistance ZnO/glass substrates and improved back reflectors are used.

Micromorph Solar Modules with VHF Technique on Large Glass Substrates

The third research area is the development of large area a-Si:H/μc-Si:H micromorph modules. The big challenge is to obtain uniformity distributions of film thickness and material quality on $0.79m^2$ size glass substrates, which need uniform electric field and gas low distributions. For the uniform gas distribution, we use a shower-head electrode. To improve the electric field uniformity excited with 40.68 MHz VHF, we designed a new power feeding method as shown in Fig.10(a), where four power feeding ports are located on the back of the

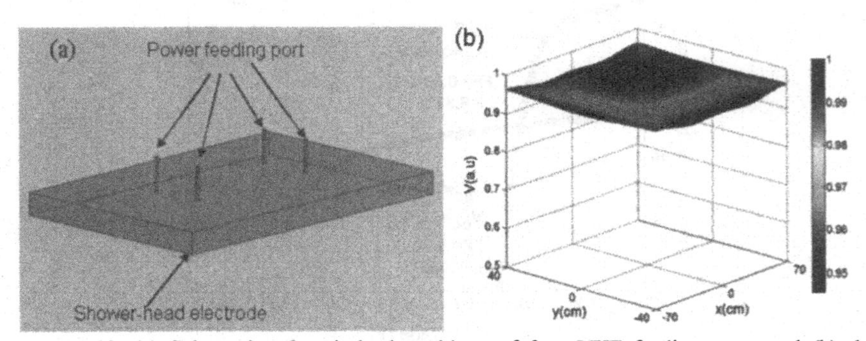

Figure 10. (a) Schematic of optimized positions of four VHF feeding ports and (b) the calculated normalized voltage distribution between shower-head electrode and grounded electrode for 40.68MHz.

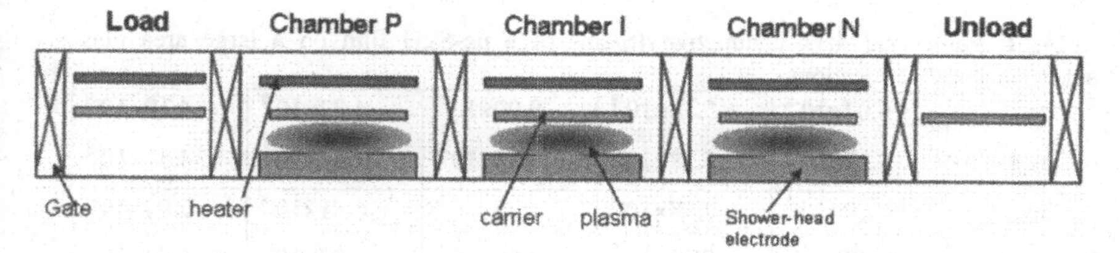

Figure 11. Schematic of the vertical in-line five chamber VHF-PECVD system, consisting of three deposition chambers, one load chamber, and one unload chamber.

shower-head electrode. After optimizing the position of the four power feeding ports for this multi-points power back feeding (MPBF) mode, the calculated non-uniformity of the voltage distribution for 40.68MHz excitation is reduced to ±2.7 % for 140×80 cm^2 electrode area as shown in Fig.10 (b).

Based on the simulation of electric field distribution as shown in Fig. 10, a five-chamber vertical in-line VHF-PECVD system was designed and built. The schematic of the system is shown in Fig. 11, where three VHF chambers (Chamber P, Chamber I, and Chamber N) with 40.68 MHz excitation frequency are used for the deposition of the p, i, and n layers. The MPBF cathode structure as shown in Fig. 10 is used in the three deposition chambers. The base pressure of the system is about 2×10^{-4} Pa. This system has been used to deposit micromorph solar module deposition on TCO/glass substrates with an area of 0.79 m^2. The μc-Si:H thin film thickness

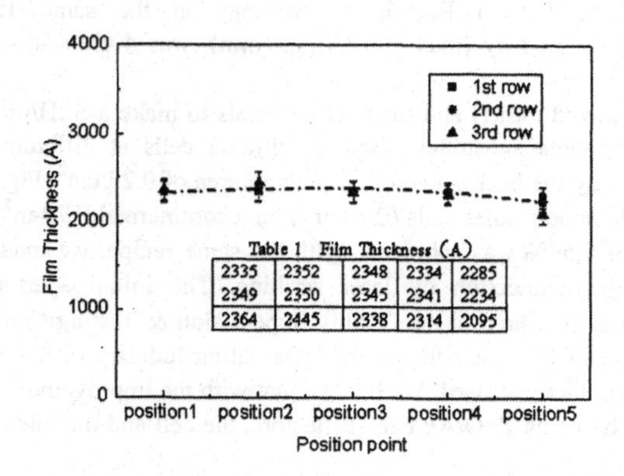

Table 1 Film Thickness (A)

2335	2352	2348	2334	2285
2349	2350	2345	2341	2234
2364	2445	2338	2314	2095

Figure 12. Thickness profile of μc-Si:H thin films on a commercial 0.79m^2 glass. The thickness was measured along three lines at 5 different positions in each line.

Table I. Photo and dark conductivity(S/cm) of a μc-Si:H film on a large area glass substrate at the 15 locations.

σ_{photo}	6.02×10^{-8}	5.18×10^{-8}	9.90×10^{-8}	4.82×10^{-8}	4.10×10^{-8}
σ_{dark}	5.22×10^{-5}	5.58×10^{-5}	9.95×10^{-5}	8.84×10^{-5}	4.51×10^{-5}
σ_{photo}	5.19×10^{-8}	7.98×10^{-8}	4.40×10^{-8}	6.81×10^{-8}	2.94×10^{-8}
σ_{dark}	5.19×10^{-5}	5.55×10^{-5}	4.80×10^{-5}	4.93×10^{-5}	5.04×10^{-5}
σ_{photo}	6.35×10^{-8}	4.60×10^{-8}	9.22×10^{-8}	7.70×10^{-8}	1.34×10^{-8}
σ_{dark}	4.36×10^{-5}	5.21×10^{-5}	4.97×10^{-5}	6.81×10^{-5}	4.34×10^{-5}

uniformity is shown in Fig, 12, where 15 locastions on the 0.79 m^2 gass substrate were measured. The 15 locations consist three rows with five points in each row. Fig. 12 shows the film thickness distribution. The non-uniformity, defined as (max-min)/2(max+min)[11], of 3.85% was achieved with 2 cm of edge exclusion, which meets the requirement of our thickness uniformity. The excellent uniformity can minimize variations of V_{oc} and J_{sc}. Electrical properties of intrinsic μc-Si:H thin films is very important and also crucial factor to monitor the quality of materials, especially for dark conductivity σ_{dark}. The medium σ_{dark} in the order of 10^{-8}s/cm as listed in Table I, where the dark- and photo-conductivity were measured at the same locations as described in Fig. 12. This result shows that the electronic properties of μc-Si:H material on the large area substrates are reasonably uniform and suitable for solar cells. In addition, the uniformity of the crystallinity was measured with Raman spectroscopy at the same 15 locations. The non-uniformity of X_C, defined by (max-min)/2(max+min), was 4.4% with the average X_C of 47.5%.

We used the optimized a-Si:H and μc-Si:H materials to make a-Si:H/μc-Si:H micromorph solar cells on the large-area substrates. First, small-area cells at different locations on the substrates were defined by the back Al contacts with an area of 0.25 cm^2. Fig. 13 shows the J-V curve of a micromorph tandem solar cells (0.25cm^2) on a commercial 0.79-m^2 SnO_2 glass, where the initial efficiency of 9.59% was achieved. With the same recipe, we made the micromorph solar modules with interconnection by laser scribing. The initial aperture-area (0.79 m^2) efficiency of 8.12% was measured at the Quality Supervision & Testing Center of Chemical & Physical Power Sources of Chinese Ministry of Information Industry (QSTC). We should point out that only Al back contact was used. We believe that with the improvement in material quality and cell structure and by using ZnO/Ag back reflectors, the cell and module efficiencies will be increased in the future.

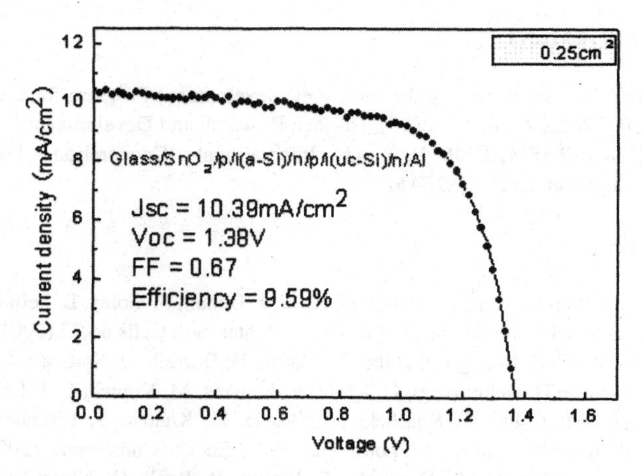

Figure 13. J-V curve of an a-Si:H/μc-Si:H tandem solar cells ($0.25 cm^2$) on a commercial $0.79 m^2$ TCO glass.

CONCLUSIONS

Generally, crystalline structural evolution in μc-Si:H films occurs under a constant deposition condition. In this paper, to resolve this issue, we first developed a high-quality seed layer by lowering silane concentration and very high frequency (VHF) power. Second, we developed a novel VHF power profiling technique to control the structural evolution in μc-Si:H i-layer. A high efficiency of 9.36% has been obtained in a μc-Si:H single-junction solar cell at average deposition rate over 10 Å/s. Using this recipe for depositing the bottom cell, an initial active-area efficiency of 11.14% for micromorph tandem solar cell was obtained.

The *p/i* interface contamination problem in μc-Si:H *pin* solar cells deposited in a single chamber system had been studied. A high crystalline interface layer was used to reduce the cross-contamination for depositing μc-Si:H solar cells in the single-chamber system. A μc-Si:H single-junction solar cell with an efficiency of 6.3% ($1.0 cm^2$) was demonstrated. With such μc-Si:H single-junction cell in a micromorph tandem solar cell, an initial efficiency of 10.1% ($1.0 cm^2$) were obtained in the single-chamber system.

We have designed and built a large-area multi-chamber system. A systematic simulation study on the cathode design has been carried out. Based on the simulation, the cathode structure was optimized. We have demonstrated a very uniform distribution of μc-Si:H deposition rates and material properties using the large-area machine with optimized deposition parameters. Micromorph tandem solar modules were fabricated on the commercial SnO_2 substrate of $0.79 m^2$. We achieved an aperture-area efficiency of 8.12% with an Al back reflector without ZnO.

ACKNOWLEDGEMENT

The work was supported by the State Key Development Program for Basic Research of China (2006CB202602, 2006CB202603), Hi-Tech Research and Development Program of China (2007AA05Z436, 2009AA050602), and International Cooperation Project between China-Greece Government (2009DFA62580).

REFERENCES

1. J. Meier, E. Vallat-Sauvain, S. Dubail, U. Kroll, J. Dubail, S. Golay, L. Feitknecht, P. Torres, S. Fay, D. Fischer, and A. Shah, Sol. Energy Mater. Sol. Cells 66,73 (2001).
2. J. Meier, U. Kroll, S. Benagli, J. Hötzel, J. Bailat, D. Borrello, J. Spitznagel, L. Castens1, E. Vallat-Sauvain,B. Dehbozorgi, O. Kluth, R. Kravets, M. Kupich, C. Ellert, S. Bakehe, H. Goldbach, M. Keller, T. Roschek, M. Gossla, H. Knauss, T. Eisenhammer, J. Henz, Proceedings of 23rd European Photovoltaic Solar Energy Conference, (2008) p.2057.
3. O. Vetterl, A. Lambertz, A. Dasgupta, F. Finger, B. Rech, O. Kluth, H. Wagner, Solar Energy Materials & Solar Cells **66**, 345 (2001).
4. B. Yan, G. Yue, J. Yang, A. Banerjee, and S. Guha, Mat. Res. Soc. Symp. Proc. **762**, 309 (2003).
5. B. Kalache, A. I.Kosarev, R.Vanderhaghen, and P. Roca.I. Cabarrocas, J. Appl. Phys. **93**, 1262 (2003).
6. M. Kondo, M. Fukawa, L. Guo, and A. Matsuda, J. Non-Cryst. Solids **266-269**, 84 (2000).
7. G Hou, X. Han, G Li, X. Zhang, N. Cai, C. Wei, Y. Zhao, and X. Geng, Technical Digest of the 17th International Photovoltaic Science and Engineering Conference (Fukuoka, Japan, 2007), p.1112.
8. U. K. Das, E. Centurioni, S. Morrison, and A. Madan, Proceedings of 3rd World Conference Photovoltaic Energy Conversion (Osaka, Japan, 2003), p.1776.
9. A. Catalano and G Wood, J. Appl. Phys. **63**, 1220 (1988).
10. S.C. Lee, J. Appl. Phys. **55**, 4426 (1984).
11. S. Klein, M. Rohde, T. Stolley, K. Schwanitz, and S. Buschbaum, Proceedings of 23rd European Photovoltaic Solar Energy Conference, (2008), p. 2088.

Mater. Res. Soc. Symp. Proc. Vol. 1245 © 2010 Materials Research Society 1245-A21-05

Study of Crystallinity in μc-Si:H Films Deposited by Cat-CVD for Thin Film Solar Cell Applications

C.H. Hsu[1], Y.P. Hsu[1], F.H. Yao[1], Y.T. Huang[1], C.C. Tsai[1], H.W. Zan[1],
C.C. Bi[2], C.H. Lu[2] and C.H. Yeh[2]
[1]National Chiao Tung University, Hsinchu, Taiwan
[2]NexPower Technology Corporation, Taichung, Taiwan

ABSTRACT

The crystallinity of the hydrogenated microcrystalline silicon (μc-Si:H) film was known to influence the solar cell efficiency greatly. Also hydrogen was found to play a critical role in controlling the crystallinity. Instead of employing conventional plasma deposition techniques, this work focused on using catalytic chemical vapor deposition (Cat-CVD) to study the effect of hydrogen dilution and the filament-to-substrate distance on the crystallinity, deposition rate, microstructure factor and electrical property of the μc-Si:H film. We found that the substrate material and structure can affect the crystallinity of the μc-Si:H film and the incubation effect. Comparing bare glass, TCO-coated glass, a-Si:H-coated glass and μc-Si:H-coated glass, the microcrystalline phase grows the fastest onto μc-Si:H surface, but the slowest onto a-Si:H surface. Surprisingly, the template effect lasted for more than a thousand atomic layers of silicon.

INTRODUCTION

The μc-Si:H with a smaller bandgap is well suited for the multi-junction solar cell applications [1] when combined with larger bandgap top-cells. The material also has the advantage of having little or no light-induced degradation effect in solar cell applications [2], and a lower cost compared to other low bandgap material such as a-SiGe:H. However, it is difficult to control the crystallinity (X_c) of the μc-Si:H due to the incubation phenomenon and the thickness dependence [3]. The μc-Si:H material is a mixture of amorphous and crystalline phases with better cells corresponding to a X_c in the range of 30-70% [4]. Compared to the conventional plasma deposition, Cat-CVD technique has the advantages of producing low H-content films with reduced Staebler-Wronski effect [5-7], and the potential of obtaining high deposition rate [8].

Many studies on the X_c of μc-Si:H focused on growing μc-Si:H on bare glass. However, in most solar cell applications, μc-Si:H films are deposited onto substrates covered by hydrogenated amorphous silicon (a-Si:H), μc-Si:H or transparent conducting oxide (TCO) [9]. Nucleation and subsequent film growth are likely to be affected by the substrate template effect [10] which can also influence the electrical and optical properties of the thin film. Therefore, in this study, the effect of the substrates on the X_c and the phenomenon of incubation were also investigated.

EXPERIMENTAL DETAILS

The μc-Si:H films were deposited from a mixture of H_2 and SiH_4 at a substrate temperature of about 210°C. Two tungsten wires (with diameter of 0.5mm) heated up to 1850 °C were used as the filament. The high temperature of the filament has a great influence on the substrate temperature due to the heat radiation. Therefore, the substrate temperature was first carefully calibrated to reach the desired temperature. A substrate heater was utilized to speed up reaching a certain stable temperature. The power supply of the filaments was then turned on in order to further heat up the surface to the process temperature. The gas mixture was introduced only after the substrate temperature was stabilized. The underlayers of 5nm a-Si:H and 20nm μc-Si:H were deposited onto glass substrate for the study on the substrate template effect.

Several analytical tools including Raman spectroscopy (with wavelength of 632.8nm), Fourier-transformed infrared spectroscopy (FTIR) and electrical measurement were used to investigate the X_c, microstructure factor, deposition rate and conductivity of the μc-Si:H. The X_c, was defined as the ratio of the integrated intensity of the 510cm^{-1} and 520cm^{-1} Raman peaks to that of the 480cm^{-1}, 510cm^{-1} and 520cm^{-1} peaks. The microstructure factor, F, was defined as the ratio of the integrated intensity of SiH_2 mode at 2100cm^{-1} to that of the SiH_2 and SiH modes at 2100cm^{-1} and 2000cm^{-1} as obtained from the FTIR spectroscopy, respectively. The sample without annealing is measured at room temperature under AM1.5G illumination in atmosphere. The film thickness was measured by the alpha-step.

RESULTS AND DUSCUSSIONS

The effect of H_2 dilution on X_c and F is shown in Fig. 1(a) and 1(b). As the hydrogen dilution increased, X_c increased due to the preferential etching of the amorphous phase on the substrate surface [11], as shown in Fig. 1(a). The X_c was found to be dependent on the film thickness. Under the same deposition condition, a thicker film of 400nm has the higher X_c than the thinner film of 200nm due to the incubation phenomenon. Compared to a-Si:H, the μc-Si:H exhibits large fraction of SiH_2 configuration, corresponding to more dangling bond and grain boundary defects. As the hydrogen dilution or X_c increased, F increased, which can be seen in Fig. 1(b). Combining the effects of film thickness and crystallinity, the microstructure factor also increases. It should also be noticed that the hydrogen dilution needed for the μc-Si:H in

Cat-CVD is much lower than in PECVD. This is due to the higher dissociation efficiency of H_2 in Cat-CVD.

The effect of the filament-to-substrate spacing, d_{fs}, on the deposition rate is presented in Fig. 2(a). The result shows that the deposition rate decreased from 0.12nm/s to 0.08nm/s as the d_{fs} increased from 50mm to 130mm. The reduction should be due to the reducing flux of the disassociate species from the filaments to the substrate. Generally, in PECVD the crystallinity should increase with decreasing deposition rate arising from more etching and relaxation time [12]. However, in Cat-CVD here, the deposition rate and X_c decrease together as d_{fs} increases, which can be seen in Fig. 2(a) and 2(b). This is probably due to the formation of polymeric species in the gas phase as d_{fs} increases. Besides, since the H-dissociation only occurs near the filament surface, the H radial density near the substrate surface is reduced as d_{fs} increases, leading to a smaller etching effect which reduces the crystallinity.

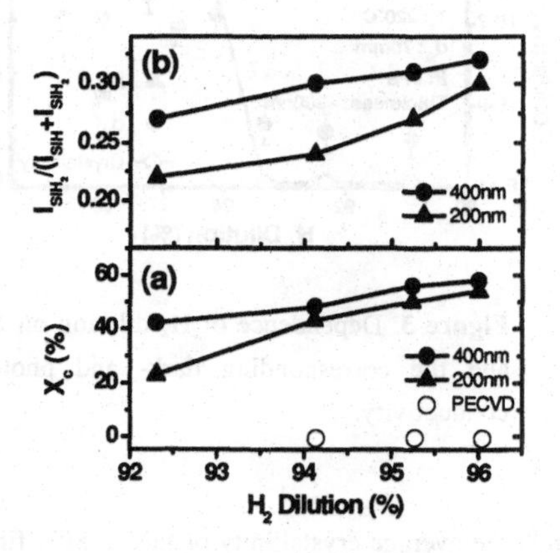

Figure 1. The X_c and the microstructure factor as a function of H_2 dilution. The filament temperature (T_f), substrate temperature (T_s), pressure (P) and the filament-to-substrate distance were 1850°C, 210°C, 5Pa and 75mm, respectively.

The dependence of F and X_c on d_{fs} is shown in Fig. 2(c). As the d_{fs} increased, the gas phase reaction increased due to the larger spacing than the mean free path [13]. This probably produced polymeric silicon hydride species which degrade the film [14]. On the other hand, increasing d_{fs} lead to a more amorphous film, which reduced F. As a result, F stayed unchanged in this study. In fact in the Cat-CVD, some SiH_4 are disassociated at the filament into $Si_{(s)}$ and $H_{(g)}$. The $Si_{(s)}$ has less mobility on the substrate surface and could degrade the film quality without the help of hydrogen [14]. At a very small d_{fs}, the film may exhibit less crystallinity but more defective.

The photo-conductivity is an indication of the ability of generating photocurrent in the absorption layer for solar cells. Different from a-Si:H film, μc-Si:H film has a higher dark conductivity arising from more defects [1]. Figure 3 shows the dependence of photo- and dark-conductivity with the X_c modulated by the H_2 dilution. As the H_2 dilution ratio is below 93.3%, the film behaves as a-Si:H which has a dark conductivity near $2x10^{-10}$S-cm^{-1}. When the H_2 dilution ratio is larger than 93.3%, the X_c rapidly increases, leading to a higher dark conductivity of about 10^{-7}S-cm^{-1}. In contrast, the photo-conductivity stay unchanged despite of the change in the structure. Beyond the transition region, the photo- and dark- conductivity remain constant even though the X_c changes from 30-60%.

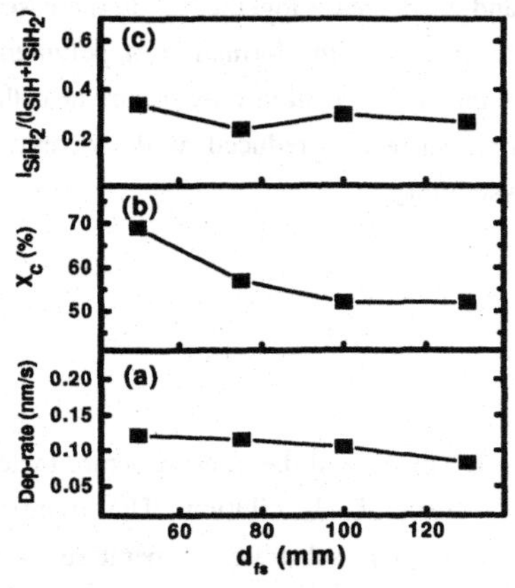

Figure 2. Effect of d_{fs} on (a) deposition rate, (b) X_c and (c) microstructure factor. The T_f, T_s, P and hydrogen dilution were 1850°C, 210°C, 5Pa and 95.2%, respectively.

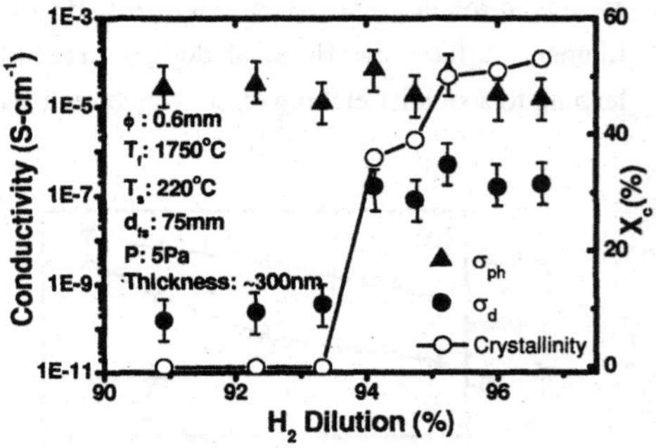

Figure 3. Dependence of H_2 dilution on X_c and the corresponding dark- and photo-conductivity.

In practical application of solar cell, although the average crystallinity of the μc-Si:H film can be controlled by changing the silane or hydrogen concentration, the incubation layer which near the interface can affect the cell efficiency significantly [15]. In this study we investigated the dependence of the crystallinity on various templates. The substrates include bare glass, non-textured TCO glass, a-Si:H-coated glass and μc-Si:H-coated glass. The a-Si:H and μc-Si:H films were deposited on glass by a PECVD system. The X_c of the μc-Si:H film deposited by PECVD is about 40% with a film thickness of 20nm. The a-Si:H film with hydrogen content of about 10% and film thickness of 5nm was used for the a-Si:H-coated glass.

Our results show that there is a strong substrate template effect on the initial film growth which extends beyond a film thickness of about 200nm. This is shown in Fig. 4(a) which exhibits the thickness dependence of incubation phenomenon. Initially, the X_c increased fast which slow down after 200nm in thickness and reach a similar X_c of 60%. It is rather surprising that the substrate effect lasted over a thousand atomic layers of silicon. Such effect can influence the performance of the thin film solar cells significantly.

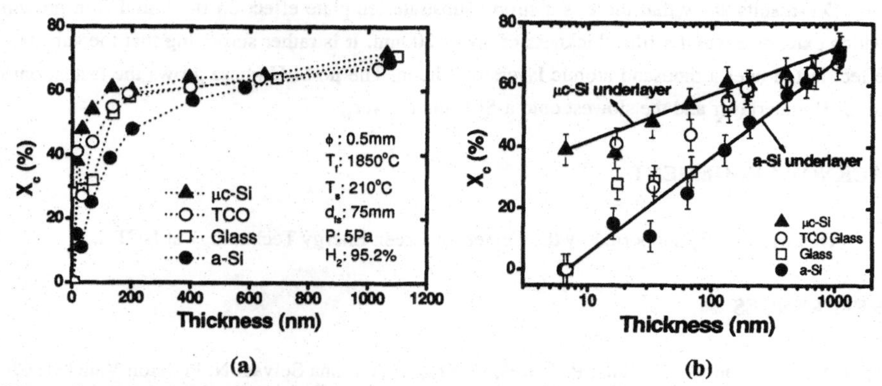

Figure 4. (a) Thickness dependence of crystallinity of μc-Si:H grown on bare glass and different underlayers of a-Si:H, μc-Si:H and TCO on glass, and (b) the same data plotted on a logarithmic scale.

Figure 4(b) shows the trend of crystallinity by plotting the film thickness on a logarithmic scale. Initially, μc-Si:H phase was only achieved on μc-Si:H underlayer, all other surfaces can only grow amorphous phase. The result shows that the μc-Si:H film deposited on μc-Si:H underlayer has the highest X_c due to the substrate template effect while on a-Si:H has the lowest. It should also be noticed that although the glass is an amorphous material, the μc-Si:H film grew onto the glass has a higher crystallinity than the a-Si:H underlayer although they both showed no X_c initially. Even after a thickness of 100nm, the crystallinity on the a-SiH underlayer still is the lowest while those on other underlayers show similar X_c. Especially for the μc-Si:H grown on glass exhibits similar X_c with the μc-Si:H or TCO underlayer. For the application as solar cell that requires high quality interface, the control of the incubation layer is important.

CONCLUSIONS

In this work, we have studied the growth of μc-Si:H phase in the Cat-CVD deposition of μc-Si:H. The hydrogen dilution required in Cat-CVD for the deposition of μc-Si:H is much lower than PECVD. Increasing d_{fs} would decrease the deposition rate but shows an opposite trend of decreasing crystallinity due to more gas phase reaction and less H etching near the surface.

Combining the effect of more amorphous phase but more gas phase products, the microstructure factor had no significant change by modulating d_{fs} in this study. With the X_c ranged in 30-60%, the photo-conductivity and dark-conductivity stayed unchanged.

Our results show that there is a strong substrate template effect on the initial film growth which extends beyond a film thickness of about 200nm. It is rather surprising that the substrate effect lasted over a thousand atomic layers of silicon. The μc-Si:H phase grows the fastest onto μc-Si:H underlayer and the slowest onto a-Si:H underlayer.

ACKNOWLEDGEMENT

This work partly supported by the Center for Green Energy Technology at NCTU.

REFERENCES

1. J. Meier, S. Dubail, R. Platz, P. Torres, U. Kroll,J. A. Anna Selvan, N. Pellaton Vaucher, Ch. Hof, D. Fischer, H. Keppner, R. Flückiger, A. Shah, V. Shklover and K. -D. Ufert, *Sol. En. Mat. Sol. Cells*, **49**, 35 (1997).
2. F. Wang, H. N. Liu, Y. L. He, A. Schweiger, and R. Schwarz, *J. Non-Cryst. Solids*, **137&138**, 511 (1991).
3. R.W. Collins, A.S. Ferlauto, G.M. Ferreira, C. Chen, J. Koh, R.J. Koval, Y.Lee, J.M. Pearce and C.R. Wronski, *Sol. En. Mat. Sol. Cells*, **78**, 143 (2003).
4. F. Meillaud, E. Vallat-Sauvain, X. Niquille, M.Dubey, J. Bailat, A. Shah and C. Ballif, *Proc. 31th IEEE Photovoltaic Specialist Conf.*, 1412 (2005).
5. D.L. Staebler and C.R. Wronski, *Appl. Phys. Lett.*, **31**, 292 (1977).
6. E. Spanakis, E. Stratakis, P. Tzanetakis and Q. Wang, *J. Appl. Phys.*, **89**, 4294 (2001)
7. A. H. Mahan and M. Vanecek, *AIP Conf. Proc.*, **234**, 195 (1991)
8. Q. Wang, *Thin Solid Films*, **517**, 3570 (2009).
9. C. Ross, Y. Mai, R. Carius and F. Finger, *Mater. Res. Soc. Symp. Proc.*, **862**, A10.4 (2005)
10. P.R. Cabarrocas, N. Layadi, T. Heitz, and B. Drevillon, *Appl. Phys. Lett.*, **66,** 3609 (1995).
11. C.C. Tsai, G.B. Anderson, R. Thompson, and B. Wacker, *J. Non-Cryst. Solids*, **114**, 151(1989).
12. S. Klein, F. Finger, R. Carius and J. Lossen, *Thin Solid Films*, **501**, 43 (2006).
13. A. Gallagher, *Thin Solid Films*, **395**, 25 (2001).
14. K. F. Feenstra, R. E. I. Schropp, and W. F. Van der Weg, *J. Appl. Phys.*, **85**, 6843 (1999).
15. Y. Mai, S. Klein, R. Carius, J. Wolff, A. Lambertz, F. Finger and X. Geng, J. Appl. Phys., **97**, 114913 (2005).

AUTHOR INDEX

SUBJECT INDEX

Printed in the United States
by Baker & Taylor Publisher Services